CONSTRUCTION 4.0

Modelled on the concept of Industry 4.0, the idea of Construction 4.0 is based on a conflu-ence of trends and technologies that promise to reshape the way built environment assets are designed, constructed, and operated.

With the pervasive use of Building Information Modelling (BIM), lean principles, digital technologies, and offsite construction, the industry is at the cusp of this transformation. The critical challenge is the fragmented state of teaching, research, and professional practice in the built environment sector. This handbook aims to overcome this fragmentation by describing Construction 4.0 in the context of its current state, emerging trends and technologies, and the people and process issues that surround the coming transformation.

Construction 4.0 is a framework that is a confluence and convergence of the following broad themes discussed in this book:

- Industrial production (prefabrication, 3D printing and assembly, offsite manufacture)
- Cyber-physical systems (actuators, sensors, IoT, robots, cobots, drones)
- Digital and computing technologies (BIM, video and laser scanning, AI and cloud com-puting, big data and data analytics, reality capture, Blockchain, simulation, augmented reality, data standards and interoperability, and vertical and horizontal integration)

The aim of this handbook is to describe the Construction 4.0 framework and consequently highlight the resultant processes and practices that allow us to plan, design, deliver, and operate built environment assets more effectively and efficiently by focusing on the physical-to-digital transformation and then digital-to-physical transformation. This book is essential reading for all built environment and AEC stakeholders who need to get to grips with the technological transformations currently shaping their industry, research, and teaching.

Anil Sawhney is Director of the Infrastructure Sector for the Royal Institution of Chartered Surveyors and Visiting Professor of Project Management at Liverpool John Moores Uni-versity. Anil is a Fellow of the Royal Institution of Chartered Surveyors and a Fellow of the Higher Education Academy. He has over 20 years' academic and industry experience inter-nationally and has over 200 academic publications. Anil is the co-editor of the *Construction*

Innovation journal and serves on the International Editorial Board of the *Journal of Civil Engineering and Environmental Systems* and *Journal of Information Technology in Construction*. He has consulted for numerous organizations including the World Bank.

Mike Riley is Pro Vice Chancellor, Engineering and Technology and Professor of Building Surveying at Liverpool John Moores University. He is Visiting Professor at the University of Malaya and at RICS School of Built Environment, Delhi, with over 25 years' academic and industrial experience, and is joint author of numerous textbooks on Construction Technology and Sustainability as well as numerous academic papers. Mike obtained his first degree from Salford University, followed by a Master of Science from Heriot-Watt University and PhD from Liverpool John Moores University. He is a Fellow of the Royal Institution of Chartered Surveyors and Fellow of the Royal Institution of Surveyors Malaysia, Senior Fellow of the Higher Education Academy, and Chartered Environmentalist.

Javier Irizarry P.E., PhD is an Associate Professor in the School of Building Construction at the Georgia Institute of Technology. A pioneer of research on Unmanned Aerial System applications in the built environment, Javier is Director of the CONECTech Lab, which aims to establish the framework for developing next generation technology enhanced solutions for construction problems by incorporating the cognitive processes of the human component of construction operations. He has over 20 years of academic and industry experience and has authored over 100 academic articles. Javier's research focuses on construction information technologies including virtual and augmented reality, HCI issues in mobile applications for AEC information access, reality capture technology, Situation Awareness driven information system design, and Unmanned Aerial Systems application in the AEC domain. He holds a B.S. in Civil Engineering from the University of Puerto Rico, Mayagüez, a Masters in Engineering Management from the Polytechnic University of Puerto Rico, and a PhD in Civil Engineering from Purdue University. Javier is also a registered Professional Engineer as well as a FAA Licensed Drone Pilot.

CONSTRUCTION 4.0

An Innovation Platform for the Built Environment

Edited by Anil Sawhney, Mike Riley and Javier Irizarry

Routledge
Taylor & Francis Group

LONDON AND NEW YORK

First published 2020
by Routledge
4 Park Square, Milton Park, Abingdon, Oxon OX14 4RN
605 Third Avenue, New York, NY 10017

First issued in paperback 2023

Routledge is an imprint of the Taylor & Francis Group, an informa business

Publisher's Note
The publisher has gone to great lengths to ensure the quality of this reprint but points out that some imperfections in the original copies may be apparent.

British Library Cataloguing-in-Publication Data
A catalogue record for this book is available from the British Library

Library of Congress Cataloging-in-Publication Data
A catalog record has been requested for this book

Typeset in Times New Roman

ISBN-13: 978-0-367-02730-8 (hbk)
ISBN-13: 978-1-03-265360-0 (pbk)
ISBN-13: 978-0-429-39810-0 (ebk)

DOI: 10.1201/9780429398100

by Swales & Willis, Exeter, Devon, UK

CONTENTS

Acknowledgements	*xiii*
Foreword	*xv*
Notes on contributors	*xvii*
List of figures	*xxviii*
List of tables	*xxxvi*

PART I
Introduction and overview of Construction 4.0, CPS,
Digital Ecosystem, and innovation — **1**

1 Construction 4.0: Introduction and overview — 3
   Anil Sawhney, Mike Riley, and Javier Irizarry

   1.1 Aims — 3

   1.2 Introduction to Construction 4.0 — 3

   1.3 Current state of the construction sector — 6

   1.4 Overview of Industry 4.0 — 7

   1.5 Construction 4.0 framework — 13

   1.6 Benefits of Construction 4.0 — 15

   1.7 Challenges to implementation of Construction 4.0 — 16

   1.8 Structure of the handbook — 17

   1.9 Conclusion — 19

   1.10 Summary — 19

   References — 19

2   Introduction to cyber-physical systems in the built environment   23
    Pardis Pishdad-Bozorgi, Xinghua Gao, and Dennis R. Shelden

    2.1 Aims   23

    2.2 Introduction   23

    2.3 Cyber-physical systems and Construction 4.0   24

    2.4 What does success look like?   25

    2.5 CPS for Smart Built Environment   30

    2.6 Conclusion   37

    2.7 Summary   38

    References   39

3   Digital ecosystems in the construction industry—
    current state and future trends   42
    Anil Sawhney and Ibrahim S. Odeh

    3.1 Aims   42

    3.2 Introduction to digital ecosystems   42

    3.3 Current state of digital technologies in construction   46

    3.4 Overview of ecosystems and platforms   47

    3.5 Digital ecosystems in construction   54

    3.6 Emerging trends and future directions—platforms and ecosystems   58

    3.7 Conclusion   60

    3.8 Summary   60

    References   60

4   Innovation in the construction project delivery networks
    in Construction 4.0   62
    Ken Stowe, Olivier Lépinoy, and Atul Khanzode

    4.1 Aims   62

    4.2 Introduction   62

    4.3 Context   64

    4.4 Opportunity for Construction 4.0 (the promise)   71

    4.5 Ramifications of Construction 4.0   76

    4.6 Challenges and considerations   81

    4.7 Conclusion   84

    4.8 Summary   87

    References   87

PART II
Core components of Construction 4.0   **89**

5   Potential of cyber-physical systems in architecture and construction   91
Lauren Vasey and Achim Menges

   5.1 Aims   91

   5.2 Introduction   91

   5.3 Towards cyber-physical construction   92

   5.4 Context: challenges and opportunities for cyber-physical
       systems within architectural production   93

   5.5 New possibilities enabled in architecture by cyber-physical systems   96

   5.6 Conclusion   107

   5.7 Summary   109

   References   109

6   Applications of cyber-physical systems in construction   113
Abiola A. Akanmu and Chimay J. Anumba

   6.1 Aims   113

   6.2 Introduction   113

   6.3 Drivers for cyber-physical systems in construction   114

   6.4 Requirements for CPCS in construction   115

   6.5 Cyber-physical construction systems   117

   6.6 Case studies   120

   6.7 Challenges and barriers   126

   6.8 Conclusion   128

   6.9 Summary   128

   References   128

7   A review of mixed-reality applications in Construction 4.0   131
Aseel Hussien, Atif Waraich, and Daniel Paes

   7.1 Aims   131

   7.2 Introduction   131

   7.3 Conclusion   139

   7.4 Summary   139

   References   140

8   Overview of optoelectronic technology in Construction 4.0   142
Erika A. Pärn

   8.1 Aims   142

Contents

8.2	Introduction	142
8.3	Fundamentals of laser scan devices	144
8.4	Modes of delivery and specifications	144
8.5	Applications in construction	146
8.6	Translating as-built progress	147
8.7	Conclusion	150
8.8	Summary	151
	References	151

9 The potential for additive manufacturing to transform the construction industry 155
Seyed Hamidreza Ghaffar, Jorge Corker, and Paul Mullett

9.1	Aims	155
9.2	Introduction	155
9.3	Additive manufacturing processes	157
9.4	Printable raw materials for construction	159
9.5	Practical and commercial challenges and opportunities for 3D printing construction	164
9.6	Future areas of research and development	180
9.7	Conclusion	182
9.8	Summary	183
	References	183

10 Digital fabrication in the construction sector 188
Keith Kaseman and Konrad Graser

10.1	Aims	188
10.2	Introduction	188
10.3	Digital fabrication in architecture and construction	189
10.4	State of research in digital fabrication	194
10.5	Research demonstrator case study: DFAB HOUSE (ETH, 2017–2019)	195
10.6	Practice case study: SHoP architects	200
10.7	Conclusion	204
10.8	Summary	205
	References	205

11 Using BIM for multi-trade prefabrication in construction 209
Mehrdad Arashpour and Ron Wakefield

11.1	Aims	209
11.2	Introduction – prefabrication as a core component of Construction 4.0	209

11.3 Background – applications of BIM for multi-trade prefabrication   210

11.4 Decision making on BIM for multi-trade prefabrication   212

11.5 Conclusion   217

11.6 Summary   218

References   218

12 Data standards and data exchange for Construction 4.0   222
Dennis R. Shelden, Pieter Pauwels, Pardis Pishdad-Bozorgi, and Shu Tang

12.1 Aims   222

12.2 Introduction   222

12.3 Elements of a data standard   223

12.4 Industry Foundation Classes: overview, application, and limitation   224

12.5 Uses and applications   229

12.6 Evolutions of building data standards   230

12.7 Conclusion   237

12.8 Summary   237

References   238

13 Visual and virtual progress monitoring in Construction 4.0   240
Jacob J. Lin and Mani Golparvar-Fard

13.1 Aims   240

13.2 Computer vision for monitoring construction – an overview   240

13.3 Review on the current state-of-the-art for computer vision
applications in the industry and research   244

13.4 Conclusions   257

13.5 Summary   259

References   259

14 Unmanned Aerial System applications in construction   264
Masoud Gheisari, Dayana Bastos Costa, and Javier Irizarry

14.1 Aims   264

14.2 Introduction   264

14.3 Unmanned Aerial Systems (UASs)   265

14.4 UAS applications   274

14.5 UAS implementation challenges   283

14.6 Conclusion   284

14.7 Summary   284

References   285

15 Future of robotics and automation in construction   289
    Borja Garcia de Soto and Miroslaw J. Skibniewski

    15.1 Aims   289
    15.2 Introduction   289
    15.3 Classification   292
    15.4 Current status/examples   295
    15.5 Main challenges and future directions   299
    15.6 Conclusion   303
    15.7 Summary   303
    References   303

16 Robots in indoor and outdoor environments   307
    Bharadwaj R. K. Mantha, Borja Garcia de Soto,
    Carol C. Menassa, and Vineet R. Kamat

    16.1 Aims   307
    16.2 Introduction   307
    16.3 Classification of robots   308
    16.4 Key fundamental capabilities   309
    16.5 Case study application   316
    16.6 Conclusion   322
    16.7 Summary   322
    References   322

17 Domain-knowledge enriched BIM in Construction 4.0:
    design-for-safety and crane safety cases   326
    Md. Aslam Hossain, Justin K. W. Yeoh, Ernest L. S. Abbott,
    and David K. H. Chua

    17.1 Aims   326
    17.2 Introduction   326
    17.3 Tacit knowledge application 1: design-for-safety knowledge
          enrich BIM for risk reviews   330
    17.4 Explicit knowledge application 2: BIM-based tower
          crane safety compliance system   340
    17.5 Conclusion   347
    17.6 Summary   347
    References   347

18 Internet of things (IoT) and internet enabled physical
    devices for Construction 4.0   350
    Yu-Cheng Lin and Weng-Fong Cheung

18.1   Aims   350

18.2   Introduction   350

18.3   Background   351

18.4   The IoT technologies   352

18.5   Applications of IoT in construction   356

18.6   Case study   359

18.7   Conclusion   367

18.8   Summary   367

References   368

19   Cloud-based collaboration and project management   370
     Kalyan Vaidyanathan, Koshy Varghese, and Ganesh Devkar

19.1   Aims   370

19.2   Introduction   370

19.3   Construction today – critical evaluation of current project
        management frameworks   372

19.4   Construction 4.0 – cloud based collaboration and evolution
        of construction information supply chain solutions   375

19.5   Transitioning to Construction 4.0   384

19.6   Conclusion   389

19.7   Summary   390

References   390

20   Use of blockchain for enabling Construction 4.0   395
     Abel Maciel

20.1   Aims   395

20.2   Introduction   395

20.3   Construction challenges in the era of BIM   396

20.4   Context and aspects of blockchain   400

20.5   Application considerations   406

20.6   Blockchain and the construction sector   408

20.7   Challenges in the adoption of blockchain   410

20.8   The fourth wave: what might happen next   410

20.9   Conclusion   411

20.10   Summary   412

References   413

PART III
Practical aspects of construction 4.0 including case studies,
overview of start-ups, and future directions   **419**

21 Construction 4.0 case studies   421
    Cristina Toca Pérez, Dayana Bastos Costa, and Mike Farragher
    21.1 Aims   421
    21.2 Case study 1: 4D BIM for logistics purposes   421
    21.3 Case study 2: the WikiHouse project   428
    21.4 Case study 3: the innovation lab   434
    21.5 Conclusion   439
    21.6 Summary   440
    References   440

22 Cyber threats and actors confronting the Construction 4.0   441
    Erika A. Pärn and Borja Garcia de Soto
    22.1 Aims   441
    22.2 Introduction   441
    22.3 The digital uprising   445
    22.4 Smart cities and digital economies   446
    22.5 Cyberspace, cyber-physical attacks and critical infrastructure hacks   448
    22.6 What motivates a cyber-attacker? Actors and incident analysis   449
    22.7 Looking at the literature   450
    22.8 Cyber-deterrence   452
    22.9 Conclusion   454
    22.10 Summary   454
    References   455

23 Emerging trends and research directions   460
    Eyuphan Koc, Evangelos Pantazis, Lucio Soibelman, and David J. Gerber
    23.1 Aims   460
    23.2 Introduction   460
    23.3 Background   463
    23.4 Emerging 4.0 trends in the AEC industry   467
    23.5 Research directions   470
    23.6 Conclusion   474
    23.7 Summary   475
    References   475

Acronyms   *477*
Index   *483*

ACKNOWLEDGEMENTS

The editors would like to thank everyone who helped with the inception, development writing, editing, and support for this handbook. This manuscript has resulted from the combined efforts of nearly 60 individual contributors from across the globe, including academics, practitioners, and industry stakeholders. This is a triumph of collaboration and teamwork!

The editors wish to express their sincere thanks to all of the individual chapter authors for their commitment, energy, and enthusiasm in co-creating this work.

Many individuals, companies, and organizations have assisted in developing the various chapters and providing illustrations and information that has allowed the various authors to produce this work. Many of these are cited in the context of specific illustrations or figures within the main text.

In addition, however, we would like to express specific acknowledgement for various chapters as follows:

In relation to Chapter 3, we would like to thank Takefumi Watanuki, a graduate student of the Construction Engineering and Management graduate programme at Columbia University. We also would like to thank Autodesk and Procore for their permission to use the illustrations. Autodesk products referenced in the chapter are registered trademarks or trademarks of Autodesk, Inc, and/or its subsidiaries and/or affiliates in the USA and/or other countries. Procore products referenced in the chapter are registered trademarks or trademarks of Procore Inc. and/or its subsidiaries and/or affiliates in the USA and/or other countries. iModelHub and iModel are registered trademarks or trademarks of Bentley Systems Inc. and/or its subsidiaries and/or affiliates in the USA and/or other countries.

In relation to Chapter 4, we would like to thank everyone who helped with the inspiration, ideas, structure, writing, editing, and support for this chapter. An incomplete list includes: Tristan Randall, Eddy Krygiel, Frank Moore, Steve Duffett, Manu Venugopal, Scott Borduin, Dustin Hartsuiker, Richard Holbrook, James McKenzie, Anil Sawhney, Alan Mossman, Zoubeir Lafhaj, Rafael Sacks, and Lauri Koskela.

The material within Chapter 13 is, in part, based upon work supported by the National Science Foundation Grant #1446765. The support and help of Reconstruct and the construction team in collecting data are greatly appreciated. The opinions, findings, and conclusions or recommendations expressed within this chapter are those of the authors and do not reflect the views of the NSF, or the company mentioned above.

The author of Chapter 20 would like to thank Professor Alan Penn for supporting his construction blockchain research at the Bartlett Faculty of the Built Environment, University College London. He would also like to thank the Construction Blockchain Consortium (CBC) team for their ongoing support and encouragement.

The authors would also like to acknowledge the kind support of WikiHouse (now Open-systemslab) and Oracle in allowing reproduction of the materials required to produce the case studies in Chapter 21.

Finally, the editors would like to express their sincere gratitude for the invaluable contribution of Cristina Toca Pérez, who has committed time, effort, and energy without hesitation. Cristina's input has been crucial to the successful completion of this work.

Co-Editors: Anil Sawhney, Mike Riley, and Javier Irizarry

FOREWORD

With increasing pressures on the built environment sector to provide the infrastructures and homes that are the key economic enablers to city growth, and as people globally are entering our cities at the rate of 3 million people a week, the heat is on construction to design, build, modify, and operate these assets to our changing needs and that of the communities whose evolving demands occupy the space provided.

Construction has continued to innovate, but not at the rate or expectations demanded of it. Therefore, we need to consider the opportunity to radically transform our methods and approaches to construction that enable it to be more efficient and effective in adopting the technologies from other sectors and services to enable it to reshape the way our built environment assets emerge now and for the future. The time is now, because the demand is there for rapid supply, balanced against the costs of the intensification for the supply of skills and resources, coupled with the desire for improved and innovative design.

You will see from reading this book that the key to unlocking the potential and pace of a more rapid "right first time" mentality is putting the physical-to-digital and digital-to-physical transformation at the heart of the delivery process.

The book is timely, as we are at the tipping point of transformative change for construction with already establishing digital practices of Building Information Modelling (BIM), Modern Methods of Construction (MMC) as well as VR/AR, AI, 3D, and IoT as rapidly evolving technologies to expedite design, deliver, and operate are all coming to the fore, backed by the essential data to feed and inform.

There is no doubt that this is an exciting time for the built environment sector and for the transformation of the construction processes that deliver it. But there is a level of pace of change that is needed now to deliver, transform, and metamorphosize the sector and this book captures well the elements necessary to deliver that change.

With this in mind, this book provides the key to unlocking the potential of the built environment sector at a time where the sector needs unlocking to gear it to transform the delivery of our infrastructure, homes, and cities. The key to unlocking the change needed sits within these pages, with digital innovation at the heart, and the power of you to drive the transformation that will re-establish construction as a key economic enabler for growth.

Amanda G Clack MSc BSc PPRICS FRICS FICE FAPM FRSA CCMI FIC CMC
Executive Director and Head of Strategic Advisory
CBRE Ltd
RICS Past President

The 2017 McKinsey Global Institute's publication "Reinventing Construction: A Route to Higher Productivity" showed that the greatest impact on productivity improvement in the construction industry is through the advancement and application of digital and technology solutions. Over the last few years there has been a significant investment by private equity funds in construction industry related digital and technology start-ups and tech companies which has fuelled tremendous growth and innovation in this part of the industry. In "Construction 4.0" the authors pull together all of the relevant elements of these essential solutions and practices and show how they will enable more effective and efficient planning, design, delivery, and operation of physical assets (i.e., capital projects) through a digital transformation. The industry has already made significant advancements over the past 2–3 years but much more is required among all members of the supply chain involved with capital projects. This publication presents a comprehensive review of these emerging solutions and systems and makes the connection of technology with people and processes. Companies and organizations that do not have a "digital strategy" will be able to understand better through "Construction 4.0" how each element complements one other and how each is able to improve performance across all phases of a capital project. While many companies in the industry have utilized BIM or VR/AR in one form or another, other advancements such as data analytics, Internet of Things (IoT) and use of artificial intelligence are shown to be significant disruptors to the traditional model of project development, design, and delivery with significant benefits to be realized by project owners, designers, and contractors. Construction performance and productivity has stalled tremendously since World War II compared to every other major industry and in order to be ready for the Fourth Industrial Revolution, industry players will need to change and "Construction 4.0" is an excellent guidebook to such transformation.

In addition to helping professionals working in the industry already, this handbook will be a useful resource for several folks in academia … undergraduate and graduate students, researchers and scholars with a keen interest in the ongoing transformation of the construction industry using the Industry 4.0 framework.

Tim McManus, Adjunct Professor, Columbia University

CONTRIBUTORS

Ernest L. S. Abbott BSc (Special), MIMA, BA (Hons), MTh, PhD is a research fellow in the Department of Civil and Environmental Engineering, Faculty of Engineering at the National University of Singapore. He has worked in a variety of industries including machine makers, the retail oil industry, shipping – port automation, the finance industry and software development. His interests are eclectic to say the least. His recent research has included productivity improvements in off-shore rig construction, BIM to productivity in construction of public housing, Design for Safety using BIM, and Prefabricated Prefinished Volumetric Construction with RFID and BIM for intelligent logistic and process productivity.

Abiola A. Akanmu is an Assistant Professor in the Myers-Lawson School of Construction at Virginia Tech. She is Director of the Automated Systems Lab whose mission is to apply intelligence to the design, construction, and maintenance of building and civil infrastructure systems, and construction workforce education using information and communication technologies. Dr Akanmu's research interests involves tightly integrating visualization and sensing systems for improving progress monitoring, construction safety, facility management, and construction education so as to enhance predictability and control.

Chimay J. Anumba BSc (Hons), PhD, DSc, Dr hc, FREng, CEng, FICE, FIStructE, FASCE is Professor and Dean of the College of Design, Construction and Planning at University of Florida. His research interests are collaborative design, construction engineering and informatics, intelligent systems, knowledge management, cyber-physical systems, and project management, with over 500 publications in these fields. He is Editor-in-Chief of the *Engineering, Construction and Architectural Management* journal and Co-Editor of the *Journal of Information Technology in Construction*.

Mehrdad Arashpour PhD, MASCE, MPMI, MRICS, MIEAust is an expert and leading researcher on automation and optimization of design, manufacturing, and assembling processes in offsite prefabrication. His research collaborations with Professor Wakefield on process integration and automation in construction have resulted in several publications in top tier journals and books. Dr Arashpour is one of the selected worldwide members of the Working

Commission on Offsite Construction (W121) and Infrastructure Task Group (TG91), established by the International Council for Building (CIB). He is the member of Editorial Boards for international journals including *Construction Innovation* and *Journal of Advances in Civil Engineering*.

Dayana Bastos Costa is Associate Professor from the Department of Structural and Construction Engineering at School of Engineering at Federal University of Bahia. She received her PhD in Civil Engineering and MSc in Civil Engineering both at Federal University of Rio Grande do Sul, Brazil. Dr Costa had opportunities to work as visiting scholar at Salford University, UK and Georgia Institute of Technology, USA. Dr Costa's research interest includes construction management and technology aiming the improvement of the industry performance. She works in business management and construction sites, involving aspects related to production, quality, safety, and sustainability, integrating with information technology and emergent technologies, such as Unmanned Aerial Systems and Building Information Modelling.

Weng-Fong Cheung has a PhD from Department of Civil Engineering of National Taipei University of Technology. His major study is about construction management and focuses on the categories of wireless sensor network (WSN), Building Information Modelling (BIM), digitalization in infrastructure management, construction safety management, and the relative integrated application in construction field. Cheung serves in the Department of Rapid Transit Systems (DORTS) of Taipei city government and acts as a specialist of the management and quality assurance of the construction project. He is also the supervisor and report signatory of the central civil laboratory of DORTS, which ensures the material quality of the metro system.

David Chua Kim Huat is a registered professional engineer in Singapore and Professor and Vice Dean at the National University of Singapore. He obtained his PhD from the University of California and has more than 35 years' research and industry experience. His research areas are lean construction, computer integrated and IT-based construction management, BIM, construction simulation, risk management, and construction safety. He has been Council Member of the Society of Project Managers and System Safety Society and a member of BCA's International Panel of Experts on BIM and Honorary Fellow of WSHi. He has served as editor for the *Journal of Construction Engineering and Management*, ASCE and other international journals and conferences.

Jorge Corker is senior researcher and Technical Director of the Granular Materials Characterization and Certification Unit lab at the Laboratory for Wear, Testing and Materials of Instituto Pedro Nunes (IPN), Portugal. He is taking his PhD in Civil Engineering at Brunel University London, and is responsible for the development of applied science research projects and specialized technical services in materials at IPN. He has participated in and coordinated several European FP7 and H2020 projects on development of advanced solutions and new insulation materials for construction. His research includes advanced materials for construction, nanomaterials, composites, and materials for energy efficiency and sustainability.

Ganesh Devkar is currently working as Associate Professor in the Faculty of Technology, CEPT University, Ahmedabad. He completed his PhD from Indian Institute of Technology Madras (IIT Madras), India, and his doctoral thesis focuses on identifying the competencies necessary for effective implementing Public Private Partnerships (PPPs) in infrastructure service delivery and developing a framework that allows urban local bodies to assess their capability to successfully deliver projects through PPPs. His research interests are in the areas of lean construction, PPPs, and urban infrastructure development.

Mike Farragher graduated from Liverpool John Moores University in 1991 with a first class degree in Architecture. He is a Chartered Architect and member of the RIBA and has won government awards for Integrated Renewable Technology. He is a Fellow of the Higher Education Academy and has nearly 20 years of industry experience before moving into academia where he has taught at Henan University, China and in the UK. He has published on zero-carbon projects in the UK and is currently researching the development of sanitation systems in sub-Saharan Africa. He has been instrumental in the inception of and enabling works for a primary school in a rural location on the border of Uganda and South Sudan.

Xinghua Gao is an Assistant Professor with the Myers-Lawson School of Construction at Virginia Tech. He received his PhD degree in Building Construction and MSc in Computational Science and Engineering from Georgia Institute of Technology in 2019, MSc in Structural Engineering from Cardiff University, UK in 2012, and BEng in Civil Engineering from Central South University, China in 2010. Prior to pursuing his PhD, he worked as a structural/Building Information Modeling (BIM) engineer at the China Institute of Building Standard Design & Research. His research interests lie in BIM and Internet of things (IoT)-enabled smart built environments that encompass automated data collection, analysis, and visualization for more efficient and effective construction and facilities management.

Borja Garcia de Soto PhD, PE is an Assistant Professor of Civil and Urban Engineering at New York University Abu Dhabi (NYUAD) and holds an appointment as Global Network Assistant Professor in the Department of Civil and Urban Engineering at the Tandon School of Engineering at New York University (NYU). He is the director of the S.M.A.R.T. Construction Research Group at NYUAD and conducts research in the areas of automation and robotics in construction, cybersecurity in the AEC industry, artificial intelligence, lean construction, and BIM. Borja has extensive experience in the industry as a structural engineer, project manager, and construction consultant. He is a Professional Engineer (PE) with licenses in California and Florida and has international experience in multiple aspects of construction projects. Borja received his PhD from ETH Zurich in Switzerland. He also holds an MSc in Civil Engineering with a concentration in engineering and project management from the University of California at Berkeley, an MSc in Civil Engineering with a concentration in structural design from Florida International University (FIU), and a BSc in Civil Engineering (graduated cum laude) also from FIU.

David Jason Gerber received his Doctor of Design and Master of Design Studies from Harvard University Graduate School of Design. He also has an M.Arch from the Architectural Association DRL in London and BA in Architecture, University of California, Berkeley. He holds a joint appointment at USC's Viterbi School of Engineering and USC School of Architecture as Associate Professor of Civil and Environmental Engineering Practice and of Architecture and is Director for the Civil Engineering Building Science undergraduate programme. He has extensive international, professional experience with practices including Zaha Hadid Architects in London, Gehry Technologies in Los Angeles, Moshe Safdie Architects in Massachusetts, The Steinberg Group Architects in California, and for Arup as Global Research Manager.

Seyed Hamidreza Ghaffar BEng (Hons), PhD, CEng, MICE is the leader of Additive Manufacturing Technology in Construction (AMTC) research group in the Department of Civil and Environmental Engineering – Brunel University London. The main area of his research is on valorising construction and demolition waste using materials science and innovative technologies for achieving circular construction. The AMTC research group focuses on the recovery of resources from waste with industrially feasible approaches towards eco-innovative manufacturing of construction products which is an important element of the transition to a low-carbon economy.

Masoud Gheisari is an Assistant Professor in the Rinker School of Construction Management at the University of Florida where he is leading Human-Centered Technology in Construction (HCTC) research group. He received his PhD in Building Construction from Georgia Tech and considers himself an educator and a researcher focusing on theoretical and experimental investigation of human-computer/robot systems in construction. Dr Gheisari's research interests are in the areas of human-computer/robot interaction, Unmanned Aerial Systems/Vehicles (UAS/UAVs), real/virtual humans in mixed reality environments, and human-technology interaction in educational settings. He also teaches various courses in the area of Building Information Modelling and technology applications in construction.

Mani Golparvar-Fard is Associate Professor of Civil Engineering, Computer Science, and Technology Entrepreneurship at the University of Illinois, Urbana-Champaign (UIUC). He received his PhD degree in Civil Engineering and MS degree in Computer Science from UIUC and MASc in Civil Engineering from University of British Columbia. He received the 2018 ASCE Walter L. Huber award for Innovation in Civil Engineering, the 2017 ENR National Top 20 under 40, 2016 ASCE Dan H. Halpin Award for Scholarship in Construction, the 2013 ASCE James R. Croes Medal for innovation in Civil Engineering, and numerous conference paper awards. He is co-founder and COO of Reconstruct Inc, offering visual data analytics to construction projects.

Konrad Graser is researcher at the Chair of Innovative and Industrial Construction at ETH Zurich. He was the project manager and lead architect of the DFAB HOUSE, a full-scale architectural demonstrator of interdisciplinary research in digital fabrication technologies at the Swiss National Center for Competence in Research Digital Fabrication. He was Project Manager Facade Engineering at Werner Sobek Stuttgart, responsible for digital design and engineering of complex building envelopes. Konrad has been a Teaching Assistant at Yale School of Architecture, Adjunct Professor at California College of the Arts, and a technical advisor for at the Institute for Lightweight Structures (ILEK) at the University of Stuttgart.

Md. Aslam Hossain is an Assistant Professor at the Department of Civil and Environmental Engineering, Nazarbayev University. He obtained his PhD from the National University of Singapore (NUS) and graduated with BSc in Civil Engineering from Bangladesh University of Engineering and Technology. Prior his current appointment, Dr Hossain worked as a post-doctoral research fellow at NUS, Assistant Professor at Presidency University and the Islamic University of Technology, Dhaka and senior lecturer at University Malaysia Pahang. His research interests include Design for Safety (DfS), BIM, Construction 4.0, productivity improvement, concurrent engineering, and lean construction. He has published his research findings in premium journals and international conferences.

Aseel Hussien is Programme Leader of Building Surveying and Facilities Management in the Department of the Built Environment (BUE) at Liverpool John Moores University (LJMU). Hussien holds a BSc Degree in Architecture Engineering, and a MSc in Computing Information Systems with Speciality in Augmented Reality. She completed her PhD from the Department of the Built Environment at LJMU in 2017, in which she developed a novel integration of Augmented Reality (AR) technologies with Agile project management philosophy to enhance collaboration and decision-making within the construction industry. She has published many research papers in many peer reviewed international conferences and high impact journals.

Vineet R. Kamat PhD is a Professor and the John L. Tishman Faculty Scholar in the Department of Civil and Environmental Engineering at the University of Michigan. He directs the Laboratory for Interactive Visualization in Engineering and co-directs the Construction Engineering Laboratory. Dr Kamat's research is primarily focused on Virtual and Augmented Reality Visualization, Simulation, Mobile Computing, Robotics, and their applications in construction.

Keith Kaseman is Assistant Professor at the Georgia Institute of Technology School of Architecture, and coordinates the Master of Science in Architecture – Advanced Production, leads design and research studios and directs the interdisciplinary Spatial Futures Lab. His research focuses on AR, VR, MR, UAS, robotics, and other digital fabrication practices in multisystem design-production workflows and experimental spatial systems. Keith has held simultaneous academic posts at the University of Pennsylvania PennDesign, Columbia University GSAPP, and University of Tennessee, Knoxville. He is a partner of KBAS, a design practice launched in 2002 with Julie Beckman and has since led the firm through a diverse array of work utilizing customized design and advanced digital fabrication protocols.

Atul Khanzode leads DPR Construction's Technology and Innovation Group, responsible for the Virtual Design and Construction (VDC), Operations and Preconstruction Technologies, strategic technology initiatives, Innovation, Research & Development and consulting. He is a Board Member at WND Ventures, responsible for managing strategic investments and providing mentorship and operational guidance to leadership teams of the portfolio companies. Atul has a PhD in Construction Engineering and Management, focused on Integrated Practice, VDC, and Lean Construction, from Stanford University and a Master's Degree in Civil and Environmental Engineering from Duke University, Durham NC.

Eyuphan Koc is a PhD student with Dr Lucio Soibelman at University of Southern California Sonny Astani Department of Civil and Environmental Engineering. He obtained his Bachelor's degrees at Bogazici University, Istanbul, in Civil Engineering with minor in Industrial Engineering. His research, funded by Caltrans, focuses on smart cities with functionalities drawn from smart and resilient infrastructures. His current focus is designing multidisciplinary frameworks investigating resilience in transportation systems to leverage data and modelling from multiple fields including earthquake engineering, transportation systems analysis, and disaster economics. He previously worked for a design and engineering consultancy, a large real estate developer and a heavy civil contractor.

Olivier Lépinoy is part of Autodesk Business Development team for the AEC industry. He has led and contributed to wide-ranging projects across architecture, landscape, urban design, real estate, land development, management consulting, digital and new technologies. At Accenture,

Olivier was part of the team in charge of French general contractors, and was in charge of strategic initiatives in engineering and construction worldwide. At IBM, he was global leader for Engineering and Construction, taking part in multiple strategic initiatives around digital reinvention worldwide. Olivier holds Master's degrees in Civil Engineering (Ecole Spéciale des Travaux Publics, Paris) and Earthquake Engineering (UCLA, USA) and is a licensed architect and an urban designer (Ecole d'Architecture de Versailles).

Jacob J. Lin is a PhD candidate in the Department of Civil and Environmental Engineering at the University of Illinois at Urbana-Champaign (UIUC). His research focuses on developing and validating computer vision and machine learning algorithms with visual data and Building Information Modelling (BIM) for construction project controls. He has developed a production management system that integrates and visualizes 4D BIM with reality models and enables proactive management of schedule delays. He has awards and recognitions from the World Economic Forum, MIT Tech Review, and the 2015 American Society of Civil Engineers International Workshop on Computing in Civil Engineering for his work on visualizing big visual data for automated UAV-driven construction progress monitoring.

Yu-Cheng Lin is a Professor of Construction Engineering and Management at the Department of Civil Engineering of the National Taipei University of Technology. He received a MS degree in the construction management programme of civil engineering from the Polytechnic University, New York, USA (1997) and a PhD from the National Taiwan University (2004). His major research interests are the application of project management and information management in construction projects. His current research interests include construction knowledge management, Building Information Modelling, web-based project management system, IT technology application, IoT and sensors, and construction automation related topics.

Abel Maciel is Architect and Senior Research Associate at University College London, focusing on Computational Design, Artificial Intelligence and Distributed Ledger Technology (Blockchain). He is Director of Design Computation Ltd. and has worked, internationally, with some of the world's leading practices, including Foster + Partners, Zaha Hadid Architects, Herzog & de Meuron, Heatherwick Studio, Arup and BuroHappold. He has advised on numerous projects including Google Campus Mountain View, Google Headquarters London, Apple Park in Cupertino, and Vessel Hudson Yards, New York City. He is a Founding Director and Academic Lead of the Construction Blockchain Consortium (CBC) and Faculty Member of the UCL Centre of Blockchain Technologies.

Bharadwaj R. K. Mantha received his Masters and PhD in construction engineering and management at the University of Michigan. He is currently working as a Post-Doctoral Research Associate in the S.M.A.R.T. Construction Research Group at the New York University Abu Dhabi. Dr Mantha's research focuses on developing technological solutions to monitor, control, and sustainably operate buildings and civil infrastructure systems. To achieve this, he explores interdisciplinary approaches using tools, methods, and frameworks from robotics, cybersecurity, data analytics, simulation, and modelling.

Carol C. Menassa PhD is an Associate Professor and the John L. Tishman Faculty Scholar in the Department of Civil and Environmental Engineering at the University of Michigan. She directs the Sustainable and Intelligent Civil Infrastructure Systems Laboratory. Dr Menassa's research focuses on understanding human-robot collaborative processes in the building sciences, modelling the impact of occupants on energy use in buildings, and developing decision frameworks to sustainably retrofit existing buildings.

Achim Menges is a registered architect in Frankfurt and professor at Stuttgart University, where he is the Founding Director of the Institute for Computational Design and Construction (ICD) and the Director of the Cluster of Excellence Integrative Computational Design and Construction for Architecture (IntCDC). In addition, he has been Visiting Professor in Architecture at Harvard University's Graduate School of Design and held multiple other visiting professorships in Europe and the United States. He graduated with honours from the AA School of Architecture, where he subsequently taught as Studio Master and Unit Master in the AA Graduate School and the AA Diploma School.

Paul Mullett BEng (Hons), CEng, FICE, FCIArb, PSE is a Chartered Fellow of the Institution of Civil Engineers, Fellow of the Chartered Institute of Arbitrators and registered Professional Simulation Engineer with NAFEMS. An accomplished professional consultant with over 25 years' experience in a variety of technical and managerial roles, Paul has led or executed services including high-level engineering consultancy, specialist engineering, forensic engineering studies, and expert advice. He is Group Engineering and Technology Director for Robert Bird Group pursuing engineering excellence through the strategic application of transformative technologies, identification and adoption of value-adding innovation, and development of staff skills necessary for the future of engineering across the global business.

Ibrahim S. Odeh is the Founding Director of the Global Leaders in Construction Management program at Columbia University. His work seamlessly merges the worlds of industry and academia, spans the global construction market, and provides leading construction firms with unique market insights and growth strategies. Professor Odeh worked as an advisory committee member at the World Economic Forum on the Future of Construction initiative and as an advisory board member at Financial Times on a similar initiative. In the fall of 2016, Professor Odeh introduced a project focused on providing Massive Open Online Courses to educate global learners about Construction Management under the Coursera platform. In the program's first three years, the number of learners registered for his courses exceeded 80,000 from 195 countries, with an additional 400 new learners joining every week! Professor Odeh has won several awards and other recognition such as Columbia University's Presidential Award for Outstanding Teaching; Top 20 Under 40 by Engineering News Records; and Top 40 Under 40 by Consulting-Specifying Engineer. Professor Odeh received his PhD in Civil Engineering from the University of Illinois at Urbana-Champaign and he also holds an MBA.

Daniel Paes is a PhD candidate in Building Construction with a minor concentration in Cognitive Psychology and a research assistant in the CONECTech Lab, Georgia Institute of Technology. His research is at the intersection of basic and applied science and is centred on visual perception, presence, and learning in Immersive Virtual Reality (IVR) environments in architectural design, construction, and workforce training applications, identifying opportunities for revolutions in both usability metrics and virtual experiences in the field. His research interests also include Building Information Modelling (BIM), Augmented Reality platforms (AR), Unmanned Aerial Systems (UAS), Human-Computer Interaction (HCI) methods, Cognitive Psychology, and Ontologies.

Evangelos Pantazis is research assistant and PhD candidate at University of Southern California, Department of Civil and Environmental Engineering, Viterbi School of Engineering. He focuses on Multi Agent Design Systems and integration of generative design, numerical analysis and digital fabrication. He holds a Master's of Advanced studies, Computer Aided Architectural Design from ETH, Zurich and Diploma in Architecture from Aristotle's University of Thessaloniki and a degree in jewellery design and manufacturing from MOKUME jewellery design school. He is a registered architect in Greece and gained international experience including Graft Architects, Berlin; Studio Pei Zhu, Beijing and BuroHappold Engineering, Los Angeles. He co-founded Topotheque, engaging computational design with tangent disciplines, including architecture, furniture, and product design.

Erika A. Pärn is a Research Associate at the Institute for Manufacturing (IfM) and Centre for Digital Built Britain (CDBB), Cambridge University. Previously whilst working in industry, she led multiple research projects for the EU commission such as two H2020 initiatives on BIM, digitalisation of the renovation sector and sustainable green construction materials. Her PhD has evaluated the use of BIM and cloud-based technologies to streamline facilities management into the as-built BIM. During her time in academia as lecturer at Birmingham City University, she taught Architectural Technology. Her research focus is upon the multidisciplinary area of "digital built environment and smart city developments" but remains actively involved in broader "construction and civil engineering management" topics such as BIM and FM integration, cybersecurity, and networked and sensor-based BIM.

Pieter Pauwels is an Associate Professor at the Eindhoven University of Technology, the Department of the Built Environment. He previously worked at the Department of Architecture and Urban Planning at Ghent University (2008–2019). His work and interests are in information system support for the building life-cycle (architectural design, construction, building operation). With expertise in computer science and software development, he is involved in a number of industry-oriented research projects on topics affiliated to AI in construction, design thinking, Building Information Modelling (BIM), Linked Building Data (LBD), Linked Data in Architecture and Construction (LDAC), and Semantic Web technologies. He has initiated the BuildingSMART LDWG and the W3C LBD Community Group. He implemented and published an IFC-to-RDF conversion service and EXPRESS-to-OWL conversion service, to retrieve complex RDF data for building models. Furthermore, he is highly active as a teacher, leading courses on Fundamentals of BIM, Parametric Design, Building Informatics, and Linked Data in Architecture and Construction.

Pardis Pishdad-Bozorgi is Assistant Professor in the School of Building Construction at Georgia Tech. She focuses on developing innovative integrated Architecture/Engineering/Construction/Facility Management solutions for smart built environments, developing cutting-edge mechanisms that address process fragmentations in delivering and operating the built environment. Her publications are on emerging fields, such as Building Information Modelling, Internet of Things, Integrated Project Delivery, Flash Tracking, Trust-Building, and Lean Construction with around 40 articles in key journals and conferences. She holds a PhD degree in Environmental Design and Planning (Virginia Tech), Master's degrees in Civil Engineering (Virginia Tech), Project Management (Harvard), and Architecture (University of Tehran). She has received multiple awards, including Construction Management Association of America National Educator of the Year.

Dennis R. Shelden is Associate Professor of Architecture and Director of the Digital Building Laboratory at Georgia Tech. He is expert in applications of digital technology to building design, construction, and operations. His experience spans education and research, technology development, and professional practice across architecture, engineering, and computing disciplines. He directs Georgia Tech's PhD in Architecture and MS in Architecture: Building Information and Systems programmes. Prior to joining Georgia Tech, he led the development of Frank Gehry's digital practice as Director of R&D and Director of Computing and later as Co-founder and Chief Technology Officer of technology spin-off Gehry Technologies. He was Associate Professor of Practice in Computation and Design at MIT and has taught at UCLA and SCIARC.

Miroslaw J. Skibniewski PhD is Professor of Civil Engineering and Construction Project Management at the University of Maryland and serves as Editor-in-Chief of *Automation in Construction*. He heads the e-Construction Group, focused on innovative applications of construction information and automation technologies, collaborating with leading American, Asian, and European academic and industry partners in construction automation and robotics R&D. He was Professor of Civil Engineering, Construction Engineering and Management at Purdue University, where he collaborated with major Japanese engineering and construction firms in construction robotics implementation.

Lucio Soibelman was Assistant Professor at University of Illinois at Urbana-Champaign and Associate Professor and Professor to the Civil and Environmental Engineering Department at Carnegie Mellon University. He joined the University of Southern California as Chair of the Sonny Astani Department of Civil and Environmental Engineering and is editor of the *Journal of Computing in Civil Engineering*. He is recipient of the FIATECH Outstanding Researcher Celebration of Engineering & Technology Innovation Award, and ASCE Construction Institute Construction Management award. He was appointed as Viterbi Dean Professor at USC and Distinguished 1,000 talent Professor at Tsinghua University. He focuses on advanced data acquisition, visualization, and mining for construction and operations of advanced infrastructure.

Ken Stowe P.E. is construction technology advisor, development strategist, and project systems coach at Autodesk, Inc. With 25 years of experience in construction management and project control on US and international projects as large as $1.4 billion; 15 years in BIM, cloud software, ROI analysis, and big analytics experience in 22 countries, and contributions to four books, Stowe works with a team at Autodesk responsible for construction business development and strategy initiatives worldwide. Together, they leverage project performance research and pursue the synergy of technologies and construction performance improvement for measurably better decisions, planning, construction execution and asset performance. Kenneth holds two degrees from Dartmouth College and Thayer School of Engineering and Professional Engineering certification.

Shu Tang is currently a PhD student in Design Computation in the School of Architecture at Georgia Tech. She is conducting research in Digital Building Laboratory. She previously worked for years as a BIM Engineer for several BIM consultancy companies. Prior to joining Georgia Tech, Shu taught and conducted research at the University of Nottingham Ningbo China.

Shu's current research is a systematic investigation of defining data interoperability requirements between smart building systems for the Internet of Things (IoT) and Building Information Modelling (BIM), trying to extend digital systems (the Internet and the Web) into the physical realm (the built environment, the transportation system, etc.)

Cristina T. Pérez is a doctoral student of Engineering at Federal University of Bahia (UFBA) – Brazil. She completed her undergraduate studies at the University of Cantabria, Spain in 2011. Cristina received her MSc in Environmental and Urban Engineering at Federal University of Bahia, Brazil, in 2015. Since 2012, Pérez works as a researcher at the Department of Structural and Construction Engineering at the School of Engineering at UFBA. Pérez had opportunities to work as visiting researcher at Georgia Institute of Technology – USA in 2018–2019 and at Federal University of Rio Grande do Sul (UFRGS) – Brazil in 2014. Her doctoral research, in 2018, was recognized in a Brazilian competition with the Academic Innovation Award. Pérez has worked as an Assistant Professor at undergraduate and graduate programs at SENAI CIMATEC School of Technology, Ruy Barbosa DeVry University and Estacio de Sá University, all of them in Salvador, Brazil.

Kalyan Vaidyanathan is co-founder CEO of Nadhi Information Technologies Pvt Ltd, incubated out of IIT Madras research park. He has nearly 20 years' experience in managing construction projects and enterprise software development. He has won many awards, including the 2015 Technologist Innovator award: Construction Industry Development Council, Government of India. He was named Entrepreneur of the Year 2014 by CII, Chennai. He has published widely on modern construction technology and lean construction both nationally and internationally. "Overview of IT Applications in the Construction Supply Chain" was his chapter in the *Construction Supply Chain Management Handbook*. He holds an MS from Cornell University, USA and a B Tech in Civil Engineering from IIT Madras, India.

Koshy Varghese is Professor at IIT Madras in Building Technology and Construction Management Division, Department of Civil Engineering. His research interest is in computer-integrated construction and he has published widely in this area. He is a member of the editorial board of the *Journal of Automation in Construction* and the *Journal of Information Technology in Construction* and the past-president of the International Association for Automation and Robotics in Construction (IAARC). Varghese has been a visiting faculty at the University of Texas and Arizona State University, USA. He completed his PhD from the University of Texas at Austin, USA and a B.E. in Civil Engineering from the College of Engineering Guindy, Chennai, India.

Lauren Vasey is Research Associate and Doctoral Candidate at the Institute for Computational Design and Construction, University of Stuttgart. Lauren focuses on adaptive robotic construction processes, with sensor feedback tightly coupled to backend computational processing. She previously researched at University of Michigan Taubman College FABLab and ETH-Zurich, Chair for Architecture and Digital Fabrication. Her work has been recognized in international competitions including Kuka Innovation Award, Fast Company Innovation By Design, and AEC Hackathons. She has lectured at venues including ACM Siggraph and Advances in Architectural Geometry (AAG), and collaborated with research partners including Autodesk and the European Space Agency. Lauren is a member of the Board of Directors of ACADIA, Chair of the scientific committee, and editor of *International Journal of Architectural Computing*.

Atif Waraich PhD is Head of Department of Computer Science at Liverpool John Moores University. He was previously Head of Division of Digital Media and Entertainment Technology at Manchester Metropolitan University and founder and Director of the Manchester Usability Laboratory. His research focuses on the use of technology to enhance learning, specifically, how game-like environments can be used to promote learning and motivate learners to engage in their studies. He researches the use of usability engineering in developing and accessing gameplay for computer games and has a growing interest in the application of usability and behaviour modification techniques (including gamification) to Augmented and Virtual Reality environments.

Ron Wakefield is Professor of Construction, Deputy Pro Vice-Chancellor; International and Dean of the School of Property, Construction and Project Management at RMIT, Australia. Ron researches and teaches process simulation and modelling, residential and commercial construction and use of information technology in construction management. Prior to joining RMIT, he was the William E. Jamerson Professor of Building Construction in the Department of Building Construction and Associate Director for Building Technology Research at the Center for Housing Research, Virginia Tech. Ron has over 25 years' experience as an international researcher, consultant, and engineer in building construction. He is Director of Launch Housing and JJR Consulting and co-opted member of the Victorian Building Practitioners Board.

Justin K. W. Yeoh is a Lecturer at the Department of Civil and Environmental Engineering, National University of Singapore. He graduated with a BEng (Civil Engineering), and subsequently obtained his PhD. His research interests lie in Applied Optimization for Construction and Facilities Management, Semantics in Building Information Models (BIM), Process Re-engineering and Machine Learning for Construction Metrology. Prior to his current appointment, Dr Yeoh worked as a post-doctoral research fellow at NUS and Carnegie Mellon University. He has also worked as a project management consultant and BIM consultant, identifying good site practices for construction.

FIGURES

1.1	Conceptual model of CPS	4
1.2	Conceptual model of a digital ecosystem	4
1.3	Construction 4.0 as a combination of CPS and digital ecosystem	4
1.4	Physical to digital and digital to physical transformation	5
1.5	Themes of Construction 4.0	6
1.6	Key challenges faced by the construction sector	8
1.7	Evolution of embedded systems	10
1.8	Industry 4.0 and its key components	11
1.9	Enabling technologies and key features of I4.0	12
1.10	Conceptual illustration of Construction 4.0 framework	14
1.11	Components of Construction 4.0 framework	15
2.1	An architecture of the envisioned Smart Built Environment in Construction 4.0	31
2.2	The basic facility data package generation	33
2.3	Data flow of the proposed screw gun with counter	37
2.4	The proposed smart robotic agents	38
3.1	Role of digital ecosystems in Construction 4.0	43
3.2	Construction 4.0 transformation based on platforms and ecosystems	45
3.3	Traditional model of digital transformation in the construction sector	46
3.4	Traditional product firm	48
3.5	Platform-based firm	48
3.6	Ecosystem-based firm	49

3.7	Classification of ecosystems and their link to platforms	51
3.8	Illustration of a digital ecosystem	52
3.9	Complementary products based on digital ecosystem	54
3.10	Autodesk Forge ecosystem	55
3.11	Forge design automation API for Revit	56
3.12	Services available through the Forge platform	56
3.13	Revit Family app using design Automation API	57
3.14	Procore App Marketplace	57
3.15	iModel.js platform from Bentley Systems	58
4.1	The concept of waste as understood in lean construction	67
4.2	COAA's Fishbone Rework Cause Classification	68
4.3	The four dimensions influence on the innovation initiative	72
4.4	The House of Construction Tech	75
4.5	Data breakdown structure for two purposes – better asset and better project	82
4.6	The innovation environment leading to beneficial adaptation to Construction 4.0	86
5.1	Components of a cyber-physical system	93
5.2	A dual robot collaborative cell with external axis for on-site timber manufacturing	98
5.3	Multi-robot collaborative set-up for the fabrication of long-span fibre composites	100
5.4	12-meter long cantilever under construction	101
5.5	Collaborative robotic workbench for timber prefabrication	103
5.6	Augmented reality interface for human-robot collaborative construction	103
5.7	Wood Chip Barn, Architectural Association's Design + Make program	105
5.8	Scanned locally sourced wood utilized in a CAD design environment, Architectural Association's Design + Make program	105
5.9	An active control-loop manipulating a network of turn buckles was connected to a back-end form-finding process to help correct deviations resulting from the physical production process	107
5.10	Textile reinforced thin-shell concrete in production, NEST HiLo	107
6.1	Key features of CPCS	118
6.2	Bi-directional coordination approach to CPCS	118
6.3	System architecture for bi-directional coordination approach to CPCS	119
6.4	Steel placement scenario	121
6.5	Bulk material tracking and control	122
6.6	Flexible postural learning environment	124
6.7	User performing a 'attach stud' subtask and virtual instructor showing his performance – safe and unsafe posture	124
6.8	Site layout management workflow	126
7.1	Daqri® AR hard hat application in facility management (upper) and projections on the hat's display (bottom)	135
7.2	The SiteVision™ AR viewer	136

7.3	Immersive design review application user interface (left) inside an HMD-based VR system (right)	137
7.4	Projection-based VR crane operator training platform	138
8.1	Exemplar of triangulation calculation method for a terrestrial laser scanner which commonly use either TOF or phase-shift methods for point capture	145
8.2	Example of Scan-to-BIM used for automated masonry recognition	149
8.3	Example of Scan-vs-BIM	149
9.1	3D printing system mechanisms	156
9.2	Printable feedstock (raw materials) formulation from waste using additives	161
9.3	Apis Cor mobile 3D printer's specification and the 3D printer in action	165
9.4	Some of the various AM project demonstrations around the world (a) WinSun in China; (b), (c), and (g) CyBe Construction B.V. in Italy, Saudi Arabia, and UAE respectively, (d) Marine Corps Systems Command in USA; (e) XtreeE project demonstrations in France; (f) WASP in Italy	166
9.5	The opportunities presented by AM for the construction industry	169
9.6	Collapse of 3D-printing under the weight of subsequent layers	178
9.7	Temporary shoring to provide stability	178
9.8	Integrated delivery of an AM system for construction	181
10.1	Component catalog from bid documents for Camera Obscura at Mitchell Park by SHoP Architects, ca 2003	191
10.2	Roof module, Heasley Nine Bridges Golf Resort by Shigeru Ban Architects in Seoul, South Korea (2008)	193
10.3	ICD/ITKE Research Pavilion 2014–2015, a composite structure utilizing robotically woven carbon fibers within a pneumatic shell	194
10.4	DFAB HOUSE by ETH as completed in February 2019	195
10.5	In situ Fabricator (IF) deploying the Mesh Mould, a fabrication system for free-form cast-in-place concrete structures, at the DFAB House	197
10.6	Spatial Timber Assemblies in production in the Robotic Fabrication Lab at ETH Zurich	199
10.7	DFAB House Innovation Objects (IOs): Lightweight Translucent Façade, Spatial Timber Assemblies, Smart Slab, Smart Dynamic Casting (SDC), Mesh Mould (listed from top to bottom)	200
10.8	Structure and envelope model for Barclays Center in Brooklyn, NY by SHoP Architects (completed 2012)	201
10.9	Installation of digitally fabricated facade component assemblies at Barclays Center	201
10.10	Lidar scan of Botswana Innovation Hub under construction	203
11.1	Case project – an educational complex that deploys MTP and involves various prefabricated components and multiple trades	210

11.2 Structural BIM model of the case project 213

11.3 BIM for MTP – diffusion of innovations within the
Construction 4.0 context 214

11.4 Number of BIM users (innovators) vs. potential BIM adopters
(imitators) 214

11.5 Simulation results – number of BIM adopters influenced by
dynamic variables 215

11.6 Training effect on BIM for MTP implementation 216

11.7 Importance of BIM training for MTP in increasing early implementation 216

11.8 Optimizing BIM training for MTP 217

12.1 Layers of internet stack 224

12.2 IFC Infra project schema organization 225

12.3 IFC standards triangle 228

12.4 The experiment framework: connecting BACnet, IFC, and CityGML
based databases 232

12.5 IFC data and schema 234

12.6 Example of RDF data model 234

12.7 Example of IFC-based RDF graph 236

12.8 Linked Building Data graph 237

13.1 Various forms of visual data and their frequency of capture 243

13.2 The top image shows MEP and structural system overlay on top of
a 360 image, the bottom left shows a 2D drawing overlay on
orthographic photo, the bottom right shows volumetric measurement
on a point cloud model 244

13.3 Typical challenges in using Standard SfM techniques for image-
based 3D reconstruction: camera misregistration due to repetitive
structure and lack of long range shots (top left); distortions in angle and
distance (top right); curved point cloud due to camera calibration failure
and lack of oblique photo capture (bottom left), and incomplete point
cloud reconstruction due to fail loop closure 246

13.4 Image-based 3D reconstruction workflow 247

13.5 Image-based 3D reconstruction where X1, X2, X3 are three back-
projected 3D points from the visual features of three Images P1, P2,
and P3 247

13.6 Surveying targets used for registration of point clouds and point cloud
to BIM 250

13.7 Progress is shown in as-built and 4D BIM models with color-coded
status superimposed together (left) (M. Golparvar-Fard et al., 2012),
laser scanned as-built (middle) and 4D BIM model (right) 252

13.8 Using patches retrieved from BIM to 2D back-projection to classify
material and performing depth test to exclude occluding area (K. Han
and Golparvar-Fard, 2015) (left); progress status is extracted by
comparing as-built and as-planned after occupancy detection and
material classification (right) 253

13.9   The first column shows progress via 4D point cloud and BIM; the second column shows that location-based 4D BIM model and work-in-progress tracking integrated with point clouds; the third column shows the 4D BIM with subcontractor responsible tasks color-coded to communicate who does what work in what location   256

13.10 The web-based system has been used on different construction site during the coordination meetings, it has been proved that it can efficiently enhance accountability and traceability, and predictive analytics improve reliability in short-term planning   256

14.1   UAS types and examples   266

14.2   UAS degree of autonomy continuum   267

14.3   Level of autonomy   269

14.4   Applying a UAS to project evaluation in an urban area   275

14.5   Applying a UAS to earth moving   276

14.6   Applying a UAS to site transportation   277

14.7   Applying UASs to chimney construction   280

14.8   Applying a UAS to steep slope roof inspection   281

15.1   Number of articles in the ISARC proceedings from 1984 to 2019   290

15.2   History of robotics in construction   291

15.3   General view of SAM and mason worker in the Delbert Day project, MO, USA in 2016   295

15.4   Single task robot to tie rebar in a bridge project   297

15.5   In situ Fabricator   298

15.6   Autonomous installation of gypsum board by HRP-5P   299

15.7   Exterior view of sky factory (a) and general view of the interior workspace (b)   300

16.1   Different subtypes of robots namely a) wheeled, b) tracked, and c) legged depending on the mechanical structure of the robot   309

16.2   Taxonomy of robots based on work environment (i.e. indoor and outdoor), task requirements (e.g. domestic tasks), and construction life cycle phase   310

16.3   Graphical node network representation consisting of 13 nodes and 14 edges   311

16.4   Two of the AprilTags from 36h11 series   313

16.5   Overview of the marker network map design process for autonomous indoor robotic navigation   314

16.6   Example representation of a few different markers   315

16.7   Components of the mobile robot used for ambient data collection in buildings   317

16.8   Marker Network Map (MNM) with virtual information regarding the location stored in each of the markers   318

16.9   Autonomous indoor treasure hunt based navigation algorithm with the help of a network of fiducial markers   320

16.10	Illustration of drift accumulation and marker to marker distance for an indoor mobile robot	320
17.1	The DIKW pyramid	329
17.2	Example of rules for DfS	331
17.3	Six-level DfS taxonomy hierarchy	332
17.4	Structure of DfS rule	332
17.5	Structure of DfS required design feature	333
17.6	List of risk narratives	335
17.7	Defining mitigation narratives ("D" stands for design suggestion and "C" stands for construction suggestion)	335
17.8	Language structure of defining Atomic Rules	336
17.9	Key components for DfS review system	337
17.10	Illustrative BIM	338
17.11	Rule-based checking for design element beam and column	339
17.12	Risk Register for the case example	340
17.13	System architecture of automated tower crane compliance system	341
17.14	A taxonomy for fixed crane spaces	342
17.15	Mapping of spaces to crane information models	343
17.16	Relationships between spaces from CIM and BIM	343
17.17	Workflow for checking safety compliance using SPARQL queries	344
17.18	Proposed layout with Crane Information Models	345
17.19	Plan view showing site boundary, CIM and deep beam position	346
17.20	Elevation view showing CIM and building heights with clearances	346
18.1	Multi-tier architecture analysis of the system	360
18.2	The developed sensor node (expanded)	362
18.3	The BIM model constructions for the system	363
18.4	Integrating mechanism of the WSN data and BIM model component	364
18.5	The design layout of the field test	365
18.6	The field test in the tunnel	366
18.7	Screenshot of the operational CPS for tunnel safety management (Location 01 detected hazardous gas and demonstrates the warning in red)	366
19.1	Construction supply chain	371
19.2	Timelines of stakeholder involvement in construction projects	373
19.3	Lean enabled integrated project controls	380
19.4	Collaborative project controls platform	381
19.5	Levels of development of CSC	385
20.1	Bitcoin block data	402
21.1	Floor plan of buildings	422
21.2	Example of a one-story building with two units	422
21.3	Improvised anchors are used for keeping up the steel formwork panels	423
21.4	Formwork is transported above a worker	424
21.5	A formwork panel is stocked in the main door	424

21.6	Transportation activity made by the telescopic handler	425
21.7	Crane being used for the formwork disassembly and telescopic handler working in the formwork assembly	425
21.8	Seminar with managers	426
21.9	Workshop with workers at work site	426
21.10	Workers are aware the sequence plan and improvisation activities are eliminated	427
21.11	Work-in-progress is reduced	427
21.12	Typical procurement route	429
21.13	Wren system	429
21.14	Typical joint assembly in the Wren system	430
21.15	WikiFarmHouse completed	431
21.16	Drawing section	431
21.17	Concrete foundation	432
21.18	CNC-milled plywood was cut in an offsite workshop	432
21.19	Structural cassettes	432
21.20	Cassettes were lifted onto the foundation	433
21.21	Water-resistant membrane application	433
21.22	Roof and cladding installation	433
21.23	Innovation Lab	434
21.24	Drones used at Innovation Lab	435
21.25	Autonomous vehicle	435
21.26	Worker sensors and alerts	435
21.27	RFID tag scanner	436
21.28	RFID tag	436
21.29	Workers sensors and alerts	437
21.30	Spot-r Clips	438
21.31	Worker data in BIM environment	438
21.32	Cloud-based data collected by drone	439
22.1	The cyber risks of networked CDE and levels of BIM	451
23.1	Chronology of industrial revolutions	461
23.2	Three types of fragmentation in the construction industry	464
23.3	Integration in the industry through the adoption of 4.0 concepts and technologies	466
23.4	Categorization of concepts and technologies in Construction 4.0 into the clusters they belong in (e.g. Smart Construction Site) and corresponding stages of diffusion within different phases of the building life cycle	468
23.5	Construction 4.0 concepts and technologies in terms of their "paths" across stages of innovation diffusion in time, and the envisioned convergence of technologies onto integrated project delivery settings in the upcoming decades	471

23.6 The flows of information (influence) between industry, academia and various project phases today and the changes envisioned to occur as the industry moves forward to a 4.0 state. Disruption waves originating in tech industries as well as the challenges of the 21st century are describing the disciplines to be integrated in the next generation curricula will be necessary   473

TABLES

1.1	Contents of the handbook	18
9.1	13 house construction 3D printers	159
9.2	Cost comparison of AM versus traditional construction	172
12.1	The common data fields between BACnet XML and IFC XML	233
12.2	The common data fields between IFC XML and CityGML	233
14.1	Main requirements of 14 CFR, Part 107 (FAA, 2017)	272
14.2	Main requirements for small UASs (20–25 kgs.) according to international regulations	273
15.1	Buildings constructed by Japanese contractors using construction automation	294
15.2	Projects in which SAM has been used since its first application in 2015	296
18.1	Comparison between Wi-Fi, Bluetooth, and ZigBee	354
18.2	The descriptions of each layer	361
18.3	Developed WSN nodes and its functions	362
19.1	The various processes in the EPC phase of construction projects that involve collaboration and project management	377
20.1	SHA-256 example	401

PART I

Introduction and overview of Construction 4.0, CPS, Digital Ecosystem, and innovation

1

CONSTRUCTION 4.0

Introduction and overview

Anil Sawhney, Mike Riley, and Javier Irizarry

1.1 Aims

- Provide an overview of Industry 4.0 and the Fourth Industrial Revolution.
- Provide a comprehensive review of the current state of the construction sector.
- Describe the overall Construction 4.0 framework.
- Articulate the purpose of Construction 4.0.
- Describe the handbook, its three parts, and its various chapters.

1.2 Introduction to Construction 4.0

With the advent of the Fourth Industrial Revolution (4IR) and the resulting framework of *Industry 4.0 (I4.0)* (MacDougall, 2014), the built environment sector also has the opportunity to leapfrog to more efficient production, business models, and value chains. Such a transformation is possible through the convergence of existing and emerging technologies that form part of the Industry 4.0 paradigm (Oesterreich and Teuteberg, 2016). This transformative framework is called the *Construction 4.0* framework in this handbook. Modelled after the concept of Industry 4.0, the idea of Construction 4.0 is based on a confluence of trends and technologies (both digital and physical) that promise to reshape the way built environment assets are designed and constructed.

In 4IR, the fundamental driver is the use of cyber-physical systems. *Cyber-physical systems* (CPS) are enabling technologies that bring the virtual and physical worlds together to create a truly networked world in which intelligent objects communicate and interact with each other (Griffor et al., 2017). A conceptual model of the CPS is provided in Figure 1.1.

The Construction 4.0 framework uses CPS as a core driver and links it with the concept of Digital Ecosystem where 'A digital ecosystem is an interdependent group of enterprises, people and/or things that share standardized digital platforms for a mutually beneficial purpose, such as commercial gain, innovation or common interest' (Gartner, 2017). The idea of a Digital Ecosystem is shown in Figure 1.2.

Construction 4.0 combines CPS and Digital Ecosystem to create a new paradigm for the design and construction of our built environment assets as shown in Figure 1.3.

CYBER-PHYSICAL SYSTEM

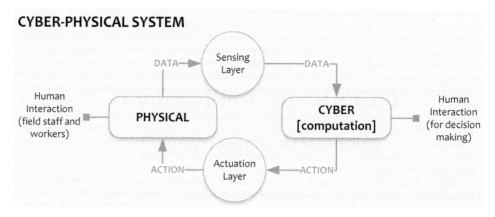

Figure 1.1 Conceptual model of CPS

Digital Ecosystem

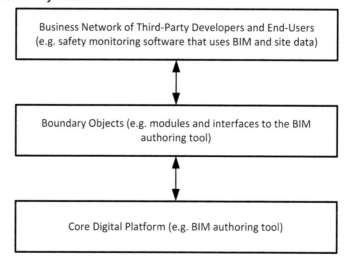

Figure 1.2 Conceptual model of a Digital Ecosystem

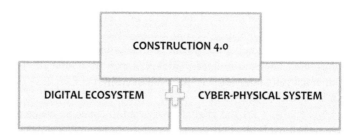

Figure 1.3 Construction 4.0 as a combination of CPS and Digital Ecosystem

Using the CPS, the cyber-physical gap that exists in the built environment can be bridged, and by concomitantly using the Digital Ecosystem the work processes to collaborate efficiently across the project delivery network to design and construct the asset can be enhanced. The Construction 4.0 framework, therefore, provides a mechanism via which we can:

a. Digitally model the built assets that already exist in our physical world.
b. Design new assets in the backdrop of what already exists or plan for the retrofit and rehabilitation of existing assets using these digital models.
c. Once these assets are digitally captured and designed, use digital and physical technologies to deliver these physical assets.

The same framework can be adopted during the operation phase of the constructed asset by using similar digital and physical technologies to support Facilities Management (FM) functions. However, the focus of this handbook is limited to the design and construction phases.

The aim of this handbook is to describe the Construction 4.0 framework and consequently highlight the resultant processes and practices that allow us to plan, design, and deliver built environment assets more effectively and efficiently by focusing on the physical-to-digital transformation and then digital-to-physical transformation. This concept is illustrated graphically in Figure 1.4.

With the pervasive use of Building Information Modeling (BIM), lean principles, digital technologies, and offsite construction the industry is at the cusp of this transformation. The critical challenge is the fragmented state of our teaching, research, and professional practice in the built environment domain. The authors and editors of this handbook aim to overcome this fragmentation by describing Construction 4.0 in the context of current state, emerging trends and technologies, and people and process issues that surround the proposed transformation.

Construction 4.0 is a framework that is a confluence and convergence of the following broad themes:

• Industrial production (prefabrication, 3D printing, and assembly, offsite manufacture).
• Cyber-physical systems (robots and cobots for repetitive and dangerous processes, and drones for surveying and lifting, moving and positioning, and actuators).

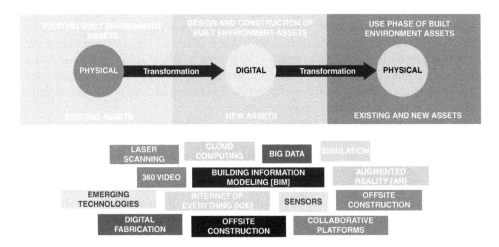

Figure 1.4 Physical to digital and digital to physical transformation

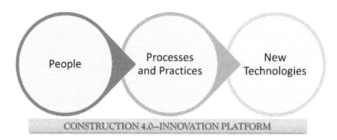

Figure 1.5 Themes of Construction 4.0

- Digital technologies (BIM, video and laser scanning, IoT, sensors, AI and cloud computing, big data and data analytics, reality capture, Blockchain, simulation, augmented reality, data standards and interoperability, and vertical and horizontal integration).

With this background and motivating factors, the handbook will address issues surrounding the key themes of people, processes and practice, and new technologies (as shown in Figure 1.5).

Modern digital and physical technologies are required to achieve the overarching vision of the 4IR (Jacobides, Sundararajan, and Van Alstyne, 2019) that underpins the Construction 4.0 framework, therefore, the framework relies on two broad paradigms: (1) cyber-physical systems and (2) Digital Ecosystems. Innovations in both cyber-physical and digital paradigms are necessary to advance the vision of Construction 4.0 in our industry.

1.3 Current state of the construction sector

Given the importance of the construction sector to their national economies, several countries have undertaken studies to identify the challenges and opportunities that the industry presents. For example, the UK has conducted several prominent studies to document the problems of the construction sector to put in place a program for improvement of the whole-of-the-sector. Sir John Egan, the chair of the Construction Task Force, published his report entitled *Rethinking Construction* in 1998 (Egan, 1998). It was instrumental in laying out a road map for the efficiency improvements within the construction industry in the UK. This came close on the heels of the report, titled 'Constructing the Team' authored by Sir Michael Latham and published in 1994. The Latham report identified inefficiencies and made recommendations for enhanced collaboration and coordination in the industry (Latham, 1994). More recently, the UK released a report by Mark Farmer entitled 'Modernise or Die' (Farmer, 2016) that used a 'strong medical process analogy'. Around the time that this study was being conducted, the UK government also released their Construction 2025 industrial strategy with a plan to commit close to £75 million in research and development.

Other countries, such as the US, Australia, Canada, Singapore, and China, have also undertaken sector-wide studies. For example, a similar exercise was conducted in the US, where Construction Users Roundtable produced a detailed report to outline a path to competitive advantage for construction users.

Several developing nations have also undertaken such studies that identify the problems faced and listed the difficulties hindering growth (Al-Momani, 1995b, 1995a; Edmonds, 1979; Manoliadis, Tsolas, and Nakou, 2006; Moavenzadeh, 1978; Moore and Shearer, 2004; Ofori, 1989, 1994, 2000). This is even more important because in developing countries the construction sector's capacity constraints impact the economic development process (Wells, 2001). These studies have also developed action points necessary for the development of the construction industry (Ofori, 1994, 2000) including the importance of developing key performance indicators (Beatham et al., 2004; Ofori, 2000). A priority-based approach was proposed to rank solutions offered by the researchers and policymakers (Ofori, 1990) with several researchers presenting an optimistic case about the improvement plans (Koenigsberger and Groak, 1978; Turin, 1973).

The repeated nature of these national studies show that there is stagnation and barring some incremental improvements; the industry as a whole has still not managed to show major improvements. The results have been mostly disappointing (Chemillier, 1988; Ofori, 1984, 1990; UNCHS, 1990). Barring a few countries, the problems have persisted over a long period despite efforts made to overcome them. This has been pointed out in the Farmer report that states 'construction has not even made the transition to "industry 3.0" status which is predicated on large scale use of electronics and IT to automate production' (Farmer, 2016; Gerbert et al., 2017). Research has pointed towards a long-term strategic approach to be followed, which is related to the socio-economic needs of the country, often overseen by a steering committee (Farmer, 2016; Ofori, 1994).

The studies described above have generally identified a standard set of challenges or problems that the industry faces. In one such study a list of ten grand challenges (shown in Figure 1.6) faced by the construction sector in India were identified (Sawhney, Agnihotri, and Paul, 2014).

The following are the key challenges that have been collated from these studies (Farmer, 2016; Gerbert et al., 2017; Global Industry Council, 2018; Sawhney and Agnihotri, 2014; Witthoeft and Kosta, 2017):

1. Low levels of research and development leading to a lack of innovation and delayed adoption of technologies.
2. Workforce issues including shortage of young talent due in part to poor industry image.
3. Informal processes and lack of process standardization leading to structural fragmentation.
4. Low levels of cross-functional cooperation and limited collaboration leading to a lack of improvement culture.
5. Low productivity, predictability, and profits.
6. Adversarial transaction-based procurement regime.
7. Insufficient knowledge transfer from project to project.
8. Cultural and mindset issues that act as a blocker to any change.

These issues require a transformational change (Farmer, 2016) in the industry, and we envision that Industry 4.0 can provide a broad framework for such a change.

1.4 Overview of Industry 4.0

During the Hannover Messe in 2011, the German Federal Government released its vision for the future of the manufacturing sector under the broad umbrella term INDUSTRIE 4.0 (Roblek, Meško, and Krapež, 2016). It became part of the 'High-Tech Strategy 2020' project that continues to grow and evolve (MacDougall, 2014). This initiative later became a globally recognized paradigm that was broadly referenced as I4.0, also seen as a precursor to the Fourth Industrial Revolution (Drath and Horch, 2014). Other terms such as smart factory, smart manufacturing, smart production, etc., have also been used to define this broad paradigm (Oesterreich and Teuteberg, 2016).

Similar initiatives have also been launched by other countries. For example, the United States developed the 'Advanced Manufacturing Partnership' in 2014 (Rafael, Jackson, and

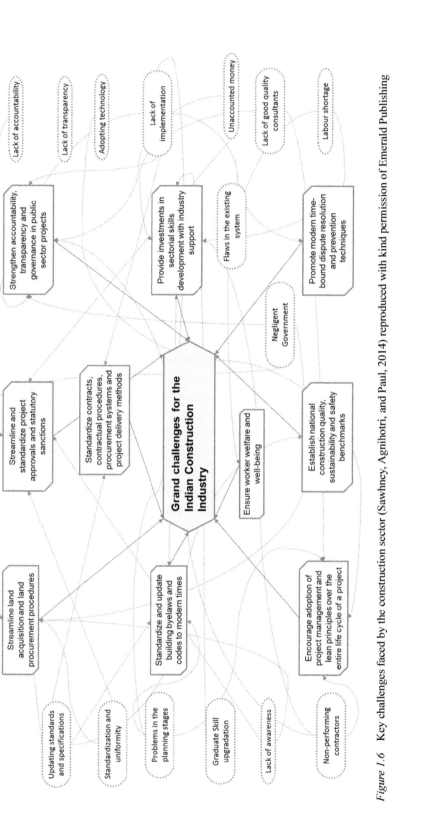

Figure 1.6 Key challenges faced by the construction sector (Sawhney, Agnihotri, and Paul, 2014) reproduced with kind permission of Emerald Publishing

Liveris, 2014) and updated it in 2016, the UK launched an initiative entitled 'Future of Manufacturing' (Foresight, 2013) and China is implementing the 'Made in China 2025' program (Liao et al., 2017).

While the First Industrial Revolution was catalyzed by steam-powered mechanical production, the second was driven by electrical-powered mass production; the third was based on electronics and automation, the Fourth Industrial Revolution has begun with the promulgation of CPS and related technologies (MacDougall, 2014; Pereira and Romero, 2017). It is envisioned that I4.0 will have far-reaching implications on the manufacturing sector that are, in turn, likely to have broad social and economic benefits for nations and societies that embrace this framework (Oesterreich and Teuteberg, 2016). Furthermore, I4.0 uses technologies such as service orientation, smart production, interoperability, cloud computing, big data analytics, and cybersecurity (Vogel-Heuser and Hess, 2016). I4.0 facilitates interconnection and computerization in traditional industries, which makes an automatic and flexible adaptation of the production chain and provides new types of services and business models of interaction in the value chain (Liao et al., 2017; Lu, 2017).

1.4.1 Definition of Industry 4.0

I4.0 is a broad term that has been presented as a 'confluence of trends and technologies promises to reshape the way things are made' (Baur and Wee, 2015). There are several definitions of I4.0 but no globally accepted one because the vision, mission, and components of I4.0 are still emerging and are being connected to more significant and broader themes such as sustainability and circular economy (Lopes de Sousa Jabbour et al., 2018; Müller, Kiel, and Voigt, 2018; Rajput and Singh, 2019).

The German government describes I4.0 as 'a new technological age for manufacturing that uses cyber-physical systems and Internet of Things, Data and Services to connect production technologies with smart production processes' (Kagermann, Wahlster, and Helbig, 2013; MacDougall, 2014) to make manufacturing smart. I4.0 has also been defined at a higher level as 'a new level of value chain organization and management across the lifecycle of products' (Hermann, Pentek, and Otto, 2016; Kagermann, Wahlster, and Helbig, 2013). It is also defined as the integration of machinery and devices with networked sensors and software that can be used to predict, control, and plan for better business and societal outcomes (Shafiq et al., 2015). In a way, I4.0 improves manufacturing organizations, business models that they use, and their production processes through the use of physical and digital technologies.

I4.0 is seen as a cross-cutting paradigm that can have broad social and economic benefits. It is seen as a way to revolutionize manufacturing and other major sectors, such as energy, health, smart cities, and mobility (MacDougall, 2014). The motivation behind this handbook is that I4.0 can also act as catalyst for the future of construction that is more industrialized and automated. We use this motivation to coin the term Construction 4.0.

1.4.2 Key components of I4.0

I4.0 is a very broad and encompassing term. Therefore, it is essential to understand the key components of I4.0. Researchers agree that the push towards I4.0 came from the evolution of embedded systems to more advanced cyber-physical systems (CPS) (Vogel-Heuser and Hess, 2016). This has also formed the basis of the vision developed by the German government. CPS is a set of technologies that connect the virtual and physical worlds together to create a genuinely networked production environment in which intelligent objects communicate and interact with each other (Kagermann, Wahlster, and Helbig, 2013a). The journey towards I4.0 began with the embedded systems and their technological evolution towards CPS and further to provide an Internet of Things (IoT), Data and Services. Figure 1.7 shows this evolution of the embedded systems to CPS.

A CPS is defined as 'a mechanism that is controlled or monitored by computer-based algorithms, tightly integrated with the Internet and its users' (Monostori et al., 2016). CPS creates a virtual copy of the physical production system that is also called the digital twin. This is the first step towards I4.0, where a physical-digital-physical loop is created (Rutgers and Sniderman, 2018). The production environment in the factory that is created through this is also known as the Cyber-Physical Production System (CPPS) (Vogel-Heuser and Hess, 2016). CPPS results in a digitalized, smart, optimized, service-oriented, and interoperable production environment upon which other components of I4.0 are built.

Once the digital twin of a manufacturing environment is created, other business and technical aspects of the production process are linked into the I4.0 framework through the Internet of Things, Data and Services. Figure 1.8 shows the key components of I4.0. The very core of I4.0 is formed by the IoT layer that connects physical objects and things, collects data from these connected objects, and allows connected objects to communicate with each other. Based on this core layer are the CPS and CPPS layers of the I4.0. CPS helps create the digital twin of the physical world, in this case, the manufacturing unit. This allows a loop in which the physical components that are connected to each other relay data that can be used for a variety of purposes including decision making. Changes to the physical world can be made via actuators thereby completing the loop.

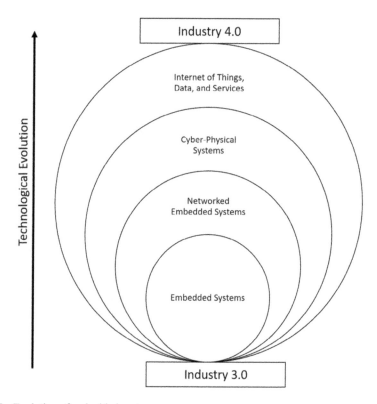

Figure 1.7   Evolution of embedded systems

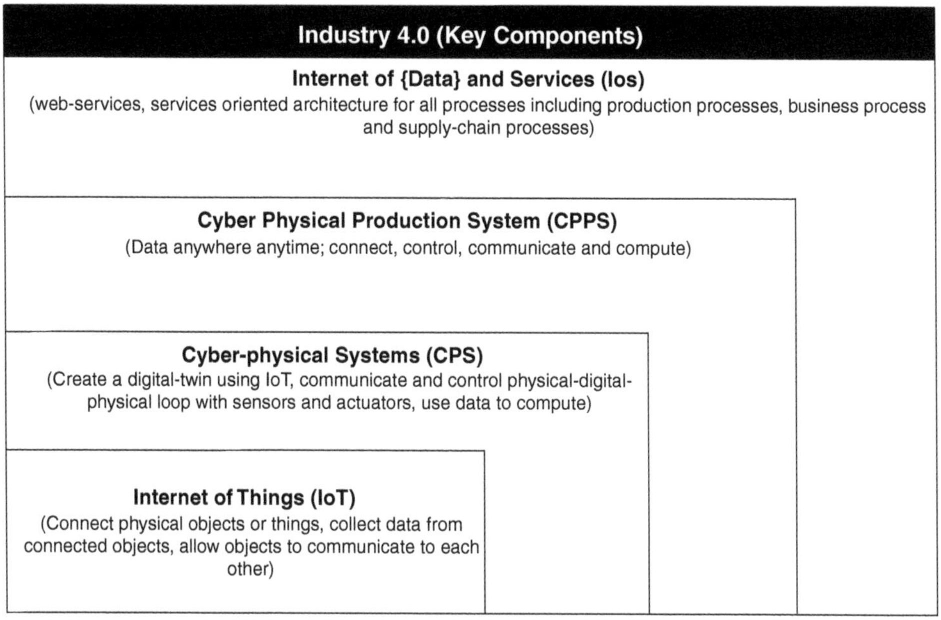

Industry 4.0 (Key Components)
Internet of {Data} and Services (Ios) (web-services, services oriented architecture for all processes including production processes, business process and supply-chain processes)

Cyber Physical Production System (CPPS) (Data anywhere anytime; connect, control, communicate and compute)

Cyber-physical Systems (CPS) (Create a digital-twin using IoT, communicate and control physical-digital-physical loop with sensors and actuators, use data to compute)

Internet of Things (IoT) (Connect physical objects or things, collect data from connected objects, allow objects to communicate to each other)

Figure 1.8   Industry 4.0 and its key components

The CPPS sits on top of the CPS layer and provides data about the physical world anywhere and anytime, and helps connect, control, communicate, and compute. CPPS provides an intensive connection with the surrounding physical world and its ongoing processes (Monostori et al., 2016). Finally, the topmost layer is the Internet of Data and Internet of Services (IoS). The IoS creates a service-oriented ecosystem and brings the end-user of customer centricity to the system (Hofmann and Rüsch, 2017). IoS allows the digital tools that support end-user functions to be available as a service on the Internet (Alcácer and Cruz-Machado, 2019). Both internal and cross-organizational services are offered and utilized by participants of the value chain (Reis and Gonçalves, 2018). The IoS helps create networks incorporating the entire manufacturing process that convert factories into a smart environment (Kagermann, Wahlster, and Helbig, 2013a).

1.4.3   Enabling technologies and key features

In addition to defining the I4.0 framework by describing its key components, the framework can also be defined by identifying its enabling technologies and key features. Liao et al. used over 224 research papers published over five years (2012–2016) to determine these technologies and key features of I4.0 (Liao et al., 2017). Figure 1.9 shows the enabling technologies and key features of I4.0 as determined by the literature review. The vision of I4.0 can be accomplished through a collective deployment of several related technologies (Alcácer and Cruz-Machado, 2019). These technologies work in conjunction with the IoT, CPS, CPPS, and IoS as identified in the previous section (Griffor et al., 2017). Based on the frequency of usage and mention in the literature these technologies are rank-ordered in Figure 1.9.

Similarly the key features of I4.0 from literature are also listed in rank order in the figure. From the literature it can be seen that both in research and practice significant attention is given to automation, integration, and collaboration. Less tractable features such as innovation, quality, and sustainability are still not prevalent.

11

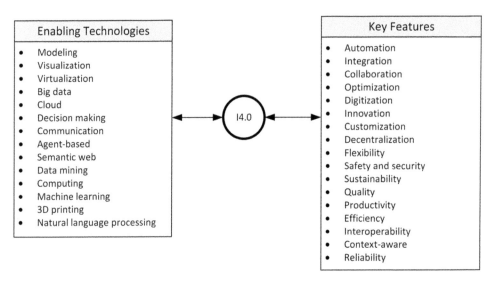

Figure 1.9   Enabling technologies and key features of I4.0

1.4.4 Interoperability and integration in I4.0

Integration and interoperability are two key drivers in the I4.0 framework (Kagermann, Wahlster, and Helbig, 2013a; Vogel-Heuser and Hess, 2016). Interoperability helps two or more systems work with each other to exchange data, information, and knowledge. Interoperability is achieved through a shared understanding of concepts, standards, languages, and relationships (Xu, Da, Xu, and Li, 2018).

I4.0 leads to the integration of processes, systems, applications, and organizations (Oesterreich and Teuteberg, 2016). It is anticipated that I4.0 will allow the following three levels of integration (Kagermann, Wahlster, and Helbig, 2013a):

* Horizontal integration through value networks.
* End-to-end digital integration of engineering across the entire value chain.
* Vertical integration and networked manufacturing systems.

1.4.5 Impact of Industry 4.0

There are several areas that can be impacted and improved by the application of I4.0 at the sector level (Kagermann, Wahlster, and Helbig, 2013; Oesterreich and Teuteberg, 2016; Rose et al., 2016):

1. Productivity improvement: I4.0 provides several improvements such as automation, real-time inventory management, and continuous optimization that lead to productivity enhancement.
2. Increased quality: ongoing monitoring and control of production allows for improved quality of products and services.
3. Increased flexibility: with a customer-centric approach, I4.0 allows manufacturing flexibility through automation and robotics.
4. Increased speed: with enhanced product life cycle management and physical-digital-physical integration, the speed of production is enhanced.

5. Safer and better working conditions: with increased automation, real-time monitoring of incidents, better-designed workstations, and enhanced work structuring, workers have safer and better working conditions.
6. Improved collaboration: as the availability of data is enhanced, and digital layer and physical layer are integrated the intra- and inter-organization collaboration is improved.
7. Sustainability: optimized use of resources, reduction in defects, and other environmental improvements make operations more sustainable.
8. Innovation: I4.0 leads to new ways of creating value and new forms of employment, for example through downstream services.

1.5 Construction 4.0 framework

Figure 1.10 shows the various layers and components of the Construction 4.0 framework. BIM and a cloud-based Common Data Environment (CDE) are central to the Construction 4.0 framework (Cooper, 2018; Oesterreich and Teuteberg, 2016). While BIM provides the modeling and simulation features that are a core component of the I4.0 framework, CDE acts as a repository for storing all the data that relates to the construction project over its life cycle.

The use of BIM and CDE creates a single platform that helps promote:

1. Integration of all phases of the project life cycle (vertical integration), all members of the project team (horizontal integration), and inter-project learning and knowledge management (longitudinal integration).
2. Linkage between the physical and cyber (digital) layer over the entire project life cycle. This allows the implementers of Construction 4.0 to utilize both physical and digital technologies in an integrated manner.

Within the Construction 4.0 framework, the following three transformational trends take place:

1. Industrial production and construction: by using prefabrication, 3D printing, and assembly, offsite manufacture, and automation, the issues and challenges caused by on-site construction techniques are significantly reduced. This type of industrialized process allows production to be digitally linked to BIM and CDE so that instructions can be directly delivered for physical production and any production-related information from the physical layer can be fed back to the digital layer.
2. Cyber-physical systems: the construction site under Construction 4.0 uses robotics and automation for production, transport, and assembly, actuators for converting digital signals into physical actions, and sensors and IoT to sense important information about physical objects (including people) from the physical layer.
3. Digital technologies: the digital transformation relies on the Digital Ecosystem that is developed in the digital layer of the Construction 4.0 framework. BIM and CDE provide the framework upon which integrated digital tools are built. With the help of video and laser scanning technology, artificial intelligence (AI) and cloud computing, big data and data analytics, reality capture, Blockchain, simulation, and augmented reality the delivery and business process is supported in the Construction 4.0 framework. While Digital Ecosystems provide the innovation needed for this support, data standards and interoperability also play an essential role in this overall transformation.

Figure 1.11 shows the components of the Construction 4.0 framework, the role they play in the framework and the layer in which they are present.

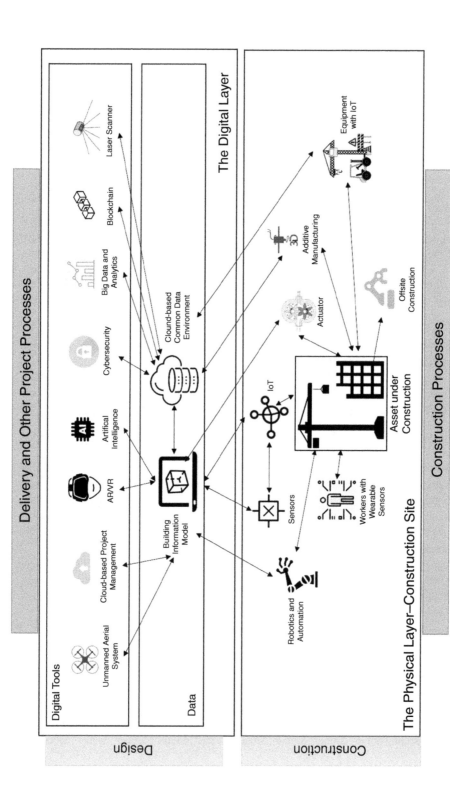

Figure 1.10　Conceptual illustration of Construction 4.0 framework

Emerging Technology or Trend		Construction 4.0 Layer	Construction 4.0 Functions
	BIM	Digital	Modeling and simulation
	CDE	Digital	Collect, manage and disseminate documentation, the graphical model and non-graphical data for the whole project team
	Unmanned Aerial Systems	Digital	Aerial image collection
	Cloud-based Project Management	Digital	Digital tools to support delivery and business processes
	AR/VR	Digital	Virtual application in all phases and for all team members
	Artificial Intelligence	Digital	Classifying, predicting, image processing, mining and problem-solving
	Cybersecurity	Digital	Securing the physical-digital-physical loop
	Big Data and Analytics	Digital	Trend analysis and business intelligence
	Blockchain	Digital	Smart contracts, building trust, and maintaining records
	Laser Scanner	Digital	Point-cloud data collection
	Robotics and Automation	Physical	Transport, assembly and production
	Sensors	Physical	Collect location, temperature, humidity, and movement information
	IoT	Physical	Connectivity of things, people and data
	Workers with wearable sensors	Physical	Collect location, temperature, humidity, and movement information
	Actuators	Physical	Convert digital interactions into physical action
	Additive Manufacturing	Physical	Print parts and products using the BIM model
	Offsite Construction	Physical	Use manufacturing to produce parts and products
	Equipment with Sensors	Physical	Assembly of parts and products in a location aware environment

Figure 1.11   Components of Construction 4.0 framework

1.6 Benefits of Construction 4.0

Several recent studies have attempted to define the I4.0 framework in general and provide a road map for research and implementation based on a detailed literature scan (Alcácer and Cruz-Machado, 2019; Liao et al., 2017; Pereira and Romero, 2017). Similar attempts are being made in regards to Construction 4.0 (Cooper, 2018; Dallasega, Rauch, and Linder, 2018; Oesterreich and Teuteberg, 2016). These studies focus on identifying the sectoral benefits of the Industry 4.0 concept in general and Construction 4.0 in particular. Based on these studies, the benefits of the Construction 4.0 framework are listed below:

1. Enabling an innovative environment: the Construction 4.0 framework may provide the right mix of enablers to allow the innovation mindset to take root in the industry. Through

an integration of the physical and digital layer, it is likely that this innovation will lead to integrated solutions that will strike at the heart of horizontal, vertical, and longitudinal fragmentation that currently dominates the industry.

2. Improving sustainability: the integrated framework of Construction 4.0 allows the industry to fully embrace a life cycle approach and ensure prudent use of resources with a significant reduction in energy usage and emissions.

3. Improving the image of the industry: the construction industry suffers from an image problem caused by several factors. It is well known for its harsh working environment and its low level of automation and digitization (Farmer, 2016; Oesterreich and Teuteberg, 2016). The digital and physical technologies of Construction 4.0 can improve the image of the industry by transforming the work, the worker, and the workplace, and make it more attractive for recruitment and retention of talent.

4. Cost savings: use of industrialized construction supported by digital technologies, BIM, and CDE, can help reduce inefficiencies and waste. Robotics and automation can result in a reduction in direct costs. Real-time access to the physical layer with abundance of data will improve decision making and provide financial incentives for project teams to collaborate and innovate.

5. Time savings: modern methods of construction like prefabrication, additive manufacturing, and on-site assembly will improve the speed of construction. With real-time access to field data, any potential delays can be avoided, resulting in time savings.

6. Enhancing safety: Construction 4.0 will enhance site safety. Augmented Reality/Virtual Reality (AR/VR) based training, wearable technologies, IoT based connectivity of objects, things, and people, image and video processing can enhance safety.

7. Better time and cost predictability: with real-time monitoring, automated site data collection, image processing, AI, and analytics tools the time and cost predictability of ongoing projects can be improved. Availability of large volumes of historical data and information can help set benchmarks for early time and cost prediction of new projects, thereby allowing longitudinal integration.

8. Improving quality: the horizontal and vertical integration resulting from the adoption of Construction 4.0 framework allows the monitoring and control of the design and production processes, thereby improving the quality of construction.

9. Improving collaboration and communication: use of cloud-based project management tools, Blockchain, central repository of information and real-time data access enhances trust among the project team members and enhances communication, coordination, and collaboration.

10. Customer and end-user centric world view: with the reduction in tedious and repetitive tasks, the project team focuses on creating value and focusing on what matters most to the customer.

1.7 Challenges to implementation of Construction 4.0

The Farmer report documented the reluctance of the construction sector to embrace technology and summarized that the industry missed the Industry 3.0 transformation (Farmer, 2016). Dallasega, Rauch, and Linder and Oesterreich and Teuteberg based on an extensive literature review developed the following list of implementation challenges the sector faces while implementing Construction 4.0 framework (Dallasega, Rauch, and Linder, 2018; Oesterreich and Teuteberg, 2016):

1. Resistance to change: the construction sector has a conservative world view when it comes to change. With a considerable resistance, the industry embraces change. Construction 4.0 requires significant change in the way we construct and is likely to face skepticism and resistance.
2. Unclear value proposition: adoption of innovation requires a clear value proposition for all stakeholders. The complex value chain and transactional nature of the sector make it difficult to document benefits and financial gains leading to a hesitation by construction companies to invest.
3. High implementation cost: Construction 4.0 may require high initial investments. This may become a barrier to adoption primarily due to unclear benefits and prediction of cost savings.
4. Low investments in research and development: historically, the construction sector has ignored investments in research and development. This may be a significant blocker of the Construction 4.0 framework.
5. Need for enhanced skills: the industry is already facing a shortage of skilled workforce. Without significant enhancements in education and training, the pipeline of skilled workforce is non-existent. Many aspects of Construction 4.0 may require new roles, new functions, and perhaps new departments or functional teams. This can be a serious threat to implementation.
6. Longitudinal fragmentation: many in the industry believe that we make the same mistakes from one project to the next. This is mainly due to the longitudinal fragmentation in the sector (Fergusson and Teicholz, 1996). To reap benefits from the Construction 4.0 framework it is essential to learn and implement best practices across projects and organizations.
7. Lack of standards: globally agreed standards for the construction sector are important in the transformative processes that are induced by the Construction 4.0 framework. While significant progress has been made on this front, there is still considerable effort needed to embrace data and process standards.
8. Data security, data protection, and cybersecurity: the I4.0 framework mandates that the physical and cyber layers are protected and secured. Lack of security and protection can become a major barrier to the implementation of Construction 4.0.
9. Legal and contractual uncertainty: the construction sector uses a transactional procurement regime that promotes adversarial relationships and limits innovation. The Construction 4.0 framework requires that delivery and procurement processes transform in sync with proposed digital and physical transformations.

1.8 Structure of the handbook

The handbook has the following three parts:

1. Part I: Introduction and overview of Construction 4.0, CPS, Digital Ecosystem, and innovation.
2. Part II: Core components of Construction 4.0.
3. Part III: Practical aspects of Construction 4.0, including case studies and future directions.

Each chapter is structured in such a way that clear and demonstrable linkages to the Construction 4.0 concept and theme of the respective section are provided. The chapters within these three parts and their links to the central theme of the handbook are summarized in Table 1.1.

Table 1.1 Contents of the handbook

Part	Chapter	Chapter title	Link to Construction 4.0
I	1	Construction 4.0: Introduction and overview	Overall vision and description
I	2	Introduction to cyber-physical systems in the built environment	Definition of CPS and its role in the framework
I	3	Digital Ecosystems in the construction industry—current state and future trends	Role of digital tools, platforms and ecosystems
I	4	Innovation in the construction project delivery networks in Construction 4.0	How Construction 4.0 leads to innovation in the industry
II	5	Potential of cyber-physical systems in architecture and construction	Use of CPS in the design and engineering processes
II	6	Applications of cyber-physical systems in construction	Role of CPS in the construction phase
II	7	A review of mixed-reality applications in Construction 4.0	VR and AR applications in Construction 4.0
II	8	Overview of optoelectronic technology in Construction 4.0	Laser scanning of ongoing construction activities
II	9	The potential for additive manufacturing to transform the construction industry	3D Printing of parts, products, and modules
II	10	Digital fabrication in the construction sector	Digital fabrication of parts, products, and modules
II	11	Using BIM for multi-trade prefabrication in construction	BIM for the design and construction phases
II	12	Data standards and data exchange for Construction 4.0	Interoperability
II	13	Computer vision and BIM-driven data analytics for monitoring construction and operation in the built environment	Prediction, forecasting, analytics
II	14	Unmanned Aerial System applications in construction	Data collection, inspection, and imaging
II	15	Future of robotics and automation in construction	Robotics and automation
II	16	Robots in indoor and outdoor environments	Dangerous and repetitive work, inspections and quality control
II	17	Domain-knowledge enriched BIM in Construction 4.0: applications in design-for-safety and crane safety compliance	Knowledge enriched processes and knowledge management
II	18	Internet of Things (IoT) and internet enabled physical devices for Construction 4.0	Machine to machine connectivity
II	19	Cloud-based collaboration and project management	Coordination, communication, and collaboration
II	20	Use of Blockchain for enabling Construction 4.0	Smart contracts and trust building
III	21	Construction 4.0 case studies	Integrated solutions using Construction 4.0 technologies
III	22	Cyber threats and actors confronting the Construction 4.0	Security of physical and digital layers
III	23	Emerging trends and research directions	Research road map

1.9 Conclusion

Construction is a globally significant industry that employs millions of people and contributes massively to the GDP of individual nations and the global economy. However, it is conservative in its approach to innovation and suffers inertia in the face of the need to change. Unlike other industrial sectors such as manufacturing, automotive, and aerospace the construction industry has failed to embrace the opportunities afforded by technology and advances in data management to enhance the efficiency and performance of the sector and the consistency and quality of its outputs. Despite numerous historic attempts to initiate and effect meaningful change the industry still suffers from fragmentation and inefficiencies in process, information flows, and collaborative working. The opportunities afforded by the concepts, principles, and components of I4.0, translated in to a strategic, tactical and operational paradigm as Construction 4.0 have the potential to truly revolutionize the sector. We now approach a potential tipping point at which the concepts generated and applied by forward thinking innovators and early adopters are accepted and applied in the mainstream sector. The effective application of cyber-physical systems and the associated technologies and practices that combine to manifest C4.0 require a new set of skills within the sector workforce. The development of such skills impacts upon the required approaches of educators, employers, and sector leaders to enable a phase shift in the very manner in which construction operates as an industrial and economic domain.

This chapter provides an introduction to the Construction 4.0 framework and its main components. It places them into context with the past and current state of the industry and describes their potential for significant change. The wide adoption of Construction 4.0 related technologies can transform the construction industry into a more efficient and transparent enterprise. Real-time progress monitoring, enhanced quality and safety, and improved communication between stakeholders are just a few of the benefits the industry can enjoy for years to come. It is incumbent on the construction industry to partner with technology innovators, academic institutions, and researchers and educators to make the implementation of Construction 4.0 a reality.

1.10 Summary

- A detailed description of Industry 4.0 and the Fourth Industrial Revolution.
- Key components, technologies, and features of I4.0.
- Benefits of I4.0.
- Description of Construction 4.0.
- Structure of the handbook.
- Future of Construction 4.0.

References

Alcácer, V. and Cruz-Machado, V. (2019) 'Scanning the Industry 4.0: A Literature Review on Technologies for Manufacturing Systems', *Engineering Science and Technology, an International Journal*, 22(3), pp. 899–919. DOI: 10.1016/j.jestch.2019.01.006.

Al-Momani, A. H. (1995a) 'Construction Practice: The Gap between Intent and Performance', *Building Research & Information*, 23(2), pp. 87–91.

Al-Momani, A. H. (1995b) 'The Economic Evaluation of the Construction Industry in Jordan', *Building Research and Information*, 23(1), pp. 30–48.

Baur, C. and Wee, D. (2015) *Manufacturing's next act, McKinsey & Company*. Available at: www.mckinsey.com/business-functions/operations/our-insights/manufacturings-next-act.

Beatham, S. et al. (2004) 'KPIs: A Critical Appraisal of their Use in Construction', *Benchmarking: An International Journal*, 11(1), pp. 93–117. DOI: 10.1108/14635770410520320.

Chemillier, P. (1988) 'Facing the Urban Housing Crisis', *Batiment International, Building Research and Practice*, 16(2), pp. 99–103. DOI: 10.1080/01823328808726874.

Cooper, S. (2018) 'Civil Engineering Collaborative Digital Platforms Underpin The Creation of "Digital Ecosystems"', *Proceedings of the Institution of Civil Engineers – Civil Engineering*, 171(1), pp. 14. DOI: 10.1680/jcien.2018.171.1.14.

Dallasega, P., Rauch, E., and Linder, C. (2018) 'Industry 4.0 as an Enabler of Proximity for Construction Supply Chains: A Systematic Literature Review', *Computers in Industry*, Elsevier 99(August 2017), pp. 205–225. DOi: 10.1016/j.compind.2018.03.039.

Drath, R. and Horch, A. (2014) 'Industrie 4.0: Hit or Hype?' *IEEE Industrial Electronics Magazine*, IEEE 8(2), pp. 56–58. DOI: 10.1109/MIE.2014.2312079.

Edmonds, G. A. (1979) 'The Construction Industry in Developing Countries', *International Labour Review*, 118(3), pp. 355.

Egan, J. and Construction Task Force (1998) *Rethinking Construction*. Available at: www.constructingexcellence.org.uk/pdf/rethinking.construction/rethinking_construction_report.pdf.

Farmer, M. (2016) '*Modernise or die: The Farmer Review of the UK Construction Labour Model*'. London: UK Government. Available at: www.gov.uk/government/publications/constructionlabour-%0Amarket-in-the-uk-farmer-review.

Fergusson, K. J. and Teicholz, P. M. (1996) 'Achieving Industrial Facility Quality: Integration Is Key', *Journal of Management in Engineering*, 12(1), pp. 49–56. DOI: 10.1061/(ASCE)0742-597X(1996)12:1(49).

Foresight. (2013) *The Future of Manufacturing: A New Era of Opportunity and Challenge for the UK*. London: The Government Office for Science. Available at www.bis.gov.uk/foresight.

Gartner. (2017) *Seize the Digital Ecosystem Opportunity*. Stamford, USA: Gartner, Inc. Available at gartner.com/cioagenda.

Gerbert, P. et al. (2017) *Shaping the Future of Construction: A Breakthrough in Mindset and Technology, World Economic Forum (WEF)*. Geneva, Switzerland: World Economic Forum. Available at: www.bcgperspectives.com/Images/Shaping_the_Future_of_Construction_may_2016.pdf.

Global Industry Council (2018) *Five Keys to Unlocking Digital Transformation in Engineering & Construction*. Available at: http://cts.businesswire.com/ct/CT?id=smartlink&url=http%3A%2F%2Faconex.com%2FDigitalTransformation&esheet=51747950&newsitemid=2018012400 5355&lan=en-US&anchor=http%3A%2F%2Faconex.com%2FDigitalTransformation&index=2&md5=7eabd45450cad87dde7ff45fc276a673.

Griffor, E. R. et al. (2017) *Framework for Cyber-Physical Systems: Volume 1, Overview*. Gaithersburg, MD: National Institute of Standards and Technology. 10.6028/NIST.SPP.1500-201.

Hermann, M., Pentek, T., and Otto, B. (2016) 'Design Principles for Industrie 4.0 Scenarios', in *2016 49th Hawaii International Conference on System Sciences (HICSS)*. IEEE, pp. 3928–3937. doi: 10.1109/HICSS.2016.488.

Hofmann, E. and Rüsch, M. (2017) 'Industry 4.0 and the Current Status as Well as Future Prospects on Logistics', *Computers in Industry*, 89, pp. 23–34. DOI: 10.1016/j.compind.2017.04.002.

Jacobides, M. G., Sundararajan, A., and Van Alstyne, M. (2019) *Platforms and Ecosystems: Enabling the Digital Economy, Briefing Paper World Economic Forum*. Available at: www.weforum.org.

Kagermann, H., Wahlster, W., and Helbig, J. (2013). *Recommendations for Implementing the Strategic Initiative INDUSTRIE 4.0 – Final Report of the Industrie 4.0 Working Group Industrie*. Frankfurt, Germany: German Academy of Science and Engineering.

Koenigsberger, O. and Groak, S. (1978) *Essays in Memory of Duccio Turin (1926–1976): Construction an Economic Development Planning of Human Settlements*. Edited by S. G. D. A. Turin, Otto H. Koenigsberger, Oxford, UK: Pergamon Press.

Latham, M. (1994) '*Constructing the Team*. London: HMSO.

Liao, Y. et al. (2017) 'Past, Present and Future of Industry 4.0 – A Systematic Literature Review and Research Agenda Proposal', *International Journal of Production Research*, Taylor & Francis 55(12), pp. 3609–3629. DOI: 10.1080/00207543.2017.1308576.

Lopes de Sousa Jabbour, A. B. et al. (2018) 'Industry 4.0 and the Circular Economy: A Proposed Research Agenda and Original Roadmap for Sustainable Operations', *Annals of Operations Research*, 270(1–2), pp. 273–286. DOI: 10.1007/s10479-018-2772-8.

Lu, Y. (2017) 'Industry 4.0: A Survey on Technologies, Applications and Open Research Issues', *Journal of Industrial Information Integration*, Elsevier 6, pp. 1–10. DOI: 10.1016/j.jii.2017.04.005.

MacDougall, W. (2014) *Industrie 4.0: Smart Manufacturig for the Future*. Berlin, Germany: Germany Trade and Investment. Available at www.gtai.de/GTAI/Navigation/EN/Invest/industrie-4-0.html.

Manoliadis, O., Tsolas, I., and Nakou, A. (2006) 'Sustainable Construction and Drivers of Change in Greece: A Delphi Study', *Construction Management and Economics*, 24(2), pp. 113–120. DOI: 10.1080/01446190500204804.

Moavenzadeh, F. (1978) 'Construction Industry in Developing Countries', *World Development*, 6(1), pp. 97–116.

Monostori, L. et al. (2016) 'Cyber-physical Systems in Manufacturing', *CIRP Annals*, 65(2), pp. 621–641. DOI: 10.1016/j.cirp.2016.06.005.

Moore, J. P. and Shearer, R. A. (2004) 'Expanding Partnering's Horizons: The Challenge of Partnering in the Middle East', *Dispute Resolution Journal*, 59(4), pp. 54–59.

Müller, J. M., Kiel, D., and Voigt, K.-I. (2018) 'What Drives the Implementation of Industry 4.0? The Role of Opportunities and Challenges in the Context of Sustainability', *Sustainability*, 10(1), pp. 247. DOI: 10.3390/su10010247.

Oesterreich, T. D. and Teuteberg, F. (2016) 'Understanding the Implications of Digitisation and Automation in the Context of Industry 4.0: A Triangulation Approach and Elements of a Research Agenda for the Construction Industry', *Computers in Industry*, 83, pp. 121–139. DOI: 10.1016/j.compind.2016.09.006.

Ofori, G. (1984) 'Improving the Construction Industry in Declining Developing Economies', *Construction Management and Economics*, 2(2), pp. 127–132. DOI: 10.1080/01446198400000012.

Ofori, G. (1989) 'A Matrix for the Construction Industries of Developing Countries', *Habitat International*, 13(3), pp. 111–123. DOI: 10.1016/0197-3975(89)90026-X.

Ofori, G. (1990) *The Construction Industry: Aspects of Its Economics and Management*. Singapore: Singapore University Press.

Ofori, G. (1994) 'Practice of Construction Industry Development at the Crossroads', *Habitat International*, 18(2), pp. 41–56. DOI: 10.1016/0197-3975(94)90049-3.

Ofori, G. (2000) 'Challenges of Construction Industries in Developing Countries: Lessons from Various Countries', in *2nd International Conference on Construction in Developing Countries: Challenges Facing the Construction Industry in Developing Countries*. Gabarone, Botswana: Botswana National Construction Industry Council (BONCIC) and University of Botswana, pp. 15–17. doi: 10.1.1.198.2916.

Pereira, A. C. and Romero, F. (2017) 'A Review of the Meanings and the Implications of the Industry 4.0 Concept', *Procedia Manufacturing*, Elsevier B.V. 13, pp. 1206–1214. DOI: 10.1016/j.promfg.2017.09.032.

Rafael, R., Jackson, S. A., and Liveris, A. (2014). *Report to the President Accelerating U.S. Advanced Manufacturing*. Washington, DC.

Rajput, S. and Singh, S. P. (2019) 'Connecting Circular Economy and Industry 4.0', *International Journal of Information Management*, Elsevier 49(March), pp. 98–113. DOI: 10.1016/j.ijinfomgt.2019.03.002.

Reis, J. Z. and Gonçalves, R. F. (2018) 'The Role of Internet of Services (IoS) on Industry 4.0 Through the Service Oriented Architecture (SOA)', in Moon, I. et al. (ed) *Advances in Production Management Systems. Smart Manufacturing for Industry 4.0*. Cham, Switzerland: Springer, pp. 20–26. doi: 10.1007/978-3-319-99707-0_3.

Roblek, V., Meško, M. and Krapež, A. (2016) 'A Complex View of Industry 4.0', *SAGE Open*, 6, pp. 2. DOI: 10.1177/2158244016653987.

Rose, J. et al. (2016). *Sprinting to Value in Industry 4.0*. Chicago, USA: Boston Consulting Group. Available at www.bcg.com/en-us/publications/2016/lean-manufacturing-technology-digital-sprinting-to-value-industry-40.aspx.

Rutgers, V. and Sniderman, B. (2018) *The Industry 4.0 Paradox: Overcoming Disconnects on the Path to Digital Transformation Milwaukee*, USA: Deloitte Consulting LLP.

Sawhney, A. and Agnihotri, R. (2014) 'Grand Challenges for the Indian Construction Industry', *Built Environment Project and Asset Management*, 4(4), pp. 317–334. DOI: 10.1108/BEPAM-10-2013-0055.

Sawhney, A., Agnihotri, R., and Paul, V. K. (2014) 'Grand Challenges for the Indian Construction Industry', *Built Environment Project and Asset Management*, Edited by D. Florence Yean Yng Ling and Dr Carlo 4(4), pp. 317–334. DOI: 10.1108/BEPAM-10-2013-0055.

Shafiq, S. I. et al. (2015) 'Virtual Engineering Object/Virtual Engineering Process: A Specialized form of Cyber Physical System for Industrie 4.0', *Procedia Computer Science*, 60, pp. 1146–1155. DOI: 10.1016/j.procs.2015.08.166.

Turin, D. A. (1973). *The Construction Industry: Its Economic Significance and its Role in Development.* London, UK: Building Economics Research Unit, University College London.

UNCHS. (1990) *People Settlements, Environment and Development.* Nairobi: United Nations Centre for Human Settlements.

Vogel-Heuser, B. and Hess, D. (2016) 'Guest Editorial Industry 4.0–Prerequisites and Visions', *IEEE Transactions on Automation Science and Engineering,* 13(2), pp. 411–413. DOI: 10.1109/TASE.2016.2523639.

Wells, J. (2001) 'Construction and Capital Formation in Less Developed Economies: Unravelling the Informal Sector in an African City', *Construction Management and Economics,* 19(3), pp. 267–274. DOI: 10.1080/01446190010020363.

Witthoeft, S., & Kosta, I. (2017) Shaping the Future of Construction. Inspiring innovators redefine the industry. In *World Economic Forum (WEF).* Retrieved from http://www3.weforum.org/docs/WEF_Shaping_the_Future_of_Construction_Inspiring_Innovators_redefine_the_industry_2017.pdf.

Xu, L. Da, Xu, E. L.,, and Li, L. (2018) 'Industry 4.0: State of the Art and Future Trends', *International Journal of Production Research,* 56(8), pp. 2941–2962. DOI: 10.1080/00207543.2018.1444806.

2

INTRODUCTION TO CYBER-PHYSICAL SYSTEMS IN THE BUILT ENVIRONMENT

Pardis Pishdad-Bozorgi, Xinghua Gao, and Dennis R. Shelden

2.1 Aims

- To introduce the concept of cyber-physical system (CPS).
- To discuss the vision for successful applications of CPS in Construction 4.0.
- To propose a CPS data framework for the Smart Built Environment.
- To propose innovative use cases of CPS implementation in Smart Built Environment.

2.2 Introduction

Cyber-physical systems (CPS) are highly interconnected and integrated smart systems that include engineered interacting networks of physical and computational components (The CPS Public Working Group, 2016). Typically, a CPS consists of the physical part – a device, a machine, or a building – and the digital or cyber part – the data, the software system, and the communication network. The cyber part of CPS represents digitally the state of the physical part and impacts it by automated control or informing people of control actions. For example, if a modern car is looked at from a CPS perspective, the physical part consists of the car's physical being, while the cyber part may involve the data from the sensors, the navigation system, and the radio system, which provide the driver with required information while driving.

The physical and digital twins of CPS are reciprocally connected and synchronized in real time through interconnected sensors and actuators. The digital twin of CPS serves as a medium to visualize, simulate, manifest, observe, and control the physical twin (Shelden, 2018). The key potential of the CPS is the infinite horizon it opens for data analytics that can be performed on the digital twin's sensory input, which is constantly "in sync" with what is happening in the physical world. The data analytics enables measuring the physical twin's performance against the design targets, which were initially simulated in the digital twin, and thus provides a closely coupled feedback loop for assessing the effectiveness of the design. In addition, machine learning tools can be utilized to make future predictions of physical performance based on historical data. Ultimately, through CPS, the digital twin could also perform autonomous control, interventions, and respond to the users' needs.

The applications of laser scanning, Unmanned Aerial Vehicles (UAV), Virtual Reality (VR), Augmented Reality (AR), and Internet of Things (IoT) devices are all examples of how

CPS can transform the construction industry by making it increasingly intelligent, digitally connected, and efficiently performed. This chapter first discusses current CPS applications in the construction industry, and then introduces a framework for achieving a scalable CPS for the Smart Built Environment – Smart Buildings, Smart Communities, and Smart Cities – that enables the utilization of the data housed in separate systems for innovative CPS use cases. The chapter outline includes an introduction to the CPS and its role in Construction 4.0, what does success look like, CPS for Smart Built Environments, and a few examples of application use cases.

2.3 Cyber-physical systems and Construction 4.0

CPS is a novel concept with various potential applications that cannot be fully described yet. It is also an umbrella concept that can represent intelligent systems such as the Internet of Things (IoT), Machine-to-Machine (M2M), Industrial Internet, Smart City, Smart Grid, etc. The CPS are systems of interconnected physical and digital duals. A digital twin is a virtual asset or simulation, running concurrently in real time with its physical twin, with digital and physical twins reciprocally connected by sensors and actuators. CPS are emerging in other industries, such as the telecommunications, the household appliances, and the automobile industry, where digital and physical assets are designed as integrated systems. CPS involve sensor data, computer models, simulation, and physical systems, connected to one another through ubiquitous networked computing (Lee and Seshia, 2016). Numerous examples of CPS are emerging, including autonomous vehicles, smart grids, robotics, and automated manufacturing (Baheti and Gill, 2011, Lee, 2008, Shi et al., 2011, Wang et al., 2015a).

The National Institute of Standards and Technology (NIST) Engineering Laboratory is leading a program to advance cyber-physical systems (NIST, 2017). In this program, the NIST's CPS Public Working Group has developed a CPS framework that presents a set of high-level concepts, their relationships, and a vocabulary for clear communication among stakeholders (e.g. designers, engineers, users) (The CPS Public Working Group, 2016). The goal of the CPS framework is "to provide a common language for describing interoperable CPS architectures in various domains so that these CPS can interoperate within and across domains and form systems of systems" (SoS) (The CPS Public Working Group, 2016).

The NIST CPS framework can be used as guidance in designing, building, and verifying CPS and as a tool for analyzing complex CPS (The CPS Public Working Group, 2016). It consists of three major components: domains, aspects, and facets. The domains are the industries that the CPS can be specialized and applied to, such as building, manufacturing, transportation, and energy. The aspects are high-level groupings of cross-cutting concerns of CPS, involving functional, business, human, data, etc. Concerns are interests in a system relevant to one or more stakeholders. The facets are views on CPS encompassing identified responsibilities in the system engineering process. They contain well-defined activities and artifacts (outputs) for addressing concerns (The CPS Public Working Group, 2016). The three facets, conceptualization, realization, and assurance, deal with three major questions, respectively, which are 1) what things should be and what things are supposed to do, 2) how things should be made and operate, and 3) how to prove things work the way they should (The CPS Public Working Group, 2016). The built environment is a rich area for application of the CPS framework. Smart buildings, cities, and infrastructure are all examples of what we may term cyber-physical environments, where the built environment becomes increasingly intelligent and digitally connected (Shelden, 2018).

CPS is the "heart" of Construction 4.0. As an important part of the Fourth Industrial Revolution, or Industry 4.0, Construction 4.0 encompasses the applications of many innovative technologies such as prefabrication, automation, 3D printing, virtual reality, Augmented Reality, Unmanned Aerial Vehicles (UAV), sensor networks, and robotics for repetitive or unsafe procedures. All of these innovations are CPS or enabled by CPS. These "smart" systems can provide construction stakeholders with the ability to collect data in an automated fashion, to simulate different scenarios, to perform advanced analysis, to visualize the simulations and analysis results, and to control the equipment. Ultimately, CPS is the key to achieve more efficient, more effective, safer, and more environmental-friendly construction projects, which are the goals of Construction 4.0.

2.4 What does success look like?

CPS are formed from integrations among a number of technology assets, some of which are already available to the building industry as by-products of digital design. Geospatial models, including BIM and GIS data, capture the geometry, spatial organization, and relationships among buildings and building elements. Simulations provide representations of building, environment, and occupant behaviors over time. Numerous facets of a building's future performance are already simulated as part of engineering design, from structural performance, air flow, and energy consumption to crowd behavior. Instrumentations are physical devices – sensors and actuators – that provide a means for bi-directionally exchanging physically measured and digitally simulated performance, along with information about device location, and other behaviors (Shelden, 2018).

We look at the construction performance from four stakeholder groups' perspectives – in chronological order of building delivery, they are the design team, the construction team, the operation team, and the users of the building. Hence, we define the performance of a construction project in these four categories: 1) realization of the design intent, 2) construction efficiency, effectiveness, and safety, 3) the building's operability and sustainability, and 4) its usability (Fischer et al., 2017). Emerging CPS enable construction practitioners to achieve improved performance in all these categories. This section discusses the vision for successful CPS applications in the era of Construction 4.0.

2.4.1 Realization of the design intent

The construction industry has been suffering from fragmentations for a long time (Fellows and Liu, 2012). Architecture's position at the beginning of a complex process of procurement and delivery separates the discipline from deep engagement in its ultimate subjects of interest: built projects and their interactions with people, organizations, and the environment (Shelden, 2018). A spectrum of technologies – from Building Information Modeling (BIM) and numerical fabrication to mixed reality and sensor networks – is both redefining and narrowing the gap between concept and realization and between the digital and physical (Shelden, 2018). Despite these capabilities, human intervention and the impacts of delivery processes result in built constructs that are often significantly different from their virtual conception, and by the time of project completion, the design model has long outlived its useful life (Shelden, 2018).

BIM is one major progress in the Architecture, Engineering, Construction, and Operations (AECO) industry. BIM technology involves the creation and utilization of digital building models. It has various applications in building design, construction, and facilities management, including 3D visualization, design checking, various building performance analysis, constructability checking, improved facility operation and maintenance, etc. (Eastman et al., 2011). Ideally, BIM can serve as the integrated data platform that incorporates all the information related to a building. It offers a clear potential as the "digital twin" of the built environment that enables architects and engineers to explicitly elaborate the design. A well-developed BIM model can reduce the design errors and omissions and mis-comprehension to the largest extent possible.

Currently, BIM is already serving as the data infrastructure for innovative CPS applications in construction. For example, VR, AR, and Mixed Reality (MR) solutions use BIM to simulate future built environment for enhanced visualization (Koch et al., 2014, Lee and Akin, 2011, Rüppel and Schatz, 2011, Schatz et al., 2014). This type of visualization provides more intuitive, efficient, and maybe more accurate descriptions of the design intent than the 2D drawing based information delivery systems, hence ensures the design intent is effectively delivered to the builders.

In the era of Construction 4.0, architects and engineers will use BIM as the sole design platform to create the digital twin of a building. The manufacturers and builders will use the developed BIM as the single source of truth to prepare building components and execute construction. In the factory, the CPS – such as robotics and automated machine tools – will fabricate precisely the building components based on the design information in BIM. On the construction site, the builders will also use CPS applications that consume BIM data to ensure the construction is the same as designed. For example, robotics will be widely used on construction sites to improve efficiency and avoid human errors. The input for the robots will be the digital twin of the building and action instructions will be generated automatically. For the construction tasks too complicated to perform by a robot, construction foremen and workers will use VR, AR, or MR to fully understand the design intent and to get support information, such as dimensions, processes, and tool instructions, to perform their work. Hence the finished construction product will be the precise realization of the digital twin.

2.4.2 *Construction efficiency, effectiveness, and safety*

The construction industry has experienced continuous digital transformation, although the adoption rate of CPS innovations is relatively low compared to other industries such as manufacturing. The emergence of CPS brings opportunities for improving the efficiency, effectiveness, and safety of construction operation and management.

2.4.2.1 *Prefabrication*

Significant efficiency improvements in construction have already been brought by prefabrication (Tatum et al., 1987, Zhong et al., 2017). In Construction 4.0, the precise automation of prefabricating construction components is a major aspect of construction efficiency that CPS can facilitate. With the required component information fully described by BIM, and robotics and machines that can read data from BIM, the future factories will be able to produce building components in a highly automated and precise manner. Prefabrication is an area in which the construction industry can directly benefit from the transformation brought by the Industry 4.0 initiative. In a future CPS-enabled factory that produces building components, critical aspects, such as quality control, process optimization, and productivity analysis and prediction, will be enhanced by CPS innovations in the manufacturing industry.

2.4.2.2 Automation on construction site

We envision that in the near future, most of the daily administrative activities on the construction site will be automated and repetitive works, such as bricklaying, will be performed solely by robots. Similar to some research studies that experimented on the automated operation of cranes (Fang et al., 2018, Lee et al., 2012), future construction machines will also be operated in an automated or semi-automated fashion. For the tasks too complicated to be solely performed by automated machines, construction workers equipped with exoskeletons will complete them efficiently, with the support of augmented or mixed visualization techniques that present the detailed component design and the construction procedure. In addition, workflow for deliveries, change orders, invoicing, scheduling, inspections, and quality control, are all optimized to keep the project on schedule, and on budget.

2.4.2.3 Construction progress monitoring

The sensing devices in the field enable CPS to track the movement of labors, materials, equipment, and thus to monitor the resource status and labor behaviors (Ding et al., 2013, Park et al., 2017, Riaz et al., 2014). In addition, the real construction progress can be captured by programed UAVs, in the form of pictures or videos (Irizarry and Costa, 2016, Irizarry and Johnson, 2014, Kim and Irizarry, 2015). With the development of deep learning technologies that can automatically process vast pictures and videos (Greenspan et al., 2016, Lecun et al., 2015, Wan et al., 2014), knowledge about the construction site will be developed by deep learning models based on the data captured by the UAVs. Hence, site inspections will be performed by CPS rather than people in the future. Then, the real construction progress can be compared with the schedule and simulated construction progress in BIM. In this way, construction progress can be monitored automatically.

With Geographic Information System (GIS) and status sensors installed on the personnel's outfits, vehicles, equipment, and assets, important information related to construction will be tracked in real time and available to relevant stakeholders. For example, the cargoes loaded on a truck (can be autopilot) will be automatically checked in when they arrive at the construction site. The project manager will have accurate estimations about the arriving time of different components, provided by the logistic system that analyzes the manufacturing progress, the vehicle's location and the traffic situation. The real-time construction data captured from sensors and mobile devices will also enable in-time communication and collaboration among various parties (Tang et al., 2019).

2.4.2.4 Construction safety and security

Health and safety have been the most critical concerns in the construction industry because it is still regarded as one of the most unsafe industries (Zhou et al., 2015). The improved safety in construction enabled by CPS will be key progress in the age of Construction 4.0. In the future, project managers and safety specialists will have access to the locations of employees and heavy equipment at all times, further ensuring the safety of workers and preventing loss of valuable assets. Moreover, structure monitoring sensors can detect the malfunctioned structural components (including temporary structures on the construction site), hence to ensure site safety (Kim et al., 2016, Yuan et al., 2016).

CPS will be able to provide detailed monitoring of human activities and field situations associated with health and safety risks, identify risky patterns, and suggest programs for correction and improvement. The locations and movements of labors, materials, and equipment will be tracked during construction. Machine learning techniques will be applied to the data collected for safety analysis, which can identify potential hazards and warn of issues before they become problems. The field workers will be equipped with portable devices for early hazard warning (Ding et al., 2013). In addition, serious games for construction health and safety training will be developed and VR will be used to simulate various scenarios during construction.

Because the location and status of important assets in the field will be tracked in real time, CPS will detect any abnormal condition or movement. Moreover, the security system around the parameter of the site will turn on after hours. Any unexpected motion will be detected, and an alarm will go off. Hence, people in charge of site security will not need to look at the video surveillance all the time, or patrol frequently.

2.4.3 Operability and sustainability

Currently, multiple building systems, such as the Building Automation System (BAS), the Building Energy Management Systems (BEMS), the Computerized Maintenance Management System (CMMS) and the security systems, are collecting a large amount of data through sophisticated sensors and emerging smart devices. These data provide the infrastructure that enables innovative CPS for improved building operation and sustainability. By facilitating automated data generation, collection, analysis, prediction, decision making, and actuator control, CPS embedded in a future building will be able to release facility managers from routine building operation tasks, such as checking equipment status, collecting data, monitoring safety, and controlling accessibility.

2.4.3.1 Maintenance

Maintenance is one of the most critical tasks to ensure buildings operate properly. By providing the BIM based 3D visualization, currently, some CPS enable facility managers and technicians to locate building components more efficiently and reduce the effort to comprehend the displayed information (Yang and Ergan, 2016). CPS – which incorporate barcodes, Radio-frequency Identification (RFID), and AR together with BIM – are developed to facilitate maintenance and repair activities by providing relevant information in a timely and intuitive fashion, such that maintenance personnel can easily comprehend it and follow the suggested procedures (Gao and Pishdad-Bozorgi, 2019, Pishdad-Bozorgi, 2017). In addition to providing information, CPS can also analyze maintenance and repair tasks such as equipment fault detection and diagnosis (FDD). Researchers have shown ways of improving maintenance and repair procedures by leveraging CPS's analysis capabilities to detect and locate system faults (Golabchi et al., 2016, Zimmermann et al., 2012) and identify failure cause-effect patterns (Asen et al., 2012, Motamedi et al., 2014).

2.4.3.2 Energy management

Currently, buildings are not operated energy-efficiently (Mills, 2011) and consume 30% to 40% of global energy annually (Sisson et al., 2009). Energy management is challenging by nature because optimizing energy consumption requires understanding the real energy needs and adjusting operation activities accordingly. Energy efficiency becomes an important consideration as early as the schematic design phase of a project, but the real energy consumption data is not available until the operation phase starts. Even though many software applications can perform energy simulation and analysis, there is always a deficit between design intent and real energy performance (Corry et al., 2015, De wilde, 2014, Shiel et al., 2018). Collecting real energy consumption data is critical to provide feedback for future design and simulation (De wilde, 2014, Shiel et al., 2018). CPS have been developed to integrate BIM with the energy data extracted from multiple building systems, such as Building Management System (BMS) (Oti et al., 2016) and BEMS (Dong et al., 2014, Jung et al., 2014, Lee et al., 2016). These CPS can provide facility managers the capability 1) to monitor the energy consumption and to analyze the performance 2) to perform energy simulation and forecast, 3) to visualize energy information through 3D model, 4) to acquire building control suggestions for energy conservation, 5) to monitor thermal condition, 6) to conduct fault detection and diagnosis, and 7) to assess sustainability (Gao and Pishdad-Bozorgi, 2019).

2.4.3.3 Emergency management

The aspects of facility emergency management that CPS can improve include path planning/finding, indoor localization, fire emergency simulation and analysis, and facility safety management. Pathfinding is a fundamental issue in an emergency evacuation situation. It generally refers to "finding the shortest path connecting two points, while avoiding collision with obstacles" (Lin et al., 2013) but pathfinding in the context of emergency is much more demanding. During an emergency, the evacuees need to be directed to safe locations while avoiding potentially dangerous areas; the first responders need to find the trapped evacuees as soon as possible while ensuring their own safety. Determining the optimal path and providing guidance to evacuees and first responders require not only comprehensive building information and real-time situation awareness, but also intelligent algorithms that can analyze the situation and support decision making. By providing both BIM-based accurate 3D geometry and semantic information on a building, CPS are developed for pathfinding and evacuation guidance during an emergency (Chen and Chu, 2015, Lin et al., 2013, Tashakkori et al., 2015).

During an emergency, the real-time locations of first responders and occupants are critical for improving the rescue efficiency and reducing fatalities and injuries (Li et al., 2014). CPS have been created based on the pathfinding and indoor localization. In the future, CPS embedded in a building will be able to identify the number of people in each room and provide the information to first responders when needed. These CPS will have the capability of evacuation assessment, escape route planning, safety education, and fire safety equipment maintenance (Wang et al., 2015b). In addition, the most common emergency for buildings is fire. Currently, serious games with VR have been created for evacuation training (Wang et al., 2014) and exploring the effect of building condition on human behavior during a fire (Rüppel and Schatz, 2011, Schatz et al., 2014). We envision that, in the future, comprehensive CPS used during building operation will have all these emergency management related functionalities and more yet-to-be-developed capabilities that can save human lives.

2.4.4 Usability

Many Smart Building initiatives have described the visions for future CPS-enabled buildings' usability (Sinopoli, 2009, Zhang et al., 2013). Future Smart Buildings will integrate and account for intelligence, enterprise, and control with adaptability at the core, "to meet the drivers for building progression: energy and efficiency, longevity, and comfort and satisfaction"

(Buckman et al., 2014). In light of the usability of current buildings, we are still a long way from the envisioned future Smart Buildings.

If the "smartness" of a building can be compared to the human nervous system, some buildings already have "senses" – vision from cameras, auditory sense from microphones, thermal sense from sensors, etc. However, they do not have a centralized "memory" yet. Currently, a typical commercial building only has some records of the video surveillance and electricity and water usage stored in separate systems. We believe a building has the potential to "know" more about its system operations and occupants' activities. Collecting the information and saving this "memory" in the "brain" of a building is a critical step in realizing a Smart Building. The "memory" of buildings can be invaluable for innovative CPS developments. If established, the "memory" of each Smart Building will provide the data infrastructure for the CPS applications in the era of Construction 4.0.

2.5 CPS for Smart Built Environment

The built environment in general is a critical component of the envisioned Construction 4.0 paradigm. Buildings systems already incorporate proprietary networks of sophisticated sensors and devices in the form of energy systems, security systems, and emerging smart home devices, albeit with limited inter-system connectivity or exposure to the larger networks of CPS devices. These building IoT networks represent potential platforms for CPS deployment, and are data sources of occupancy and space that can provide significantly enhanced value to innovative CPS (Gao et al., 2018).

In this section, we propose a scalable data acquisition framework for the smart built environment – Smart Buildings, Smart Communities, and Smart Cities. The framework enables the utilization of the data housed in separate building systems for innovative CPS use cases. It will help CPS stakeholders understand their common data needs from buildings and identify the overlaps between the data protocols used by different building systems.

The framework is presented in a "top-down" sequence. An architecture of the envisioned Smart Built Environment in Construction 4.0, which is based on the proposed data acquisition framework, is presented in Section 5.1. The proposed conceptual framework to establish the facility data infrastructure is described in Section 5.2. Several examples of application use cases are presented in Section 5.3.

2.5.1 *The envisioned Smart Built Environment in Construction 4.0*

A conceptual Smart Built Environment architecture from the perspective of a Smart Building network is proposed and shown in Figure 2.1. It is our vision of future Smart Cities in the era of Construction 4.0.

In the Smart Built Environment we proposed, as Figure 2.1 shows, multiple Smart Buildings form a community. Multiple Smart Communities – such as residential community, campus, healthcare, commercial, office, and government – form a Smart City. Each facility will be "smart" enough to provide a certain amount of data to the IoT-enabled Smart City network in real-time. The data flow generated in each building is collected by CPS and sent to the data hub of each community, and then connected to the city-level IoT network. The data contents can vary based on the facility type but some of them are universal. We name the data that will be provided by all Smart Buildings "the basic facility data package". The basic facility data package provides the fundamental data for the Smart Building network and enables innovative CPS related to buildings.

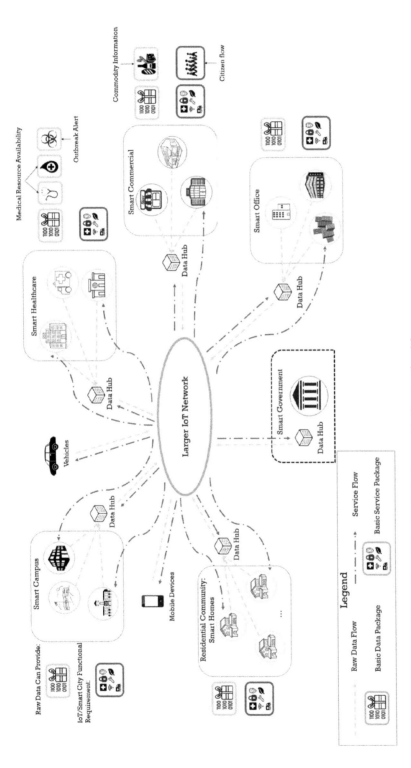

Figure 2.1   An architecture of the envisioned Smart Built Environment in Construction 4.0

Besides the basic facility data package, different data will be provided by certain types of facilities and we name them "extra data". For example, healthcare facilities can provide information pertaining to medical resource availabilities, such as the doctors' schedules and the blood bank inventory. They can also send outbreak alerts to the Smart City network if an infectious disease case is identified. Another example of extra data is that the supermarket in the smart commercial community can provide real-time commodity information to the Smart City network so that citizens can locate the commodities they need. This is particularly crucial when natural disasters, such as hurricanes, tsunamis, and sandstorms, threaten the city and citizens are hoarding necessities.

The Smart Buildings in the proposed architecture not only provide data to the network but also require services from it. The service requirements may vary based on the facility types but there are some common services required by all. We name them "the basic service package". Some examples of the basic service package involve security, emergency assistance, data connection, operation and maintenance, etc. Besides the basic service package, unique services may be requested by certain types of facilities and we name them "extra service". For example, a shopping mall may request real-time citizen flow information from the Smart City network to predict the customer flow (Figure 2.1).

2.5.2 *The establishment of the facility data infrastructure*

Currently, multiple building systems, such as BAS, BEMS, CMMS and the security systems, are collecting a large amount of data through sophisticated sensors and emerging smart devices. However, the inter-system data interoperability is limited, and the data formats vary based on different vendors. Our hypothesis is that the evolving building systems already contain many valuable data for building related CPS innovations but are not effectively exploited because they are not connected, available to analysts and developers in a consumable way. We can establish the federated and integrated building data foundation and enable CPS innovations in future Smart Built Environment by extracting relevant data from multiple building systems, storing them in cloud databases, and connecting these databases through customized Extensible Markup Language (XML) schemas, derived from the data mapping between different data standards.

Figure 2.2 illustrates the high-level framework to establish the basic facility data package. Separate building systems with different data protocols generate various types of raw data, such as the surveillance system generate videos, the Heating, Ventilation, and Air Conditioning (HVAC) system generates temperature information, the security and access control system generates access record, etc. The raw data can be processed by certain techniques – data mining, pattern recognition, Artificial Intelligence, etc. – to produce the "basic data", such as space occupancy rate, people counting, and electricity usage.

Although a huge amount of building data is being generated every hour, the systems that generate these data are established based on various data standards and protocols. Currently, each type of building data is only used for a single purpose. For example, electricity consumption is for energy monitoring, temperature data is for HVAC system control, etc. The lack of a comprehensive usage of these building data is hindering the innovative CPS applications of the Smart Built Environment. A data framework that can provide integrated and comprehensive building data by connecting multiple databases, which are based on various standards and protocols, is the basis for realizing the innovative CPS use cases.

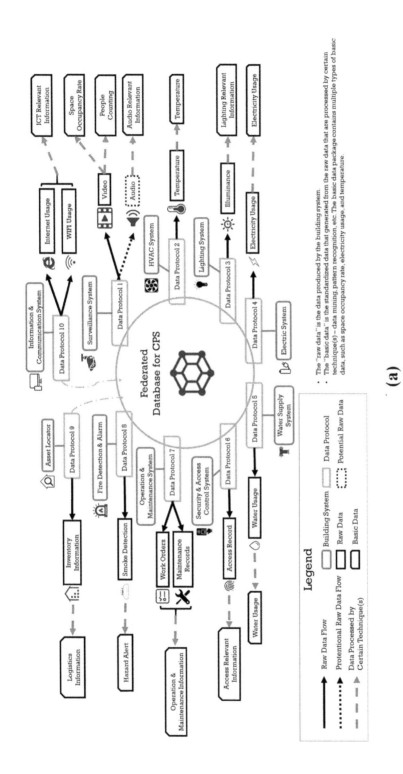

Figure 2.2 The basic facility data package generation

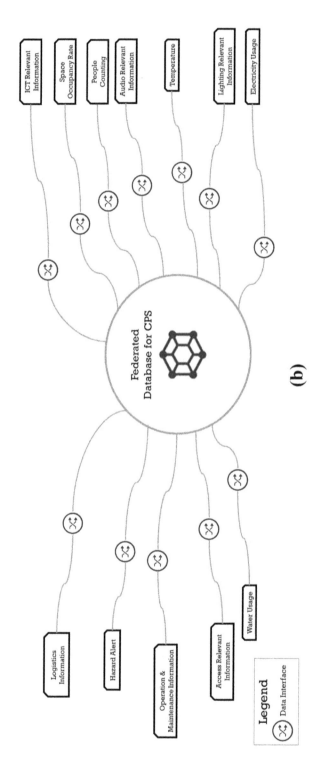

Figure 2.2 (Cont.)

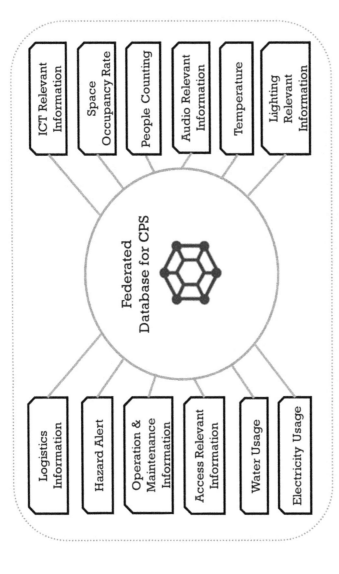

The Basic Facility Data Package

(c)

Figure 2.2 (Cont.)

2.5.3 CPS use cases

Several potential CPS use cases enabled by the proposed basic facility data package are identified. Each of these systems uses a subset of the basic facility data package to enable new CPS.

2.5.3.1 Use case 1: a life-cycle cost analysis (LCCA) system for facilities management decision making

An organization's facility management (FM) department usually faces decision making challenges because the estimated life-cycle cost of a building component or system is not available to facility managers. For example, if the energy performance of a building's HVAC system is worse than expected, it is difficult to determine which of the following options is the most economical one: hiring a vendor to identify the cause and repair it if possible, updating the HVAC system by replacing the old energy-intensive equipment such as the heaters, or just running the system as usual and bearing the extra energy cost. The life-cycle cost analysis (LCCA) can be complicated because if renovation or installation work is required, the facility managers should consider not only the purchasing expense and the actual energy consumption but also the cost of labor, building closure, business interruption, and the influences on the future maintenance.

Facility managers are making important decisions that affect the operation of organizations but, usually, they do not have the life-cycle cost information of facility components to support their decision making because the building systems lack connectivity, even though these systems already contain the information that can be used to derive the life-cycle cost. It is critical to develop a CPS to enable facility managers to perform LCCA efficiently during their decision-making process by connecting the separated building systems, thereby optimizing FM budget allocations.

2.5.3.2 Use case 2: a real-time occupancy information visualization and recording system

This is a web-based CPS that can show the real-time occupancy information of each space in the buildings of an organization. Through color schemed 3D building models, the web shows whether there are people in a certain space or not, and the web server has been recording the historical data of the occupancy information. This CPS can be used for the following purposes:

1. The facility management department analyzes the statistical data to support the decision making in space management: identifying underused and overused spaces, forecasting space requirements, and managing the new or renovation construction projects accordingly.
2. The facility management department can use it to control energy consumption. For example, corrective measures can be taken when a room is unoccupied, such as turning off the computers and air conditioner remotely.
3. During an emergency (fire, shooting, etc.), the police department can use the web to understand which buildings/rooms close to the incident site still have people in them, and dispatch police force and first aid accordingly.
4. The police department can also use the web as a supplementary security system. If the access control system shows there should not be anyone in the building/room but the web shows there is, the police can identify a potential problem.

2.5.3.3 Use case 3: counter in screw guns

This proposed CPS is a smart screw gun with the sensor that can count how many screws have been drilled into wallboards and can recognize who is using it (Gao et al., 2018). The project manager would like to know the statistic of how much material (of various types) is installed and by whom. The goal of this product is to know the productivity of an individual laborer. The users involve carpenters, foremen, project managers, and executives. Sub-devices involve a fingerprint gun trigger lock and the counting mechanism/sensor. Probably, the user's smartphone can work as the data collection device. Besides the general construction project information, additional data are needed, including the time stamp, the location (floor, room), tool status (e.g. how many cycles before failure, proactive and preventative maintenance, output), problems, such as productivity drops (Figure 2.3).

2.5.3.4 Use case 4: automated operation and maintenance request generation (smart floor)

This proposed CPS uses simple robotic agents (Figure 2.4) to 1) monitor building conditions, 2) perform basic cleaning, and 3) generate maintenance requests automatically (Gao et al., 2018). For example, the agent can detect a coffee spill on the floor and clean it. If the agent cannot handle the contamination, it will generate a maintenance request and notify the building operator.

2.6 Conclusion

Cyber-physical systems (CPS) consist of interconnected and integrated smart systems that include both physical and digital or cyber parts. The physical and digital twins of CPS are reciprocally connected and synchronized in real time through interconnected sensors and actuators. The digital twin of CPS serves as a medium to visualize, simulate, manifest, observe, and control the physical twin. The key potential of the CPS is the infinite horizon it opens for data analytics that can be performed on the digital twin's sensory input. The applications of laser scanning,

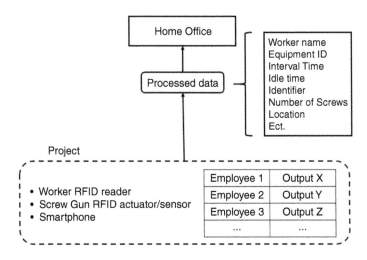

Figure 2.3   Data flow of the proposed screw gun with counter

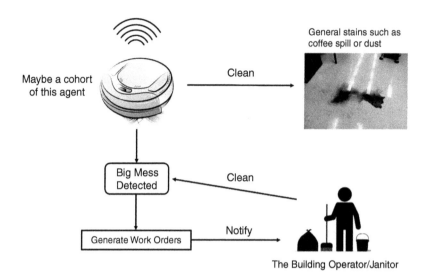

Figure 2.4 The proposed smart robotic agents

Unmanned Aerial Vehicles (UAV), Virtual Reality (VR), Augmented Reality (AR), and Internet of Things (IoT) devices are all examples of how CPS can transform the construction industry by making it increasingly intelligent, digitally connected, and efficiently performed. CPS is the "heart" of Construction 4.0. In this chapter, we discussed our vision for successful applications of CPS in Construction 4.0 spanning different phases of a project life cycle from design intent to construction and operation. During design, BIM technologies can be used to create a digital twin and the cyber part of CPS. BIM together with Augmented Reality (AR), Virtual Reality (VR), and Mixed Reality (MR) can be used for enhanced visualization and communication of design intent among different stakeholders. During construction, BIM models can be used to supply data required for automated fabrication, and robotics to implement standardized and repetitive procedures. CPS can also be applied for construction progress monitoring using an interconnected system of technologies, such as UAVs, BIM, and GIS to monitor, capture, and compare the real-time construction progress with planned schedules in BIM. Other applications include health and safety monitoring of workers in the field and construction site security. During operation, CPS can incorporate barcodes, Radio-frequency Identification (RFID), and AR with BIM to easily locate assets and to facilitate maintenance. CPS can also be used for energy monitoring and management through an integrated and interconnected system including BIM, and Building Energy Management Systems. Other CPS applications include emergency management enabled by an interconnected system involving BIM, GIS, pathfinding algorithms, and sensors. Finally, we proposed a data framework for the envisioned Smart Built Environment and provided a few examples of CPS use cases enabled by the proposed framework.

2.7 Summary

- Cyber-physical systems consist of interconnected and integrated smart systems that include physical and digital or cyber parts.
- Physical and digital twins of CPS are reciprocally connected and synchronized as a medium to visualize, simulate, manifest, observe, and control the physical twin based on the digital twin's sensory input.

- CPS can transform construction using digitally connected, intelligent devices such as laser scanning, Unmanned Aerial Vehicles (UVA), Virtual Reality (VR), Augmented Reality (AR), and Internet of Things (IoT).
- CPS is the "heart" of Construction 4.0 with applications at all phases of a project life cycle from design intent to construction and operation.
- An effective and integrated data framework is essential for the benefits of CPS to be liberated in construction.

References

Asen, Y., Motamedi, A., and Hammad, A. 2012. BIM-based visual analytics approach for facilities management. 14th International Conference on Computing in Civil and Building Engineering. Moscow, Russia.

Baheti, R. and Gill, H. 2011. Cyber-physical systems. *The Impact of Control Technology*, 12, 161–166.

Buckman, A., Mayfield, M., and Beck, B. M. 2014. What is a smart building? *Smart and Sustainable Built Environment*, 3, 92–109.

Chen, A. Y. and Chu, J. C. 2015. TDVRP and BIM integrated approach for in-building emergency rescue routing. *Journal of Computing in Civil Engineering*, 30(5), C4015003.

Corry, E., Pauwels, P., Hu, S., Keane, M., and O'Donnell, J. 2015. A performance assessment ontology for the environmental and energy management of buildings. *Automation in Construction*, 57, 249–259.

The CPS Public Working Group 2016. Framework for cyber-physical systems. Available: https://pages.nist.gov/cpspwg/ Accessed on April 19, 2019.

De wilde, P. 2014. The gap between predicted and measured energy performance of buildings: a framework for investigation. *Automation in Construction*, 41, 40–49.

Ding, L., Zhou, C., Deng, Q., Luo, H., Ye, X., Ni, Y., and Guo, P. 2013. Real-time safety early warning system for cross passage construction in Yangtze Riverbed Metro Tunnel based on the Internet of Things. *Automation in Construction*, 36, 25–37.

Dong, B., O'Neill, Z., and Li, Z. 2014. A BIM-enabled information infrastructure for building energy fault detection and diagnostics. *Automation in Construction*, 44, 197–211.

Eastman, C., Teicholz, P., Sacks, R., and Liston, K. 2011. *BIM Handbook: A Guide to Building Information Modeling for Owners, Managers, Designers, Engineers and Contractors*, 2nd Edition. Hoboken, NJ: Wiley, 1–648.

Fang, Y., Cho, Y. K., Durso, F., and Seo, J. 2018. Assessment of operator's situation awareness for smart operation of mobile cranes. *Automation in Construction*, 85, 65–75.

Fellows, R. and Liu, A. M. 2012. Managing organizational interfaces in engineering construction projects: addressing fragmentation and boundary issues across multiple interfaces. *Construction Management and Economics*, 30, 653–671.

Fischer, M., Ashcraft, H., Reed, D., and Khanzode, A. 2017. *Integrating Project Delivery*, Hoboken, NJ: Wiley, 1–450.

Gao, X. and Pishdad-Bozorgi, P. 2019. BIM-enabled facilities operation and maintenance: a review. *Advanced Engineering Informatics*, 39, 227–247.

Gao, X., Tang, S., Pishdad-Bozorgi, P., and Shelden, D. 2018. Foundational research in integrated building Internet of Things (IoT) data standards. Available: https://cdait.gatech.edu/sites/default/files/georgia_tech_cdait_research_report_on_integrated_building_-_iot_data_standards_september_2018_final.pdf Accessed on April 19, 2019.

Golabchi, A., Akula, M., and Kamat, V. 2016. Automated building information modeling for fault detection and diagnostics in commercial HVAC systems. *Facilities*, 34, 233–246.

Greenspan, H., Van ginneken, B., and Summers, R. M. 2016. Guest editorial deep learning in medical imaging: overview and future promise of an exciting new technique. *IEEE Transactions on Medical Imaging*, 35, 1153–1159.

Irizarry, J. and Costa, D. B. 2016. Exploratory study of potential applications of unmanned aerial systems for construction management tasks. *Journal of Management in Engineering*, 32, 05016001.

Irizarry, J. and Johnson, E. N. 2014. Feasibility study to determine the economic and operational benefits of utilizing unmanned aerial vehicles (UAVs). Available: https://smartech.gatech.edu/handle/1853/52810?show=full Accessed on April 19, 2019.

Jung, D.-K., Lee, D., and Park, S. 2014. Energy operation management for smart city using 3D building energy information modeling. *International Journal of Precision Engineering and Manufacturing*, 15, 1717–1724.

Kim, K., Cho, Y., and Zhang, S. 2016. Integrating work sequences and temporary structures into safety planning: automated scaffolding-related safety hazard identification and prevention in BIM. *Automation in Construction*, 70, 128–142.

Kim, S. and Irizarry, J. 2015. Exploratory study on factors influencing UAS performance on highway construction projects: as the case of safety monitoring systems. Conference on Autonomous and Robotic Construction of Infrastructure, Ames, IA, 132.

Koch, C., Neges, M., König, M., and Abramovici, M. 2014. Natural markers for augmented reality-based indoor navigation and facility maintenance. *Automation in Construction*, 48, 18–30.

Lecun, Y., Bengio, Y., and Hinton, G. 2015. Deep learning. *Nature*, 521, 436.

Lee, D., Cha, G., and Park, S. 2016. A study on data visualization of embedded sensors for building energy monitoring using BIM. *International Journal of Precision Engineering and Manufacturing*, 17, 807–814.

Lee, E. A. Cyber physical systems: design challenges. 2008 11th IEEE International Symposium on Object and Component-Oriented Real-Time Distributed Computing (ISORC), 2008. IEEE, 363–369.

Lee, E. A. and Seshia, S. A. 2016. *Introduction to Embedded Systems: A Cyber-physical Systems Approach*, Cambridge, MA: MIT Press.

Lee, G., Cho, J., Ham, S., Lee, T., Lee, G., Yun, S.-H., and Yang, H.-J. 2012. A BIM-and sensor-based tower crane navigation system for blind lifts. *Automation in Construction*, 26, 1–10.

Lee, S. and Akin, Ö. 2011. Augmented reality-based computational fieldwork support for equipment operations and maintenance. *Automation in Construction*, 20, 338–352.

Li, N., Becerik-Gerber, B., and Soibelman, L. 2014. Iterative maximum likelihood estimation algorithm: leveraging building information and sensing infrastructure for localization during emergencies. *Journal of Computing in Civil Engineering*, 29, 04014094.

Lin, Y.-H., Liu, Y.-S., Gao, G., HAN, X.-G., Lai, C.-Y., and Gu, M. 2013. The IFC-based path planning for 3D indoor spaces. *Advanced Engineering Informatics*, 27, 189–205.

Mills, E. 2011. Building commissioning: a golden opportunity for reducing energy costs and greenhouse gas emissions in the United States. *Energy Efficiency*, 4, 145–173.

Motamedi, A., Hammad, A., and Asen, Y. 2014. Knowledge-assisted BIM-based visual analytics for failure root cause detection in facilities management. *Automation in Construction*, 43, 73–83.

NIST. 2017. *Cyber-Physical Systems*. Available: www.nist.gov/el/cyber-physical-systems, Accessed on April 19, 2019.

Oti, A., Kurul, E., Cheung, F., and Tah, J. 2016. A framework for the utilization of building management system data in building information models for building design and operation. *Automation in Construction*, 72, 195–210.

Park, J., Chen, J., and Cho, Y. K. 2017. Self-corrective knowledge-based hybrid tracking system using BIM and multimodal sensors. *Advanced Engineering Informatics*, 32, 126–138.

Pishdad-Bozorgi, P. 2017. Future smart facilities: state-of-the-art BIM-enabled facility management. *Journal of Construction Engineering and Management*, 143, 02517006.

Riaz, Z., Arslan, M., Kiani, A. K., and Azhar, S. 2014. CoSMoS: a BIM and wireless sensor based integrated solution for worker safety in confined spaces. *Automation in Construction*, 45, 96–106.

Rüppel, U. and Schatz, K. 2011. Designing a BIM-based serious game for fire safety evacuation simulations. *Advanced Engineering Informatics*, 25, 600–611.

Schatz, K., Schlittenlacher, J., Ullrich, D., Rüppel, U., and Ellermeier, W. 2014. Investigating human factors in fire evacuation: a serious-gaming approach. In K. Schatz, J. Schlittenlacher, D. Ullrich, U. Rüppel, and W. Ellermeier (eds) *Pedestrian and Evacuation Dynamics 2012*, New York: Springer, 600–611.

Shelden, D. 2018. Cyber-physical systems and the built environment. *Technology|Architecture+ Design*, 2, 137–139.

Shi, J., Wan, J., Yan, H., and SUO, H. A survey of cyber-physical systems. 2011 International Conference on Wireless Communications and Signal Processing (WCSP), 2011. IEEE, 1–6.

Shiel, P., Tarantino, S., and Fischer, M. 2018. Parametric analysis of design stage building energy performance simulation models. *Energy and Buildings*, 172, 78–93.

Sinopoli, J. M. 2009. *Smart Buildings Systems for Architects, Owners and Builders*, Oxford: Butterworth-Heinemann.

Sisson, W., Van-Aerschot, C., Kornevall, C., Cowe, R., Bridoux, D., Bonnaire, T. B., and Fritz, J. 2009. Energy efficiency in buildings: transforming the market. Available: www.commonenergyproject.eu/uploads/biblio/document/file/244/WBCSD_Green_Construction.pdf Accessed on April 19, 2019.

Tang, S., Shelden, D., Eastman, C., Pishdad-Bozorgi, P., and Gao, X. 2019. A review of building information modeling (BIM) and Internet of Things (IoT) devices integration: present status and future trends. *Automation in Construction*, 101, 127–139.

Tashakkori, H., Rajabifard, A., and Kalantari, M. 2015. A new 3D indoor/outdoor spatial model for indoor emergency response facilitation. *Building and Environment*, 89, 170–182.

Tatum, C., Vanegas, J. A., and Williams, J. 1987. *Constructability Improvement Using Prefabrication, Preassembly, and Modularization*, Bureau of Engineering Research, University of Texas, Austin, TX, USA.

Wan, J., Wang, D., Hoi, S. C. H., Wu, P., Zhu, J., Zhang, Y., and Li, J. Deep learning for content-based image retrieval: a comprehensive study. Proceedings of the 22nd ACM International Conference on Multimedia, 2014. ACM, 157–166.

Wang, B., LI, H., Rezgui, Y., Bradley, A., and Ong, H. N. 2014. BIM based virtual environment for fire emergency evacuation. *ScientificWorldJournal*, 2014, 589016.

Wang, L., Törngren, M., and Onori, M. 2015a. Current status and advancement of cyber-physical systems in manufacturing. *Journal of Manufacturing Systems*, 37, 517–527.

Wang, S.-H., Wang, W.-C., Wang, K.-C., and Shih, S.-Y. 2015b. Applying building information modeling to support fire safety management. *Automation in Construction*, 59, 158–167.

Yang, X. and Ergan, S. 2016. Design and evaluation of an integrated visualization platform to support corrective maintenance of HVAC problem-related work orders. *Journal of Computing in Civil Engineering*, 30(3).

Yuan, X., Anumba, C. J., and Parfitt, M. K. 2016. Cyber-physical systems for temporary structure monitoring. *Automation in Construction*, 66, 1–14.

Zhang, D., Shah, N., and Papageorgiou, L. G. 2013. Efficient energy consumption and operation management in a smart building with microgrid. *Energy Conversion and Management*, 74, 209–222.

Zhong, R. Y., Peng, Y., Xue, F., Fang, J., Zou, W., Luo, H., Ng, S. T., Lu, W., Shen, G. Q., and Huang, G. Q. 2017. Prefabricated construction enabled by the Internet-of-Things. *Automation in Construction*, 76, 59–70.

Zhou, Z., Goh, Y. M., and Li, Q. 2015. Overview and analysis of safety management studies in the construction industry. *Safety Science*, 72, 337–350.

Zimmermann, G., Lu, Y., and Lo, G. 2012. Automatic HVAC fault detection and diagnosis system generation based on heat flow models. *HVAC&R Research*, 18, 112–125.

3

DIGITAL ECOSYSTEMS IN THE CONSTRUCTION INDUSTRY—CURRENT STATE AND FUTURE TRENDS

Anil Sawhney and Ibrahim S. Odeh

3.1 Aims

- Provide an overview of ecosystems and platforms.
- Provide a comprehensive review of digital ecosystems in general and in construction in particular.
- Discuss the role of the digital ecosystem in the Construction 4.0 framework.
- Articulate the purpose of the digital ecosystem in promoting innovation in the industry.
- Describe examples to explain the implementation of digital ecosystems in construction.

3.2 Introduction to digital ecosystems

A digital ecosystem is a complex intermeshing of an interdependent group of organizations, people, products, and things that work on a shared digital platform for a mutually beneficial purpose and value creation (Tiwana, Konsynski, and Bush, 2010; Gartner, 2017). Digital ecosystems have been popularized by the success of software ecosystems such as Firefox browser, Apple iOS (Tiwana, Konsynski, and Bush, 2010), and by high-tech businesses like Google, Intel, Cisco (Gawer and Cusumano, 2014), and many similar initiatives.

Digital ecosystems are the key drivers of innovation as is evident from other sectors of the economy, e.g. ride-sharing, mobile phone apps, social networking, etc. They help drive innovation both within and outside the organizations that participate in the ecosystem (Gawer and Cusumano, 2014). For example, the Apple iOS operating system is a digital ecosystem that has allowed innovation to take place in the mobile phone business by bringing organizations (Apple, app developers, hardware manufacturers, etc.), people (mobile phone users, designers, app developers, etc.), products (iPhone, iPad, iPod, etc.), and things (add-on hardware and software). Rather than selling iOS operating systems to phone manufacturers as a product, Apple has created this digital ecosystem which has driven innovation, resulting in higher profits, and value and productivity gains for users of mobile devices. There are some downsides to this phenomenon also, e.g. tight control of Apple over its devices, higher prices of iPhone, etc.

While this is a simple description of digital ecosystems, there are several other formal definitions of digital ecosystems that will be introduced in section 4 of this chapter.

3.2.1 Construction 4.0 and digital ecosystems

New digital and physical technologies are required to achieve the overarching vision of the Fourth Industrial Revolution (Jacobides, Sundararajan, and Van Alstyne, 2019) that underpins the Construction 4.0 (C4.0) framework. As described in the introduction chapter, the C4.0 framework relies on two broad paradigms: (1) cyber-physical systems (CPS) and (2) digital ecosystems. Innovations in both cyber-physical and digital paradigms are necessary to advance the vision of Construction 4.0 in our industry.

Figure 3.1 shows the role of digital ecosystems in the C4.0 framework. In C4.0, CPS connects the physical layer, i.e. the production space on a construction site to the digital layer. The digital layer uses the Internet of Data and Services to provide a layer consisting of a virtual model of what is being constructed and cloud-based storage to act as a repository of data and information. The top-most layer in this representation of C4.0 is the digital ecosystem that consists of a core digital platform and digital add-in tools that are known as complementary digital tools. For example, in this representation of C4.0 the construction site of building projects is the physical layer. Inside the physical layer technologies such as sensors, actuators, robotics, etc. are placed to connect this layer to the digital layer which in turn may consist of a Building Information Model (BIM) and a cloud-based Common Data Environment (CDE). Within the digital layer a BIM-based digital ecosystem exists. Organizations that are part of this BIM-based digital ecosystem work together to produce innovative solutions for various design, construction, and delivery tasks. For example, a construction company within the ecosystem can work with trade contractors and third-party software developers to develop a tool for real-time monitoring of construction workers to track the safety, productivity, and worker well-being.

At this stage, it is important to briefly discuss the broad contours of an ecosystem, especially the idea of platforms as part of ecosystems. The term 'platform' used in Figure 3.1 is

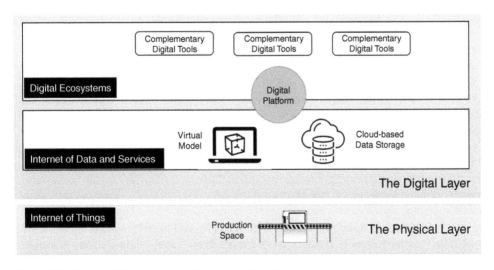

Figure 3.1 Role of digital ecosystems in Construction 4.0

crucial to the discussion about the importance of the digital ecosystem in the Construction 4.0 framework. Platforms, as will be described later, are core to the concept of ecosystems. Sometimes the terms ecosystem and platform are used interchangeably. In a digital ecosystem, software platforms that are the extensible codebase (Tiwana, Konsynski, and Bush, 2010) become the main drivers of innovation, emergent behavior of ecosystems, and creation of value. However, the idea of platforms is not new and is not specific to the software industry. In addition to digital, platforms can be business-centric platforms, product (physical) platforms, brand platforms, etc. (Sawhney, 1998).

3.2.2 *Role of digital ecosystems in construction 4.0*

The authors envision the co-emergence of digital ecosystems (e.g. BIM software-based ecosystems) and product ecosystems (e.g. modular or offsite product based ecosystems) in the Construction 4.0 innovation journey. This concept is illustrated in Figure 3.2. We believe that the Construction 4.0 framework is possible with a combination of three transformative processes (each illustrated as a vertical pillar in the figure):

1.   Product transformation
2.   Digital transformation
3.   Transformation in project delivery processes and related business processes

While the digital transformation is driven by the concept of digital ecosystems, there is a need for the simultaneous evolution of product-platform based ecosystems to reap the full benefits of Construction 4.0. The transformation journey on the digital front has already begun. As an industry, we are moving away from a non-model based approach to a more model-centric approach (BorjeGhaleh and Sardroud, 2016; Sawhney, Khanzode, and Tiwari, 2017). The next step in this digital journey is the evolution towards an integrated, synchronous, and collaborative model-centric approach (World Economic Forum, 2018). Once this step is achieved the digital systems used by the industry will integrate with cyber-physical systems that, for example, will help in creating digital twins.

On the physical front, the transformation journey is also moving away from 'stick-built' on-site methods towards modular and offsite construction. In the physical transformation of the construction industry, product platforms can play a significant role (Bryden Wood, 2018).

As shown in the figure, the digital transformation for our industry is facilitated by digital ecosystems; the product transformation is underpinned by product platforms but is incomplete without considering changes to the current delivery and procurement regime. Therefore, concomitant with these two transformations, the industry must also transform the delivery processes and commercial terms from a transactional to enterprise-centric and integrated one (Construction Leadership Council, 2018).

For an industry that is generally regarded as a laggard when it comes to technology adoption, digital ecosystems can help it leapfrog the digital transformation and lead to sector-level productivity improvements (Cooper, 2018).

In this chapter, digital ecosystems are defined in more detail by tracking their brief history and by considering their applications in other sectors. Physical platforms and ecosystems are also briefly described as these are covered elsewhere in the handbook in chapters related to additive manufacturing and offsite construction.

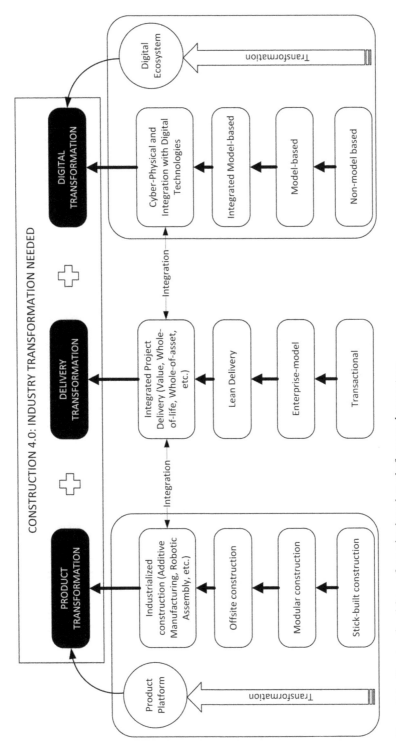

Figure 3.2 Construction 4.0 transformation based on platforms and ecosystems

3.3 Current state of digital technologies in construction

The industry agenda formulated by the World Economic Forum in February 2017 summarized that 'the construction industry has been slow to adopt new technologies and processes and over the past 50 years has undergone no fundamental change' (Gerbert et al., 2017). This may not be for long, however. As the construction sector is seeing significant investments in new technology startups that are in the process of developing new digital tools for the industry— with some reports estimating total expenditures of $10 billion in the 18 months beginning January 2018 (Putzier, 2018), it seems that there are some changes on the horizon. However, the sector has not shown much improvement in terms of investments into information technology—68.2 per cent of companies reporting investments of 1 per cent or less as a percentage of annual sales volume in 2017 compared to 64.1 per cent of companies stating the same level of investments in 2018 (JBKnowledge, 2018).

It is now clear that the construction industry must modernize as a sector and fully embrace a digital transformation. To undertake the change, the authors feel that the traditional model of transformation popular in the industry must be documented alongside the deficiencies it generates. In Figure 3.3, we have developed a high-level illustration of the traditional model of adoption of digital technologies in construction.

Under the traditional approach, the following four challenges exist:

1. *Take-one-at-a-time approach*: most organizations consider each innovation, including digital technologies 'one-at-a-time' and therefore, do not gain the maximum benefits possible due to the interconnected nature of these innovative ideas (Cone, 2013). The root of this issue lies in the fragmented view of the environment-related improvements

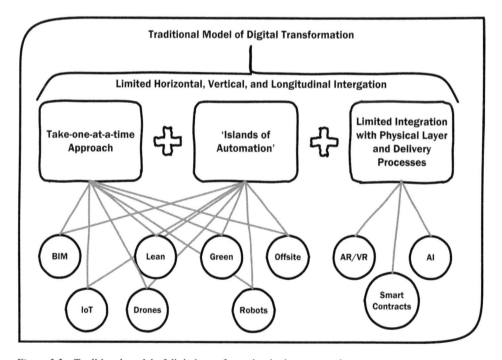

Figure 3.3 Traditional model of digital transformation in the construction sector

(green initiatives), digital-related improvements (BIM) and the process-related (lean principles-based initiatives) improvements by the construction entities (Cone, 2013; Ahuja, Sawhney, and Arif, 2018).

2. *Islands of automation* (Hannus, 1996): with limited desire to adopt digital technologies, the construction sector has taken information technology as an isolated development of functional, departmental, or organizational solutions (Hannus, 1996; Bowden et al., 2006). This has led to a situation where 'islands of automation' have developed requiring a need for 'bridges' of interoperability. Ad-hoc development of isolated, function-driven solutions (Bowden et al., 2006) for the sector is therefore not desirable.

3. *Limited integration with physical layer*: the information technology implementations currently undertaken by entities in the construction sector do not consider the physical-digital-physical loop (Rutgers and Sniderman, 2018). Real-time data from the physical layer, i.e. the asset being constructed and the surrounding space is not connected back to the digital layer. This limited integration makes digital transformation in our industry difficult. Limited and ad-hoc use of sensors and edge computing leads to further fragmentation.

4. *Incremental improvements in delivery processes*: physical and digital transformation of the industry is usually hampered by the transactional nature of the delivery processes. The transformative initiatives in the industry do not account for the delivery processes and procurement regimes that are currently in play.

The traditional model described above leads to low uptake of digital solutions. This leads to the limited influence of the digital solutions across the life cycle stages of the projects (vertical fragmentation), across the project team members and project supply chain (horizontal fragmentation), and across ongoing and new projects (fragmentation) (Fergusson and Teicholz, 1996).

The Construction 4.0 framework in general and digital ecosystems, in particular, provide an impetus to overcome these fragmentations and therefore are crucial in the digital transformation journey.

3.4 Overview of ecosystems and platforms

To understand ecosystems and platforms, it is essential to understand the emerging shift from the traditional product-based firms where production, and to a large extent innovation, only happens inside the firm (Jacobides, Sundararajan, and Van Alstyne, 2019) to firms that use platform and ecosystem. For example, a widget-maker makes widgets (e.g. a Building Information Modeling (BIM) authoring software) and sells these widgets to the customers at a profit using the mantra of selling more and more, for less and less (as shown in Figure 3.4). Platform firms, on the other hand, use a core product (or idea or concept) and invert the firm by opening up certain portions of their product's design to their partners and customers (Sawhney, 1998). Two sides of the market, e.g. in the case of a bidding platform shown in Figure 3.5, i.e. project sponsors and bidders come together to use the platform and extend the platform in specific (limited) ways to enhance the usage of the platform itself thus making it valuable for both sides.

Therefore, platforms are resources, computing or otherwise, that connect different stakeholder groups and derive benefits from others participating in the platform. In a platform-based approach, not only the product adds value for the respective players, but also the participation of the two sides creates additional value for these participants.

A product firm is concerned with selling a, typically standalone, product. The goal of the firm is to sell the product to as many customers as possible at the best price to maximize profits. Although the product may have to be integrated with other systems, the product firm only needs to focus on a single stakeholder group, i.e. the buyers. A platform firm, on the other hand, seeks to commercialize in a multi-sided market where, in the simplest case, suppliers and consumers transact to exchange value by using the platform. This commercialization means the platform firm needs to balance the interests of multi-sided stakeholders with their interests in a way that leads to the best value creation environment.

In Figure 3.6, this idea is extended further to introduce the concept of ecosystems. In an ecosystem, a core product, concept, or design is used as a platform to develop a business network such that entities participating in the network create additional complementary products, concepts, and designs to enhance the value for all stakeholders (Sawhney, 1998). In an ecosystem, the network effect takes over, and other firms, end-users, and a combination of firms

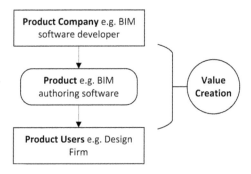

Figure 3.4   Traditional product firm

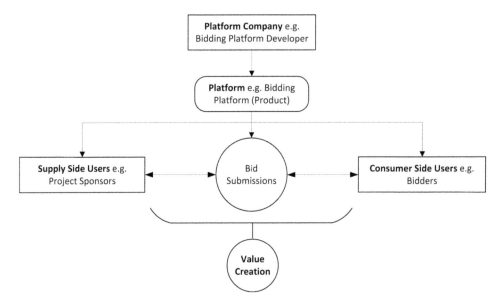

Figure 3.5   Platform-based firm

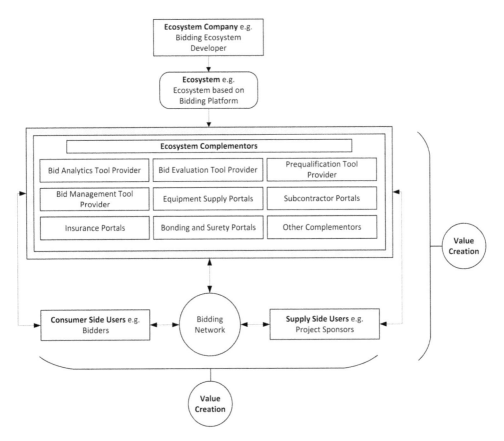

Figure 3.6 Ecosystem-based firm

and end-users start developing complementary products to serve various aspects of their business (see the example of Apple iOS in section 2). The production gets inverted—happening
both inside and outside the originating firm (Jacobides, Sundararajan, and Van Alstyne, 2019).
Finally, a firm organized around an ecosystem has many similarities with a platform company,
especially if it owns, operates, and evolves the platform in the ecosystem. However, there are
some critical differences in that functionality provided by the platform, the functionality provided by complementary products, and customer-specific functionality built by customers, are
all in the same domain and center on the evolution of functionality. For instance, the platform
in an ecosystem needs to continuously incorporate new features to stay valuable and avoid
commoditization (Baldwin and Woodard, 2008). In this context, the firm that owns, operates,
and evolves the platform needs to be very careful to ensure a balanced setting for all participants in the ecosystem governance. As shown in Figure 3.6, the bidding platform (a product)
is used to not only develop bid submission functionalities but is also opened up to partners
and end-users so that an ecosystem develops. Several developers use the bidding platform
and extend it by developing additional complementary products, e.g. bidding analytics, bid
management, bid evaluation, etc. This brings into play the network effect where users seeing
more value in the ecosystem are attracted to it, thereby bringing in new users (Jacobides, Cennamo, and Gawer, 2018). Potentially the business network that surrounds the platform grows,
and new constituents are added to the ecosystem, e.g. equipment and materials supply portals,
sub-contractor portals, are added to the ecosystem.

In this ecosystem, the bidding platform provider opens the core platform for other firms, including customers, to develop additional customer-centric products. This development allows value creation in multiple ways and for various stakeholders. Value is created by the use of the core bidding platform and by the addition of complementary products. So project sponsors gain as more and more bidders participate. Complementors derive value by selling customer-centric products. The original firm wins more ecosystem participants as the number of players in the ecosystem increase, and as a result, more complementary products are added.

3.4.1 Types of ecosystems

Platforms and ecosystems are closely related terms that are used interchangeably (Altman and Tushman, 2017); there is no satisfactory resolution to the underlying lack of boundary between the two in research or practice. The critical challenge in demarcating and defining the two terms is that there are significant philosophical and practical overlaps. Based on the published literature, we feel that generally, platforms are a sub-set of ecosystems. While it is possible that platforms can exist without an explicit or clearly defined ecosystem that surrounds them in most instances, platforms form the core of an ecosystem. Generally speaking, a platform is the core 'design, concept, idea, pattern or model' (Baldwin and Woodard, 2008), while the business network consisting of a core organization that owns or governs the platform interacting to various degrees with external entities to generate value (Gawer and Cusumano, 2014; Altman and Tushman, 2017) is the ecosystem. We have used the taxonomy discussed by Sawhney (Sawhney, 1998); Baldwin and Woodard (Baldwin and Woodard, 2008); and Altman and Tushman (Altman and Tushman, 2017) to classify ecosystems and define both platforms and ecosystems for this handbook. Figure 3.7 shows this classification. In the literature, ecosystems can be primarily classified into:

a. Business ecosystems;
b. Innovation ecosystems; and
c. Platform-based ecosystems (also known as platform ecosystems).

There are other classifications (Gawer and Cusumano, 2014) also, but we will discuss these later in the chapter.

A business ecosystem is the network of organizations that includes suppliers, distributors, customers, government agencies, and others—involved in the delivery of a specific product or service through both competition and cooperation (Moore, 2006).

An innovation ecosystem is based on the core organization's use of a community of external entities, including crowd members and users to innovate (Altman and Tushman, 2017). Typically, an innovation ecosystem consists of volunteer contributors and innovation communities in which value creation happens via open innovation and open coordination (Chesbrough and Appleyard, 2007).

Platform-based ecosystems are driven by a core platform (physical or digital) upon which the ecosystem is based. Physical or product-based platforms first originated out of a need to address the high-variety challenge, i.e. the need of a firm to offer a wide variety of products in a cost-effective manner (Sawhney, 1998; Bryden Wood, 2018). Software-based or digital-platform ecosystems have an extensible software product at its core (Tiwana, Konsynski, and Bush, 2010; Um, Yoo, and Wattal, 2015). In this handbook, we focus on the platform-based ecosystems and more importantly, on digital (platform-based) ecosystems.

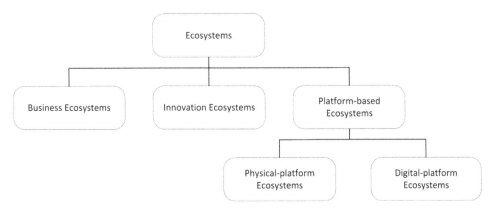

Figure 3.7 Classification of ecosystems and their link to platforms

Some researchers have also classified ecosystems as both internal and external. Internal eco-systems are generally firm-centric that allow the firm to work by itself or with a close network of suppliers to develop a family of related products and components (Gawer and Cusumano, 2014). External ecosystems are platform based in which outside firms act as complementors (Nalebuff and Brandenburger, 1997) and provide new products and components that generate value due to network effects (Baldwin and Woodard, 2008; Gawer and Cusumano, 2014).

3.4.2 Digital platforms and ecosystems

A digital platform-based ecosystem consists of the following main elements:

1. *Digital platform*: the product, software, concept, idea, or thinking that a company chooses to open to others including their end-users and other firms in the hope that new products emerge and value is created manifold (Gawer and Cusumano, 2014) as compared to the value generated by a product-firm that keeps the product closed. For digital ecosystems, the platform consists of a core software product or a digital tool.
2. *Boundary objects or modules*: the artifacts that connect external complementary products to the platform to add functionality and features, some of which emerge as the ecosystem grows and adds new users. These also include interfaces that provide the specifications and design rules that describe how the platform and modules interact (Um and Yoo, 2016).
3. *Complementary products*: new products that the network of external developers devel-ops based on the platform and by using the boundary objects available to the ecosystem participants. The complementary products add new functionality and are highly custom-er-centric (Tiwana, 2014). They become the key drivers of innovation.
4. *Keystone company*: this is the principal member of the ecosystem that owns, operates, governs, and evolves the platform (Iansiti and Levien, 2004). Other terms are also used to reference these companies (Gawer, 2015), e.g. keystone firms are also known as platform leaders (Gawer and Cusumano, 2002) and hubs (Dhanaraj and Parkhe, 2006).
5. *Complementors*: the developers of complementary products are called complemen-tors. They develop complementary products to add value to the platform, the keystone company, and the end-users of the ecosystem (Nalebuff and Brandenburger, 1997; Tiwana, 2014).

51

The concept of digital ecosystems is illustrated graphically in Figure 3.8. A digital platform or product becomes the core component of the ecosystem. The entire ecosystem is based on the platform that provides the extensible codebase (Um and Yoo, 2016) for the core functionality of the ecosystem. Other parts of the ecosystem are the digital components, e.g. the software development kit (SDK) and the application programming interface (API) (Um, Yoo, and Wattal, 2015; Bonardi et al., 2016). Using the platform and these boundary objects (Islind et al., 2019) third-party developers, customers, and end-users come together to create and co-create add-on digital products that solve a particular problem innovatively and create value for all involved (Gawer, 2009).

For example, a BIM authoring tool can act as the core digital platform for the construction sector. The creator of the authoring tool, the keystone company, provides the boundary objects or modules that are interfaces to access the functionality of the core product. Users of the BIM authoring tool and a set of external complementors add new add-on digital products (e.g. apps) to develop new customer-centric functionality. These add-on tools or products are then made available to the broader network or ecosystem. The BIM authoring tool developer, the end-users, and the app developers all participate in this ecosystem to contribute and derive value from the ecosystem. In the platform-based ecosystem, interactions between two distinct groups (the two 'sides') is crucial (Tiwana, 2014). The platform's value to a user depends on the number of adopters on the other side, i.e. growth in the number of complementors and the number of end-users must be interlinked.

3.4.3 Characteristics of digital ecosystems

A digital ecosystem has several characteristics that are primarily determined by the platform architecture (Tiwana, Konsynski, and Bush, 2010). The platform architecture is defined as the conceptual framework that describes how the ecosystem is divided into a

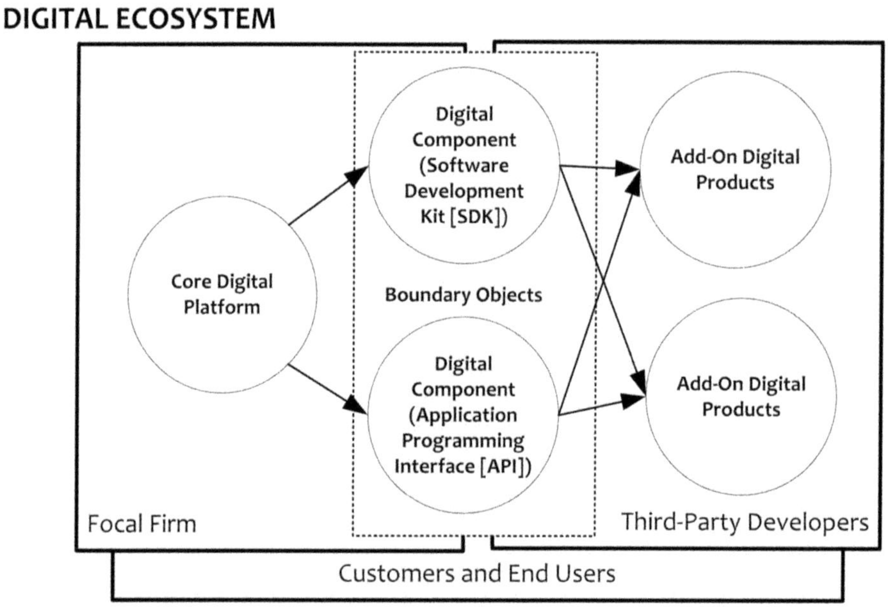

Figure 3.8 Illustration of a digital ecosystem

stable core platform and a complementary set of modules or boundary objects. In addition, the architecture of the platform also consists of the rules that are binding on the platform as well as for the modules and are managed through the ecosystem's governance mechanism. It is the platform's architecture that positions the platform and the modules within the ecosystem. The architecture generally allows low variety and high reusability in the platform and wide variety and low reusability in the complementary modules (Baldwin and Woodard, 2008).

The challenge lies in the design of the platform architecture as this is often irreversible and must accommodate changes unforeseen at the time that the platform was created (Baldwin and Woodard, 2008). The following three distinctive perspectives can be used to define the platform architecture (Tiwana, Konsynski, and Bush, 2010):

1. Decomposition: it explains how the form and function of the digital ecosystem are broken down into subsystems (Tiwana, Konsynski, and Bush, 2010). It defines which subsystems and functionality are part of the platform codebase and which ones reside outside of it in modules and other types of boundary objects. A platform in the ecosystem is often decomposed hierarchically into smaller subsystems (Um and Yoo, 2016). Decomposition minimizes interdependence among the evolution processes of components of the ecosystem, supporting change and variation, and it also helps cope with complexity. However, it often comes with an upfront design cost and may also irreversibly constrain or overwhelmingly expand the scope and span of an ecosystem's components.
2. Modularity: in a digital ecosystem this is crucial because it limits the impact of changes within one subsystem or module on the behavior of other parts of the ecosystem (Baldwin and Woodard, 2008; Tiwana, 2014). The modular structure of the ecosystem is a subjective choice made upfront by the keystone firm. Too few modules make the platform challenging to extend, and too many modules can leave the platform in a fragmented shape.
3. Design rules: design rules define how the ecosystem will be governed and what rules the ecosystem stakeholders will follow as the network evolves. These are set by the keystone company sometimes jointly by other players in the ecosystem. Design rules ensure interoperability with the rest of the ecosystem. Design rules must be stable and versatile (Tiwana, Konsynski, and Bush, 2010).

3.4.4 Governance of digital ecosystem

A digital ecosystem cannot be left to govern itself. Typically, the keystone firm develops implicitly or explicitly a governance structure of the digital ecosystem, where rules are set regarding the decision-making process (Tiwana, 2014). Digital ecosystem governance influences how the platform, boundary objects, complementary products and other stems of the ecosystem evolve (Gawer, 2009). The governance mechanism controls the emergent properties of the digital ecosystem. The key design feature of the governance mechanism is to strike a balance between sufficient controls over the platform to ensure the integrity of the platform while democratizing the control sufficiently to encourage innovation by the complementors. Governance of the digital ecosystem is all about sharing responsibilities and authority, governance by aligning incentives, and governance by sharing stakes (Tiwana, Konsynski, and Bush, 2010).

3.5 Digital ecosystems in construction

BIM and collaborative project management platforms are pushing the industry towards the use of digital ecosystems to address the fragmentation challenges it faces (Cooper, 2018). While the formal recognition of the idea of digital ecosystems is difficult to trace, in a recent report by the Global Industry Council, a framework based on digital ecosystems is advocated (Global Industry Council, 2018). The report suggests a digital ecosystem-based digital transformation to consolidate fragmented platforms, standardize processes, and attract digital-savvy talent.

The idea of platforms and ecosystems in the construction sector may be roughly traced to the use of procurement and bidding platforms, and materials and equipment supply portals (Noelling, 2016). These ecosystems did not address the root causes of the horizontal, vertical, and longitudinal fragmentation in the industry, but they did demonstrate the value of an ecosystem approach in the sector. Figure 3.9 illustrates the various complementary tools or applications that can be developed using the digital ecosystem concept.

More recently, major software companies with significant stakes in the construction sector have adopted the digital ecosystem approach in the hope of providing more value to their

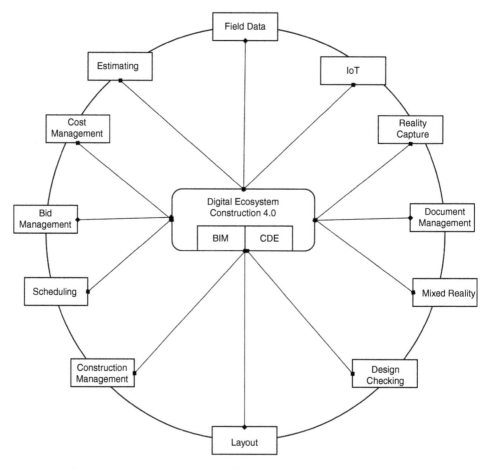

Figure 3.9 Complementary products based on digital ecosystem

customer by bringing on board complementors especially tech startups who are willing to innovate in a customer-centric ecosystem. In the next section, we briefly provide an overview of three such construction focused digital ecosystems primarily to illustrate the concept and its benefits. Several other ecosystems are also available, but these are not covered here due to lack of space.

3.5.1 Autodesk Forge and BIM360 ecosystem

Autodesk uses the Autodesk Forge platform to promote the usage of digital ecosystems in the industry, including the construction sector. Autodesk Forge is a platform of web service APIs that allow developers to integrate Autodesk software-as-a-service (SaaS) products (such as AutoCAD, Fusion, BIM 360, etc.) into their workflows and to embed some of the components used in those products into their complementary web or mobile applications. Figure 3.10 shows the Forge ecosystem.

In early 2019, Autodesk made available the Forge Design Automation API that exposes the Revit's engine for users and complementors to develop complementary products (see Figure 3.11). Using these APIs, users can create and modify BIM models and extract and analyze model data via an external app.

As shown in Figure 3.12, the Forge platform offers the following services externally: Authentication; Data Management; Design Automation; BIM 360; Reality Capture; Model Derivatives; Viewer; and Web-hooks.

Several external organizations have used the Forge ecosystem to develop complementary products. Autodesk provides an example of a Revit Family creation app that uses the Design Automation API of Revit for creating and editing Revit Family anytime, anywhere, without any installation of Revit software (Figure 3.13).

Autodesk Forge APIs are also available for BIM 360 Docs, BIM 360 Team, and BIM 360 Admin. Using these APIs, several complementary products are available on the Autodesk App Store.

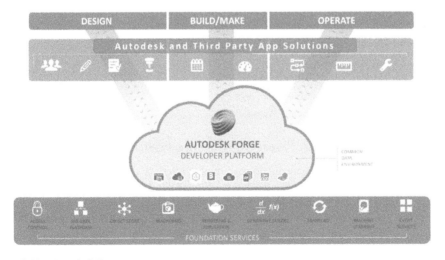

Figure 3.10   Autodesk Forge ecosystem
(Reproduced with the kind permission of Autodesk)

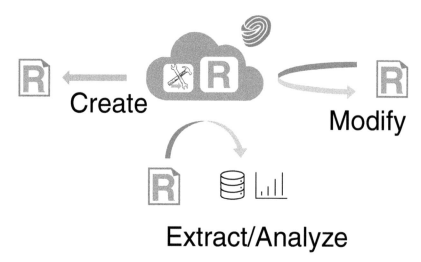

Figure 3.11   Forge Design Automation API for Revit
(Reproduced with the kind permission of Autodesk)

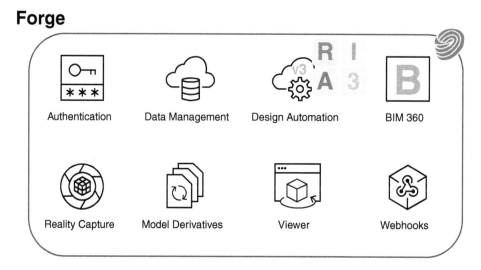

Figure 3.12   Services available through the Forge Platform
(Reproduced with the kind permission of Autodesk)

3.5.2 Procore developers ecosystem

Procore's (the platform) open Application Programming Interface (API) is another example of a construction industry digital ecosystem that has recently gained popularity. It provides the underlying framework for developing applications and custom integrations between Procore and other software tools and technologies. Customers and complementors can expand the functionality of Procore by leveraging existing integrations available in their App Marketplace (shown in Figure 3.14), or by developing new applications and customized connections using the Procore API.

The Procore API allows the ecosystem participants to leverage Procore resources within the Procore cloud using a simple architecture. The endpoints provided by Procore API are intuitive

Figure 3.13 Revit Family App using Design Automation API
(Reproduced with the kind permission of Autodesk)

Figure 3.14 Procore App Marketplace
(Reproduced with the kind permission of Procore)

and powerful, enabling the developer to easily make calls to retrieve information or execute actions on the various resources in Procore. The Procore App Marketplace serves as a repository for applications and integrations developed by the ecosystem partners using the Procore API.

These offerings allow Procore clients to integrate Procore with their existing tools and workflows. Integrations currently available in the App Marketplace expand project management possibilities for Procore clients in the areas of Analytics, Business Intelligence, Accounting, Estimating, Building Information Modeling (BIM), and others. The two primary developer personas that interact with the Procore API are Procore Clients and Procore Technology Partners.

3.5.3 *Bentley iModel.js model server ecosystem*

Bentley Systems offers an open platform for infrastructure digital twins called the iModel.js (a registered product of the company with additional information available at https://imodeljs. github.io/iModelJs-docs-output/). As shown in Figure 3.15, iModel.js is a platform that can be used by end-users and third-party developers to develop new products. For example, the platform can be used for creating, accessing, leveraging, and integrating infrastructure digital twins. The core artifacts in the platform are the iModelHub and Base Infrastructure Schema (BIS). iModelHub acts as a model server while the BIS provides a data standard for storing BIM models known as iModels. Several boundary objects are supplied as part of the platform: (a) iModel.js service; (b) iModel Sync Service; and (c) iModel Web SDK. With the help of these boundary objects, new iModels can be created from various sources of engineering data that may be stored in a Common Data Environment. The platform can be used for storing, managing, and saving changes to the iModel and other related functions.

3.6 Emerging trends and future directions—platforms and ecosystems

Digital ecosystems in construction are beneficial in several ways. In addition to driving the digital transformation that is part and parcel of the Construction 4.0, digital ecosystems can provide the following direct benefits to the industry (Global Industry Council, 2018):

1. Integration across heterogeneous internal and external systems: construction companies rely on internal and external systems and tools that are heterogeneous and fragmented. With the help of digital ecosystems organizations seamlessly combine these multiple systems and tools.
2. Rationalization and standardization: digital ecosystems can help promote rationalization of processes and practices and allow the use of standardization in data and information flow.

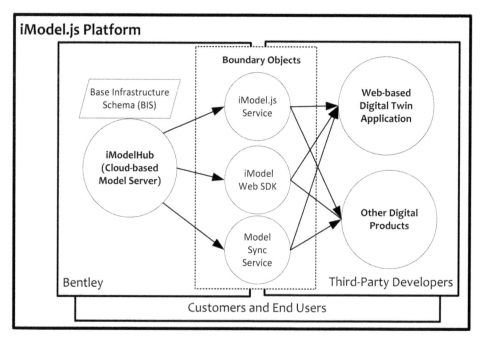

Figure 3.15   iModel.js Platform from Bentley Systems

3. Adoption of digital technologies: adoption of digital tools is a crucial challenge for most construction organizations due to the reluctance to change. Digital ecosystems can help overcome this reluctance by developing add-ons and additional complementary products that provide interfaces and artifacts that are more acceptable to project participants.
4. Establishing a convincing value proposition: digital ecosystems help showcase benefits of the digital transformation, making a strong value proposition to key decision-makers and stakeholders in construction organizations.

Broadly these benefits stem from the following three scenarios:

1. Software out-of-the-box does not work in construction: a significant irritant in the adoption saga of digital tools in the construction sector has been the mindset that off-the-shelf software does not work for the highly specialized processes and practices prevalent in the industry. Given that these processes and practices are generally not standardized in the companies that want to adopt digital tools, this may seem like a convincing blocker of technology. With digital ecosystems based transformation, this concern can be largely overcome. The boundary objects available in a digital ecosystem can be used to extend a given digital platform to meet the needs of a particular user or group of users. In theory, these extensions and add-on products can provide limitless flexibility to the users of the ecosystem, thereby making a strong case for adoption.
2. Budget is a blocker: knowing that the construction sector does not spend enough to embrace digital technologies it becomes challenging to convince construction companies to deploy resources to develop custom software or to fully standardize their processes and practices to use off-the-shelf software. Digital ecosystems can help overcome this hurdle. While acting alone is too demanding for most construction companies, working collaboratively—to complement, adjust, and support joint efforts that are essential to leveraging digital ecosystems (Jacobides, Sundararajan, and Van Alstyne, 2019) can make innovation affordable.
3. Heterogeneous software systems and data streams: construction projects depend on a number of collaborating disciplines where the work is interconnected, with a plethora of asynchronous decisions and changes made over the life of the project. While we cannot escape the use of heterogeneous digital tools, the data for the construction project can be streamlined and brought into a Common Data Environment (CDE). Use of CDE can be further enhanced by the availability of a digital ecosystem that supports the development of add-on products.

The application of digital ecosystems in the construction sector is still in its nascent stage. Several issues remain unresolved. Moving forward, the industry, as a whole, must address these issues. Researchers and scholars must build on the theory and knowledge on platforms and ecosystems available in the literature, mostly in other sectors of the economy to promote the use of digital ecosystems in the construction sector. Some of the questions that need to be answered are:

1. How should platform-based ecosystems be designed to address the challenges faced by the industry?
2. How will digital ecosystems transform our industry and contribute to horizontal, vertical, and longitudinal integration?
3. What are the interlinkages between existing data standards such as Industry Foundation Classes (IFC) and digital ecosystems?
4. How can competing digital ecosystems co-exist?
5. How should researchers develop the theory for digital ecosystems in construction?
6. How do digital platforms affect our projects and our profession and influence the future of work in construction?

3.7 Conclusion

Digital ecosystems can lead to the Construction 4.0 transformation and help construction companies become true digital enterprises. With platforms at the core, augmented by digital add-on products and data-enabled innovative tools, digital ecosystems can provide the much needed horizontal, vertical, and longitudinal integration. These digital enterprises can work together with owners, design and engineering firms, suppliers, and other stakeholders in industrial digital ecosystems.

3.8 Summary

- A detailed description of ecosystems and platforms.
- Key characteristics of ecosystems.
- Digital ecosystems in construction.
- Example digital ecosystems in construction.
- Future trends in digital ecosystems and their implication for Construction 4.0

References

Ahuja, R., Sawhney, A. and Arif, M. (2018) 'Developing Organizational Capabilities to Deliver Lean and Green Project Outcomes using BIM', *Engineering, Construction and Architectural Management*, 25(10), pp. 1255–1276. DOI: 10.1108/ECAM-08-2017-0175.

Altman, E. J. and Tushman, M. L. (2017) *Platforms, Open/User Innovation, and Ecosystems: A Strategic Leadership Perspective, Advances in Strategic Management*. DOI: 10.1108/S0742-332220170000037007.

Baldwin, C. Y. and Woodard, C. J. (2008) 'The Architecture of Platforms: A Unified View', *SSRN Electronic Journal*. DOI: 10.2139/ssrn.1265155.

Bonardi, M. et al. (2016) 'Fostering Collaboration through API Economy', in *Proceedings of the 3rd International Workshop on Software Engineering Research and Industrial Practice - SER&IP '16*. New York: ACM Press, pp. 32–38. DOI: 10.1145/2897022.2897026.

BorjeGhaleh, R. M. and Sardroud, J. M. (2016) 'Approaching Industrialization of Buildings and Integrated Construction Using Building Information Modeling', *Procedia Engineering*, 164, pp. 534–541. DOI: 10.1016/j.proeng.2016.11.655.

Bowden, S. et al. (2006) 'Mobile ICT Support for Construction Process Improvement', *Automation in Construction*, 15(5), pp. 664–676. DOI: 10.1016/j.autcon.2005.08.004.

Bryden Wood. (2018) *Bridging the Gap Between Construction + Manufacturing*. London: Bryden Wood.

Chesbrough, H. W. and Appleyard, M. M. (2007) 'Open Innovation and Strategy', *California Management Review*, 50(1), pp. 57–76. Available at: https://pdxscholar.library.pdx.edu/cgi/viewcontent.cgi?article=1021&context=busadmin_fac.

Cone, K. (2013) *Sustainability + BIM + Integration, a Symbiotic Relationship, AIA*. Available at: www.aia.org/practicing/groups/kc/AIAB081071 (Accessed: 20 June 2016).

Construction Leadership Council. (2018) *Procuring for Value*. London: Construction Leadership Council. Available at: http://constructionleadershipcouncil.co.uk/wp-content/uploads/2018/07/RLB-Procuring-for-Value-18-July-.pdf.

Cooper, S. (2018) 'Civil Engineering Collaborative Digital Platforms Underpin the Creation of "Digital Ecosystems"', *Proceedings of the Institution of Civil Engineers – Civil Engineering*, 171(1), pp. 14. DOI: 10.1680/jcien.2018.171.1.14.

Dhanaraj, C. and Parkhe, A. (2006) 'Orchestrating Innovation Networks', *Academy of Management Review*, 31(3), pp. 659–669. DOI: 10.5465/amr.2006.21318923.

Fergusson, K. J. and Teicholz, P. M. (1996) 'Achieving Industrial Facility Quality: Integration Is Key', *Journal of Management in Engineering*, 12(1), pp. 49–56. DOI: 10.1061/(ASCE)0742-597X (1996)12:1(49).

Gartner (2017) *Seize the Digital Ecosystem Opportunity*. Available at: gartner.com/cioagenda.

Gawer, A. (2009) *Platforms, Markets and Innovation, Platforms, Markets and Innovation*. Cheltenham, UK: Edward Elgar Publishing. DOI: 10.4337/9781849803311.

Gawer, A. (2015) 'What Drives Shifts in Platform Boundaries? An Organizational Perspective', *Academy of Management Proceedings*, p. 13765. DOI: 10.5465/ambp.2015.13765abstract.

Gawer, A. and Cusumano, M. A. (2002) *Platform Leadership: How Intel, Microsoft, and Cisco Drive Industry Innovation*. Boston: Harvard Business School Press.

Gawer, A. and Cusumano, M. A. (2014) 'Industry Platforms and Ecosystem Innovation', *Journal of Product Innovation Management*, 31(3), pp. 417–433. DOI: 10.1111/jpim.12105.

Gerbert, P. et al. (2017) *Shaping the Future of Construction: A Breakthrough in Mindset and Technology, World Economic Forum (WEF)*. Geneva. Available at: https://bcgperspectives.com/Images/Shaping_the_Future_of_Construction_may_2016.pdf.

Global Industry Council (2018) *Five Keys to Unlocking Digital Transformation in Engineering & Construction*. Available at: http://cts.businesswire.com/ct/CT?id=smartlink&url=http%3A%2F%2Faconex.com%2FDigitalTransformation&esheet=51747950&newsitemid=20180124005355&lan=en-US&anchor=http%3A%2F%2Faconex.com%2FDigitalTransformation&index=2&md5=7eabd45450cad87dde7ff45fc276a673.

Hannus, M. (1996) 'Islands of Automation in Construction', in Z. Turk (ed) *Construction on the Information Highway*. Ljubljana: CIB, pp. 20–24. Available at: http://fagg.uni-lj.si/bled96/.

Iansiti, M. and Levien, R. (2004) *The Keystone Advantage: What the New Dynamics of Business Ecosystems Mean for Strategy, Innovation, and Sustainability*. Boston: Harvard Business School Press.

Islind, A. S. et al. (2019) 'Co-Designing a Digital Platform with Boundary Objects: Bringing Together Heterogeneous Users in Healthcare', *Health and Technology*, 9(4), pp. 425–438. DOI: 10.1007/s12553-019-00332-5.

Jacobides, M. G., Cennamo, C., and Gawer, A. (2018) 'Towards a Theory of Ecosystems', *Strategic Management Journal*, 39(8), pp. 2255–2276. DOI: 10.1002/smj.2904.

Jacobides, M. G., Sundararajan, A., and Van Alstyne, M. (2019) Platforms and Ecosystems: Enabling the Digital Economy, *Briefing Paper World Economic Forum*. Available at: www.weforum.org.

JBKnowledge (2018) *The 7th Annual Construction Technology Report*. Bryan. Available at: https://jbknowledge.com/2018-construction-technology-report-survey.

Moore, J. F. (2006) 'Business Ecosystems and the View from the Firm', *The Antitrust Bulletin*, 51(1), pp. 31–75. DOI: 10.1177/0003603X0605100103.

Nalebuff, B. J. and Brandenburger, A. M. (1997) 'Co-opetition: Competitive and Cooperative Business Strategies for the Digital Economy', *Strategy & Leadership*, 25(6), pp. 28–33. DOI: 10.1108/eb054655.

Noelling, K. (2016) *Digitization in the Construction Industry: Building Europe's Road to 'Construction 4.0'*. Munich, Germany: Roland Berger GMBH. Available at: file:///C:/Users/bueasawh/Downloads/tab_digitization_construction_industry_e_final.pdf.

Putzier, K. (2018) 'Momentum Builds for Automation in Construction', *The Wall Street Journal*, 2 July, p. 1. Available at: https://wsj.com/articles/momentum-builds-for-automation-in-construction-11562073426.

Rutgers, V. and Sniderman, B. (2018) *The Industry 4.0 Paradox: Overcoming Disconnects on the Path to Digital Transformation*.

Sawhney, A., Khanzode, A. R., and Tiwari, S. (2017) 'Building Information Modelling for Project Managers', *RICS Insight Paper*. London: RICS. Available at www.rics.org/globalassets/rics-website/media/knowledge/research/insights/bim-for-project-managers-rics.pdf.

Sawhney, M. S. (1998) 'Leveraged High-Variety Strategies: From Portfolio Thinking to Platform Thinking', *Journal of the Academy of Marketing Science*, 26(1), pp. 54–61. DOI: 10.1177/0092070398261006.

Tiwana, A. (2014) 'The Rise of Platform Ecosystems', in A. Tiwana (ed) *Platform Ecosystems: Aligning Architecture, Governance, and Strategy*. New York: Elsevier, pp. 3–21. DOI: 10.1016/B978-0-12-408066-9.00001-1.

Tiwana, A., Konsynski, B., and Bush, A. A. (2010) 'Research Commentary—Platform Evolution: Coevolution of Platform Architecture, Governance, and Environmental Dynamics', *Information Systems Research*, 21(4), pp. 675–687. DOI: 10.1287/isre.1100.0323.

Um, S. and Yoo, Y. (2016) 'The Co-Evolution of Digital Ecosystems', in *2016 International Conference on Information Systems, ICIS 2016*, pp. 1–15.

Um, S., Yoo, Y., and Wattal, S. (2015) 'The Evolution of Digital Ecosystems: A Case of Wordpress from 2004 to 2014', in *2015 International Conference on Information Systems: Exploring the Information Frontier, ICIS 2015*, pp. 1–29.

World Economic Forum (2018) *An Action Plan to Accelerate Building Information Modeling (BIM) Adoption*. Available at: http://weforum.org/docs/WEF_Accelerating_BIM_Adoption_Action_Plan.pdf.

4

INNOVATION IN THE CONSTRUCTION PROJECT DELIVERY NETWORKS IN CONSTRUCTION 4.0

Ken Stowe, Olivier Lépinoy, and Atul Khanzode

4.1 Aims

- Explore the four major influential forces that will shape innovation initiatives in Construction 4.0.
- Describe a strategic pathway to innovative and sustained analytical excellence for project delivery networks adapting to Construction 4.0.
- Suggest an organizational structure for the transforming organizations adapting to Construction 4.0.

4.2 Introduction

What do the most effective leaders today have in common? They wake up every morning and ask themselves the same questions: "What world am I living in? What are the biggest trends in this world? And how do I educate my citizens about this world and align my policies so more of my people can get the best out of these trends and cushion the worst?

… The second thing the best leaders understand is that in a world of simultaneous accelerations in technology and globalization, keeping your country as open as possible to as many flows as possible is advantageous for two reasons: You get all the change signals first and have to respond to them, and you attract the most high-I.Q. risk-takers, who tend to be the people who start or advance new companies.

(Thomas L. Friedman – Opinion Columnist – New York Times, April 2, 2019)

In his *New York Times* column above, Thomas Friedman describes what the best leaders do to create a vibrant economy … get the signals, respond, excel, and as a result, attract the best and most courageous professionals to join you. He focuses in this column on political leaders and citizens, but we are inspired to think of construction leadership having the same urgent assignment to identify trends and keep their construction culture as receptive open and agile as possible. In the same column, he describes cutting-edge "information flows" as a resource that must be treasured and protected. Innovation is also a resource that, like knowledge, will be prized and nurtured by the best leaders in our industry.

In this chapter, we describe a future of construction that no author can assuredly predict. In our earnest effort, we examine the current state of the industry, observe the trajectories of technical, methodological, market pressures, and cultural changes over the last several years, and attempt to describe the environments for innovations that will shape Construction 4.0. We address the opportunity with a global perspective and relate them to trends in other industries.

We present a strategic pathway for leaders of project delivery networks to inspire and guide their stakeholders to achieve success through the integration of technology, market savvy, cultural identity, and principled purpose. Along the path, the culture must enjoy "small wins" that sustain the commitment to long-term transformation to analytical excellence, growth, new services, and an inviting and innovative environment for attracting and retaining talent.

We hope the chapter stimulates, inspires, and guides the reader to enjoy and thrive during the transformation that is assuredly ahead of us.

We break the discussion into five sections:

Context – Our industry has been at work and juggling traditional behaviors with new opportunities to meet competitive pressures. We read about new trends of collaboration, principles, and contract methodologies. We consume new technologies and integrate them into redesigned workflows. We anticipate the future when we see signals of a transformation that is assuredly coming.

Opportunity – We discuss the nature of innovations and motivations for various stakeholder in various project delivery networks. We assert that a pathway that integrates trends and technology will create investment in pursuit of competitive advantage. We attempt to forecast the early adopting disciplines and stakeholders. We divide the patterns into technology, culture, principles, and market pressures.

We try to address a range of topics – from data integrity to the emotional impact of root cause analysis and transparency.

Ramifications – We anticipate the ramifications of Construction 4.0. Firms will need to continue to balance winning work with their current business models, and to invest in generating new forms of wins, and earning wins for new reasons. There will be an expanded appreciation for innovation and agility. The cultures, contracts, organizations, and institutions will evolve to provide new value and eliminate waste. We anticipate which stakeholders will be the early beneficiaries, and what will happen to the laggard firms that move slowly to adapt.

Challenges – Our innovators and practitioners have work to do. We describe that work and the obstacles and friction that their efforts face. The barriers are categorized as cultural, technical, behavioral, contractual, and educational. Leaders that are either unprepared or fearful or both will reinforce those obstacles.

Closing thoughts – There will be positive outcomes for the industry. Some firms will transform successfully and create a sustained winning formulation. We anticipate some of the adverse outcomes ... shifts in job roles, vulnerabilities to certain industry stakeholders, and the impact of transparency. And finally, we present some personal convictions from the authors.

Addressing innovation in Construction 4.0 means anticipating a convergence of the major forces that will shape this transformation. We endeavor to help the reader brace, strategize, invest, and adapt to take advantage of the opportunities.

Our innovators will benefit from lessons from manufacturing, finance, politics, retail, entertainment in their journey to Industry 4.0. We can anticipate that in our multi-trillion-dollar industry, some have the vision and wisdom to make the intellectual leap to see power, scout the opportunities, invest, persist, and enjoy early successes and rewards that Construction 4.0 promises.

We hope that the reader finds the chapter to be stimulating and bestows a perspective that guides their strategic planning as Construction 4.0 emerges and allows them to further their research, careers, and their company or institutional goals and helps to advance our industry and our planet.

4.3 Context

Construction 4.0 has begun appearing in isolated networks, and only foreshadowing what we will observe and experience. The global construction industry has often been described as big and fragmented and viewed as a laggard in the areas of investing, innovation, research, experimentation, and reaping productivity gains from technology. The fragmentation is reflected in the different stakeholder purposes, motivations, systems, behaviors, and siloed data.

The industry is segmented into project delivery networks (PDNs), each with its own contracts, culture, reward systems, and appetite for innovation and technology. Universal are the pervasive problems of risk (injury, deaths, and failed projects), uncertain forecasts, wasted time and money, and misalignment.

4.3.1 Market context

Industry segmentation

Essential to our topic of innovation in Construction 4.0 is the variety of funding sources: public, private, and public-private partnerships. Within these segments is another division, a contractual one: design-bid-build, design-build, construction management (at-risk and for fee), public-private-partnerships, and Integrated Project Delivery. Finally, there is a division by built-asset type: transportation, industrial, residential, pharmaceutical, high-tech, etc. The arc of transformation to Construction 4.0 will vary for each of the many environments. For some PDNs, the funding source will resist new methodologies, either due to personal or political risks. For other PDNs, the funding source will mandate use of data capture that will produce the data deliverables needed for analytical excellence.

For this chapter, we will navigate through the segments with an eye to the "universal" potential improvements in the traditional construction workflows, who currently pays for the waste, who is initiating efforts to recognize and address the waste, and who will soon be culturally and technically developed to a strategic maturity level when they will invest.

Market pressures to improve

There are thousands of new products and services that are created by innovative parties eager to compete at the forefront of Construction 4.0. Developed by major vendors, third-party start-ups, or by large stakeholders, these products will form the components for integrated solutions that respond to market pressures. Those pressures to improve in the current environment include:

1. Reduced risk.
2. Predictability.
3. Reduction of waste.
4. Shortening project duration.

5. Higher quality.
6. Improved safety.
7. Operational efficiency of the built asset.
8. Contractors' general conditions and overhead.
9. Pursuit win-rate.
10. Owner/customer experience.
11. Attracting management and trade personnel.

4.3.2 Cultural context

Cultures

Naturally, not all the construction cultures are prepared to invest in innovative methods or leading-edge technologies like machine learning, lean principles, or even BIM. For firms that find investing in innovation too difficult or risky, sustaining the day-to-day operations dominates their planning – focusing on winning work and executing successfully.

Anecdotal evidence indicates that some executives are concerned about their firms' readiness for the challenges on the horizon. The authors have heard statements like: "We know that Big Data is going to require us to transform our firm to be innovative and agile, but we don't know how we should proceed."

Project metrics and measurement

The project success rates measured by the traditional metrics (on time, in budget, safe, quality, and satisfaction) still indicate that there is much that needs improving. New success metrics will be critical to inspiring innovation.

As Construction 4.0 evolves, additional success metrics will emerge, and the attention to Key Performance Indicators (KPI) selection will intensify. Among the success metrics to consider are: return customer rate, retained employees, data quality, accurate reporting, insights, latency, trade contractor success, publishable results, and asset performance during operations (predicted vs actual). Currently, KPIs are mostly financial and lagging indicators and rarely lead to actionable intelligence.

Risk and trust

With a fragmented delivery model, with low levels of trust between owners and suppliers, collaboration is limited, and each party protects its margin, inhibiting information-sharing. Project waste results from tension between the self-interest of the parties and the common good of the project. A win by one stakeholder often means a loss for another. Because of this waste, the construction industry is perceived to be underperforming the productivity improvements seen in other sectors of the economy.

Alignment between the physical and digital worlds

Each PDN culture has two domains: the physical domain and the digital domain, each with specialists. Some individual contributors provide services in both domains, but the division between the two is a source of cultural, intellectual, and behavioral conflict that inhibits innovation.

Those individuals in the traditional (physical world) roles may resist the help from the digital world experts. The business network of technology companies can be more closely aligned with the physical world practitioners and can help the project networks ease those conflicts, and to develop strategies to deliver new value and solve persistent and pervasive problems.

Technology firms (software, consulting, startups, etc.) often have a perception problem. Perceived as delivering only hype, they are often seen as disruptive and overselling an impossible dream. That said, however, there is evidence that vital strategic partnerships between technology providers and traditional construction stakeholders are emerging. These partnerships sometimes include academic contributors and require attention to cultural fit, to the process innovation, and the knowledge shared (intellectual property).

The current data environment

When we explore the data that the PDNs regularly capture, utilize, and archive, we find documents (sometimes BIM), requests for information, schedules, estimates and cost management (with unit costs forecast and actual), safety records, and punch-lists. The fragmentation of the data *storage and access* correlates to the fragmentation of the industry. Some PDNs create lean project data, such as commitments, defects, planned percent complete, latency, etc. (Kam, 2018).

Many PDNs have data that is not centralized nor utilized in an integrated way to be mined affordably to provide digital twins, insights, or alerts. Often this data is found to be of poor quality and needs to be normalized. Labor productivity data information is captured, tracked and "locked away" in trade contractors' siloed systems, and unavailable to other stakeholders. The root causes are tracked in one system, and the impacts are monitored in another. The abundance of poor-quality data and stranded data provides a clear and present obstacle to Construction 4.0 and analysis across projects.

Current technologies and digital maturity

A steady stream of new technologies is providing the conditions for a profusion of decisions for leaders. Cyber-physical systems (sensors, robots, drones, actuators, laser scanners, mobile devices, augmented/virtual reality tools, etc.) have improved and become affordable. They have proven the value they can deliver for the industry. In parallel, Digital and Computing technologies (Machine Learning, Big Data, Cloud, Reality Capture technologies, Building Information Modeling, Internet of Things, and Blockchain) are suggesting opportunity for potential performance advancements. Add real-time tracking of physical work, connected BIM, advanced work packaging, smart contracts, and expert systems. These technical capabilities are enticing the industry's decision-makers, making managing very complicated, and result in pressure on decision-makers for fast and accurate responses to opportunities. This management challenge invites an innovation strategy, balancing structured decision-making and agility.

4.3.3 Prevailing trends and influences

Stakeholders and integration

The stakeholders in a PDN include owners (and representatives), designers and sub-consultants, general contractors, trade contractors, material vendors, and equipment suppliers. The technology currently used by the PDNs vary greatly, but design (CAD and BIM), scheduling, estimating, project controls, and field engineering layout/control constitute most of the day-to-day use of technical tools. Some level of integration (depending on digital maturity of the PDN) of the tools exists. This integration is reflected in advances like model-to-cost integration, resource-loading, 4D construction simulations, etc.

Current technical adoption

We can confidently state that many PDNs have adopted BIM, even if not by entire teams, and not yet delivering on BIM's total potential across all phases. Mobile technologies have also become mainstream for some teams, working in the field and connecting stakeholders in the project delivery networks. The integration of BIM and mobile technologies is young and promising, but adoption is not globally widespread at this writing.

Persistent and pervasive challenge: waste

There is one indicator of the persistent and widespread challenge in the industry: the value/waste/support ratio. Waste (process and physical) account for more than 50% of construction time (Figure 4.1).

Researchers have explored the pervasive problem of rework in construction and its root causes. A 2003 study commissioned by the Construction Owners of Alberta (COAA) established a framework for the root causes of rework. The $599.6M Syncrude Aurora 2 Project was analyzed by Amineh Fayek using the fishbone diagram (see Figure 4.2). The study revealed that Engineering and Review was the root cause of 55.41% of the rework identified and that 55.41% could be parsed into 34.48% Errors and Omissions, 6.93% Scope Changes, and 10% Late Design Changes (Fayek et al., 2003). Construction 4.0 promises to make root cause analysis commonplace and rework causes transparent, which will have wide implications discussed later in Section 4.

The "avalanche of waste"

Douglas Chelson's thesis asserts: "The most significant factors causing rework are poor design and coordination" (Chelson, 2010). Stowe later asserted that

> Projects often suffer rework that can be viewed as an 'avalanche of waste', starting small and gathering speed and volume, and leading to a large and destructive force. Design Errors and omissions cause requests for information, which result in design

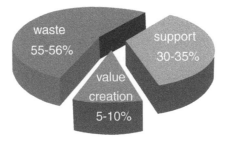

Figure 4.1 The concept of waste as understood in lean construction (with permission from Alan Mossman/Lean Production, 2009)

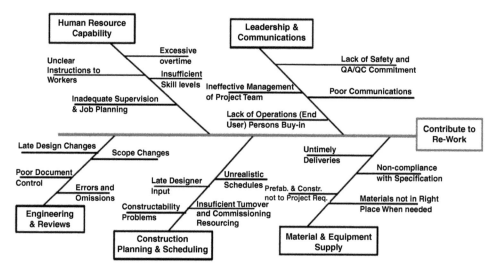

Figure 4.2 COAA's Fishbone Rework Cause Classification (last updated October 2002) with permission – © Canadian Science Publishing

changes, cost changes, and project delays. Overtime, 2nd shift, and other measures to regain the lost time create an accelerated and chaotic setting often resulting in quality failure, and worker fatigue amplifies the risk of injuries. Sometimes, the final destructive element of waste in the avalanche is litigation.

(Stowe, 2018).

4.3.4 Innovations that foreshadow Construction 4.0

There are a few inspiring examples of innovative contractors and other stakeholders that may foretell what we may expect to become mainstream systems and behaviors in Construction 4.0. These innovators and AI users likely have a strong track record of digitization. McKinsey research finds that "companies with a strong track record of digitization are 50 per cent more likely to generate profit from using AI" (Blanco, 2018).

Machine learning

Some of these early and successful adopters of digitization have invested in AI and machine learning that are enjoying positive gains and sometimes publishing and or presenting the results. Layton Construction (Richard Holbrook) of Sandy, Utah presented in Las Vegas in 2018. He reported that with the use of AI, Layton had reduced their lost-time accident rate from 2017 to 2018 by 38%. Swinerton Construction (Dustin Hartsuiker) at that same seminar reported that at Swinerton they "only use API-enabled vendors" and "AI highlighted issues that could have been overlooked" (Autodesk, 2018).

Issue tracking/analysis – linked to the 3D model

SWECO is a European engineering consultancy that aims to plan and design the cities of the future. To meet their customers' demands, they developed a third-party application built on one of their software provider's platform. They report; "At any one time, the team tracks approximately 3,000 issues – each detailed at the task level and linked to the 3D model" (Autodesk, 2018).

Design innovations

Inspiration also comes from design stakeholders that leverage computing power and point to life-cycle operational performance opportunities for Construction 4.0. The H.D.S. Beagle 1.0 tool was developed that

> enables the generation of design alternatives according to user-defined param-
> eter ranges; automatically gathers the energy analysis result of each design
> alternative; automatically calculates three objective functions; and uses Genetic
> Algorithm to search intelligently, rank, select, and breed the solution space for
> decision making.
>
> *Gerber et al. (2012)*

New hardware and promising peripherals

The project delivery networks have faced investment decisions in new systems that each involved return on investment decisions: drones, virtual reality, and augmented reality goggles, robotics, sensors, smart helmets and vests, etc. Decision-makers are repeatedly faced with a plethora of investment choices and can easily be overwhelmed with evaluating, testing, and implementing new solutions.

The smart asset becomes the "learning asset"

Some PDNs have invested in digital twin, machine learning, cloud computing, and sensors for simulation, optimization, and efficiencies in operations and maintenance. General Electric offers a technology that "adjusts the angle of turbine blades by changes in temperature or atmospheric pressure" and estimates that the system may "raise power output by around 5% and lower maintenance costs by 20%" (Nikkei, 2017).

BIM justification, adoption, and its natural succession

In the early phases of BIM implementation, many stakeholders searched for an excellent way to justify the investment in technology. In response, Stowe created a BIM return-on-investment workshop in 2008 designed to help project stakeholders calculate the investment and forecast the savings from what he called "All-in BIM" Stowe (2018).

During a span of ten years, Stowe delivered the workshop in 22 countries. Over that span of years, Stowe found the participants more and more interested in *performance measurement* and inquiring about KPIs. The most advanced of those participant stakeholders wanted to discuss KPI priorities, data capture, platforms, analytics, and the challenges to create a "data-centric" culture within their organizations.

BIM has catalyzed several ingredients that set the stage for innovation. It changed the workflow, inspired PDNs to move from document-centric to data-centric thinking, provided a potential core for data structure, and spawned hundreds of innovative third-party applications by developers.

Pockets of innovation

There are "pockets" of innovation where the promise of Construction 4.0 has been appealing enough for some PDNs to invest and prove that higher performance is achievable. A look at the emergence of Building Information Modeling (BIM) will be instructive. For Rafael Sacks (2019), BIM has become a foundational component of the development of construction technologies. We witnessed a small percentage of innovators in the 1990s and early 2000s that began to use BIM, not just for design, but also for logistics planning, clash detection, 4D construction simulation, mockups, and visualization. This is an example of integration of technology, collaboration, and data-sharing. That innovative adaptation often increased construction site productivity (Chelson, 2010).

At the other end of the adoption curve, some contractors waited to invest in BIM until it was mandated contractually by their clients (in pursuit of savings and predictability) or until standards and detailed BIM execution plans were available.

4.3.5 *Methodologies that foreshadow Construction 4.0*

Mandated project deliverables

Since the 1970s, governments have contractually mandated more and more digital deliverables. Examples include logistics plans; schedules; BIM deliverables (Australia calls it "Digital Engineering" mandates) in the UK, Singapore, Japan, US agencies, airports, and New York City's "3D Site Safety Modeling" requirements (New York City Department of Buildings, 2019).

Now some agencies are going beyond BIM and mandating a common data environment for data and metadata deliverables from other systems to set the stage for insights, alerts, simulation, and digital twin. They are evaluating the standards such as ISO 19,650, which address the "common data environment" and the "project information model". As these emerge, they will have enormous implications for new processes, federated data, deliverables, and the innovation environment.

Advanced Work Packaging

Advanced Work Packaging (AWP) is one of the methodologies that foreshadow how Construction 4.0 will emerge. Combined with data-centric building modeling and moving from the era of documentation the Construction Industry Institute defined AWP as:

> a construction-driven process that adopts the philosophy of "beginning with the end in mind." The work packaging and constraint management process removes the guesswork from executing at the workface by tightly defining the scope of all work involved, and by ensuring that all things necessary for execution are in place.
>
> *(CII, 2019a)*

This is an example of technology integrated with a methodology developed for "improving craft labor productivity and predictability" (CII, 2019b).

*The intersection of lean construction and Construction 4.0
and becoming value-centric*

Aptly stated, lean construction is "the quest for synchronized and continuous improvements in design, procurement, construction, operations and maintenance processes to deliver value to the entire project delivery network".

(Sawhney, 2011)

The Lean Project Delivery System (LPDS) advocates the application of the Lean Production principles to the delivery of construction projects. Construction projects are conceptualized as production systems, and work structuring is managed using the lean principles by minimizing waste and improving workflow reliability. BIM and Virtual Building technologies include the application of symbolic models of Product, Organization, and Process using 3D and 4D CAD applications and help accomplish visualization of work sequences and early planning. Sawhney (2011) advocates a focus on value: "Value, especially the way it is perceived by the user, should be used as a driver for the design and construction of our built environment."

Last Planner

Production control deals with putting a system in place that will help control the process so that the desired outcome of the lean project is achieved. The Last Planner system is used for production control on lean projects, and increasingly has a digital form that can be an essential part of a common data environment for the PDN.

4.4 Opportunity for Construction 4.0 (the promise)

What is the promise of Construction 4.0 and what are the core influences that will push and pull us to achieve that promise?

… In this world, growth increasingly depends on the ability of yourself, your community, your town, your factory, your school and your country to be connected to more and more of the flows of knowledge and investment – and not just rely on stocks of stuff.
(Thomas L. Friedman – Opinion Columnist – New York Times, *April 2, 2019)*

Friedman adeptly describes the Construction 4.0 challenge with just four words that refer to the innovation setting. We will be living and working amid changing "*flows of knowledge and investment*". Knowing that the flows will be accelerating, we might call them *floods* of knowledge, and anticipating that innovation teams will be overwhelmed, considering the magnitude and quickening of the challenges for construction decision-makers. Again we are reminded that our innovation initiative must balance structure and agility.

4.4.1 Practitioner's path to success in Construction 4.0

Vision and pressures

The optimal path to success in Construction 4.0 will not be ad-hoc discovery of technological value, but form an advantageous integrated response to four influences:

- Technologies – emerging at an accelerating pace, and straightforward to implement.
- Market pressures – owners demanding higher project performance that is measured and proven, and digital requirements and project analytics.
- Principles:
    o openness and trust
    o collaboration
    o focus on value
    o reward systems
    o retraining.
- Culture – company strategies amplifying consideration of technology in their strategic initiatives, creating partnerships and a paced cultural shift toward data-centric thinking and pursuing analytical excellence, with priorities and rationale clear.

During the transformation to Construction 4.0, companies will evolve from their customary business model, while maintaining their profits to fund the initiatives that form their reshaped company (Figure 4.3).

We categorize the challenges that stakeholder face into three time-horizons:

1. *Day-to-day:* Pursuing, winning, and executing projects is the perennial day-to-day challenge, which generates profit, enables promotion and hiring, and sustains the reputation for high performance.
2. *On the horizon:* Responding to the convergence of digital opportunities, principles, market pressures, and innovative culture is a focus first on *value (creating unity in the firm)*, and then the "on the horizon" challenges, when innovation will flourish. Addressing these challenges are the trials that occur in the medium-term future that integrate the trends, secure the continuation of competitive position in existing PDNs to win, and capture the knowledge to execute.

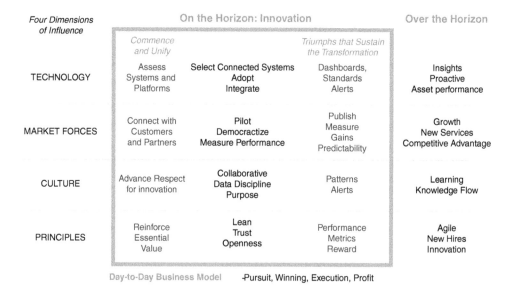

Figure 4.3   The four dimensions of influence on the innovation initiative

3. *Over the horizon*: With the learning and successes (dashboards, patterns, alerts, measured gains, and metrics-oriented thinking from the "on-the-horizon" initiatives, the team will generate opportunities for "over-the-horizon" success: new business models, growth into new markets and services, an agile learning organization, and analytical excellence. This successful execution only happens with essential contributions of cultural transformation and reward for innovation.

We suggest that some firms should follow in the path of some leading companies: create a new leadership role, e.g. Chief Transformation/Innovation Officer, who may need four individuals to take on the four dimensions of influence: technological, market pressure, principles/methodology, and cultural transformation.

4.4.2 Opportunity for cultural gains

In addition to embracing Sacks' analogy for building successful technology, discussed below in Section 4.4.3, we want to explore how to build cultural readiness for innovation, investment in people, motivation, and tolerance for trial and error (while recognizing that entrepreneurship is related and is one of Sacks' pillars). Indeed, the cultural readiness of some contractors and designers have led to the development of their own "in-house" technologies, some of which have been acquired by leading technology firms.

Not just digitization, but data maturity

Like McKinsey's assertion that those firms that have a "strong record of digitization" are more likely to profit from AI, we may go more in-depth with the suggestion that those companies that not only successfully *digitized* tasks over the years but also invested in systems with an appreciation of the power of relational databases are becoming data-centric organizations.

In the span of 40 or 50 years, estimating went from paper and calculator to spreadsheet, to databased methodologies, with a similar story for scheduling, estimating, and for design media among other tools. The early successful project networks/cultures that benefit from Construction 4.0 will likely come from strong respect for data and integration with a focus on analytical excellence, early warnings, and optimization.

For every role – adaptation and adoption – proof of excellence

Construction 4.0 holds the promise of improving the performance of every member of the PDN team, from tradespeople and supervisors through executives. It follows that the innovations will propagate from every role. This is not a new concept that innovation does not develop entirely in a lab or experimental setting. Not only must the innovations be adopted by the managers and workers, but also the innovations must be adapted to the various roles, and many of the innovations must *originate* with those contributors. Those "here/now" contributors (pursuing, planning, and executing work) will be innovators and the early winning firms in Construction 4.0.

Some executives in the stakeholder organizations will be inspired to develop a long-term strategy with over-the-horizon goals. Some will look for ways for technology to benefit the "day-to-day" business model. Just like some executives saw BIM as a vehicle to win in interviews and proposals, they will be expecting owners will be drawn to performance improvement methodologies, the promise of the digital twin, proof of analytical excellence, and continuous improvement.

Discovery and human vision

In every redefined Construction 4.0 role, patterns can be discovered, evaluated, and actions considered. For every discovery, human vision combined with digital pattern recognition is vital.

> Humans have a remarkable pattern recognition system in our heads, and the ability for knowledge discovery in data-driven science depends critically on our ability to perform effective and flexible visual exploration. This may be one of the key methodological challenges for data-rich science in the 21st century.
>
> *Donalek et al. (2014)*

Project environments that cultivate innovation

Project environments that have embraced collaboration, digitization, integration, and life-cycle thinking will likely be settings for the innovations needed for Construction 4.0. These PDN environments will have cultural readiness and will have proven themselves to be competitive. These PDNs may have competed and won in Public-Private Partnerships, Design-Build-Finance-Operate, Integrated Project Delivery, and other networks that prize working together, combining wisdom from various disciplines and multiple phases. It is natural to think these integrated project teams will exploit the power of integrated data earlier than other cultures. They are also positioned to broadly measure performance (project performance and asset performance), simulated and actual, like few different project environments. This setting also makes the creator of the digital twin for design simulation the benefactor of the savings in construction, operational excellence, and the efficiencies of the "learning asset".

Indicators of success in Construction 4.0

We consider technology "adoption" to be only one of the vital ingredients that will shape the success metrics that define Construction 4.0. We assert that integration is also an essential ingredient. Succar's BIM maturity curve describes an advance that incorporates technology, process, culture, and perhaps contracts. For Succar (2009), this intermediate point, creating an integrated project team is a *step on the way* to creating new value that will indicate we are succeeding. Our indicators of successful transformation signify the tangible value of these advances:

- Improved project life-cycle outcomes.
- Timely alerts and analytical insight.
- Vibrant worker culture of analysis and learning.
- Agile response to market dynamics.
- Project forecast accuracy.
- Improved value/waste ratio.
- Learning assets.

4.4.3 Pathway of innovations

3D as a core influence

In his publication "Path Creation in Architecture and Construction" – Richard J. Boland, Jr. (Boland et al., 2003) introduces us to the term "3D induced innovation" and conveys the notion that innovations happen along a "path" and that 3D may spark a pathway of innovations (which he calls "moments").

> There are multiple levels of potential path creation and path dependence in the use of 3D representations in architecture and construction. They include:
>
> (1) the development of the 3D software tools and underlying domain models (the digital design path),
> (2) the changing work practices of architects using the digital representations,
> (3) the kinds of projects that are designed with digital representations, and
> (4) the appropriation of digital technologies by contractors and subcontractors.
>
> All these levels are interrelated and involve changes in technology use, work practices, knowledge assets, and organizational structures and strategies, which we are trying to understand holistically in the study of 3D induced innovation as separate moments of path creation.
>
> *Boland et al. (2003)*

Technological pathway (with BIM foundation)

Rafael Sacks gives us a visual depiction of the progression of successful *technological* advances: his "House of Construction Technology" (Figure 4.4) which culminates in "Adoption".

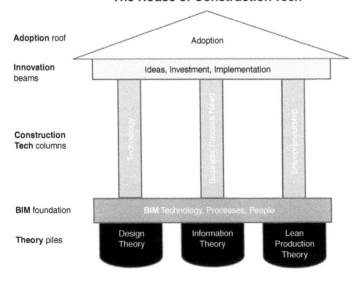

Figure 4.4 The House of Construction Tech
(reproduced with the kind permission of Rafael Sacks)

Sacks' underlying message is that advances in technology:

1. must be underpinned by three critical theories (Design, Information, and Lean Production);
2. will rely on BIM advances of technology, processes, and people;
3. must be borne by technology, business process (needs), and entrepreneurship creating the "superstructure";
4. on which is capped by what we may call an "innovation plank", with ideas, investment, and implementation; and
5. leading to the final success metric: adoption.

We may deduce that for Sacks (and perhaps Boland), BIM data is at or near the "center of data gravity" of the new integrated data environment. Probably implied in the figure, the business need is fulfilled with implementation to the extent that renders the technology *worthy* of adoption at a wide scale (Sacks, 2019).

4.4.4 BIM, integration, adoption, and alignment

The final step in Sacks' sequence is adoption. We can deliberate on how it relates to Succar's final step, *Integrated Project Delivery*, which he describes as the "long term goal" of BIM implementation. These two views represent a vital confluence of contractual, cultural, and technical progress that must proceed with common purpose.

Sacks and Succar together inspire us to think about a new industry *maturity path*. Considering those two assertions, BIM may be a "pillar" as we build toward Construction 4.0. It may lead us to Integrated Project Delivery.

In her research paper on business model renewal, Karoliina Rajakallio suggests that Integrated Project Delivery may lead us to new *alignment* of value drivers and mechanisms. "The findings suggest that the shift towards more Integrated Project Delivery models causes significant changes in the alignment of value drivers and value appropriation mechanisms across the network of firms" (Rajakallio et al., 2017). BIM may be simultaneously at the core of new spin-off technologies *and* new integrated behaviors.

4.5 Ramifications of Construction 4.0

4.5.1 The vision

Project data is affordably captured without encumbering the project teams. Projects use a common data platform with quality data from many stakeholders. Primary and third-party applications crunch that data for different project roles, employing machine learning, with algorithms producing insights and alerts, developed from thousands or hundreds of thousands of project experiences.

Teams produce an ever-improving digital twin for simulation, visualization, and planning during the project. During operations and maintenance, the digital twin continues to learn from sensors and to process how to improve preventive maintenance, safety, and operational efficiency.

The business models for retail, music, transportation, hospitality, movies, banking, and others have been disrupted, and for some firms, their business models were devastated. Construction is the next sector to be disrupted. Some firms will be devastated. But what are disruptors doing when they overthrow markets and even industries? They are releasing new forms of value by using technology. They are also leveraging the digitization of just about everything – documents, news, music, photos, video, maps, personal updates, social networks, requests for information and responses to those requests, data from all kinds of sensors. The list goes on. For some industries, the "middle man" has vanished. For us in the construction industry, the middle man and perhaps middle management will be at risk – any role or activity that adds cost, but not significant value is vulnerable.

Since the 1990s, we experienced the internet's emergent power and how it changed many of our tasks, our contracts, our workflows, and behaviors. It took about 30 years from the development of the World Wide Web to mature to our current capabilities in construction. The web is still changing our industry, but now we see it as just one of the technologies that is vital to our project workflows. We can expect Construction 4.0 similarly to take years to be embraced by the industry. It will be a massive shift, but a gradual one.

4.5.2 From stocks of knowledge to knowledge flow

Knowledge flow

Over centuries ... business has been organized around stocks of knowledge as the basis for value creation. The key to creating economic value has been to acquire some proprietary knowledge stocks, aggressively protect those knowledge stocks and then efficiently extract the economic benefit from those knowledge stocks and deliver them to the market. The challenge in a more rapidly changing world is that knowledge stocks depreciate at an accelerating rate. In this kind of society, the critical source of economic value shifts from stocks to flows.

The companies that will create the most economic value in the future will be the ones that find ways to participate more effectively in a broader range of more diverse knowledge flows that can refresh knowledge stocks at an accelerating rate.

(Hagel, 2016)

Our opportunities are arriving at an accelerating rate. Refreshing our knowledge stocks will require a vibrant culture – inspired and rewarded.

Insights

Analytics will produce insights. Root causes of waste will be transparent. Waste will diminish. Stakeholders will continuously analyze the data for early warnings, be diligent to avoid being exposed as the root cause of waste and rework, and seek to be associated with successfully satisfied commitments. Individual contributors will be identified as better partners, higher performers and find themselves in demand by customers and competitors. Equipment choices and logistics decisions will be evaluated with a learning machine.

Transparency and private data

The power of having access to the data creates a crucial question in Construction 4.0. Sharing of some data that is currently private, sensitive, and part of a stakeholder's distinct competitive advantage will be too menacing. It is likely that the successful platforms will allow each stakeholder to have a "shared data space" *and* a "private data space". Owners may mandate data deliverables and a secure common data environment that has heretofore been private. This means dissolving some of the walls that would obfuscate collaboration and analytical efforts in their current project environments.

Key Performance Indicators and data integrity

Performance measurement will become commonplace. Performance charts will line the walls of the main offices and field offices of the projects. Project Infographics will be prized. Project professionals will respect the power of both leading and lagging indicators. Data integrity will be of paramount importance to the company and institutional executives. Those individuals and teams that adhere to data integrity standards, identify risks early, and produce insights will be recognized and rewarded.

Power shifts

In each project delivery network, there is a "power equation" assigning risk, reward, function, and decision-making authority. The assignment of that power is related to risk, relationships, and leadership skills. PDNs have leaders that influence the big cost/schedule/logistics decisions, control the pace and behaviors. Some stakeholders will aspire to earn that authority, as they get better and better at reducing risk in Construction 4.0. They also will want the power to influence the project *systems*, access, and deliverables. With insights that will be derived from the data analytics, the stakes will be high. Those who gain power will control the data (even if temporarily), will have access to the data for analytics, get the alerts from the "always learning" machine, and manage the multiple uses of the digital twin. They will decide the question of "insight sharing" (who is enabled with access and analytical tools).

In the design-build networks, one party leads the project team and governs the interactions with the owner. That party has more at risk and more influence. In the design-bid-build networks, the owner attempts to fix the scope, price, and schedule; but takes on risk of negative cost and schedule impacts of scope changes. With Integrated Project Delivery, the risk is shared. We assert that *shared risk will promote shared data, shared access, and shared insights*, giving those PDNs an advantage. That advantage will justify the investment in transformation to new delivery models over time.

Adaptation

We are in a race to adapt as individuals compete, as firms compete for the individuals, as firms create more competitive nations. Within stakeholder firms, professionals will be selected and evaluated for their ability to adapt. Leaders will be charged with motivating their teams to embrace new systems, behaviors, and workflows. Changes requiring new behaviors and attitudes of the workforce will need to be appropriately paced to sustain the enthusiasm, energy, a sense of purpose, and continuity. The kind of transformation that Construction 4.0 requires will create friction, and like friction in mechanical systems, it creates stress that must be dissipated. Successful leaders will assiduously consider ways to reduce and vent that emotional and psychological tension.

Company leadership

With increasing computing power and descending costs of data storage, firms will face transformation challenges. Middle management positions will be scrutinized for their costs and their contribution to successful transformation to the new environment. The current list of organizational roles/titles will change, making data decisions and transformation central to company strategy. There will be "organizational flip-flops" when technologically advanced workers begin to deliver more value for the money than their predecessors with more years of experience in the old business models, but less adaptability.

An edge in the race

Some stakeholder groups are positioned to lead in the "race" to transform successfully … to see the opportunity, be inspired to invest, and gain leadership authority within each of the PDNs. They will be first to create value from data aggregation and insights.

At-risk general contractors, EPCs, and trade contractors certainly feel the pain of wasting time and money in the current environment. Some will be agile enough to move quickly, affordably and successfully. We will see some savvy owners seize the opportunity first, but the transformation will be too complex and daunting for others. Owners have the authority to demand the lead role because they control the contract language, team selection, and funding in many PDNs. Architects and engineers may anticipate the opportunities and make a play to be the leaders of project data strategy. They will need to be data-savvy and mindful of the *owner's* potential gains from analytics, alerts, and digital twins. In many PDNs today designers, construction managers, and program managers play the role of owner's representative. If they don't move to play a leadership role in Construction 4.0, they may face lower fees, less frequent wins, and suffer loss of reputation and authority.

Trust

In many PDNs, companies will struggle with the question of trust. They will question whether any association, service provider, or other firm is trustworthy enough to host their data and "co-derive" insights. These PDNs may sense that they have proprietary information that serves as their competitive advantage that might be exposed to their rivals. They may choose to build their platform. They may justify the time, cost, and effort to develop and maintain their new systems with the promise of deriving tightly held insights and control.

4.5.3 Impacts on the construction market

Smart permitting and public pressure

The trend toward digital submissions and rule-based smart permitting will continue. It will mean shorter overall project durations. Public review of project scope and impacts (environmental, public disruption during construction, and the contribution to smart cities) will advance so that projects will produce less unexpected disruption, public resistance, and community disappointments.

Contracts

Just as with BIM mandates, owners will develop data mandates, requiring not only BIM deliverables, but also data, metadata, digital twins, simulation, and analytical reporting. Also like the global spread of institutions adopting BIM standards, there will be the early adopters of innovative new data mandates, and followers, that will study those new legal documents and adapt them to the local PDNs, be they government, university, high-tech, medical, etc. There will be successful "lighthouse" projects, perhaps funded by governments and owner associations that publish results and prove the value of the new contract language. Those projects will serve as case studies and enjoy global recognition.

Project outcomes

Project performance will improve. Forecasts will be more accurate. Completed asset performance will be measured, analyzed, and will improve steadily. Selection criteria to win work will include evidence of analytical excellence and proven project performance gains.

Adaptation strategy

To sustainably adapt to the opportunities that leaders encounter with their strategic teams, companies need to adopt a new approach to create a formulation for agile adaptation and transformation. They must tend to the existing businesses, and shape new businesses. Many leaders will seek new partnerships, some with high-tech companies, and some with other stakeholders to expand their services and range. Companies which manage to constantly reimagine themselves build higher investment capacity by revitalizing their core business.

Emotions and courage

We expect varying emotional responses from professionals in the industry as they live *with* the emergent Industry 4.0 and *live in* Construction 4.0. For some, it will be sad, perhaps fearsome. For some it will be thrilling, maybe life-changing. It is a playground for discovery and a storm of disruption.

Courageous leaders will invest, tolerate experimentation that innovation requires, steer their efforts with "guardrails", and allow for setbacks. These leaders will strive to develop the "optimal environment" for innovation. The mix of essential roles includes sponsor, re-thinker, tinkerer, mindset-changer, producer, evaluator, promoter, and standardizer.

Existing facilities

Owners will not only create *new* learning assets; they will also digitize and monitor *existing* assets with reality capture and sensor technologies, creating digital twins. These models will not be static models for newly efficient buildings, but "learning buildings", "learning campuses", and "learning cities" to generate operational efficiencies and preventive (versus reactive) maintenance.

Industry associations

Many industry associations and standards-setting institutions will create strategies to help their members through a successful transformation to mature Construction 4.0. They will educate their members, help them prepare for innovation, establish data standards, calculate benchmarks, perhaps even attempt to create a platform to federate projects and serve their members with co-funded machine learning.

4.6 Challenges and considerations

For those investing time, money, and energy to embrace technologies, hire and train people, test partnerships, let innovators innovate, or advance their education, we offer some considerations.

> Often, the greatest hurdle to implementing such solutions is the one-time backward reconciliation of data. Most firms have collected lots of information over the years, but it's stored in disparate systems and inconsistent formats. As such, the first step should be to take stock of what they have – many companies will find they have a lot more data than they realize …
>
> *(Hovnanian et al., 2019)*

A substantive challenge to assessing and addressing data integrity includes rationalizing the data capture methodologies. Stakeholders participate in many varying segments of the industry and varying geographies (and sometimes languages) that have different terminology and information needs. Some good news at this writing is that we find that some construction industry executives are stating in clear terms that data integrity is on their minds, and discussion of data capture, prioritization of KPIs, and strategic planning is timely.

4.6.1 Cultural challenges

Introducing and sustaining the data integration challenge

The task for the PDNs of integrating high-quality project data capture into a project environment can be onerous and is overwhelming for many industry professionals and leaders. It may be described as *transformation fatigue*. It is, however, possible and helpful to explain the big data initiative *simply* – as associated with just two purposes (to answer two questions): *What* are we going to build and manage? *How* are we going to build and manage it?

Much like projects benefit from the clarification that stems from Work Breakdown Structure and Organizational Breakdown Structure; a big data initiative will benefit from a Data Breakdown Structure (Figure 4.5). Stakeholders in design and operations are likely to be intrigued by the possible outcomes of integration of the "what" elements of data/analytical excellence: BIM, simulation, design optimization. Stakeholders from the parties with the challenge of planning, pricing, contracting, fabricating, constructing, and turnover will be interested in the "how" elements: RFIs, changes, delays, labor productivity, logistics, overtime. This clarification of purposes disaggregates the data-capture and data-integrity challenges. It offers a setting to illuminate prioritization of purpose for stakeholders and PDNs, so they invest their time and money aligned to their unique project environment and values.

Proactive leadership

CEOs of Real Estate, Engineering, Construction, and Facility Management firms are becoming more proactive in tackling the disruptive forces that are changing their industry. But many still struggle to pivot their organizations to new opportunities decisively or sustainably. Some

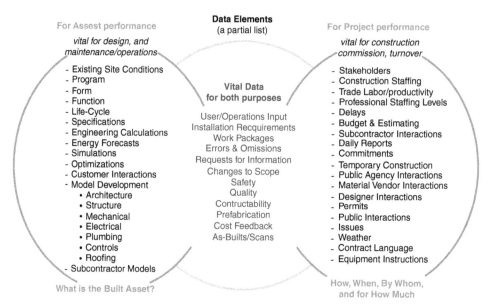

Figure 4.5   Data breakdown structure for two purposes – better asset and better project

companies remain overly focused on their legacy business such that they are unable to pursue new opportunities. Others neglect their core business to rush to the new without the required investment capacities (Moore, 2015).

Adaptation – too complex? Too slow? Too political?

Large companies in this sector will adapt to Construction 4.0 but may choose to do so slowly. Several factors will inhibit them from accelerating their journeys. These include capital-intensive infrastructures, contractual agreements, retaining outdated technology, and relentless devotion to legacy products, services, and brands. Some strategic decision-makers will believe they are less vulnerable to disruption because construction is a local industry and relies on local relationships. Some will think they're secure because this industry has a high degree of visibility in politics and the regulatory bodies will not allow disruption. Besides, this industry has a very fragmented supply chain; the business processes at stake are very complex. Legal, social, environmental, technical, regulatory constraints can anchor the industry to traditional systems.

4.6.2 Market challenges

Data scientists and education

The primary software developers, third-party developers, and those firms that elect to build their platforms will compete for a limited supply of data scientists. Even some of the stakeholders that elect to work with a commercial common data platform will choose to hire data scientists to help them pursue analytical excellence and create reports and dashboards. Educators charged to train the next generation of designers and construction professionals will include data analytics in their curricula. Data science will be another "tool in the toolbox" for PDNs.

Vulnerability

Although some companies will be shielded from the pressure to adapt to Construction 4.0, that protection will be impermanent. In the more demanding and technically advanced markets, some companies are very vulnerable to disruption. This is the case when they either fall behind in performance, rely too much on regulators to derail insurgents, or when they won't allocate enough capital to grow new initiatives. In some cases, companies serve customers/owners that fall behind their competitors that are demanding higher performance from service providers. Those firms that serve laggard customers will fail to create new business models and may find themselves years behind in the marathon to provide excellence in analysis, execution, and service.

4.6.3 *Technical decisions & challenges*

Big decisions

Project delivery networks will be faced with the selection of one or more shared data platforms, and whether to *build* a proprietary platform or *join* a commercial or institutional platform. Building and maintaining a proprietary platform may offer desirable privacy and control but may be revealed to be prohibitively expensive over the long term. Commercial platforms will likely have a time-to-market advantage, have machine-learning with many thousands of reference projects, and offer dozens or hundreds of third-party applications. Still, three questions persist:

- Will the stakeholders trust their data to be hosted by a private company, whether it be a startup or publicly traded enterprise?
- Will the stakeholders choose to keep insights from *their data* so tightly guarded that they build their platform, and reject the opportunity to co-invest?
- Will owners assert their authority to mandate (by contract) structured data deliverables like the way that they have begun to mandate BIM deliverables?

Data standards

As stakeholders from the PDNs, institutions, and commercial companies invest in the potential of common data environment, they will struggle to identify, integrate, embrace, and perhaps customize workflows with data standards in mind. A company may have 50,000 employees, with thousands of projects. A commercial platform may serve millions of construction professionals with hundreds of thousands of projects in the data library. Precise definitions of terms like error, omission, and request for information, change order, delay, etc. will be vital for data integrity. With knowledge of the three decades that it has taken to evolve the many current BIM standards, this challenge will be daunting, but it is not a sufficient reason to delay the innovations.

Network effects

The more users who adopt a platform, the more valuable the platform becomes to the owner and the users. Among technology providers, there will be a high-stakes global competition to be a platform leader, with the "dominant design", and with "core and peripheral systems". Some academic researchers and providers are betting on BIM as a core system. Some innovators are decidedly focused on platforms. They are betting that there will be "keystone firms" that have a strong starting point, whose current data specialties create an advantage if they are sufficiently open (Baldwin and Woodard, 2008). "The interfaces around the platform should be sufficiently 'open' to allow outside firms to 'plugin' complements as well as innovate on these complements and get a worthwhile return from their investments" (Gotsopoulos, 2015).

Multi-sided market

Our industry is a multi-sided market, defined as providing services "to several distinct groups of customers, all of whom need each other" (Gawer, 2014). Currently siloed data needs to be integrated. Owners need contractors. Contractors need trade contractors. Designers need owners. In our PDNs, all stakeholders need new levels of cooperation from each other to perform at the highest level. Implementing BIM proves the essential nature of this cooperation.

That cooperation is made more difficult because the stakeholders likely have different priorities in term of benefits from Construction 4.0. Designers may consider generative design and optimization their highest priority. Contractors may focus on prefabrication or change order and rework reduction. Owners may prize the digital twin and "asset learning" pursuing decades of operational excellence after turnover.

The most successful PDNs will intentionally target the highest price/performance benefits of the myriad possible returns. They will surmount the diverse and competing perspectives.

Proof of efficacy and academic contributions

This need for evidence introduces the "chicken and egg" dilemma in technology adoption. Many firms need evidence to justify investing. But what firms want to spend, measure benefits, and then publish their methods and results so competitors can close the performance gap? We will watch firms investing, partnering, proving value, and deciding to what extent (and how) to make their performance gains visible, at least, to their customers. Some of those firms will endeavor to obscure their gains and methodologies from alerting their competitors for as long as possible.

4.7 Conclusion

Construction 4.0 will invite innovation and generate positive outcomes for the industry. We assert this with firm conviction but acknowledge that the task of forecasting is multifaceted and the transformation is so disruptive that it demands great humility. We are forecasting *what* will happen and led by *whom*.

We anticipate some negative consequences ... shifts in job roles, vulnerabilities for certain industry players, the complex (positive and negative) impact of transparency. Finally, we present some personal convictions from the authors.

With owners compelling the project delivery networks to do better, the pull of the competitive nature of the stakeholders, and the inevitability of Construction 4.0 technologies to play a positive and strategic role, project outcomes and productivity will improve. Projects will be faster, leaner, safer, of higher quality, and more predictable. The built asset will be intelligent and learning, creating maintenance efficiencies and superior environmental performance.

Owners will choose to either: 1) invest, own, mandate, and manage the common data, or 2) demand the higher performance that results from data discipline and analytical excellence from their service providers. Contracts will evolve to capitalize on quality data, integrated systems, integrated thinking, "learning digital twins", and analytical deliverables.

Construction practitioners will better understand their essential part of the value chain, and continuously innovate to trim waste and low-value-added activities. The top performers will differentiate themselves with their analytical excellence and their skills to bring high performance to their customers. Those top performers will expand their service offerings. Top performers will acquire some of the firms that lose competitive ground in Construction 4.0 out of neglect to move quickly enough to seize the opportunity.

New executive-level positions will emerge. A new strategic role will emerge (Chief Transformation Officer, or Chief Innovation Officer) with the assignment to create an innovative environment, identify and evaluate the choices for investment in Construction 4.0, and implement them with short-term wins, leading to long-term, and sustained competitive advantage (Figure 4.6).

The technologies will be *one* of the vital ingredients that will power the expansion of IT-led innovation and include field-led discovery and innovation. Indeed, a new mandate for leaders and IT will be to create the digital and cultural environment that empowers and rewards innovation in every role. The "tiger teams" that steer and lead the organizational transformation will be wise to consider the four dimensions of successful transformational initiatives: culture, market pressure, principles, and technology.

Partnerships will flourish. Some firms will partner with like-minded organizations, perhaps with software companies. Some will partner with large consulting organizations. Some will partner with academia or institutions.

Academic institutions will help some firms or projects to enjoy early success and publish results. Some of those projects will be "lighthouse" projects that will be widely published and illuminate the systems, workflow, motivations, and cultural modifications that are needed for higher performance. The new students readied for Construction 4.0 will have analytical skills and will experience continuous learning during employment from the valued accessible data.

Finally, we expect to see substantial productivity gains across the industry, after years of lagging other sectors that have enjoyed proven benefits with technology. As stakeholders, we all have decisions about how we will approach innovation in this era of Construction 4.0. It will be a journey, a challenge, a marathon, a team effort, and full of discovery. Innovators will create value. Innovative and courageous executives in our industry will excel at the intersection of construction science, data science, and new hardware systems. Leaders will inspire and guide their organizations, transform and nourish their businesses, and make the ascendancy in Construction 4.0 fun, challenging, and rewarding. They will clarify the essential value that their organizations produce and create a transformation initiative with innovation as a treasured asset. Writing this chapter has been a privilege and inspiring for us. We sincerely hope it helps the readers.

> The only true voyage of discovery, the only fountain of Eternal Youth, would be not to visit strange lands but to possess other eyes, to behold the universe through the eyes of another, of a hundred others, to behold the hundred universes that each of them beholds, that each of them is.
>
> *(Proust, 1923)*

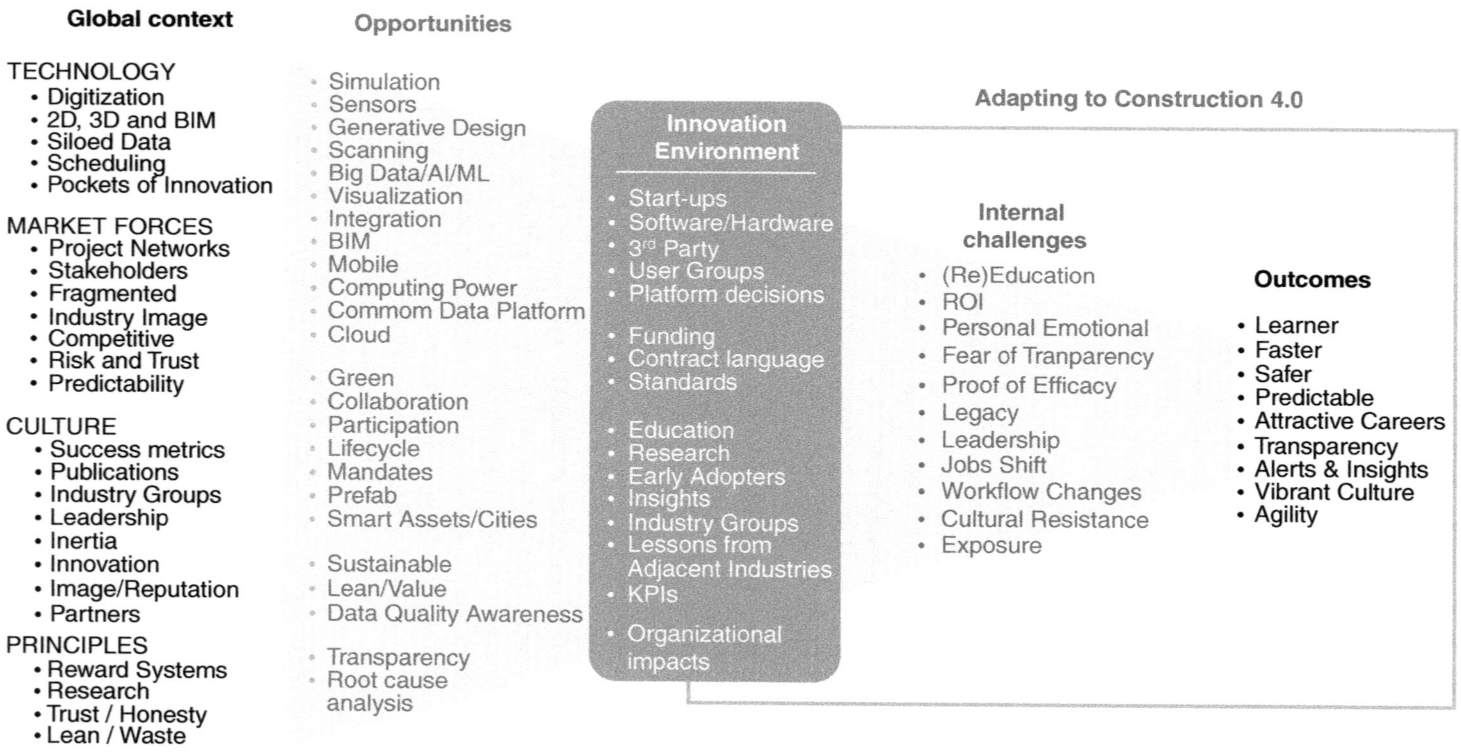

Global context

TECHNOLOGY
• Digitization
• 2D, 3D and BIM
• Siloed Data
• Scheduling
• Pockets of Innovation

MARKET FORCES
• Project Networks
• Stakeholders
• Fragmented
• Industry Image
• Competitive
• Risk and Trust
• Predictability

CULTURE
• Success metrics
• Publications
• Industry Groups
• Leadership
• Inertia
• Innovation
• Image/Reputation
• Partners

PRINCIPLES
• Reward Systems
• Research
• Trust / Honesty
• Lean / Waste

Opportunities

• Simulation
• Sensors
• Generative Design
• Scanning
• Big Data/AI/ML
• Visualization
• Integration
• BIM
• Mobile
• Computing Power
• Commom Data Platform
• Cloud

• Green
• Collaboration
• Participation
• Lifecycle
• Mandates
• Prefab
• Smart Assets/Cities

• Sustainable
• Lean/Value
• Data Quality Awareness

• Transparency
• Root cause analysis

Innovation Environment

• Start-ups
• Software/Hardware
• 3rd Party
• User Groups
• Platform decisions

• Funding
• Contract language
• Standards

• Education
• Research
• Early Adopters
• Insights
• Industry Groups
• Lessons from Adjacent Industries
• KPIs

• Organizational impacts

Adapting to Construction 4.0

Internal challenges

• (Re)Education
• ROI
• Personal Emotional
• Fear of Tranparency
• Proof of Efficacy
• Legacy
• Leadership
• Jobs Shift
• Workflow Changes
• Cultural Resistance
• Exposure

Outcomes

• Learner
• Faster
• Safer
• Predictable
• Attractive Careers
• Transparency
• Alerts & Insights
• Vibrant Culture
• Agility

Figure 4.6 The innovation environment leading to beneficial adaptation to Construction 4.0

4.8 Summary

- Construction 4.0 will invite innovation and generate positive outcomes for the industry.
- There may be some negative consequences such as role shift, the vulnerability of some stakeholders, and complex impact of transparency.
- As project delivery networks improve driven by needs and responses of stakeholders, the positive impact of Construction 4.0 technologies will improve.
- Projects will be faster, leaner, safer, of higher quality, and more predictable.
- The built asset will be intelligent and learning, creating maintenance efficiencies and superior environmental performance.
- Owners will choose effective strategies to maximize the utility and benefit of data and contracts will evolve to capitalize on quality data, integrated systems, integrated thinking, "learning digital twins", and analytical deliverables.
- Construction practitioners will understand their part of the value chain, innovate to trim waste and low-value-added activities.
- Top performers will differentiate themselves through analytical excellence and skills to bring high performance to their customers and will expand their service offerings.
- Top performers will acquire firms that lose competitive ground in Construction 4.0.
- New executive-level positions will emerge to create an innovative environment for competitive advantage.
- Construction 4.0 will liberate substantial productivity gains across the construction industry.

References

Autodesk (2018). Enhancing Construction Project Management Through AI. www.autodesk.com/autodesk-university/class/Enhancing-Construction-Project-Management-Through-AI-2018 (Accessed September 7, 2019).

Autodesk (2019). Forge Links Agile Issue Tracking to 3D Models. https://forge.autodesk.com/customer-stories/sweco (Accessed September 7, 2019).

Baldwin, C. Y. & Woodard, C. J. (2008). The Architecture of Platforms: A Unified View. *Harvard Business School Finance Working Paper No. 09-034*. September 8, Available at SSRN: https://ssrn.com/abstract=1265155 or 10.2139/ssrn.1265155.

Blanco, J. L. (2018). Artificial Intelligence: Construction Technology's Next Frontier. April 2018 www.mckinsey.com/industries/capital-projects-and-infrastructure/our-insights/artificial-intelligence-construction-technologys-next-frontier.

Boland, R. J. et al. (2003). Path Creation with Digital 3D Representation: Networks of Innovation in Architectural Design and Construction. *Sprouts: Working Papers on Information Environments. Systems and Organizations*, Vol. 3, Winter.

Chelson, D. (2010). The Effects of Building Information Modeling on Construction Site Productivity. *Dissertation submitted to the Faculty of the Graduate School of the University of Maryland*, College Park, in partial fulfilment of the requirements for the degree of Doctor of Philosophy 2010.

CII (2019a). Transforming the Industry: Making the Case for AWP as a Standard (Best) Practice. www.construction-institute.org/resources/knowledgebase/best-practices/advanced-work-packaging/topics/rt-319 (Accessed September 7, 2019).

CII (2019b). Enhanced Work Packaging: Design through WorkFace Execution (Best Practice). www.construction-institute.org/resources/knowledgebase/best-practices/advanced-work-packaging/topics/rt-272 (Accessed September 7, 2019).

Donalek, C. et al. (2014). *Immersive and Collaborative Data Visualization Using Virtual Reality Platforms*, Conference paper, October 2014. DOI: 10.1109/BigData.2014.7004282.

Fayek, A. et al. (2003). Measuring and Classifying Construction Field Rework. *Presented to: Construction Owners Association of Alberta (COAA) Field Rework Committee*, May 2003.

Friedman, T. L. (2019). The United Kingdom Has Gone Mad. *New York Times*, April 2, 2019. www.nytimes.com/2019/04/02/opinion/brexit-news.html.

Gawer, A. (2014). Industry Platforms and Ecosystem Innovation. *Journal of Product Innovation Management* 31(3): 417–433. Product Development & Management Association. DOI: 10.1111/jpim.12105.

Gerber, D. et al. (2012). Design Optioneering: Multi-disciplinary Design Optimization through Parameterization, Domain Integration and Automation of a Genetic Algorithm. *Simulation Series*. 44.

Gotsopoulos, A. (2015). Dominant Designs. *International Encyclopedia of the Social & Behavioral Sciences* (Second Edition). New York: Elsevier, pp. 636–640. www.sciencedirect.com/topics/computer-science/dominant-design.

Hagel, J. (2016). Flows, Fragility and Friction. *Edge Perspectives with John Hagel*, October 26, 2016. https://edgeperspectives.typepad.com/edge_perspectives/2016/10/flows-and-friction.html.

Hovnanian, G. et al. (2019). How Analytics can Drive Smarter Engineering and Construction Decisions. January 2019. www.mckinsey.com/industries/capital-projects-and-infrastructure/our-insights/how-analytics-can-drive-smarter-engineering-and-construction-decisions.

Kam, C. (2018). Key Performance Metrics from Lean Construction Perspective. *20th LCI Congress*. October 15–19, Orlando, FL. www.lcicongress.org/pdfs/2018/THA6-Kam.pdf.

Moore, G. A. (2015). *Zone to Win*. New York: First Diversion Books.

Moore, G. A. (1991, 1999–2002). *Crossing the Chasm*. New York: HarperCollins.

Mossman, A. (2009). Creating Value: A Sufficient Way to Eliminate Waste in Lean Design and Lean Production (2009) *Lean Construction Journal* 2009: 13–23. https://leanconstructionblog.com/The-Concept-of-Waste-as-Understood-in-Lean-Construction.html.

New York City Department of Buildings (2019). 3D Site Safety Plans. www1.nyc.gov/site/buildings/safety/3d-site-safety-plans.page](Accessed September 7, 2019).

Nikkei (2017). AI to Propel Wind Farm Efficiency in Japan. https://asia.nikkei.com/Business/AI-to-propel-wind-farm-efficiency-in-Japan (Accessed September 7, 2019).

Proust, M. (1923). *Remembrance of Things Past* (published in French in 1923, and first translated into English by C. K. Moncrief).

Rajakallio, K. et al. (2017). Business Model Renewal in Context of Integrated Solutions Delivery: A Network Perspective. *International Journal of Strategic Property Management* 21: 72–86. DOI: 10.3846/1648715X.2016.1249533.

Sacks, R. (2019). Construction Tech: Innovation Founded on Lean Theory and BIM Technology. *Constructs: Digital Innovation in the Built Environment, 2018–19 Keynote Lecture Series*. The Bartlett School of Construction and Project Management, University College London, London. February 2019.

Sawhney, A. (2011). Modelling Value in Construction Processes using Value Stream Mapping. *Masterbuilder*, October 2011/88.

Stowe (2018). The Avalanche of Waste. www.linkedin.com/pulse/avalanche-waste-ken-stowe?articleId=6524639178163003392#comments-6524639178163003392&trk=public_profile_article_view (Accessed September 7, 2019).

Succar, B. (2009). Building Information Modelling framework: A Research and Delivery Foundation for Industry Stakeholders. *Automation in Construction* 357–375. DOI: 10.1016/j.autcon.2008.10.003.

Part II

Core components of Construction 4.0

5

POTENTIALS OF CYBER-PHYSICAL SYSTEMS IN ARCHITECTURE AND CONSTRUCTION

Lauren Vasey and Achim Menges

5.1 Aims

- Identify and describe the potentials for cyber-physical systems in architecture and construction.
- Contextualize current opportunities and challenges for cyber-physical systems within stereotypical design workflows and production processes.
- Describe the critical technical components of a cyber-physical system and the enabling technologies that facilitate their deployment in AEC.
- Identify areas where cyber-physical systems can be applied utilizing specific case studies.
- Conclude with a discussion of the additional facilitating technologies and methodologies that would enable and enhance the impact of cyber-physical systems in AEC.

5.2 Introduction

Cyber-physical systems present enormous potentials in architecture and construction, enabling a reconsideration of the highly fragmented nature of the industry through increased possibilities for integration and collaboration as well as a re-examination of the role of simulation and standards within stereotypical design and production processes. This chapter considers the greater context of architectural design, approval and realization, including conventions that have historically limited the typologies and materiality of architecture, and structures that can be designed, simulated, and realized.

Through cyber-physical systems, physical fabrication and construction projects can be tightly coupled to background computational monitoring mechanisms; data gathered during physical processes can be iteratively processed, and feedback can be iteratively exchanged between digital environments and physical processes. With intelligent and reactive handling protocols, cyber-physical systems can enable an entirely new set of material systems, including non-standard, anisotropic, and differentiated material systems, as well as unpredictable material systems which resist prediction with existing models of analysis and reductive calculations.

These possibilities enable a shift between formally compartmentalized processes, such as digital design processes and physical production processes, and a shift towards process flexibility and autonomy: where robots, machines, and construction equipment can leverage sensor feedback and closed-loop control to respond to unpredictable material systems within unstructured environments. Inter-device communication can connect multiple parallel processes, and new modes of networked and collective construction are possible, where teams of machines, robots, and workers can work collaboratively on common construction tasks.

These possibilities are not without challenges, and require additional facilitative technologies, including domain-specific hardware and software tools developed for the needs and constraints of construction, in addition to integrated co-design processes, where new building systems, machines, and processes are co-developed, their constraints mutually influencing interdependent development. Similarly, there is a need for new building codes and regulations to codify these opportunities, reconsidering, for example, the use of structural analysis as an integral part of production and construction, opening the door for validation of mechanical properties rather than conservative pre-calculation. Such developments could truly enable the adoption and the full potentials of cyber-physical systems into the greater architecture and building industry eco-systems.

5.3 Towards cyber-physical construction

A cyber-physical system is a tight coupling of physical devices, sensors, and back-end computational processes (Rajkumar et al., 2010). While the concept of a cyber-physical system is similar to both the Internet of Things and embedded systems, a few critical distinctions can be made. Embedded systems are engineered systems that a couple of physical processes with computing. However, embedded systems tend to be hermetic and do not typically expose their inner functionalities to external entities or enable manipulation at run-time. Cyber-physical systems, in contrast, are complex networks which offer opportunities for interaction (Schirner et al., 2013). In addition, a cyber-physical system includes several parallel computational processes acting at multiple levels of hierarchy, enabling computational processes to affect physical processes and vice versa. Thus cyber-physical systems are a new generation of interconnected systems which leverage IOT and embedded systems but are much more complex and stochastic.

5.3.1 Components of a cyber-physical system: computing, communicating, sensing, and controlling

A cyber-physical system includes several layers and components dedicated towards tasks in computing, communicating, sensing, and controlling, illustrated in Figure 5.1. Networked communication protocols connect multiple devices, hardware elements, computational resources, and sensors. Each physical entity, including actuators, machines, robots, and devices, may have its on-board computational resources responsible for local decision making and control, while more centralized processes collect and process data from many sources. Control algorithms may be utilized to maintain process-specific parameters at the local level of a device or machine, while higher-level computational processes may be responsible for autonomous or semi-autonomous decision making at the system level. Symbolic 3D representations of the system can enable sensor data to be related temporarily and spatially to a "digital twin," or 3D model representing the current system state. Within the domain of architecture and construction, 3D data structures are of particular importance, to

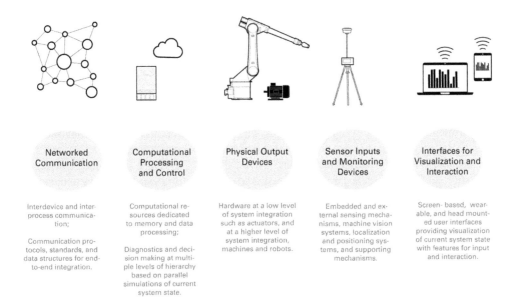

Figure 5.1   Components of a cyber-physical system

represent both design intentions as well as the as-built status. A cyber-physical system may also include multiple input and monitoring devices, including user interfaces and displays, depicting diagnostic information.

5.4 Context: challenges and opportunities for cyber-physical systems within architectural production

In contrast to other industries, several characteristics of the AEC industry make the implementation of cyber-physical systems challenging. Compartmentalization separating processes and professions have historically created barriers to collaboration and fluid data exchange. Similarly, the prevalence and reliance on standardization acting across many scales are extremely limiting. Recognizing and overcoming these challenges could be a significant step towards increased levels of digitalization within the context of Construction 4.0.

5.4.1 Existing production chains in architecture and construction

5.4.1.1 Fragmented and compartmentalized design processes

Within the AEC industry, several professional entities spread across different organizations must collaborate towards the design, approval, and production of a building. This includes, but is not limited to architects, building owners, building operators, construction professionals, contractors and sub-contractors, civil and structural engineers, and material suppliers. In a common design delivery process, particularly pre-dating integrated project delivery, these roles were functionally separated. An architect would deliver a set of drawings representing design intent, later to be interpreted by contractors and engineers. Similarly, static dimensional documents and representations served as the primary means of exchange during collaboration, utilized for approval and tendering processes. Thus, historically, design-to-construction processes have been highly linear, inflexible, and compartmentalized, with little feedback occurring between professions or between construction phases.

5.4.1.2 Specialized and incompatible software for different building services

Compartmentalization between professions has been exacerbated by incompatibility between software environments for different building services (Poirier et al., 2014). The development of end-to-end solutions has proved difficult, as there have not been features such as common data structures or shared APIs, facilitating fluid data exchange between parties. Interoperability has arguably reinforced the industry's reliance on static and inflexible systems of representation. Similarly, conversion between file types has also been a necessary consequence of software interoperability, often causing a loss of critical metadata.

5.4.1.3 Standardization of materials, processes, and systems

Standards within AEC operate at different points during a construction and realization process, functioning as a necessary mechanism for quality control between disconnected processes. Historically, standardization has been essential because of processes of exchange, whereby designs, documents, and components could be passed from one entity to another with consistency and predictability. Today, the dependence on standardization needs to be critically examined for its underlying limitations. In particular, the common practice and codification of utilizing standardized components, predominant since modernization, hinders the development of new legislation and codes which would enable differentiated building components, where each building component can be manufactured uniquely in response local mechanical needs or design intentions.

Standardization is also operative at the level of material processes, whereby naturally variable and anisotropic materials, including wood, often undergo rigorous processing steps to transform into quasi-isotropic materials. In this case, the technical means of acting, reacting, and verifying non-standard, anisotropic material has not been feasible within highly linear and inflexible production chains.

5.4.1.4 Conservative calculations based on known typologies

Limitations are also operative due to the compartmentalized and conservative nature of structural engineering within stereotypical design workflows. As Jan Knippers noted, "the paradigm of 'calculability' has been pre-dominant in structural engineering since the mid-nineteenth-century until today: what cannot be calculated cannot be built" (Knippers, 2017). Many existing calculation methods, including the working stress method, are based on simplified assumptions and require manual calculations that are usually extremely conservative (Mufti et al., 2005). The need to pre-calculate structures according to an inherently limited range of approved methods for deformations, strain and internal forces, limits the types of structures which can be verified and validated (Knippers, 2017). Thus, many of the potentials of digital fabrication and sensor integrated production, including the invention and verification of new material systems, cannot be realized in the context of conservative industry practices.

5.4.2 Enabling technologies for cyber-physical systems in AEC

Technological advances in sensor feedback, inter-device communication, and real-time data processing enable cyber-physical systems in AEC, facilitating a rethinking of compartmentalized, linear production chains within architecture. Increased computational power, which doubles predictably every two years according to Moore's Law (Koomey et al., 2011, Waldrop, 2016), is now enhanced through increased communication reliability, whereby the speed and consistency of wireless communication are increasing with 5G networks, known for low latencies and high connection densities (Agiwal et al., 2016).

Arguably, many of these enabling technologies are not new, but rather, are newly accessible to multiple actors within the industry through increased ubiquity and usability. Open-source initiatives and community-based forums for exchange and development, including Grasshopper for McNeel and Autodesk's Dynamo, provide common development platforms where tools for simulation, design generation, and manufacturing can be integrated into one workflow.

However, open-source initiatives operating purely through software are insufficient for cyber-physical systems and do not provide the critical path necessary from digital design workflows to physical production processes. Increasingly, the open-source ethos has spread to hardware, where common off-the-shelf input and output devices, sensors and actuators, increasingly support inter-device communication through add-on APIs (Bonvoisin et al., 2017; Powell, 2012). Micro-processors and even micro-computers with significant processing power are now readily available at a low cost. Additionally, the decreasing cost of sensors and actuators and the simultaneous increase in their accuracy has improved the economic feasibility of sensor integration into common production workflows (The Atlas, 2016).

Robust data processing techniques utilizing machine learning can also be leveraged to make existing sensors more accurate. Emerging platforms, including Paperspace, Amazon Web Services, and Google's TensorFlow, are now working towards making complex data processing techniques increasingly accessible to non-expert users, facilitating their application to a variety of complex AEC problems (Paperspace 2019; Abadi et al., 2016). Cloud computing is an additional service for purchase through vendors, enabling on-demand access to computing power beyond what may have historically been available on-site.

There is additionally a shift in accessibility in production and manufacturing technologies, due to propriety and black-box algorithms coming to the end of their terms of protection. Computer-controlled technologies for rapid prototyping, including milling machines and 3D printers, first emerged in the 1960s through the 1980s (Baudisch & Mueller, 2017). However, patents protected the first generation of rapid prototyping technologies, hindering research and development for several decades. In 2009, however, the first major patent expired, thereby initiating a significant uptake in both commercial and research activities (ibid.).

The usability of manufacturing and production equipment is also enhanced through software functionalities for data exchange and online control. Manufacturing, production, and robotics companies face increasing incentives to produce software add-ons to enable real-time control and communication between manufacturing equipment and external computing resources, features customarily excluded or minimized due to safety concerns. These software add-ons can simplify the process of establishing customizable data exchange, enabling online feedback and direct manipulation of the actions of physical machines and robots at run-time.

5.4.2.1 Communication protocols: towards interoperability between software and hardware

To establish a cyber-physical system connecting multiple digital and physical processes, custom communication protocols must connect disparate platforms, devices and workstations which may have been developed by disparate parties or entities. Thus, protocols for iterative data exchange are vitally important, and 3D data structures not tied to any individual proprietary CAD program are advantageous.

Several developments within academic and research contexts recognize the fundamental importance of such tools and are moving towards real-time collaborative data sharing. Recent projects include Speckle, an open-source (MIT) initiative for developing an extensible Design and AEC data communication and collaboration platform (Speckle Works, 2019). Important aspects of Speckle include the ability to share model information in real-time, where collaborators can mutually work on the same CAD file simultaneously.

Additionally, the COMPAS library, spearheaded at the ETH Zurich by the Block Research Group, is an "open-source, Python-based computational framework for collaboration and research in architecture, engineering and digital fabrication" (Mele et al., 2017). Key features of this library are that it is front-end and back-end agnostic, and meant to function entirely independently from CAD software. The library supports a variety of common data structures, simulation techniques, and geometric processing and manipulation methods for fabrication, construction, and simulation (ibid.).

Increasingly, major software companies, including Autodesk, are expanding their suite of products, shifting in some cases away from individual software programs, to platforms and services. Though early initiatives for a common CAD scripting language faced difficulties (Aish & Mendoza, 2016), recent initiatives include Autodesk Forge, a set of APIs that expose functionality across multiple programs (Autodesk, 2018). These APIs are compatible with multiple programming languages, web, and NET frameworks, making them increasingly accessible to devices and hardware, and thus physical processes. Similarly, McNeel, the developer of Rhinoceros, has made their core geometric library accessible without an instance of their software running, a factor that could enable geometry and processing libraries to be more fully integrated into production processes and deployed in end-to-end solutions (McNeel, 2019).

These computational frameworks and tools provide the necessary communication infrastructure to develop integrated cyber-physical systems for construction processes, whereby computational workflows and custom data structures used during design development can continue to be used in production, challenging compartmentalization between digital and physical processes.

5.5 New possibilities enabled in architecture by cyber-physical systems

5.5.1 Intelligent, interconnected, and adaptive production

The implementation of cyber-physical systems into architecture and production can challenge linear and fragmented workflows and production chains, and facilitate a transfer of many of the developments of intelligent manufacturing into the building industry, allowing gains in productivity, efficiency and quality control.

5.5.1.1 Online process monitoring and control

Cyber-physical systems can establish communication between production equipment and sensors, facilitating automated data acquisition (ADA), monitoring and control of production processes at run-time. By enabling sensor feedback to influence motion toolpaths, unpredictable temperature and pressure sensitive processes can be handled through reactive protocols operating within machine control code (Vasey, 2015). While closed-loop control is often applied in industrial and manufacturing contexts for localization and positioning, in architecture it can be used geometrically, leveraged to enable a robot or machine to closely approximate a target geometric condition despite process inconsistencies. Geometric-based closed-loop control has been utilized in 3D printing applications to correct aggregated tolerances (Cohen & Lipson, 2010), and also in applications for the robotic fabrication of on-site rebar for concrete reinforcement (Dörfler et al., 2016).

Scanning, digitally recording, processing, construction and fabrication data is an important component of a cyber-physical system for AEC, as it provides the opportunity to iteratively compare as-built conditions to intended digital design files, where tolerances or quality issues can be detected without latency, and changes and adjustments can be made on-the-fly through correctional logics. While many automation technologies have been developed to identify quality issues by measuring the inclination, alignment and evenness of components, or detecting problematic features such as cracks, a cyber-physical system could enable those features to be subsequently corrected through the actions of physical output devices.

5.5.1.2 Just-in-time and on-demand production

The ability to sense and scan what has previously built opens up the possibility of implementing on-demand fabrication in an on-site context. Just-in-time manufacturing is a technique appropriated from the manufacturing industry, where tolerances are incrementally measured, and parts files can be digitally regenerated and produced just before installation to minimize aggregated tolerances (Cheng, 1996).

To enable this type of adaptive workflow, critical transportable manufacturing and production infrastructure may be brought in strategically onto the construction site. Many initiatives in the industry are pursuing mobile fabrication and construction platforms for this purpose. Autodesk has developed a transportable shipping container containing a collaborative robotic work cell for directed energy deposition (DED) for the production of metal components (Peters, 2018). ODICO, a publicly traded robotics company based in Denmark, has developed concepts for "Factory-on-the-Fly," a portable container that can be deployed in a scenario such as a construction site (ODICO, 2019). A collaboration between the University of Stuttgart and BEC GmbH has resulted in a fully transportable mobile platform with a dual robotic cell for on-site timber manufacturing, shown in Figure 5.2 (Alvarez et al., 2019).

5.5.1.3 Tracking, localization, and positioning of building components, people, and construction equipment

There are additionally immense potentials for utilizing tracking and positioning technologies for the monitoring of people, material, and equipment in a scenario such as a construction site. Many different types of tracking and positioning technologies are available with different constraints, price-points, and specifications, making their application suitable for different environments and industry needs. Visual tag-based systems provide a relatively cheap solution for the tracking of components, requiring limited additional hardware. To implement such a

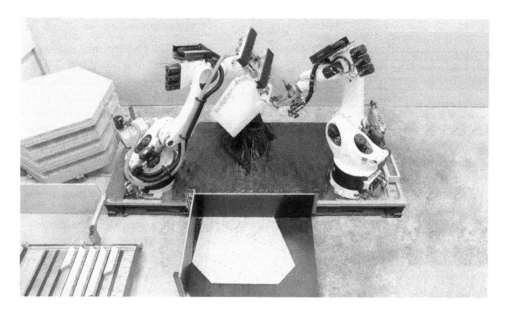

Figure 5.2   A dual robot collaborative cell with external axis for on-site timber manufacturing
ICD/ITKE, University of Stuttgart

system, printed identifiers can be attached to parts and equipment, which can then be scanned periodically, enabling the position of that element to be digitally recorded with a time stamp.

Shop Architects, an early pioneer of digital technologies and professional integration, exhibited the benefits of this approach already in 2012 for the management and construction of Barclay's center. The team developed a tagging system and custom iOS app to tag, organize and track the status of customized weathered steel building components (Nobel, 2016). A scan of an individual component enabled immediate access to the component's metadata, demonstrating that many of the added organizational and logistical challenges incurred through digital workflows can be minimized with cyber-physical systems and user interfaces.

Visual fiducial tagging systems appropriated from robotics contexts can also be used for real-time localization of equipment, machinery and people (Olson, 2011). Such systems are extremely low cost but require several cameras to be precisely placed and calibrated according to a global coordinate system. Other commonly used localization technologies, including GPS and RFID, radio-frequency identification, have previously been widely implemented in construction contexts but face challenges in some scenarios (Ergen et al., 2007). GPS can be hindered by large physical barriers, while the signal strength of RFID can be manipulated within proximity to metals (Lu et al., 2007). Alternative wireless network systems utilizing Wi-Fi, Ultra-Wideband, ZigBee, and Bluetooth, apply RSSI, a received signal strength indicator, to localize components based on the estimated distance or angle measured by a system of beacon and gateway locations (Shen et al., 2008). Collectively, these technologies could enable cyber-physical systems to gather real-time information about the location of components, people, and even production equipment in a construction site.

5.5.1.4 *Integrated diagnostics and up-to-date project scheduling*

The ability to track building components and monitor manufacturing and building processes suggest the possibility of integrated process diagnostics, where process metrics and efficiencies can be evaluated, quantified, and accessed in real-time. These capabilities also imply the future possibility of more integrated supply chains, whereby quantities of materials and supplies can be intelligently monitored, and digital mechanisms can distribute materials or equipment according to demand.

Applying cyber-physical systems towards coordination and management could potentially imply a significant shift in business practices in AEC. In other industries, the shift from product and service-based business models to platform-based business models has been profound. For example, crowdsourced ride-sharing has disrupted transportation, while entities like Airbnb have disrupted hospitality (Kenney and Zysman, 2016). Thus intelligent and augmented platforms could open up the possibility of connecting, managing, and redistributing resources, equipment or even construction and contracting services.

5.5.2 Collaborative and networked construction

Design, fabrication, and construction processes involve a number of disparate parties, entities, and processes. Thus, another immediate possibility suggested by cyber-physical construction is leveraging digital mechanisms to orchestrate and coordinate complex construction processes, where multiple tasks can be planned and dispatched to disparate entities utilizing networked systems of control.

5.5.2.1 Computationally mediated task distribution and logistics

In research contexts in Human-Machine Interaction (HMI) it has been demonstrated that multiple users can be coordinated in large-scale building processes when guided by instructions delivered via wearable devices. One such project was a proof-of-concept exhibition, the Hive Pavilion, developed at Autodesk University in 2015. For this exhibition, users with no previous construction or robotics experience were led through the process of interacting and building a modular architecture-scale system with a team of collaborative robots (Lafreniere et al., 2016). To enable the process, a computational back-end engine, the "foreman engine," played a role similar to a foreman on a construction site, iteratively tracking build status, as well as the status of each of the robotic stations, users, and completed modules. Beacons were used to track the geospatial locations of users within the space, and the foreman engine could also monitor each user's progress within the app in order to forward just-in-time instructions to the robotic workstations and devices (ibid.). Image-based machine vision measured the deflections of each part before production, and the robots' toolpaths could be adaptively regenerated at run-time in response to measured deviations (Vasey et al., 2016).

As a proof-of-concept, this system illustrated the immense potentials of cyber-physical coordination systems in practice. Similar coordination systems in construction could be applied either in co-located on-site conditions or between multiple disparate but concurrent processes.

5.5.2.2 Multi-machine and multi-robot collaboration

With the ability to connect multiple devices, machines, robots, and users in an interconnected cyber-physical system comes the ability to enable digitally truly collaborative workflows: where heterogeneous robot teams work in tandem on common construction goals. Interconnected and collaborative processes offer many advantages, as a system of robots, devices, and machines can overcome many of the limitations of a single robot.

One such demonstrator, the ICD/ITKE Research Pavilion 2016/2017, designed and built at the University of Stuttgart, investigated a collaborative building process for the construction of a long-span composite structure, illustrated in Figure 5.3. A custom UAV drone was developed to transport an exchangeable robotic end-effector from one industrial robotic station to another, enabling a continuous fibrous structure to be constructed between the two stationary robots (Felbrich et al., 2017).

The project illustrated the necessary hardware and software components which can enable such a process. Custom hardware facilitated task-trading through the physical exchange of a winding gripper. Electro-magnetic actuation enabled the drone to carry the gripper: after landing, it could be re-gripped pneumatically by the robot. The process was augmented with additional sensors for feedback and localization, enabling the drone to localize itself relative to its target in the scene, and the enabling the robot to locate and grip the end-effector after a successful drone flight. Secondary mechanical systems and online tension feedback regulated the amount of tension in the system.

To orchestrate and coordinate the actions of each of the robots and devices in the system, a set of custom computational tools complied the desired set of fibre paths into a coordinated task-list of high-level machine instructions. A distributed communication network developed using the Robot Operating System (ROS) iteratively dispatched these tasks to each of the robots and devices in the scene, coordinating their execution based on the current production status. The robots were programmed with flexible sub-routines: an incoming data-package would indicate which robotic sub-routine to run, with additional inputs for process-specific variables (ibid.).

Such a process embodies the potentials of using a cyber-physical system to manage a complex network of robots, machines, and devices, where each entity or actor in the scene can be used for its relative strengths and specifications. By combining machine and robots with different specifications, in mobility, payload, and accuracy, the complexity and scale of building components and material systems can potentially increase. In this case, a cyber-physical collaborative fabrication process enabled the construction of a cantilever over 12 meters in length (Figure 5.4), significantly longer than the working space of either of the industrial robots. Simultaneously, the project illustrates techniques that can enable the realization of a collaborative process: in particular, distributed and networked communication systems, and flexible and autonomous programming strategies that enable dynamic and sequence-dependent execution of robotic routines.

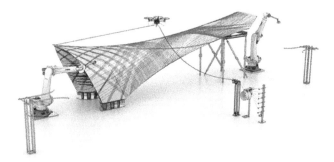

Figure 5.3   Multi-robot collaborative set-up for the fabrication of long-span fibre composites, ICD/ITKE University of Stuttgart

Figure 5.4   12-meter long cantilever under construction, ICD/ITKE University of Stuttgart

5.5.2.3 Semi-autonomous machines, robots, and processes

As the above case study illustrated, cyber-physical fabrication implies a shift in the language of machine and robot instruction. In contrast to executing predefined and static control code, machines and robots can be programmed with flexible sub-routines, where signal processing can be configured to affect robotic actions at run-time. The spectrum of autonomy can be quite varied according to project, material, or system demands. Thus one potential for cyber-physical systems is to begin to apply developments from autonomous systems research into processes of production, whereby construction equipment and fabrication machines can utilize state estimations and machine learning algorithms to achieve high-level goals within unstructured environments.

While on-site construction equipment has historically been teleoperated by human operators, increasingly, construction companies, including Built Robotics, Komatsu, John Deere, and Caterpillar are beginning to develop autonomous construction equipment, including autonomous excavators, tractors, and trucks (Built Robotics, 2019, Gershgorn, 2019). Researchers have also developed autonomous vehicles for terrain manipulation and landscape design (Hurkxkens et al., 2017).

With enhanced functionalities for handling unstructured sites and tasks, the characteristic roles that robots and other automated machines play within production and construction processes can expand. To date, robotics and other fabrication machines have had relatively compartmentalized and isolated roles within the production of buildings: in early attempts towards automation in construction, primarily occurring in Japan in the 1980s, robots were chiefly deployed to automate single tasks within an overall workflow (Bock & Linner, 2016). A new push towards automation is emerging through the use of accessible and generic industrial robots. However, today, large-scale buildings utilizing advanced fabrication technology, including Landesgartenschau Exhibition Hall, have primarily used robots as precise rather than autonomous entities, producing predefined parts or executing predefined control code (Schwinn & Menges, 2015).

When augmented with sensor feedback, robots can do more than process predefined parts: robots are increasingly being used for more dexterous tasks in a large-scale spatial assembly in non-standard wood and metal systems (Parascho, 2016; Willmann et al., 2016). Other emerging roles for robots and autonomous machines on construction sites include inspection and monitoring for quality control (Yan et al., 2019). With features including axis torque monitoring, robots can gain localized force feedback during production, enabling them to act as generic sensors for testing structural or assembly sequences (Calvo Barentin et al., 2018). Collectively, recent endeavors indicate that increased sensing and data processing capabilities can significantly expand the complexity and types of tasks which machines and robots can successfully execute.

5.5.3 Increased possibilities for human-in-the-loop construction

While robotic automation is one strategy for enhancing process efficiencies within construction, a second strategy is a strategic collaboration, where humans and machines are used for their relative strengths in a collaborative workflow, and cyber-physical systems are leveraged to digitally coordinate task trading, task sharing, and safety monitoring.

5.5.3.1 Human-Robot Collaboration

Human-Robot Collaboration (HRC) is emerging as a relatively important field of research, as many barriers to automation in construction exist due to the relative dependency on human-level dexterity and cognition in many construction tasks. Thus, one possibility towards accelerating process efficiencies in the short term can be enabled by human-robot and human and machine collaboration, whereby users can work cooperatively on construction tasks alongside human-safe collaborative robots or human safe construction equipment. In this case, cyber-physical systems can manage and orchestrate collaboration and task sharing and utilize sensor networks and iterative data processing for safety monitoring.

Collaborative robots, classified simply as a typology of a robot that is human safe through torque sensing in the joints, active vibration damping, sensitive collision detection, as well as compliant control, have emerged in the last decade, developed for deployment in the manufacturing sector (Albu-Schäffer et al., 2007). In this domain, they have proved to be effective in improved joint task performance while reducing the workload on their human co-workers (Scassellati & Hayes, 2014). Within construction, collaborative workflows can enable a smart division of labor and tasks, whereby users such as project managers, fabricators, and construction workers can leverage process knowledge, intuition, and craft to execute tasks which are still elusive or cost-prohibitive for robots. Similarly, robots and machines can be used to execute precise, monotonous, repetitive, and dangerous tasks.

One such case study was a collaborative robotic workbench developed for deployment in wood prefabrication, pictured in Figure 5.5 (Kyjanek et al., 2019). To enable users with various expertise to interact with a robotic construction process, an augmented reality scene served as a control layer and interface, enabling parts to be selected, planned, and then dispatched to the robot. Diagnostic feedback from the robotic work cell, including axis torques and construction progress, could be superimposed on the user's view, displayed in Figure 5.6. The construction workflow was based on collaboration and exchange, a user executed dexterous and cognitive tasks including fixing, screwing, and planning, while a Kuka collaborative robot placed and held parts according to a digital model.

5.5.3.2 Enhancing usability and accessibility through user interfaces

To facilitate collaboration in a construction context, many principles from the field of Human-Machine Interaction (HMI) are relevant for consideration. User interfaces can serve as a bi-directional flow of information, where critical project and system information can be visualized and delivered to users. In parallel, interfaces can enable feedback and information

Figure 5.5   Collaborative Robotic Workbench for timber prefabrication, ICD, University of Stuttgart

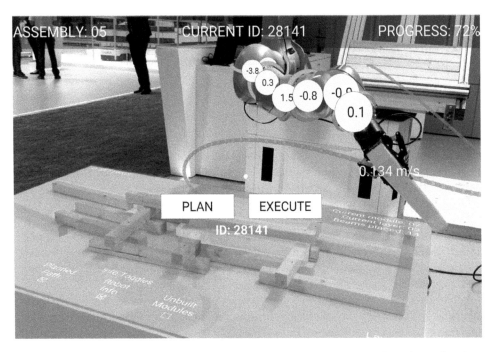

Figure 5.6   Augmented reality interface for human-robot collaborative construction, ICD, University of Stuttgart

to be gathered, collected, and collated from multiple distributed users, digitizing feedback that resists quantification through other means.

As many of the above projects illustrated, process-specific digital interfaces can also serve to enhance usability and accessibility to multiple users and project stakeholders. Fabrication equipment and construction processes require a high degree of expert knowledge. Thus one potential of cyber-physical systems is to lower barriers for participation, whereby users with different skill sets and expertise can intervene and interact with an ongoing construction process. Through the processing and display of data, a cyber-physical system can augment users with diagnostic information, providing insights into the functionality and performance of the system, enabling users to more effectively achieve their tasks.

Various interface technologies can serve different purposes within construction, while augmented reality enables the overlay of information in 3D in addition to hands-free motion, it is cost-prohibitive, and still has significant hardware limitations in terms of battery life and field-of-view. Screen-based interfaces have an added advantage of ubiquity and low-cost. Advances in browser-based 3D visualization suggest that the web is a viable solution for collaboration and exchange to address interoperability in AEC (Pauwels et al., 2017).

5.5.4 New material and construction system opportunities

Linear and fragmented production chains and highly precise and notational software environments have contributed to an approach to tolerances in AEC which is defined by predefinition, pre-calculation, and rigid control, where a design is completely modeled and defined a priori, preceding any stages of production construction (Menges, 2015). Such an approach to tolerances necessitates that materials have a high degree of predictability, favoring the use of standardized materials and prefabrication workflows, where parts are produced off-site in a structured environment. A fundamental paradigm shift emerges through the new technological possibilities afforded by cyber-physical systems, where robust data processing methods and autonomous systems can be utilized to operate robustly with non-standard geometries or imprecise material processes.

5.5.4.1 Material and construction systems which utilize non-standardized elements, irregular geometries, or upcycled materials

Sensor feedback can be coupled with computational and procedural logics of design generation and production to enable the use of material with irregular geometries. For example, Wood Chip Barn, a project emerging from the Architectural Association's Design + Make program at Hooke Parke, London, utilized irregular timber sourced from a local forest (Figure 5.7). Preceding construction, the local woodland was first sourced for trees with appropriate dimensions through photographic surveying techniques (Self & Vercruysse, 2017). A computational toolset digitalized the scanned geometries, and each unique tree branch could then be positioned relative to one another in the CAD design of a structure, shown in Figure 5.8. Custom joints could then be computationally generated and subsequently machined.

Within this workflow, much of the computation and design decision making was done offline and is thus not a cyber-physical system by definition. However, there are increasing abilities to couple real-time scanning and analysis with online physical production processes. Other cyber-physical processes have utilized scanning and adaptive robotic control to enable robotic processing to respond to wood grain directionalities and material properties (Brugnaro et al., 2019, Wood et al., 2016).

Figure 5.7   Wood Chip Barn, Architectural Association's Design + Make program

Figure 5.8   Scanned locally sourced wood utilized in a CAD Design Environment, Architectural Association's Design + Make program

Collectively, recent research illustrates that material and geometric variation can be handled with intelligent correctional logics operating within a cyber-physical system. Thus there is increasing opportunity to rethink the reliance on standardization within stereotypical work-flows, circumventing in some cases processing steps whereby inherently variable materials, such as wood, are processed into isotropic material systems.

5.5.4.2 Reconsidering simulation: from pre-calculation to verification

Structural simulations within a design and development process typically serve as a means of verification and approval: where the load-bearing capacities of final static elements are verified ahead of production. By connecting precision instruments and localization in a cyber-physical system, precisely placed digitally monitored forces can be applied to structures in real-time, either during production or before installation. The combination of mechanic precision, force feedback, and measured live deflections, suggests the possibility of real-time structural analysis, whereby a structures load-bearing capacities can be reverse-engineered and revealed through gathered data and coupled simulations. These possibilities lend themselves to application in additive based manufacturing and production processes, where additional layers of material can be applied incrementally during fabrication, and the structure or parts can be tested until a desired stiffness or performance is reached. Such a model expands the role of structural engineering within a production chain; instead of a verification step proceeding production, analysis can become fully integrated into processes of construction.

The benefits of integrating iterative structural testing within production can be found in material efficiency, whereby verification of structural properties of built prototypes circumvent the need for the reduction known and standardized structural typologies, and calculations which utilize unnecessarily high factors of safety. Additional possibilities exist for embedding sensors during the lifetime of a built structure, where online monitoring can reveal the structure's current performance and health. This emerging field, called civionics, combines electrical engineering, computer engineering, photonics, and civil engineering, and has been deployed so far towards the monitoring of civil infrastructure including bridges and roads (Mufti et al., 2005).

5.5.4.3 Cyber-physical form-finding and active control

Cyber-physical systems can also be utilized in other highly engineered design and construction processes to expand the complexity and scope of feasible architectural and structural forms. Form-finding is an iterative design process for finding a structurally performative form in static equilibrium for a given set of constraints, loads, and boundary conditions (Adriaenssens et al., 2014). Cyber-physical systems can augment physical form-finding processes, enabling an ideal geometry to emerge out of a feedback-driven construction process where networked actuators can be digitally controlled, and sensor feedback can benchmark how close a form is to an ideal structural state. The Block Research Group at the ETH Zurich has developed an augmented physical form-finding process developed for the production of textile reinforced thin-shell concrete (Liew et al., 2018). Closed-loop control, connected to a set of numeric form-finding processes shown in Figure 5.9, was leveraged to manipulate a network of digitally controlled turn-buckles acting at the perimeter of a cable-net structure, serving as the formwork for a lightweight concrete shell. The lengths of the turn-buckles could be iteratively manipulated until the desired stress state, and the geometric condition was reached. Ultimately, this cyber-physical system enabled an ideal structural form to physically materialize through a coupling of sensor feedback, digitally controlled actuators, and back-end numeric form-finding methodologies, Figure 5.10.

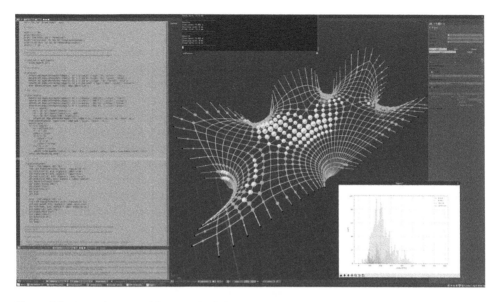

Figure 5.9   An active control-loop manipulating a network of turn buckles was connected to a back-end form-finding process to help correct deviations resulting from the physical production process

Figure 5.10   Textile reinforced thin-shell concrete in production, NEST HiLo, Block Research Group, ETH Zurich

5.6 Conclusion

Cyber-physical systems present many opportunities for the AEC industry. The adoption of cyber-physical systems would challenge many ingrained conventions, including tendering, approval, and calculation processes based on static, two-dimensional representations. In addition, many of the potentials of cyber-physical fabrication, including semi-autonomous robots and processes, can only be realized if physical systems are redesigned for the contingencies and constraints of robotic systems. In the past, automation technologies were applied to existing building systems such as masonry and timber systems (Bock & Linner, 2015). Though robot oriented design was suggested already as a concept in the 1990s, the application of robotic technologies to existing building systems faced many barriers. The failure of these early initiatives points to the need to develop new integrated design methodologies, enabling the design of new building systems to be informed by the constraints of robotic systems.

Simultaneously, new hardware tools are needed for the unique constraints and contingencies of construction. The construction industry and the architecture design field has historically been an industry where innovation is driven by appropriation: early CAD technologies such as CATIA were appropriated from the shipbuilding and aerospace industries (Szalapaj, 2005) while industrial robots were appropriated from manufacturing. The ramifications of appropriation and domain transfer become manifest through dominant constraints. For example, the kinematics of industrial robots are highly constrained to producing spherical geometries, their kinematics suited to pass parts between robotic stations. The appropriation of production equipment from structured industrial environments also reinforces the industry's linear and rigid approach to tolerances, encouraging robots to be used as precision instruments and not as autonomous, dexterous, thinking machines.

Thus, there is a need to question what hardware and software tools which would enhance the impact of cyber-physical systems, and how to target the development of such tools through fundamental research and design. Necessary developments in hardware might include physical robots and tools developed for unstructured construction sites. Like collaborative robots, an ideal construction robot would support HRC through torque monitoring and collision detection but may have additional specifications including kinematics derived from the proportionality of buildings and payloads derived from the weight of building components. Similarly, new control logic and programming paradigms for robots are needed to enable dynamic and sequence-dependent execution of high-level construction tasks. Process specific user interfaces can serve to increase usability, lowering barriers for participation for multiple stakeholders.

Additionally, there is a need to question mechanisms that might incentivize fundamental rather than applied research for domain-specific tools for the construction industry. While AEC hackathons have emerged recently to cultivate an entrepreneurial culture (AEC Hackathons, 2018), little funding or venture capital has traditionally existed for research and design in the construction industry (Azevedo, 2019). Research efforts emerging in academia, including the NCCR, Center of Digital Fabrication at the ETH Zurich, the newly established Excellenz Cluster at the University of Stuttgart, and the Center for Construction Robotics at RWTH Aachen exemplify the need to approach this research within a multidisciplinary framework, simultaneously engaging across industry, practice, and academia.

Interoperability must also be addressed in iterations of future software. Machines, robots, and devices must be able to access not only a static geometric model but also the embedded relationships implicit in its parametric generation, enabling relationships to be maintained despite tolerances or process inconsistencies. Formerly isolated software programs for various building services must begin to develop common data structures for fluid data exchange, enabling digital representations of the design and its current construction state to be accessible and editable to all collaborators.

These pursuits engender larger questions about trade-offs between standardization, specifications, and design freedom: every design environment, program, or data structure has inherent limitations and a limited range of applicability. While BIM is emerging as a much-needed single source of truth between multiple stakeholders, the limitations of BIM as a representational system must be considered. Though efforts are in place to make BIM compatible with IoT (Dave et al., 2018), BIM is highly notational and not compatible with the relative imprecision of production processes. Thus, one question is whether industry standards can learn from research practices, where geometric data representations are chosen to correspond to the accuracy of physical production processes. Similarly, tendering, verification, calculation, and bidding processes need to be restructured to utilize flexible design definitions that innately embody a range of acceptable variation.

Ultimately, cyber-physical systems with the AEC industry represent a convergence of multiple developments in robotics, digital fabrication, simulation, software development, and even building codes and legislation. Recognizing these relationships and overcoming these challenges could be a significant step towards increased levels of digitalization and automation within the context of Construction 4.0.

5.7 Summary

- Cyber-physical systems closely couple physical construction processes and computational processes.
- Several characteristics hinder the adoption of cyber-physical systems in architecture and construction, including standardization, compartmentalization, and a lack of interoperability.
- Increased computational power, increased communication stability, open-source hardware, and inter-device communication enable cyber-physical systems in AEC. Increased accessibility of manufacturing equipment and data processing techniques enable cyber-physical systems to be integrated into construction workflows.
- Cyber-physical systems in AEC can be deployed towards collaborative and networked fabrication, flexible and autonomous construction, as well as non-standard, adaptive fabrication.
- Additional facilitating technologies would enhance the impact of cyber-physical systems in AEC, including domain-specific tools and integrated design processes.

References

Abadi, M., Barham, P., Chen, J., Chen, Z., Davis, A., Dean, J., Devin, M., Ghemawat, S., Irving, G., Isard, M., & Kudlur, M. (2016). "Tensorflow: A system for large-scale machine learning." *12th {USENIX} Symposium on Operating Systems Design and Implementation ({OSDI} 16)*, pp. 265–283.

Adriaenssens, S., Block, P., Veenendaal, D., & Williams, C. (2014). *Shell structures for architecture: Form finding and optimization*. Abingdon: Routledge.

AEC Hackathon. (2018) https://aechackathon.com/(accessed 7.17.19).

Agiwal, M., Roy, A., & Saxena, N. (2016). "Next generation 5g wireless networks: A comprehensive survey." *IEEE Commun. Surv. Tutorials* 18, 1617–1655. doi: 10.1109/COMST.2016.2532458.

Aish, R. & Mendoza, E. (2016) "DesignScript: a domain specific language for architectural computing." *Proceedings of the International Workshop on Domain-Specific Modeling – DSM 2016*. Amsterdam, Netherlands: ACM Press, pp. 15–22.

Albu-Schäffer, A., Haddadin, S., Ott, C., Stemmer, A., Wimböck, T., & Hirzinger, G. (2007). "The DLR lightweight robot: Design and control concepts for robots in human environments." *Industrial Robot: An International Journal* 34, 376–385. doi: 10.1108/01439910710774386.

Alvarez, M., Wagner, H., Groenewolt, A., Krieg, O., Kyjanek, O., Aldinger, L., Bechert, S., Sonntag, D., Menges, A., & Knippers, J. (2019). "The BUGA Wood Pavilion – Integrative Interdisciplinary Advancements of Digital Timber Architecture." *Proceedings of the 39th Annual Conference of the Association for Computer Aided Design in Architecture (ACADIA)*. ISBN: 978-0-578-59179-7.

The Atlas. (2016). "The average cost of IoT sensors is falling." www.theatlas.com/charts/BJsmCFAl (accessed 6.17.19).

Autodesk. (2018). Autodesk forge APIs. https://forge.autodesk.com/en/docs/(accessed 6.12.19).

Azevedo, M. (2019). "Investor momentum builds for construction tech." *TechCrunch*. http://social. techcrunch.com/2019/02/16/investor-momentum-builds-for-construction-tech/(accessed 7.17.19).

Baudisch, P. & Mueller, S. (2017). "Personal Fabrication." *Foundations and Trends® in Human-Computer Interaction* 10(3–4), 165–293.

Bock, T. (2008). "Construction automation and robotics," in C. Balaguer and M. Abderrahim (eds), *Robotics and automation in construction*. InTech doi: 10.5772/5861. www.intechopen.com/books/robotics_and_automation_in_construction/construction_automation_and_robotics   (accessed 16.11.19).

Bock, T. & Linner, T. (2015). *Robot-oriented design: Design and management tools for the deployment of automation and Robotics in construction*. Cambridge; New York: Cambridge University Press.

Bock, T. & Linner, T. (2016). *Construction Robots: Volume 3: Elementary Technologies and Single-task Construction Robots*. Cambridge; New York: Cambridge University Press.

Bonvoisin, J., Mies, R., Boujut, J.-F., & Stark, R. (2017). "What is the 'source' of open source hardware?" *Journal of Open Hardware* 1, 5. doi: 10.5334/joh.7.

Brugnaro, G., Figliola, A., & Dubor, A. (2019). "Negotiated materialization: Design approaches integrating wood heterogeneity through advanced robotic fabrication," in F. Bianconi and M. Filippucci (eds), *Digital wood design. New York: Springer International*. pp. 135–158. doi: 10.1007/978-3-030-03676-8_4.

Built Robotics (2019). www.builtrobotics.com/ (accessed 7. 16.19).

Calvo Barentin, C., Van Mele, T., & Block, P. (2018). "Robotically controlled scale-model testing of masonry vault collapse." *Meccanica* 53, 1917–1929. doi: 10.1007/s11012-017-0762-6.

Cheng, T. & Podolsky, S. (1996). *Just-in-time manufacturing: An introduction*. New York: Springer Science & Business Media.

Cohen, D. L. & Lipson, H. (2010). "Geometric feedback control of discrete-deposition SFF systems." *Rapid Prototyping Journal* 16, 377–393.

Dave, B., Buda, A., Nurminen, A., & Främling, K. (2018). "A framework for integrating BIM and IoT through open standards." *Automation in Construction* 95, 35–45. doi: 10.1016/j.autcon.2018.07.022.

Dörfler, K., Sandy, T., Giftthaler, M., Gramazio, F., Kohler, M., & Buchli, J. (2016). "Mobile robotic brickwork," in D. Reinhardt, R. Saunders, & J. Burry (eds), *Robotic fabrication in architecture, Art and design 2016*. Cham: Springer, pp. 204–217.

Ergen, E., Akinci, B., & Sacks, R. (2007). "Tracking and locating components in a precast storage yard utilizing radio frequency identification technology and GPS." *Automation in Construction* 16, 354–367. doi: 10.1016/j.autcon.2006.07.004.

Felbrich, B., Prado, M., Saffarian, S., Solly, J., Vasey, L., Knippers, J., & Menges, A. (2017). "Multi-Machine fabrication: An integrative design process utilising an autonomous UAV and industrial robots for the fabrication of long-span composite structures." *Proceedings of the 37th Annual Conference of the Association for Computer Aided Design in Architecture (ACADIA)*. ISBN 978-0-692-96506-1.

Gershgorn, D. (2019) "After trying to build self-driving tractors for more than 20 years, John Deere has learned a hard truth about autonomy." *Quartz*. https://qz.com/1042343/after-trying-to-build-self-driving-tractors-for-more-than-20-years-john-deere-has-learned-a-hard-truth-about-autonomy/ (accessed 7.16.19).

Hurkxkens, I., Girot, C., & Hutter, M. (2017). "Robotic landscapes: Developing computational design tools towards autonomous terrain modeling." *Proceedings of the 37th Annual Conference of the Association for Computer Aided Design in Architecture (ACADIA)*. ISBN 978-0-692-96506-1, pp. 292–297.

Kenney, M. & Zysman, J. (2016). *The rise of the platform economy. Issues in science and technology*. National Academies of Sciences, Engineering, and Medicine, The University of Texas at Dallas Arizona State University.

Knippers, J. (2017). "The limits of simulation: Towards a new culture of architectural engineering." *Technology|Architecture + Design 1* 155–162. doi: 10.1080/24751448.2017.1354610.

Koomey, J., Berard, S., Sanchez, M., & Wong, H. (2011). "Implications of historical trends in the electrical efficiency of computing." *IEEE Annals Hist. Comput.* 33, 46–54. doi: 10.1109/MAHC.2010.28.

Kyjanek, O., Al Bahar, B., Vasey, L., Wannemacher, B., & Menges, A. (2019). "Implementation of an augmented reality ar workflow for human robot collaboration in timber prefabrication." *Proceedings of the 36th International Symposium on Automation and Robotics in Construction*. doi.10.22260/ISARC2019/0164.

Lafreniere, B., Grossman, T., Anderson, F., Matejka, J., Kerrick, H., Nagy, D., Vasey, L., Atherton, E., Beirne, N., & Coelho, M. H. et al. (2016). "Crowdsourced fabrication," in: *Proceedings of the 29th Annual Symposium on User Interface Software and Technology*. ACM, pp. 15–28.

Liew, A., Stürz, Y. R., Guillaume, S., Van Mele, T., Smith, R. S., & Block, P. (2018). "Active control of a rod-net formwork system prototype." *Automation in Construction* 96, 128–140. doi: 10.1016/j.autcon.2018.09.002.

Lu, M., Chen, W., Shen, X., Lam, H.-C., & Liu, J. (2007). "Positioning and tracking construction vehicles in highly dense urban areas and building construction sites." *Automation in Construction* 16, 647–656. doi: 10.1016/j.autcon.2006.11.001.

McNeel. (2019). Rest Geometry Server. *McNeel*. https://github.com/mcneel/compute.rhino3d (accessed 7.12.19).

Mele, T. V., Liew, A., Mendéz, T., & Rippmann, M., et al. (2017). *COMPAS: A framework for computational research in architecture and structures*. https://block.arch.ethz.ch/brg/tools/compas-computational-framework-for-collaboration-and-research-in-architecture-structures-and-digital-fabrication.

Menges, A. (2015). "The new cyber-physical making in architecture: Computational construction." *Architectural Design* 85, 28–33. doi: 10.1002/ad.1950.

Mufti, A. A., Bakht, B., Tadros, G., Horosko, A. T., & Sparks, G. (2005). "Are civil structural engineers 'risk averse'? Can civionics help?" in F. Ansari (ed), *Sensing issues in civil structural health monitoring*. Berlin/ Heidelberg: Springer-Verlag, pp. 3–12. doi: 10.1007/1-4020-3661-2_1.

Nobel, P. (2016). "SHoP: Make it whole – on valuing means and ends in the practice of SHoP," in M. U. Hensel and F. Nilsson (eds), *The changing shape of practice: Integrating research and design in architecture*. Abingdon: Routledge, pp 74–85.

Odico. (2019). Factory-on-the-Fly. www.odico.dk/technologies#factory-on-the-fly (accessed 7.15.19).

Olson, E. (2011). "AprilTag: A robust and flexible visual fiducial system," in: *Proceedings of the IEEE International Conference on Robotics and Automation (ICRA)*. IEEE, pp. 3400–3407.

Paperspace. (2019). www.paperspace.com/ml (accessed 11.12.19).

Parascho, S., Gandia, A., Mirjan, A., Gramazio, F., & Kohler, M. (2017). "Cooperative fabrication of spatial metal structures." *Fabricate 2017*. pp. 24–29.

Pauwels, P., Zhang, S., & Lee, Y.-C. (2017). "Semantic web technologies in AEC industry: A literature overview." *Automation in Construction* 73, 145–165. doi: https://doi.org/10.1016/j.autcon.2016.10.003.

Peters, T. (2018). "Autodesk University 2018 challenges the future of work, again." *Architect Magazine*. www.architectmagazine.com/technology/autodesk-university-2018-challenges-the-future-of-work-again_o (accessed 7.12.19).

Poirier, E. A., Forgues, D., & Staub-French, S. (2014). Dimensions of interoperability in the AEC industry, *Construction Research Congress 2014*. Presented at the Construction Research Congress 2014, American Society of Civil Engineers, Atlanta, Georgia, pp. 1987–1996. doi.10.1061/9780784413517.203.

Powell, A. (2012). "Democratizing production through open source knowledge: from open software to open hardware." *Media, Culture & Society* 34, 691–708. doi: 10.1177/0163443712449497.

Rajkumar, R., Lee, I., Sha, L., & Stankovic, J. (2010). Cyber-physical systems: The next computing revolution, in *Design Automation Conference*. IEEE, pp. 731–736.

Scassellati, B. & Hayes, B. (2014). "Human-robot collaboration." *AI Matters* 1, 22–23. doi: 10.1145/2685328.2685335.

Schirner, G., Erdogmus, D., Chowdhury, K., & Padir, T. (2013). "The future of human-in-the-loop cyber-physical systems." *Computer* 46, 36–45. doi: 10.1109/MC.2013.31.

Schwinn, T. & Menges, A. (2015). "Fabrication agency: Landesgartenschau Exhibition Hall." *Architectural Design* 85, 92–99. doi: 10.1002/ad.1960.

Self, M. & Vercruysse, M. (2017). Infinite variations, radical strategies, in: *Fabricate 2017 Conference Proceedings*. London: UCL Press, pp. 30–35.

Shen, X., Cheng, W., & Lu, M. (2008). "Wireless sensor networks for resources tracking at building construction sites." *Tsinghua Science and Technology* 13, 78–83. doi: 10.1016/S1007-0214(08)70130-5.

Speckle Works (2019). Data rich design. https://speckle.works (accessed 4.19.19).

Szalapaj, P. (2005). *Contemporary architecture and the digital design process*. Architectural Press, Amsterdam; Boston.

Vasey, L., Baharlou, E., Dörstelmann, M., Koslowski, V., Prado, M., Schieber, G., Menges, A., & Knippers, J. (2015). "Behavioral design and adaptive robotic fabrication of a fiber composite compression shell with pneumatic formwork," *Computational Ecologies: Design in the Anthropocene*, Proceedings *of the 35th Annual Conference of the Association for Computer Aided Design in Architecture (ACADIA).* pp. 297–309.

Vasey, L., Nguyen, L., Grossman, T., Kerrick, H., Schwinn, T., Benjamin, D., Conti, M., & Menges, A. (2016). "Human and robot collaboration enabling the fabrication and assembly of a filament-wound structure," *ACADIA//2016: Posthuman Frontiers: Data, Designers, and Cognitive Machines, Proceedings of the 36th Annual Conference of the Association for Computer Aided Design in Architecture.* pp. 184–195.

Waldrop, M. M. (2016). "The chips are down for Moore's law." *Nature* 530, 144–147. doi: 10.1038/530144a.

Willmann, J., Knauss, M., Bonwetsch, T., Apolinarska, A. A., Gramazio, F., & Kohler, M. (2016). "Robotic timber construction—Expanding additive fabrication to new dimensions." *Automation in Construction* 61, 16–23.

Wood, D. M., Correa, D., Krieg, O. D., & Menges, A. (2016). "Material computation—4D timber construction: Towards building-scale hygroscopic actuated, self-constructing timber surfaces." *International Journal of Architectural Computing* 14, 49–62. doi: 10.1177/1478077115625522.

Yan, R.-J., Kayacan, E., Chen, I.-M., Tiong, L. K., & Wu, J. (2019). "QuicaBot: Quality inspection and assessment robot." *IEEE Transactions on Automation and Engineering* 16, 506–517. doi: 10.1109/TASE.2018.2829927.

6

APPLICATIONS OF CYBER-PHYSICAL SYSTEMS IN CONSTRUCTION

Abiola A. Akanmu and Chimay J. Anumba

6.1 Aims

- Give an overview of the triggers for cyber-physical systems in the construction industry.
- Describe the key requirements for cyber-physical system applications in construction.
- Discuss the role of bi-directional coordination between virtual models and the physical construction/constructed facility in facilitating cyber-physical systems.
- Describe a system architecture that illustrates the key technologies and sub-systems needed for facilitating bi-directional coordination.
- Describe deployment scenarios to show the potential benefits of cyber-physical systems in construction.

6.2 Introduction

Recent developments in information and communication technologies have resulted in increased availability and proliferation of sensors, data acquisition, and control technologies. The innovations invoked by these technologies in several industry sectors have been pushing the academic and practice communities of the construction industry towards exploring their potentials. In the past two decades, construction automation has been one of the fastest-growing research areas in the construction engineering and management discipline. Significant progress has been made in the application of wireless sensing and visualization technologies for supply chain management (Demiralp et al., 2012; Irizarry et al., 2013), construction progress management (Golparvar-fard et al., 2012; Kim et al., 2013; Teizer, 2015), safety management (Cheng and Teizer, 2013; Li et al., 2018) and workforce training (Li et al., 2015; Teizer et al., 2013). More recently, the cyber-physical system has emerged as a promising direction to improve on existing efforts in the area of real-time monitoring, predictability, and control of project outcomes. Cyber-physical systems are an integration of physical systems with cyber (e.g. sensors, algorithms, and actuators) techniques and technologies to form an integrated system that intelligently responds to changes of the real world. By integrating cyber-physical systems with construction processes, opportunities exist to transform the construction industry into the future Industry 4.0 factory with significant economic potential (Dallasega et al., 2018; Oesterreich and Teuteberg, 2016). For instance, a report by Industry 4.0 Market

Research, a division of Human Sciences Research Council, forecasted that Industry 4.0 would boost the United States gross value by approximately 214 billion dollars by 2030 (HSRC), I.M.R.A.D.O., 2018).

To distinguish cyber-physical systems in the construction industry from other industry sectors, this chapter uses the term 'cyber-physical construction systems' to describe cyber-physical systems designed for the construction industry. Since cyber-physical construction systems (CPCS) are still in the infancy stage in the construction industry sector, this chapter describes a structure and framework for the implementation of cyber-physical systems in the construction industry. The framework includes requirements for cyber-physical systems, system architecture, some early developed and implemented scenarios and highlights of challenges and barriers to achieving and implementing future CPCS are presented.

6.3 Drivers for cyber-physical systems in construction

6.3.1 Construction progress management

Construction progress management involves periodically tracking and comparing the actual status of construction projects with the planned progress. Construction projects are complex, as every project generally comprises of several phases, with each phase requiring a diverse array of specialized services and the involvement of numerous participants. In addition to the diversity of expertise, the success of construction projects is influenced by factors such as design and scope changes, resource constraints and project dynamics. How information (e.g. updates and changes) is managed amongst project participants is important as this could result in deviations from the as-planned and low productivity. The ability of project teams to understand what is going on at every stage of a project is critical to the successful completion of construction projects. Every team member needs timely and accurate information about the project progress, and status in comparison to the initial plans so that they can react quickly to site problems. Also, when design changes occur, they need to be shared accurately, consistently and in a timely manner between the project team members (at the construction site) so as to reduce risks such as time and cost overruns.

When construction projects commence, it is not self-regulating and requires real-time monitoring if events are to conform to set standards or plans. To successfully capture the progress of work, there is need for strategies for effectively tracking and relaying progress information to the concerned parties. Over the years, there have been tremendous efforts to automate the progress management process of construction projects. These have ranged from isolated tracking of construction materials (Jang and Skibniewski, 2008; Song et al., 2006) to more integrated solutions involving the integration of virtual models and the physical construction (El-omari and Moselhi, 2011; Turkan et al., 2012). However, the dynamics and uncertainties inherent in construction projects have not been adequately modeled in the existing solutions; as such, there are limited opportunities to use the solutions for predicting project uncertainties and control.

6.3.2 As-built documentation

As-built models are extremely important to owners to maintain and renovate building systems and infrastructure. These models typically represent the contractors' certified record of what has been constructed in terms of the types of systems and installed locations. Complete as-built models need to account for the design intent (or as-planned) as well as any design and scope changes. Over the years, studies have demonstrated the potential of component and image-based sensing systems for improving the capturing and documenting project as-builts. Unfortunately, these improvements have only achieved incremental success with applications to elemental segments/systems of buildings and limited potential for capturing the complete as-built. In fact, there are still limited opportunities for instrumenting the construction and maintenance phases to achieve more automated and sustainable solutions to capturing project as-builts and enhancing life-cycle management of buildings. This has resulted in un-correlated collections of inaccurate, incomplete information with limited utility for describing exactly what was built. This also results in the loss of opportunity to use continuously updated as-built models to manage ongoing work and identify deficiencies early enough to avoid rework.

6.3.3 Health and safety management

The construction industry is one of the industries with the largest labor force in the United States, yet the industry continuously struggles with fatality, injury, and health issues, resulting in premature exit of workers from the workforce – this should be alarming to both practitioners and researchers. Compared with workers in other industry sectors, construction workers encounter the highest risk of occupational injuries and illnesses. In fact, the US Bureau of Labor Statistics (Statistics, 2016a) showed that the construction industry account for over 38% of all injury and illnesses encountered in all industry sectors. Furthermore, in 2016, the construction industry recorded more fatalities than in any other major industry and the number of fatalities has been on the rise since 2011 (Statistics, 2016b). These safety and health issues have been attributed to several factors including the dynamics of the site layout, worker behavior, poor coordination, the physically demanding and repetitive nature of construction tasks, safety policy, economic pressure, management training, and safety culture. While existing studies have come a long way in addressing safety and health issues in the construction industry, the interaction between the different facets of the built environment and the influencing factors of safety and health risks are still not understood, e.g. leveraging data on the influencing factors of health and safety risks of different construction projects to generate predictive models capable of achieving 'zero safety and health' risk designs and projects. Another example of the design and construction of buildings for reduced or zero safety and health risk involves the development of models for simulating the worker-structure interaction during design and construction. The models will also be tied to the actual physical construction process for predicting safety and health risks during construction and suggesting real-time interventions.

6.4 Requirements for CPCS in construction

6.4.1 Real-time capability

The real-time capability of CPCS means that sensed information or data must be transmitted in a timely manner to meet the real-time control requirements of construction processes (Lee, 2008). As a result of the dynamic nature of the construction projects, monitoring of construction activities should be in real-time so that potential issues or risks can be quickly identified and resolved. Real-time acquisition of information pertaining to the key facets of construction projects such as supply chain, installations, resource status are critical for decision making and control of construction projects. Being able to effectively capture this information and relay it in real-time to the project stakeholders is significant. Researchers have identified the benefits of real-time monitoring for supply chain management (Irizarry et al., 2013), site layout management (Akanmu et al., 2016), progress monitoring (Azimi et al., 2011; Ranaweera et al., 2012), inventory management (Kasim et al., 2012), and safety management (Cheng and Teizer, 2013; Giretti et al., 2008).

6.4.2 Adaptability

CPCS should be able to capture and handle changes, uncertainties, and abnormalities that typically occur during construction projects. Whether the anomaly is naturally, accidentally, or maliciously induced, in terms of a sensor default, failure of any of the part of the system, or the infrastructure malfunction, the CPCS must adapt to whatever situations they encounter (Wan, Cai, and Zhou, 2015; Xia et al., 2013). The CPCS must virtually reorganize, reconfigure, and heal to yield the best possible system performance. Such systems should also be scalable and tolerant to risks triggered by these uncertainties. In addition, to cope with uncertain and emerging situations, CPCS should be self-aware and context-aware. For example, be able to anticipate needed information and deliver these based on the context of the beneficiary.

6.4.3 Networked

There is a co-dependency relationship between activities in construction projects and systems within buildings and civil infrastructure systems. This has been the rationale for Bulbul, Anumba, and Messner (2009)'s advocating for a systems of systems approach to designing and analyzing construction operations. Construction activities are highly dependent on one another in terms of information, space, resources, and personnel. These activities share a number of allocations such as workspaces (Chavada et al., 2012), equipment, and temporary structures. For instance, for construction projects to be successful, workers must have enough or required knowledge to not only perform their tasks but also to collaborate with other trades (Coakes et al., 2008). To enable execution of work efficiently and effectively, each worker will depend on information provided by other workers involved in the construction project (Ochoa et al., 2011). Understanding the dependencies of information between participants is critical for building an effective network of communication. Furthermore, traditional simulation of construction operations assumes that associated tasks are discrete or continuous and provides control to the operations. In a CPCS, construction tasks will need to be mediated by networks and software that neither have discrete nor continuous behavior.

6.4.4 Predictability

The dynamic nature and uncertainties inherent in construction projects make it difficult to predict project outcomes – these make construction projects a candidate for CPS. Predictability refers to the ability and degree to which the behavior and state of construction processes and resources can be predicted qualitatively or quantitatively (Lee, 2006; Sangiovanni-Vincentelli et al., 2012). Predictable construction activities will need to assure the outcomes of the system's behavior whenever operational while also meeting all system requirements. In CPCS, sensing systems as well as control technologies should be well adapted to construction processes, anticipate changes in the physical process, risks posed to projects and amend their features based on the context of the construction environment. Some current construction automation solutions allow for real-time monitoring and control but are insensitive to project uncertainties (e.g. changing environment, project scope and delays) and time-predictability.

6.4.5 Human-in-the-loop

The construction industry is a labor intensive sector. Applicable CPCS must consider human participants as a significant component. CPCS must be instinctive and integrate with workers' behaviors in construction environments (Munir et al., 2013). This is important for communicating needed information and assisting decision processes of construction workers. Although there are significant data and experience on the performance of existing automated systems deployed in the construction industry, the extent to which human behavior has been modeled and integrated into the design of systems has varied. Many of the anticipated goals of CPCS, in terms of zero fatalities and injuries, can be achieved through the design of systems that have complete understanding of human behavior under diverse conditions such as emergency situations (Munir et al., 2013).

6.4.6 Reliability

Reliability refers to the degree of accuracy to which a CPCS performs its function (Mitchell and Chen, 2013). Certifying the capabilities of CPCS as being able to accurately perform their functions does not mean that they actually can. CPCS are expected to operate reliably in dynamic and uncertain environments. For construction safety applications involving equipment or postures, a slight change or deviation from planned performance may result in fatalities (Yuan et al., 2016). Uncertainties in knowledge, characteristics, potential errors, and outcomes in construction processes should be quantified during the design of CPCS.

6.5 Cyber-physical construction systems

6.5.1 Key features

A CPCS approach involves bridging the cyber world with the physical world through sensors/data acquisition technologies and actuators to form a closed loop or feedback system (Chen et al., 2015; Dillon et al., 2011). The need for closed loop in construction activities has been identified by (Navon and Sacks, 2007; Turkan et al., 2012). CPCS consist of two key elements: a 'physical to cyber' bridge and a 'cyber to physical' bridge (Bordel et al. 2017; Xia et al., 2011). These are described as follows (see Figure 6.1):

- Physical to Cyber Bridge: This involves sensing the elements of construction processes and activities using sensors and data acquisition technologies. Information about the condition and dynamics of construction process are captured and related to other cyber elements for processing.
- The Cyber to Physical Bridge: This represents the actuation which indicates how the sensed information affects the system. Actuation is taken to mean conveying the right information to enable quick decision making with the sensed information and/or using the sensed information to physically control construction activities, resources and components.

6.5.2 CPCS via bi-directional coordination

The need to maintain consistency between the as-planned design and as-built models have resulted in the development of bi-directional coordination approach to CPCS (Anumba et al., 2010). The bi-directional coordination approach involves integrating virtual models and the physical construction/constructed facility for consistency maintenance. Unlike traditional

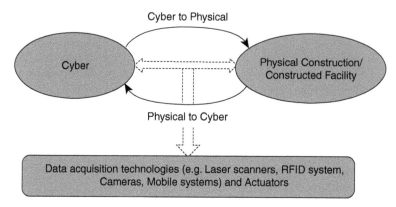

Figure 6.1 Key features of CPCS

integration approaches which only passively monitor the physical world, this approach enables tight interaction and coordination between the virtual and the physical world. Early attempts at implementing CPCS via bi-directional approach include construction component tracking and placement (Akanmu et al., 2013b, 2013a), site layout management (Akanmu et al., 2016), and postural monitoring and control. These are represented in Figure 6.2.

6.5.3 System architecture for bi-directional coordination approach

The bi-directional coordination approach is illustrated in the system architecture shown in Figure 6.3. The system architecture brings together the functionality and the key enabling technologies necessary for implementing bi-directional coordination between the virtual models and physical representations. The architecture is based on five layers, which are explained in the following sub-sections.

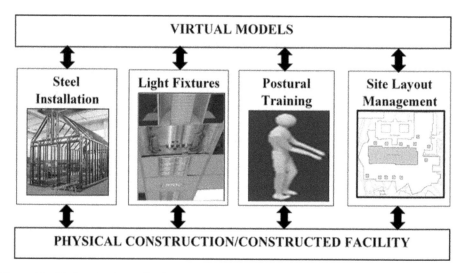

Figure 6.2 Bi-directional coordination approach to CPCS

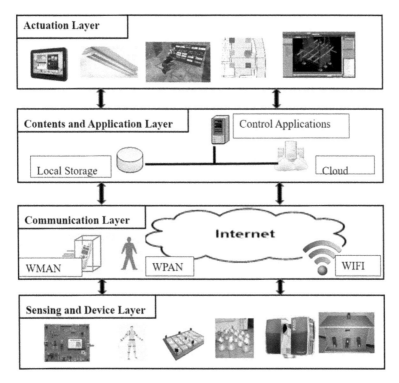

Figure 6.3 System architecture for bi-directional coordination approach to CPCS (Akanmu et al., 2013b; Kan et al., 2019)

6.5.3.1 Sensing layer

The sensing layer consists of sensors and other data acquisition technologies. The sensors monitor different aspects of the construction process/constructed facility, e.g. radio frequency identification sensing systems for tracking construction resources' status. Depending on the type of sensor used, this layer can also provide the construction personnel access to control decisions (e.g. information captured in the radio frequency identification (RFID) tag memory or information that can be accessed through the RFID tag ID). Other data acquisition technologies such as smartphones can also be used by construction personnel on site to physically extract project status information. This layer serves two purposes: it provides access to sensed data from the sensing layer, and enables the entry of information through the user interface.

6.5.3.2 Communication layer

The communication layer serves as the data processing and communication unit. This layer converts the data obtained from the sensing layer into formats readable by the contents and application layer. The communication layer contains the internet and wireless communication networks e.g. wide area networks and local area networks. These communication networks also connect the sensor and data acquisition technologies to allow for networking and information sharing between workers at the job site and the field office.

6.5.3.3 Contents and application layer

The contents and application layer contains the local database, database server, and the control applications. The type of control application depends on the context application to which the system is put. For example, the control applications could be machine learning algorithms. This layer stores, analyses, and is constantly updated with information collected from both the communication and actuation layers. The stored information includes project management data (such as resource type and status). The control applications use the sensed data from the database to make control decisions which can be either visualized using the virtual prototype in the actuation layer or used to control the physical system such as equipment and postures.

6.5.3.4 Actuation layer

Depending on the type of control needed, the actuation layer could provide access to critical information needed for decision making or physically control the physical environment using actuators. The actuation layer contains the virtual prototype which is accessed through the user interface or a mixed reality environment. The virtual prototype enables the user to visualize how the sensed information (from the contents and storage layer) affects the system. The user interface enables the user to visualize and monitor the sensed information from the contents and storage layer. The user can also embed control decisions into the virtual prototype through the user-interface to be accessed in the device layer.

6.6 Case studies

6.6.1 Steel installation tracking and control

This scenario describes how bi-directional coordination between the virtual model and the physical construction will enhance the installation of structural steel members. Each member is tagged using active RFID tags with embedded information on the erection sequence for the installation of the steel members. This is intended to avoid the problems that can occur when a member is in line for installation before the supporting members have been erected. The virtual model reflects the status of the steel erection (i.e. it can highlight which members have and have not been installed to date). The model also contains information about dependencies (i.e. which members need to be installed before others). The following numbers illustrate the sequence of actions and correspond to the numbers in Figure 6.4.

- After steel members have been fabricated, a change is made to the design (or a mistake is found). The structural engineer corrects this change, but it has an impact on the sizes of several other members.
- Information regarding the changed/affected member sizes is communicated to the construction site and written to the RFID tags: The information from the model is initially captured in the database server, transferred through the internet and read by a fixed RFID reader (which has read and write capability). The reader/writer writes the information to the RFID tags of the affected members so that the contractor knows which pieces can be installed and which need to be replaced with larger members. Since the virtual model also contains dependencies, if the original installation sequence changes, the new sequence is also communicated to the RFID tags of the affected members.

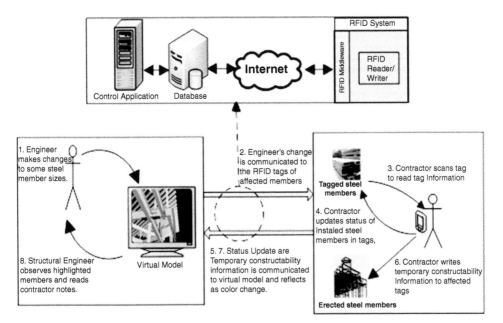

Figure 6.4 Steel placement scenario (Akanmu, Anumba, and Messner, 2013b)

- Before erecting each steel member, the steel erector scans the RFID tag with a portable device (having an in-built reader with read and write capability) and reads off embedded information.
- As each steel member is erected on site, the steel erector updates the status of the RFID tags (from 'on-site' to 'installed' or 'erected') on the members by writing to the tags using the portable device.
- The status change is detected by the fixed reader and communicated through the internet to a project database. A control application (running on the remote office computer) loops through the database for new information and updates the virtual model with the new status, resulting in the affected member being highlighted (e.g. by a color change). This color change in the virtual model enables the model coordinator/project manager to monitor the progress of the structural steel erection from the remote office.
- If large mechanical equipment needs to be lowered into the building later during construction and this requires some steel members to be left out, the steel erector can identify the steel members and write this change to their tags.
- The tag information is detected by the fixed reader through the radio frequency from the tags and communicated through the database to the virtual model, which is then updated with the new information, resulting in that member being highlighted. The role of the color change is to inform the structural engineer of the updated information relating to a steel member.
- The structural engineer reads the information in the model, and evaluates whether the remainder of the structure is adequate to support the loads temporarily.

6.6.2 Light fixture tracking and control

The locations of individual items of bulk materials (such as light fixtures) within a building are not easily differentiable; as such, facility managers cannot control each item. This scenario (see Figure 6.5) proposed and implemented by Akanmu et al. (2013a) focuses on identifying and tracking the installed locations of light fixtures for the purpose of improving energy management in buildings. The key steps are highlighted below:

- On arrival at the site, the light fixtures are tagged with passive RFID tags (if the light fixtures are not already tagged) to identify and differentiate each component.
- As each tagged fixture is installed, the electrical contractor scans the fixture tag using a portable device with an integrated RFID scanner.
- The portable device has a virtual model of the facility. On scanning the tag on the fixture, the electrical contractor binds the tagged fixture with the corresponding virtual fixture in the model and changes the status to 'installed' in the model. Binding the physically tagged fixture with the virtual fixture (in the model), creates opportunities for the electrical contractor to create individual controls for the fixtures. This virtual model can be shared by the model coordinator, who monitors the progress of work in the site office and can identify which components have been installed and uninstalled.
- During the building life cycle, the facility managers and owners can remotely use the virtual model to identify and distinguish the locations of each fixture within a physical space to enhance access to individual lighting units and control the energy performance of buildings.
- When the light fixtures are controlled remotely, control messages are sent over the internet using the TCP/IP protocol to a device server. The device server sends the control messages in the form of an IP address to the bus-master. This bus-master filters and sends the control messages to the appropriate light fixture.

Facility managers can also remotely observe and query the status (ballast failure, lamp failure, power failure, device type) of each or group of fixtures remotely through the model. For example, the status and specification of defective fixtures can be communicated to the facility

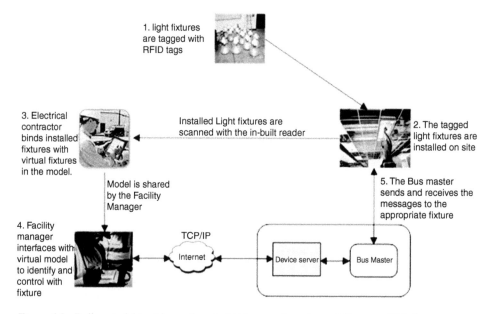

Figure 6.5 Bulk material tracking and control (Akanmu, Anumba, and Messner, 2013a)

managers (in real-time) through the model so that they can replace them. This is particularly important as problems can be identified and diagnosed early, thus, reducing the need for routine maintenance checks and enabling the owner/facility manager control over the facility. This concept can be applicable to urban infrastructure street lights, electrical billboards and similar installations, where the installation of the individual street lights can be monitored remotely for the purpose of distinguishing each fixture and controlling each fixture over the infrastructure life cycle. Being able to digitally address each light fixture provides an opportunity to digitally address each fixture from the virtual model and to communicate any changes in the status of the physical fixture (e.g. defective) to the virtual model for operational and maintenance purposes.

6.6.3 *Postural training and control*

This scenario describes how bi-directional coordination between the virtual and the physical postures/movements can enhance competences of construction workers in performing work in safe postures. This is intended to reduce the rate of occurrence of work-related musculoskeletal disorders in the construction industry by providing construction workers with a flexible learning environment where they can improve their motor skills. This scenario leverages virtual reality to provide workers with repetitive practice of construction tasks together with salient recommendations/feedback. The virtual environments have the potential to optimize motor learning by controlling practice conditions that unequivocally engages motivational, cognitive, and motor control mechanisms. A virtual reality construction environment consisting of a construction site and resources for executing construction activities and tasks, is developed. Each activity and task are simulated via discrete event simulations, with each task consisting of suggested safe postures. The virtual reality construction site also contains a virtual instructor which delivers the postures, evaluates the learner's performance and provides real-time feedback. The virtual instructor is capable of diagnosing knowledge and skill gaps of individual trainees in achieving optimal work behaviors on construction sites, and adapting instructional contents and delivery methods according to the need of individual trainees. The diagnostic and adaptive capabilities are two essential qualities of a good teacher in maximizing learning. The construction activities, tasks, corresponding safe postures and instructional contents are stored in a database. Each physical body part or segment is tagged with inertia measurement units (IMU), consisting of accelerometers, gyroscopes, and magnetometers. The IMU tracks the movement or rotation of each of the tagged body parts. The key steps are as follows:

- The worker selects an activity he wishes to train on from the virtual instructor. On selecting the activity, a list of tasks required for executing the activity is extracted from the database and shown to worker via the virtual instructor.
- The worker performs each of the tasks and the IMU captures the real-time movement of each body part.
- The posture is deduced from the movement of the body parts and reflected on the virtual instructor or any platform the worker is gazing at when performing the activity. Each body part represented in the posture on the virtual instructor shows different color codes (Figure 6.6), with each code representing the level of ergonomic exposure identified in literature (Andrews et al., 2012).
- The posture assumed by the worker is compared with the suggested postures for performing the tasks (obtained from the database) and displayed on the virtual instructor

(Figure 6.7). Presenting this information to the worker in real time will improve his/her awareness of the risk associated with his approach to work, thus encouraging him/her to take necessary precautions.

- For every unsafe posture assumed by the worker in executing each of the tasks, the virtual instructor provides instructional materials such as hints and video lessons on how to assume the appropriate or safe postures. Reinforcement learning algorithm is used to define the state of learning of the workers with respect to the instructional contents administered via the virtual instructor. The reinforcement learning algorithm will choose the type of instructional content to adapt the virtual instructor when the postures are deemed unsafe ergonomically.

- If the worker shows posture improvements, the reinforcement algorithm issue a reward for that given state of learning. If the posture continues to be unsafe, the reinforcement algorithm will issue a punishment for that state.

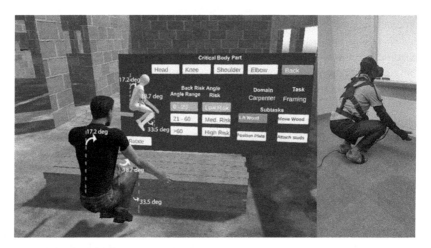

Figure 6.6   Flexible postural learning environment

Figure 6.7   User performing a 'attach stud' subtask and virtual instructor showing his performance – safe and unsafe posture

6.6.4 Site layout management

This scenario describes how bi-directional coordination between virtual site models and the physical construction site environment can enhance effective monitoring and utilization of site spaces. This addresses the issue associated with time wasted and safety risk associated with searching for available spaces to store materials. Successful site layout management is dependent on a number of factors including the project schedule, status of construction work, and real-time knowledge of the context of site resources. In this scenario, a middleware is developed which coordinates between the virtual site model and the physical construction site, for generating and managing site layouts. The middleware (Figure 6.8) integrates (i) the project information from BIM, (ii) the status of construction resources obtained from radio frequency identification real-time location sensing (RFID-RTLS) system, and (iii) genetic algorithm for optimizing the placement of site resources. The RTLS system consists of tags, readers, and a location engine. The RTLS system determines the real-time location of tagged materials/components within spatially mapped zones on the construction site. The key steps are described below:

- Coordinates of the boundary of the construction site and the as-planned facility, are received from the project manager and represented in the virtual site model. This is important to determine available spaces for staging materials and equipment.
- The middleware collects the site from the virtual site model, divides the available site area into grids and stores these information in the database.
- Within the BIM model, the middleware provides a platform for the project manager to insert the type of temporary facilities or site spaces required for executing each activity in the schedule. For the first phase of the project, the project manager assigns temporary facilities to each activity to be executed in the first phase. The middleware collects and stores these information into the database.
- Where an activity requires more than one temporary facility or site space, during the first phase of the project, the project manager inserts an estimated frequency of trips between the temporary facilities. The frequency of trips represents the rate at which a tagged worker interacts with or transitions between associated temporary facilities to complete work.
- The database sniffer from the middleware triggers the optimization algorithm to generate the site layout. Genetic algorithm uses the information contained in the database such as the coordinates of possible locations, type of activities in the first phase of the project, the temporary facilities or spaces needed for the activities and the frequency of trips between the facilities, to generate the optimum site layout. Other factors such as productivity and safety could be considered.
- The middleware stores the first site layout in the database and also displays it on the virtual site model.
- For the subsequent project phases, the middleware tracks the actual frequency of trips made by tagged workers between the temporary facilities or assigned spaces, the availability of the spaces (by tracking if there are any tagged materials within the spaces), the type of activities and required spaces, to generate the site layouts.
- If there are no tagged materials within the spaces, this means that the space is vacant and the middleware checks the schedule within BIM to determine if the task/activity associated the temporary facilities is completed.
- If the task is completed within the current phase, the facilities are registered as 'movable' and can be considered as free spaces in the next site layout generation. If not, the facilities are registered as 'fixed'.

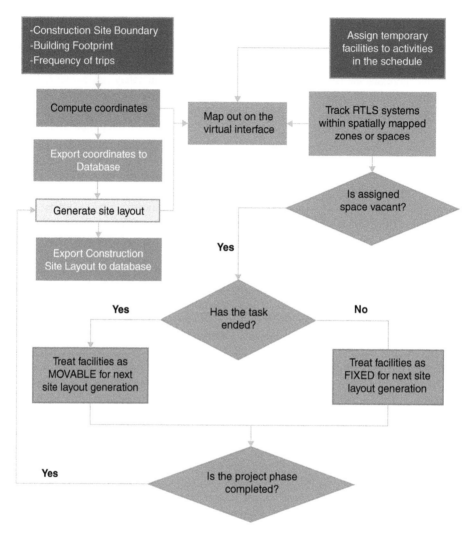

Figure 6.8 Site layout management workflow (Akanmu et al., 2016)

6.7 Challenges and barriers

6.7.1 Future workforce competency

Studies have shown that one of the ways of invoking changes in industry sectors is by equipping the future workforce with the required competencies (Breiner et al., 2012; Dede, 2007; Mcmasters, 2004; McWilliam, 2008). Existing educational frameworks are not adequate to equip construction engineering and management students with CPCS competencies. CPCS traditionally involves several academic disciplines, requiring courses that combine, for example, electrical and control theory, computer science and computer engineering. Rapid and dramatic changes are needed in the engineering education to situate CPCS in construction engineering and management curriculums. At present, only few institutions offer CPCS related education in the construction engineering and management programs. A knowledgeable and skilled workforce is therefore crucial for creating an awareness and sustaining CPCS based applications in the construction industry.

6.7.2 Training

Construction workers are typically not accustomed to technology, so considerable training (costs to train) would be required in order to get them on board with CPCS. In a study by Akanmu et al. (2013b), industry practitioners claim that it would cost a lot to educate the building managers on how to use the technologies involved in the light fixture tracking in Section 6.6.2. Although, they also acknowledged that the benefits of the CPCS outweighs the cost of training building managers. It is also worth noting that the cost of training the building managers is a one-time cost.

6.7.3 Project delivery methods

Industry practitioners will likely be concerned about the effect that bi-directional coordination would have on the current process of tracking and conveying changes to concerned parties during construction. The current practices require that the contractors raise change orders if there is a change in the scope of work, design, or conditions on site before the change is implemented. Future CPCS will need to be tightly integrated into existing practices and designed around the potential users for such systems to be accepted and sustainable. In the present form, CPCS will more likely be suitable for the design-build and integrated project delivery methods. This is because there is better communication between the design team and the field personnel in these delivery methods, thus enhancing the effective flow of changes/model updates to the site and the update of progress or as-built information.

6.7.4 Technology cost

There will likely be additional cost to construction projects as a result of implementing CPCS. This raises the question of 'who bears the cost? And if shared, how will these cost be shared? The tagging of building components requires that the use of smartphone and data acquisition technologies, thus increasing the project budget. Depending on the scale of the CPCS and type of sensing system, the cost of the sensors could be high – there will be a considerable loss if the sensors are stolen, lost, or damaged.

6.7.5 Security

There is a growing concern in other industry sectors about cyber-attacks on CPCS. In CPCS, these attacks could be costly and fatal. For example, abnormalities in CPCS employed in hospitals and automobiles could endanger lives (Prittopaul et al., 2015; Wang et al., 2012; Wang et al., 2010). The construction industry is not an exception.

6.7.6 Reliability

Several construction-related factors may threaten the life and performance of sensing systems, e.g. the environmental condition, ruggedness, and interference. These could affect the accuracy of the sensed data and also pose physical damages to the sensors.

6.8 Conclusion

The benefits of the applications of cyber-physical systems in the construction industry are unprecedented, and the impact of the solutions are far-reaching; particularly in enhancing predictability and control of project outcomes. One of the early approaches of cyber-physical systems in construction is in enhancing bi-directional coordination between virtual models and the physical construction, so as to maintain consistency between both environments. This approach has been demonstrated through the development of use-case scenarios relating to tracking the installation and control of steel and light fixtures, and equipment and posture monitoring and control. In spite of these potentials, opportunities exist in the area of how to design construction systems to achieve zero safety and health risk projects; how to design construction projects that provide information needed by the project team, yet requires minimal development and maintenance cost, software, and is reliable. Responding to these challenges requires significant advances in the state-of-the-art. A tremendous multidisciplinary approach is required to understand and develop methods, science, models, architectures, and solutions for construction cyber-physical systems. This presents a very fertile area for ground-breaking research and technologies.

6.9 Summary

- Cyber-physical systems enhance consistency and maintenance of virtual models and physical construction, and this offers considerable scope for new tracking and communication mechanisms in project control.
- Bi-directional coordination between virtual models and the physical construction has potential opportunities for improving project monitoring and control of construction process/constructed facility, tracking changes/model updates and communicating these to site in real-time, controlling equipment and worker's posture, documentation of as-built status and sustainable practices.
- The developed scenarios indicate that a cyber-physical systems approach can play an important role in enhancing bi-directional coordination between virtual models and the physical construction.
- Workforce competency, training, adjustment to existing practices, cost of the technology, security, and reliability are some of the challenges and barriers to the uptake of cyber-physical systems in the construction industry.

References

Akanmu, A., Anumba, C., & Messner, J. 2013a. Active monitoring and control of light fixtures during building construction and operation: Cyber-physical systems approach. *Journal of Architectural Engineering*, 20, 04013008.

Akanmu, A., Anumba, C., & Messner, J. 2013b. Scenarios for cyber-physical systems integration in construction. *Journal of Information Technology in Construction (ITcon)*, 18, 240–260.

Akanmu, A., Olatunji, O., Love, P. E., Nguyen, D., & Matthews, J. 2016. Auto-generated site layout: an integrated approach to real-time sensing of temporary facilities in infrastructure projects. *Structure and Infrastructure Engineering*, 12, 1243–1255.

Andrews, D. M., Fiedler, K. M., Weir, P. L., & Callaghan, J. P. 2012. The effect of posture category salience on decision times and errors when using observation-based posture assessment methods. *Ergonomics*, 55, 1548–1558.

Anumba, C. J., Akanmu, A., & Messner, J. 2010. Towards a cyber-physical systems approach to construction. Construction Research Congress 2010: Innovation for Reshaping Construction Practice, 528–537.

Azimi, R., Lee, S., Abourizk, S. M., & Alvanchi, A. 2011. A framework for an automated and integrated project monitoring and control system for steel fabrication projects. *Automation in Construction*, 20, 88–97.

Bordel, B., Alcarria, R., Robles, T., & Martín, D. 2017. Cyber–physical systems: Extending pervasive sensing from control theory to the Internet of Things. *Pervasive and Mobile Computing*, 40, 156–184.

Breiner, J. M., Harkness, S. S., Johnson, C. C., & Koehler, C. M. 2012. What is STEM? A discussion about conceptions of STEM in education and partnerships. *School Science and Mathematics*, 112, 3–11.

Bulbul, T., Anumba, C., & Messner, J. 2009. A system of systems approach to intelligent construction systems. 2009 ASCE International Workshop on Computing in Civil Engineering, 22–32.

Chavada, R., Dawood, N., & Kassem, M. 2012. Construction workspace management: The development and application of a novel nD planning approach and tool. *Journal of Information Technology in Construction*, 17, 213–236.

Chen, N., Xiao, C., Pu, F., Wang, X., Wang, C., Wang, Z., & Gong, J. 2015. Cyber-physical geographical information service-enabled control of diverse in-situ sensors. *Sensors*, 15, 2565–2592.

Cheng, T. & Teizer, J. 2013. Real-time resource location data collection and visualization technology for construction safety and activity monitoring applications. *Automation in Construction*, 34, 3–15.

Coakes, E. W., Coakes, J. M., & Rosenberg, D. 2008. Co-operative work practices and knowledge sharing issues: A comparison of viewpoints. *International Journal of Information Management*, 28, 12–25.

Dallasega, P., Rauch, E., & Linder, C. 2018. Industry 4.0 as an enabler of proximity for construction supply chains: A systematic literature review. *Computers in Industry*, 99, 205–225.

Dede, C. 2007. Transforming education for the 21st century: New pedagogies that help all students attain sophisticated learning outcomes. *Commissioned by the NCSU Friday Institute, February*.

Demiralp, G., Guven, G., & Ergen, E. 2012. Analyzing the benefits of RFID technology for cost sharing in construction supply chains: A case study on prefabricated precast components. *Automation in Construction*, 24, 120–129.

Dillon, T. S., Zhuge, H., Wu, C., Singh, J., & Chang, E. 2011. Web of things framework for cyber–physical systems. *Concurrency and Computation: Practice and Experience*, 23, 905–923.

El-omari, S. & Moselhi, O. 2011. Integrating automated data acquisition technologies for progress reporting of construction projects. *Automation in Construction*, 20, 699–705.

Giretti, A., Carbonari, A., Naticchia, B., & De Grassi, M. 2008. Advanced real-time safety management system for construction sites. Proceedings of the 25th International Symposium on Automation and Robotics in Construction, Vilnius, Lithuania, 2008, 26–29.

Golparvar-fard, M., Peña-mora, F., & Savarese, S. 2012. Automated progress monitoring using unordered daily construction photographs and IFC-based building information models. *Journal of Computing in Civil Engineering*, 29, 04014025.

(HSRC), I.M.R.A.D.O. 2018. Industry 4.0 Market, *Technologies & Industry 2019–2023*, 3, 425. https://industry40marketresearch.com/reports/industry-4-0-market-technologies/.

Irizarry, J., Karan, E. P., & Jalaei, F. 2013. Integrating BIM and GIS to improve the visual monitoring of construction supply chain management. *Automation in Construction*, 31, 241–254.

Jang, W. S. & Skibniewski, M. J. 2008. A wireless network system for automated tracking of construction materials on project sites. *Journal of Civil Engineering and Management*, 14, 11–19.

Kan, C., Anumba, C. J., & Messner, J. I. 2019. A framework for CPS-based real-time mobile crane operations. *Advances in Informatics and Computing in Civil and Construction Engineering*, Springer, pp. 653–660.

Kasim, N., Liwan, S. R., Shamsuddin, A., Zainal, R., & Kamaruddin, N. C. 2012. Improving on-site materials tracking for inventory management in construction projects. International Conference of Technology Management, Business and Entrepreneurship, 2012.

Kim, C., Son, H., & Kim, C. 2013. Automated construction progress measurement using a 4D building information model and 3D data. *Automation in Construction*, 31, 75–82.

Lee, E. A. 2006. Cyber-physical systems-are computing foundations adequate. Position Paper for NSF Workshop on Cyber-Physical Systems: Research Motivation, Techniques and Roadmap, 2006. Citeseer, 1–9.

Lee, E. A. 2008. Cyber physical systems: Design challenges. 2008 11th IEEE International Symposium on Object and Component-Oriented Real-Time Distributed Computing (ISORC), 2008. IEEE, 363–369.

Li, H., Lu, M., Chan, G., & Skitmore, M. 2015. Proactive training system for safe and efficient precast installation. *Automation in Construction*, 49, 163–174.

Li, X., Yi, W., Chi, H.-L., Wang, X., & Chan, A. P. 2018. A critical review of virtual and augmented reality (VR/AR) applications in construction safety. *Automation in Construction*, 86, 150–162.

McMasters, J. H. 2004. Influencing engineering education: One (acrospace) industry perspective. *International Journal of Engineering Education*, 20, 353–371.

McWilliam, E. 2008. *The creative workforce: How to launch young people into high-flying futures*, Sydney: UNSW Press.

Mitchell, R. & Chen, R. 2013. Effect of intrusion detection and response on reliability of cyber physical systems. *IEEE Transactions on Reliability*, 62, 199–210.

Munir, S., Stankovic, J. A., Liang, C.-J. M., & LIN, S. Cyber physical system challenges for human-in-the-loop control. Presented as part of the 8th International Workshop on Feedback Computing, 2013.

Navon, R. & Sacks, R. 2007. Assessing research issues in automated project performance control (APPC). *Automation in Construction*, 16, 474–484.

Ochoa, S. F., Bravo, G., PINO, J. A., & Rodríguez-Covili, J. 2011. Coordinating loosely-coupled work in construction inspection activities. *Group Decision and Negotiation*, 20, 39–56.

Oesterreich, T. D. & Teuteberg, F. 2016. Understanding the implications of digitisation and automation in the context of Industry 4.0: A triangulation approach and elements of a research agenda for the construction industry. *Computers in Industry*, 83, 121–139.

Prittopaul, P., Sathya, S., & Jayasree, K. 2015. Cyber physical system approach for heart attack detection and control using wireless monitoring and actuation system. 2015 IEEE 9th International Conference on Intelligent Systems and Control (ISCO), 2015. IEEE, 1–6.

Ranaweera, K., Ruwanpura, J., & Fernando, S. 2012. Automated real-time monitoring system to measure shift production of tunnel construction projects. *Journal of Computing in Civil Engineering*, 27, 68–77.

Sangiovanni-Vincentelli, A., Damm, W., & Passerone, R. 2012. Taming Dr. Frankenstein: Contract-based design for cyber-physical systems. *European Journal of Control*, 18, 217–238.

Song, J., Haas, C. T., & Caldas, C. H. 2006. Tracking the location of materials on construction job sites. *Journal of Construction Engineering and Management*, 132, 911–918.

Statistics, B. O. L. A. 2016a. Nonfatal Occupational Injuries and Illnesses *Requiring Days Away from Work and Musculoskeletal Disorders, 2015*. Bureau of Labor Statistics. https://data.bls.gov/cgi-bin/dsrv?cs.

Statistics, U. S. B. O. L. 2016b. *Census of Fatal Occupational Injuries*. Bureau of Labor Statistics. www.bls.gov/iif/oshcfoi1.htm.

Teizer, J. 2015. Status quo and open challenges in vision-based sensing and tracking of temporary resources on infrastructure construction sites. *Advanced Engineering Informatics*, 29, 225–238.

Teizer, J., Cheng, T., & Fang, Y. 2013. Location tracking and data visualization technology to advance construction ironworkers' education and training in safety and productivity. *Automation in Construction*, 35, 53–68.

Turkan, Y., Bosche, F., Haas, C. T., & Haas, R. 2012. Automated progress tracking using 4D schedule and 3D sensing technologies. *Automation in Construction*, 22, 414–421.

Wan, J., Cai, H., & Zhou, K. 2015. Industrie 4.0: Enabling technologies. Proceedings of 2015 International Conference on Intelligent Computing and Internet of Things, 2015. IEEE, 135–140.

Wang, E. K., Ye, Y., Xu, X., Yiu, S.-M., Hui, L. C. K., & Chow, K.-P. 2010. Security issues and challenges for cyber physical system. IEEE/ACM International Conference on Green Computing and Communications & International Conference on Cyber, Physical and Social Computing, 2010. IEEE, 733–738.

Wang, Y., Tan, G., Wang, Y., & Yin, Y. 2012. Perceptual control architecture for cyber–physical systems in traffic incident management. *Journal of Systems Architecture*, 58, 398–411.

Xia, F., Hao, R., Li, J., Xiong, N., Yang, L. T., & Zhang, Y. 2013. Adaptive GTS allocation in IEEE 802.15. 4 for real-time wireless sensor networks. *Journal of Systems Architecture*, 59, 1231–1242.

Xia, F., Vinel, A., Gao, R., Wang, L., & Qiu, T. 2011. Evaluating IEEE 802.15. 4 for cyber-physical systems. *EURASIP Journal on Wireless Communications and Networking*, 2011, 596397.

Yuan, X., Anumba, C. J., & Parfitt, M. K. 2016. Cyber-physical systems for temporary structure monitoring. *Automation in Construction*, 66, 1–14.

7

A REVIEW OF MIXED-REALITY APPLICATIONS IN CONSTRUCTION 4.0

Aseel Hussien, Atif Waraich, and Daniel Paes

7.1 Aims

- To provide an overview of state-of-the-art mixed-reality technologies in the construction industry.
- To describe virtual reality and augmented reality technologies including their principles, differences, operation, latest devices, main applications and major impacts in the construction industry.
- To characterize current adoption levels and future expectations concerning mixed-reality applications in Construction 4.0.

7.2 Introduction

The term Industry 4.0 has been introduced to describe the trend towards digitization and automation to improve productivity and quality in the manufacturing industry (Oesterreich and Teuteberg 2016). When brought into the construction industry context, the term becomes Construction 4.0, encompassing all technologies that promote digitization and automation in the construction environment (Woodhead et al. 2018).

Currently, interest in the use of Industry 4.0 technologies permeates several construction practices. The construction industry is undergoing massive changes due to the growing adoption of such technologies, which can contribute to support, integrate, and optimize construction processes throughout a project's life cycle. Regardless of the noticeable number and diversity of technologies available, one of the contributing factors to today's industry poor performance continues to be the unsuccessful communication and information sharing among stakeholders, which is largely based on often inaccurate and time-consuming methods. In late 2014, the advent of powerful mixed-reality software and devices – backed by remarkable advances in computer processing power, low latency tracking and display technologies – has enabled the development and spread of Virtual Reality (VR) and Augmented Reality (AR) solutions – previously limited to the military and manufacturing sectors – across the Architecture, Engineering, Construction, Operations and Facility Management (AECO-FM) industry. As part of the VDC (Virtual Design and Construction) approach, VR and AR solutions have been implemented across the industry to improve the efficiency, quality and safety of construction and infrastructure. The Startup Europe Partnership has identified VR and AR as key technologies for Digital Construction and Industry 4.0 (SEP 2018), while the Hannover Fair has listed them among the top 20 most potentially transformative technologies (IoT-Analytics 2019). A recent report by Digi-Capital (2019) shows that the mixed-reality market is expected to reach $90 billion in revenue by 2023. In 2018, the World Economic Forum in conjunction with the Boston Consulting Group (BCG 2018) released a report in which mixed-reality is listed among the ten most promising technologies to improve productivity in construction. Many research institutions and industry practitioners worldwide are exploring the benefits and investigating the challenges associated with VR and AR implementation in various construction activities. This chapter focuses on VR and AR technologies in Construction 4.0, describing their: a) principles and differences, b) operation and latest devices, and c) main applications and major impacts.

7.2.1 *Principles and differences*

VR and AR are mixed-reality technologies for advanced visualization and simulation. They can enhance the cognitive capabilities of users who perform tasks that rely on or benefit from access to visual information, and can be used to improve the understanding, communication and sharing of that data. Efficient mixed-reality solutions enable users to effectively and intuitively navigate the virtual or augmented space while interacting with, generating, modifying and extracting data. Ultimately, mixed-reality technologies are used to create alternative visual information environments where users have direct and intuitive access to relevant data and consequent increased decision-making and task performance abilities. As per the KPMG Global Construction Survey (KPMG 2016), visualization is the future of decision-making in capital projects.

Mixed-reality is a spectrum of technology-mediated experiences located between the boundaries of the virtuality continuum developed by Milgram and Kishino (1994), which ranges from the completely real world (physical/concrete reality) to the completely virtual, computer-generated world. While AR blends the virtual and real worlds by overlapping real-world views with virtual information, VR is a completely computer-generated visual experience – there is nothing else in it other than virtual elements. A clear-cut distinction between VR and AR might be challenging given the unclear boundaries between virtual and real input in mixed-reality environments (such as in augmented virtuality). For this reason, many developers and scholars in the field adopt the term mixed-reality to refer to either VR or AR systems. Although they are primarily advanced visualization technologies, VR and AR environments may not be limited to visual input only as they can also encompass haptic and/or auditory stimuli. Despite their differences, these technologies share three main components, namely: the content (information being conveyed), the medium tools (software and hardware), and the user.

VR's definition is associated with one's technology-mediated experience of presence – a psychological state – in a virtual or remote environment (Steuer 1992). The VR system's apparatus aims at producing a strong sense of presence in a virtual/remote environment providing the user with intuitive and natural ways of viewing and interacting with the simulation. VR is not the computer simulation per se, but a human experience of immersion and involvement – defining factors of presence – in the simulation, which results from the interaction among hardware, software and human senses (Witmer and Singer 1998). Although full immersion in the digital world accompanied by total loss of awareness of one's physical/real surroundings seems unlikely, some VR systems are deemed "fully-immersive" to the extent that they are able to orient the user's attention and cognitive processes towards virtual stimuli only, separating her/him from the physical environment.

In turn, AR is associated with the idea of augmenting one's perceptions of the world (Azuma et al. 2001). AR is a visualization technology able to enrich information about one's surrounding environment by adding an extra layer of information and meaning onto images of real objects and spaces. It captures and overlays views of a real object or space (photos, video recordings, and real-time videos) with different types of digital content, modifying one's perceptions of that object or space and delivering an augmented experience of reality. As opposed to VR, AR not only allows but also relies on one's awareness and interaction with the real world. It does not create an imitation of reality as VR does. As per Azuma et al. (2001), AR supplements the physical reality rather than completely replacing it. In fact, in most AR applications it is critical that augmentations are distinguishable from real objects in a real-world view. Typically, augmentations are deliberately not realistic and do not intend to deceive the observer in any way so that she/he is able to distinguish virtual data from real objects within her/his augmented visual field.

7.2.2 *Operation and devices*

The context of application and the demands of the task-at-hand define a mixed-reality system's setup and specifications, including its display type, tracking sensors, interface devices, and digital content or simulation. Interfaces are particularly critical components of mixed-reality platforms as they determine the quality of interactions between user and digital content. Interfaces are basically interaction devices that capture users' entries and gestures allowing them to navigate through and interact with a VR or AR environment. They can be of different types, either tangible (keyboard, mouse, touch screen, controller, etc.) or multimodal interfaces (speech or gesture recognition, eye tracking, etc.).

VR systems can have different configurations and levels of complexity. The most basic setups comprise a 3D (three-dimensional) animation with which the user can interact using a computer consisting of a display and interaction/input devices (e.g. keyboard and mouse). High-end systems allow users to literally walk through the virtual space, manipulate virtual objects using hand gestures, and shift viewpoints with head movements as if they were in the real world. To enable such high levels of interactivity, these systems employ stereoscopic displays such as head mounted displays (HMD) and virtual retinal displays (VRD), body and head movement tracking sensors, data gloves or 3D controllers for hand gesture-based object manipulation, and treadmills. For many years, a particular type of projection-based VR system named CAVE™ (Cave Automatic Virtual Environment) had been the most common across industry segments and research groups (Cruz-Neira et al. 1993). The increase in computer processing power over the past two decades allowed for the development and spread of several high-end HMD models. Currently, enthusiasts, researchers, and industry personnel have access to off-the-shelf devices like the Oculus Rift™ and the HTC Vive™. Such platforms require very little computational and programming skills and are able to delivery quite powerful VR experiences.

AR systems can be of different types as well. These include fixed, mobile, indoor, and outdoor setups. Most AR applications in construction utilize mobile systems – Mobile Augmented Reality (MAR) systems – made of wireless technologies able to process, superimpose, and display digital data onto real-world scenes, allowing users to walk freely through an augmented environment (Carmigniani et al. 2011). AR-enabled mobile devices are becoming more powerful and less expensive at a very rapid pace. A mobile AR system consists of the integration of a few essential components, which are typically found in most smartphones, tablet computers, and high-end smartglasses such as the Microsoft® HoloLens™, the GlassUp

F4, and the Daqri Smart Glasses®. Those components include gyroscopes, accelerometers, Wi-Fi antennas, digital compasses, Global Positioning Systems (GPS), a processing unit, a conventional video camera to capture live scenes or a special depth-sensing camera to scan the shape of the surroundings, a software application that registers and overlays real-world data (in the format of images or scans) with virtual data, the virtual data per se, and a display. The display method alternates between conventional touch screens of handheld mobile devices and see-through/transparent screens of high-end smartglasses that, instead of capturing real-time videos, are able to scan the shape of the environment and place augmentations in the correct scale, location and orientation onto the real-world view. In either case, it is crucial that augmentations align properly with the real world as misalignment can lead to inaccuracies and subsequent inefficiencies in most contexts of use. Alignment is achieved through different methods, depending on whether an AR system is marker-based or markerless. While marker-based AR makes use of physical markers distributed through the real environment to anchor the augmentations, markerless AR utilizes other types of positioning technologies such as GPS (Behzadan et al. 2008, Cheng et al. 2017). Marker-based AR systems require prior registration of the real environment onto which digital data would be added, and this is probably the main issue of marker-based AR applications (Azuma et al. 2001). Registration involves capturing (registering) data from the physical environment in which physical markers such as quick response barcodes are placed. These markers serve as anchor points for subsequent overlay of digital data following real-world data registration. The position and orientation of augmentations are defined in relation to those markers. In markerless AR systems, GeoSpots can be used to create geo-reference points of latitude and longitude that are connected to the digital information. In this case, augmentations are displayed relatively to those GeoSpots, thus eliminating the need for physical markers that would not be practical in many AR applications in construction (Irizarry et al. 2013).

7.2.3 *Applications and impacts*

Mixed-reality solutions can be used for different purposes, in different phases of a construction project's life cycle – from programming and planning through operation and demolition or renovation. While VR solutions tend to be most relevant in situations where the visualization of non-existing buildings and artifacts is needed or advantageous – such as in the planning and design phases – AR technology seems most suitable to the phases where a building – or parts of it – already exists such as in the construction and operation phases, which include tasks that involve workers' interaction with built structures.

In any case, mixed-reality systems are fundamentally advanced visualization tools, and so, they can improve construction activities that rely on or benefit from access to visual data, such as design review, safety training, construction, and facility management, among others. Three-dimensional, interactive, and intuitive visualization of building information can improve the conditions for collaborative decision-making and problem solving in various construction activities. Positive impacts of different kinds and at different levels can be expected from visualization techniques like VR and AR, which enable the finalization of design decisions earlier in a project's life cycle, improve learning in safety training programs, or facilitate on-site construction monitoring and/or safety inspection. Naturally, as mixed-reality technologies are tested and implemented in construction they tend to become more effective. Understanding the usability and human factors involved in their contexts of use is crucial to the development of user-centered and increasingly effective systems that would better respond to users' demands, improving their cognitive capabilities during task performance.

7.2.3.1 AR applications

Most AR applications in construction comprise visualization systems capable of overlapping live world scenes with information extracted from Building Information Modeling (BIM) databases. In the design phase, for instance, AR can be used to display 3D interactive scale models – often built with BIM tools – of floorplans or entire buildings, giving the user the ability to "hover" above the displayed prototype, in a bird-view mode. Construction managers can use mobile AR at the jobsite to perform construction progress monitoring through visual comparisons of planned against actual completion of construction, in which case a 4D BIM model (one comprising schedule information) is overlaid on live videos of construction works. More recently, AR systems have become accurate and wearable enough to assist field workers in equipment operation, building construction, and facility maintenance by displaying interactive visual overlays with instructions on how to perform their tasks. Mobile AR is particularly helpful in the facility management phase, providing maintenance personnel with visual diagrams, instructions, and technical information overlaid on a real control panel for instance, allowing a repair worker to locate, identify, and access its various components and complete maintenance and repairing tasks efficiently (Figure 7.1).

Figure 7.1 Daqri® AR hard hat application in facility management (upper) and projections on the hat's display (bottom).

In the past few years, researchers and industry practitioners have made great progress in the development of technology required to allow for seamless operation of AR systems. Irizarry et al. (2013) developed and tested a prototype of a mobile AR application for facility management. Their prototype aimed at providing facility managers with a tool for visualizing indoor environments added with context-relevant virtual information. Moreira et al. (2018) developed and tested a mobile LAR (Living Augmented Reality)-based interactive building manual. Traditionally, a building manual consists of instructions in hard copy format used by facility managers and residents to perform simple repairs. Their LAR-based solution displays a step-by-step instructional animation of repairing procedures on top of live views of building components such as toilets. The Trimble® SiteVision™ AR viewer (Trimble 2019) is an AR device for construction progress visualization, and utilizes satellite signals to overlay a full-scale 4D BIM model on top of a real construction site, allowing managers to check the construction progress by visual comparison (Figure 7.2). In 2017, Skanska started field-testing Daqri® AR hard hats, also known as smart helmets, which besides the usual protection also provide thermal imagery and BIM data overlaid onto the helmet's visor (CIOB 2019).

7.2.3.2 VR applications

Although the causes and extent of impacts of the adoption of VR systems in construction are yet to be known, evidence of significant improvements in communication, decision-making, and problem solving is strong and recurrent. Scholars in the field argue that the benefits of VR systems over traditional simulation and representation methods are due to its immersive visualization feature, which appears to support great levels of involvement, increased attention, and improved spatial perception (Paes et al. 2017). VR-based methods have been explored across the entire construction industry. From collaborative immersive design review to VR-based training sessions, the possibility of experiencing hypothetical scenarios is an advantageous

Figure 7.2   The SiteVision™ AR viewer.

strategy in different construction areas. These include construction safety training, disaster evacuation training, heavy equipment operation training, human-building interaction studies, and immersive design review, among others.

A BIM-enabled VR system allows for the development of more integrated design solutions because it facilitates the processes of communication, review, and evaluation of the implications of solutions developed over the design process (Figure 7.3). In fact, many construction companies and design offices worldwide are exploring the benefits of immersive VR platforms for collaborative design review such as Gilbane, Mortenson Construction, and Perkins and Will. It is important to highlight that VR technology is not deemed a tool for the creative process of ideation. A VR system does not solve design problems by itself, but allows for the representation and critical assessment of design solutions – a collaborative process that usually involves multidisciplinary teams.

During collaborative VR-based design review meetings, stakeholders have the ability to examine together a design by looking at and walking through the model. Collaborative and immersive review meetings promote the development of designs that better meet end-users' needs and expectations, to the extent that when present in the meetings they can better understand the design representation and provide valuable feedback. In general, immersive design review can better support stakeholders in the assessment of design solutions that encompass issues of ergonomics, operability and maintainability of facilities, constructability, installation feasibility, and critical sight lines. For instance, virtual environments can assist design teams in the identification of end-user lighting preferences through continuous testing of alternative design solutions in order to increase end-user satisfaction while reducing the electricity waste associated with lighting systems (Heydarian et al. 2017). SHoP Architects design office (SHoP Architects 2019) is using immersive VR platforms to supplement traditional representation techniques, improving client communication and speeding up design review. Immersive design review also allows the design team at SHoP Architects to take bigger risks with clients

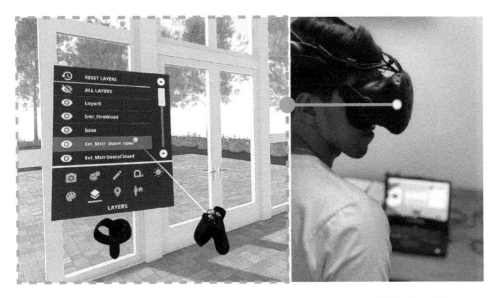

Figure 7.3   Immersive design review application user interface (left) inside an HMD-based VR system (right).

in proposing complex spaces and effects otherwise hard to convey, knowing that everyone involved in the project would understand exactly what is being proposed.

VR solutions are also helpful for safety planning in the pre-construction phase, through which potential hazards are identified and addressed by construction safety personnel before actual work at the jobsite starts. Traditionally, hazard areas are identified by imagining construction tasks at the jobsite with the help of 2D drawings, schedules, safety rules, and the managers' prior experience (Guo et al. 2017, Li et al. 2018). VR offers safety managers the possibility to get close to and experience hazard situations in a virtual construction site as if they were there, allowing for a more reliable identification of potential hazards and better safety planning.

Moreover, construction workers and heavy equipment operators can undergo hands-on training in immersive simulations that mimic their experiences in a real construction site without exposing them to actual danger (Figure 7.4). Through VR-based training, workers can better understand and retain knowledge on risk exposure and respective preventive measures. Studies on the topic suggest that immersive environments are able to improve the trainees' ability to identify and codify potential hazards. Motivation, engagement, and attention have been reported as the main causes of improved learning experiences (Sacks et al. 2013). In 2017, Bechtel Corporation launched VR-based training programs to prepare

Figure 7.4 Projection-based VR crane operator training platform.

its construction workers and crane operators before sending them out on the job. Bechtel's VR-based training leverages realistic simulations to train workers and operators in dealing with hazardous situations at the jobsite, allowing them to practice challenging skills such as unloading a beam from a crane at high altitudes on a foggy day. VR-based training data is collected and analyzed against geographic information, safety records and regulations to optimize both virtual and real-world training approaches and practices (Bechtel Corporation 2017, Construction Dive 2016).

Alternatively, multiuser VR platforms can be employed in all of the aforementioned applications, allowing for users to interact with each other and collaborate inside a virtual environment, whether participating in training simulations, immersive design review or safety planning meetings. Multiuser systems can deliver immersive simulations to team members simultaneously and regardless of their location, offering an alternative cost-effective solution to conventional methods in many contexts of application.

7.3 Conclusion

Mixed-reality systems are advanced visualization tools, and so they can improve construction activities that rely on or benefit from access to visual data, such as design review, safety training, construction and facility management, among others. Their primary goal is to improve communication and information sharing among people. Therefore, they can support collaborative decision-making and problem solving in many construction activities. In fact, mixed-reality technology has already started to revolutionize the way construction professionals perform many of their tasks, with great impacts on design environments, training programs, construction and facility management. Nonetheless, continued research in the field is critical to provide an up-to-date picture of the actual benefits, effectiveness, and limitations of such technologies in the construction domain. User-centered research is particularly important to ensure that mixed-reality develops with adequate concern for its users, investigating and addressing its limitations from the user standpoint towards increasingly effective solutions. Despite the limited number, diversity, functionality and reliability of currently available mixed-reality platforms, it is crucial that higher education institutions and training centers start educating the future workforce on such emerging technologies, given that the most disruptive devices are yet to come, and a truly widespread adoption seems on its way.

7.4 Summary

- Virtual reality and augmented reality are mixed-reality technologies for advanced visualization and simulation and can improve the understanding, communication, and sharing of visual-spatial building information.
- Mixed-reality is listed among the most potentially transformative technologies to improve productivity in construction, and its market is expected to grow rapidly in the coming years.
- In the past few years, researchers and industry have made great progress in the development of technology required to allow for seamless operation of mixed-reality systems – the features and functionality of the latest devices allow for reliable performance in most contexts of use.
- Construction companies, design offices, and research institutions across the globe are already benefiting from the implementation of mixed-reality solutions in different construction activities such as design review, safety training, construction progress monitoring, facility maintenance and operation, among others.

References

Azuma, R., Baillot, Y., Behringer, R., Feiner, S., Julier, S., & MacIntyre, B. 2001. Recent advances in augmented reality. *IEEE Computer Graphics and Applications*, 21(6), pp. 34–47.

BCG – Boston Consulting Group (2018). Available at: www3.weforum.org/docs/Future_Scenarios_Implications_Industry_report_2018.pdf.

Bechtel Corporation (2017). Available at: www.bechtel.com/newsroom/releases/2017/09/bechtel-and-iti-to-broaden-vr-training.

Behzadan, A. H., Timm, B. W., & Kamat, V. R. 2008. General-purpose modular hardware and software framework for mobile outdoor augmented reality applications in engineering. *Advanced Engineering Informatics*, 22(1), pp. 90–105. DOI: 10.1016/j.aei.2007.08.005.

Carmigniani, J., Furht, B., Anisetti, M., Ceravolo, P., Damiani, E., & Ivkovic, M. 2011. Augmented reality technologies, systems and applications. *Multimedia Tools and Applications*, 51(1), pp. 341–377.

Cheng, J., Chen, K., & Chen, W. (2017). Comparison of marker-based and markerless AR: A case study of an indoor decoration system. *Proceedings of the LC3 2017 – Lean and Computing in Construction Congress*, Heraklion, Greece, July 4–7, 483–490. DOI: 10.24928/JC3-2017/0231.

CIOB – Chartered Institute of Building (2019). Available at: www.bimplus.co.uk/news/skanska-tria7l-aug6mented-rea5lity-hard-hats.

Construction Dive (2016). Available at: www.constructiondive.com/news/bechtel-rolls-out-virtual-reality-safety-training/426674.

Cruz-Neira, C., Sandin, D. J., & DeFanti, T. A. (1993). Surround-screen projection-based virtual reality: The design and implementation of the CAVE. *Proceedings of the 20th Annual Conference on Computer Graphics and Interactive Techniques*, Anaheim, pp. 135–142. DOI: 10.1145/166117.166134.

Digi-Capital (2019). Available at: www.digi-capital.com/news/2019/01/for-ar-vr-2-0-to-live-ar-vr-1-0-must-die/.

Guo, H., Yantao, Y., & Skitmore, M. (2017). Visualization technology-based construction safety management: A review. *Automation in Construction*, 73, pp. 135–144. DOI: 10.1016/j.autcon.2016.10.004.

Heydarian, A., Pantazis, E., Wang, A., Gerber, D., & Becerik-Gerber, B. (2017). Towards user centered building design: Identifying end-user lighting preferences via immersive virtual environments. *Automation in Construction*, 81, pp. 56–66.

IoT-Analytics (2019). Available at: https://iot-analytics.com/top-20-industrial-iot-trends-hannover-messe-2019/.

Irizarry, J., Gheisari, M., Williams, G., & Walker, B. (2013). InfoSPOT: A mobile augmented reality method for accessing building information through a situation awareness approach. *Automation in Construction*, 33, pp. 11–23. DOI: 10.1016/j.autcon.2012.09.002.

KPMG – Global Construction Survey (2016). Available at: https://assets.kpmg/content/dam/kpmg/xx/pdf/2016/09/global-construction-survey-2016.pdf.

Li, X., Yi, W., Chi, H. L., Wang, X., & Chan, A. P. (2018). A critical review of virtual and augmented reality (VR/AR) applications in construction safety. *Automation in Construction*, 86, pp. 150–162.

Milgram, P. & Kishino, F. (1994). A taxonomy of mixed reality visual displays. *IEICE Transactions on Information and Systems*, 77(12), pp. 1321–1329.

Moreira, L. S., Ruschel, R. C., & Behzadan, A. H. (2018). Building owner manual assisted by augmented reality: A case from Brazil. *Proceedings of the CRC 2018 – Construction Research Congress*, Louisiana State University, April 2–5. DOI: 10.1061/9780784481264.045.

Oesterreich, T. D. & Teuteberg, F. (2016). Understanding the implications of digitization and automation in the context of Industry 4.0: A triangulation approach and elements of a research agenda for the construction industry. *Computers in Industry*, 83, pp. 121–139.

Paes, D., Arantes, E., & Irizarry, J. (2017). Immersive environment for improving the understanding of architectural 3D models: Comparing user spatial perception between immersive and traditional virtual reality systems. *Automation in Construction*, 84, pp. 292–303. DOI: 10.1016/j.autcon.2017.09.016.

Sacks, R., Perlman, A., & Barak, R. (2013). Construction safety training using immersive virtual reality. *Construction Management and Economics*, 31(9), pp. 1005–1017. DOI: 10.1080/01446193.2013.828844.

SEP–StartupEuropePartnership(2018).Researchhighlights:DigitalconstructionandIndustry4.0.Available at: https://startupeuropepartnership.eu/wp-content/uploads/2018/03/SEP-ResearchHighlights_Industry4.0.pdf.

SHoP Architects (2019). Available at: www.shoparc.com.

Steuer, J. (1992). Defining virtual reality: Dimensions determining telepresence. *Journal of Communication*, 42(4), pp. 73–93. DOI: 10.1111/j.1460-2466.1992.tb00812.x.

Trimble (2019). Available at: https://sitevision.trimble.com/.

Witmer, B. G. & Singer, M. J. (1998). Measuring presence in virtual environments: A presence questionnaire. *Presence*, 7(3), pp. 225–240. DOI: 10.1162/105474698565686.

Woodhead, R., Stephenson, P., & Morrey, D. (2018). Digital construction: From point solutions to IoT ecosystem. *Automation in Construction*, 93, pp. 35–46. DOI: 10.1016/j.autcon.2018.05.004.

8

OVERVIEW OF OPTOELECTRONIC TECHNOLOGY IN CONSTRUCTION 4.0

Erika A. Pärn

8.1 Aims

- Introduce advanced image/laser-based reality computing technologies of construction 4.0.
- Focused review of laser scanning and photogrammetry computing.
- Discuss the hierarchy of the modes of delivery for laser scan devices.
- Provide a thematic analysis of 3D terrestrial laser scan technology applications in Construction 4.0.

8.2 Introduction

The construction sector is currently confronted by the Fourth Industrial Revolution, known as 'Construction 4.0' where cyber-physical production systems (CPS), automation, data exchange, and manufacturing technologies coalesce (Perez et al., 2016). To keep ahead of these advancements, professionals in the industry must look outward to see where technological advancements have flourished or failed in other industries (Koutsoudis et al., 2014; Zou, Kiviniemi, and Jones, 2017). Construction 4.0 is set to emulate many of the more advanced sectors with its image-sensing and optoelectronic technologies. Substantial advancements in image-based sensing and optoelectronic technologies have been achieved in more technologically advanced sectors of aerospace, manufacturing, automotive, and security industries, allowing for the feasible transfer of these same technology into Construction 4.0. Embryonic exemplars of such have been reported and demonstrated upon in research, for instance on suitable automated image/laser based reconstruction to manage and update the as-built BIM.

Optoelectronic technology refers to electronic devices used for emitting, modulating, transmitting, and sensing light and have the capability to gather, store, process, and display all gathered information. Among the hierarchy of optoelectronic technology available in various industries, optical laser scanners have emerged to become the most versatile and commodified of these devices. Laser scanning devices are frequently used for the rapid automation of millimeter precision measurements and reconstruction of physical elements via processed optical signals from reflected light (Blais, 2004; Thiel and Wehr, 2004). Outside of construction we have seen the application of these devices in large swathes of industries such as: aerospace when monitoring structural health, law enforcement when reconstructing crime scenes virtually, agriculture when monitoring crop growth, archaeology when reconstructing artefacts and site details, and manufacturing for quality assurance purposes. Because of this widespread demand for optoelectronic devices in laser scanning, it is expected to exceed over 5.90 billion USD by 2022 as an industry (Markets, 2016).

Such an increase in demand for optoelectronic devices has boomed across nearly all industries, and is set to become an important tool for Construction 4.0 as a sensing mechanism to construct a digital twin. Presently the demand has incrementally increased in the architecture, engineering, construction, and the owner-operated (AECO) sector. This is set to increase with the current pace of digitization of the construction site and a growing reliance on advanced image/laser-based reality capture technologies set to proliferate which include laser scanning and photogrammetry.

Optoelectronics found in 3D laser distance and ranging (LADAR) devices are widely utilized the AECO sector for rapidly constructing point cloud data to measure large volumes of site elements, creating multiple iterations of reflection data recorded by the scanners over time. Although the optoelectronic technology used in 3D LADAR devices offers sophisticated and solid state functionality it is not without its inherent limitations. Inherent limitations in the expansion in laser scanning technology include:

- High equipment costs.
- Discrepancies in spatial information.
- Timely calibration of scanning equipment.
- Limited point cloud capture from occlusions.
- Excess time consumed with point cloud data processing.

Construction 4.0 is slowly shifting to an ever presence of ubiquitous computing and sensing on the construction site resulting in new applications for laser scanning devices. Most notably, laser scanning has been integrated with BIM for model validation and snagging purposes. Yet this integration is beset with its own limitations, as laser scanning applications integrated with BIM ostensibly lack automation in the recognition of attributes in scanned object data (Godin, Borgeat, Beraldin, and Blais, 2010; Rushmeier, 2002). Construction 4.0 is set to address many of such limitations, through a recent surge of research and industry advocating the coupling of image-based sensing technologies with laser scan technology (Golparvar-Fard et al., 2011). Image-based sensing technologies refers to machine vision (aka 'industrial vision', or 'vision systems') or most prominently known as computer vision. Computer vision technologies are set to enhance many of the shortcomings of laser scan capabilities by improvements to object recognition, algorithmic image processing, and pattern recognition (Gao, Akinci, Ergan, and Garrett, 2015).

This chapter will provide a review of laser scanning, photogrammetry computing and provide an in-depth description of how they function together in a Construction 4.0 setting. It will also be prospecting the future trajectory of augmented optoelectronic technology within the Construction 4.0 context through the use of post-processing.

8.3 Fundamentals of laser scan devices

Since early inception and development of optoelectronic scanning devices in the early 1970s, 3D laser scanning trailed behind 2D scanning until the emergence of microcomputers. The exponential improvements in computational processing power meant an increased demand for 3D scanners, nearing the performance of their 2D counterparts. Stimulating new development of cost-effective, automated, contactless 3D range scanners for a variety of industries (Hornberg, 2006; Kim, Cheng, Sohn, and Chang, 2015a). Three contactless methods of ranged measurement emerged: triangulation, phase shift, and time of flight (TOF). More recently an emergence of hybrid technologies, incorporating all three of these methods has been witnessed in contemporary 3D laser scanning devices used in construction.

8.3.1 Triangulation, phase shift, and time of flight

Of these three methods, triangulation is best utilized for short to medium length measurements, between 5 and 10 meters. Triangulation devices use an angled light line that that is projected onto objects, using the deformation of the line to track changes in the surface of the objects thus recording the image (Hornberg, 2006). Phase shift devices calculate the difference between overlapping sent and received signals within a certain wavelength (Lindner, 2016). TOF devices are ideal for long-distance range scanning. This method utilizes multiple single laser point reflections, taking millions of laser pulses and recording the time taken for the pulses to return to the source (Blais, Beraldin, and El-Hakim, 2000). A wide range of hybrid devices have been developed that allow for short-, medium-, and long-range measurements by using two or more of these modes simultaneously (Beraldin et al., 2000b; Creath and Wyant, 1992). The purpose is to counteract some of the limitations of each method. The triangulation method for instance, while visually and mechanically simple in an effort to minimize cost, has a limited field of view. Phase shift method often suffers from interference of point cloud data due to speckle noise. While the accuracy and speckle noise are still an issue for devices relying on TOF, the ability to quickly calculate large surface areas to millimeter level accuracy and precise spatial resolution far outweigh these limitations (Figure 8.1) (Blais, Beraldin, and El-Hakim, 2000; El-Omari and Moselhi, 2009; Shoji, 2013).

8.4 Modes of delivery and specifications

The largest hurdle for the widespread adaptation of 3D laser scanning devices in the AECO sector Construction 4.0 is the exorbitant cost of such devices, albeit the cost of scanners has seen an incremental decrease in recent years (Lindner, 2016). Three primary modes of conducting short, mid, and long range TOF laser scanning during construction and post-construction phases are: airborne, mobile, and terrestrial. Commercial costs for mobile and terrestrial scanners fluctuate based upon the technical specifications of the device, namely: range of scan, weight, ranging error, optical resolution, field of view, and laser class. Range of scan refers to the minimum measurable distance in an indoor or outdoor environment in an upright incidence to a surface with 90% reflective material. Ranging error is the degree of conformance between a measurement of an observable quality and a recognized specification that indicates a true value of the quantity (Blais, Beraldin, and El-Hakim, 2000). Optical resolution denotes the minimum dimensions of an object feature that the 3D scanner is able to acquire (Faro, 2004). Field of view is determined by the angle of laser scanning capabilities of the device,

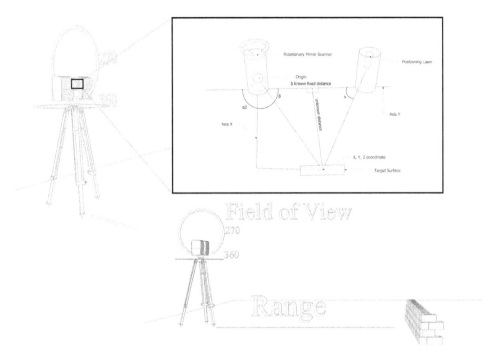

Figure 8.1 Exemplar of triangulation calculation method for a terrestrial laser scanner which commonly use either TOF or phase-shift methods for point capture

between 20–360 degrees vertically and/or horizontally. Laser class is defined by International Electrotechnical Commission (IEC) document 60,825–1 I, which assigns lasers into one of four hazard classes (1, 1M, 2, 2M, 3, 3M, 4, in ascending order – 1 being the least harmful) depending on the potential for causing biological damage.

8.4.1 Airborne Laser Scanning

Airborne Laser Scanning (ALS) devices measure the range of reflected objects on the earth's surface, including landmarks like railroad tracks, waterways, and electrical transmission lines (Wehr and Lohr, 1999). Airborne scanners have evolved from their early applications of aerospace satellites to encompass more widely adopted geological and agricultural surveys (Andujar, Ribeiro, Fernandez-Quintanilla, and Dorado, 2011; Hoffmeister, 2016; Lehtomäki et al., 2015). ALS uses LADAR systems to track geo-referenced points from the overall terrain (Cointault et al., 2016). Unmanned aerial vehicles (UAVs), otherwise known as drones are one of the more generally known ALS devices for the Construction 4.0 sector. Their main purpose is to monitor construction site progress on a small scale. Currently they lack precision of ALS, however, due to the addition of image sensing and photogrammetry techniques, which require moving photographic data in conjunction with automated reconstruction software more accurate surveys can be conducted. In a Construction 4.0 context the proliferation of remote surveying via UAVs is set to grow in demand (Irizarry and Johnson, 2014) whilst accuracy discrepancies will be tackled using improvements to post-processing techniques and image based reconstruction of site data.

8.4.2 Mobile Laser Scanning

Mobile Laser Scanning (MLS) devices are used to capture large amounts of data while in motion (Longstreet, 2010). These devices are either fitted to vehicles or other moving objects to capture large quantities of point data (Breuckmann, 2014). On construction sites currently, there are options for either a harness or hand-held MLS device to help scan data of the site as the operator moves. Hand-held devices, while cheaper and more compact, only have a very limited range, with attaining a 1-millimeter accuracy is only attained at a 1-meter range from the target object (Faro, 2004). Being trained to use these devices is also a costly endeavor for construction site operators. MLS devices can also provide machine vision, allowing for the use of navigating and orienting autonomous robots (UKIVA, 2007; Hornberg, 2006; Lindner, 2016; Perez et al., 2016). As these devices are rare in the industry, their absence reflects this. Future research endeavors for Construction 4.0 in the development of digital twins has sought to eliminate the need for manual scanning precisely by using autonomous robots to scan the construction site on a room-by-room basis. In more confined spaces of a construction site robots can offer continuous monitoring and data capture of site quality and progress via autonomous MLS.

8.4.3 Terrestrial Laser Scanning

Terrestrial Laser Scanning (TLS) technology is the current preferred method of scanning in construction of ground-based discrete objects because it captures static 3D objects with high accuracy (Bosché et al., 2015). While TLS technology is sought after, its limitations lie with the time needed to perform a single high-resolution scan, the number of scans needed to capture an entire site and the high cost of equipment (Barazzetti et al., 2015; Bosché et al., 2015; Brilakis et al., 2010; Faro, 2004). Lower specification models are available, more suited for as-built and topographical high-definition surveys, however, they are more harmful to humans, receiving a laser classification of 3R. The largest limitation for TLS technology is the stationary nature of the devices themselves. This limits both the scanning capacity with a limited field of view and potential loss of data via occlusions (Godin, Borgeat, Beraldin, and Blais, 2010).

8.5 Applications in construction

There are three broad applications of optoelectronic technologies presently used in construction namely: progress tracking, quality control and assessment; structural health monitoring and development of as-built data. In Construction 4.0 the advent of the Internet of Things (IoT) means that the use of optoelectronic devices with machine vision for progress, quality, and structural monitoring will enable inspection data to be transferred to data centers or cloud based solutions for further analysis.

8.5.1 Progress tracking

Progress tracking is a prerequisite requirement for effective and efficient cost management of labor, plant and equipment, and materials (El-Omari and Moselhi, 2009). Traditionally, progress tracking on sites have been achieved through processes such as barcode scanning, radio frequency identification (RFID) (Meadati, Irizarry, and Akhnoukh, 2010), digital photography or video footage (Riaz, Edwards, and Thorpe, 2006). More recently, the process of tracking has evolved to utilize LADAR to track structural work, progress, and quality. The data collected is able to reproduce 3D information with millimeter precision (Matthews et al., 2015) into BIM representation.

8.5.2 Quality control and assessment

Quality control and assessment have been adopted predominantly to determine the structural integrity of the existing structural elements in the AECO sector (Kim, Cheng, Sohn, and Chang, 2015a). Using TLS, structural engineers can validate the integrity of steel frames and precast concrete structural elements (Zeibak-Shini, Sacks, Ma, and Filin, 2016). The limitations caused by redundancies in the system due to the excessive time taken in the production of the point clouds has led to the development of automated quality assessment methods and procedures (Park, Lee, Adeli, and Lee, 2007). The future of construction 4.0 quality assessment will more than likely include a combination of remote monitoring capabilities in MLS and UAVs. Manufacturing setting has quickly adopted machine vision and laser scanning as a means to monitor production quality. In a similar vein it can be anticipated that construction 4.0 will become increasingly dependent on post-processing of inspection data captured in a larger scale of machine vision.

8.5.3 Structural health monitoring

Monitoring the structural health of site and machines with optoelectronics is used throughout numerous industries, including aerospace engineering (Derriso, McCurry, and Schubert Kabban, 2016), structural engineering, geological engineering, archaeological surveying. The accuracy of TLS point cloud data means that structural monitoring is used for defect detection for concrete structural elements (Kim, Sohn, and Chang, 2015b). These devices are chosen specifically for their higher accuracy and ability to record surface quality defects (Kim, Son, and Kim, 2013). For the Construction 4.0 to fully embrace defect detection with optoelectronics accuracy issues will need to be overcome with fast paced cloud based post-processing algorithms similarly applied in a fast paced moving production line in manufacturing.

8.6 Translating as-built progress

New digital technologies have emerged and coalesced with existing ones, namely BIM and laser scanning to form new methods for capturing the status of existing and new building stock. The adoption of digital technologies for semantic data recognition has equally opened the opportunity to increase the stored building data in BIM (Alvarez-Romero, 2014; Love et al., 2014; McArthur, 2015; Patacas, Dawood, and Kassem, 2015). The main benefits of as-built progress capture in BIM via laser scan technology are early detection of nonconformities to circumvent high rework costs (Bosché et al., 2015), faster approval of contractors' work (Klein, Li, and Becerik-Gerber, 2012), and the handover of contemporary as-built BIM data or a 'digital twin' to provide a partial resolution to rework cases during construction (Matthews et al., 2015).

Although numerous studies demonstrate successful implementation of laser scan technology for the purposes of site reconstruction (Bosché, 2010; Kim, Cheng, Sohn, and Chang, 2015a; Wang, Cho, and Kim, 2015; Zhang and Arditi, 2013), the process of creating an accurate representation is still bogged down with manual tasks and timely work. Creating semantically rich BIM models continue to elude the AECO sector due to the difficulties of manually transforming raw point cloud data into semantically rich BIM, as it is laborious and prone to human error (Xiong, Adan, Akinci, and Huber, 2013). Current methods of integrating point cloud data with BIM are unable to acquire semantic detail that is vital for building assets maintenance during their operational stages (Pătrăucean et al., 2015). Recent research suggests the need for automatic 3D object recognition to automate the integration with the as-built BIM, with only partial fulfillment of such functionality readily available commercially through semi-automated tools.

Most prominent research has thus far focused on automated 3D modeling as used in construction stage related activities and phases (Chen et al., 2015). While this is a step in the right direction, more attention should be given on creating semantic detail in building models for the purpose of accurately representing physical assets and their attributes. These scans can create a geometric representation between the physical and virtual space (Akanmu and Anumba, 2015; Tang et al., 2010).

8.6.1 *Post-processing information retrieval*

Post-processing algorithms and applications overcome this by delivering automated 3D laser scanning of physical assets into BIM with some degree of semantic information recognition (Sturm and Triggs, 1996; Szeliski, 2010). These processes are vital in tunnel cross section dimensional quality assessment, deformation measurement of existing concrete structures (Gordon and Lichti, 2007), and volume loss estimation for in situ concrete bridges. The algorithms are capable of expediting processing point cloud data capture into existing workflows such as BIM. Such post-processing tools have automated the generation of new 3D objects into the BIM, particularly for use in complex mechanical, electrical, and plumbing (MEP) projects (Barazzetti et al., 2015; Brilakis et al., 2010; Son and Kim, 2016). It can be envisaged that Construction 4.0 will become increasingly reliant upon a growing library of post-processing algorithms used to identify site related semantic detail from machine vision and point cloud scans.

Bosché et al., (2015, 2010) first coined the term 'Scan-vs-BIM' technique where point cloud data is compared to as-designed models (Bosché et al., 2015) (Figure 8.3). The definition assumes that the BIM merely acts as an additional reference for point cloud data (Son, Kim, and Kim, 2015a). The three-step process to this manual workflow consists of multiple scans captured from multiple source points before generating point cloud data from these scans are integrated together before being compared to BIM data which is employed to model new objects using the point scans as reference (Monserrat and Crosetto, 2008; Wang and Cho, 2015). In rare cases, these processes can be expedited using software, but the procedure lacks full automation (Bosché, 2010; Hyojoo Son and Kim, 2015). In the 'Scan-to-BIM' method, semi-automatic post-processing software is incorporated, enabling faster model reproduction times (Figure 8.2). While BIM represents a 'central information hub', it cannot become a static information repository otherwise it becomes 'blind' to the ongoing changes of the site. This generates a stable representation of assets during construction (Matthews et al., 2015), and post-construction phases (Volk, Stengel, and Schultmann, 2014).

8.6.2 *Augmenting laser scanning with machine vision*

Machine vision provides automation in the mundane manufacturing tasks of inspection and fault detection. The broader applications lie in automating collision detection in autonomous driving vehicles, robotics, and off-highway machinery (Riaz, Edwards, and Thorpe, 2006). Given the rapid emergence of laser scanning, machine vision technology applications within

Figure 8.2 Example of Scan-to-BIM used for automated masonry recognition (image courtesy of cyber build lab and Prof. Frédéric Bosché)

Figure 8.3 Example of Scan-vs-BIM (image courtesy of cyber build lab and Prof. Frédéric Bosché)

the AECO industry has simultaneously emerged. Early machine vision developments in the AECO sector have integrated laser scan technology (Hyojoo Son and Kim, 2015) and photogrammetry (Scaioni et al., 2014). One example of the practical application of the integrated solution is merging 3D TLS data with photogrammetry for asset inspection purposes (Monserrat and Crosetto, 2008; Perez et al., 2016).

For instance, in the research realm El-Omari and Moselhi (2009) have demonstrated improved site progress capture and monitoring with the combination of photogrammetry, laser scanned point cloud data and post-processing software, called PHIDIAS. This hybrid development monitored and reported upon construction progress with an impressive 75% reduction in the time needed to scan a site (ibid.). However, Golparvar-Fard et al. (2011) revealed that such work could be enhanced by integrating accurate geo-referenceable site coordinates with photographic content. BIM provides a much-needed reference point for a semantic link between photogrammetry and point cloud data (ibid), thus enabling remote decision making by contractors based on the as-designed information (Koutsoudis et al., 2014; Safa et al., 2015).

Photogrammetry techniques are heavily founded upon the Structure-from-Motion (SfM) and Dense Multi-View 3D Reconstruction (DMVR) post-processing algorithms. Newly developed photogrammetry applications are now commercially available and more accessible to construction due to a proliferation of with mobile devices used on site and lower costs in comparison to laser scan devices. SfM and DMVR algorithms automate 3D modeling generation from an unordered image collection of the target object from multiple viewpoints (Sturm and Triggs, 1996). The difference in SfM lies in the 'calibration' process, where the internal camera geometry parameters, external camera position, and orientation are automatically adjusted using correspondences between the image features (Peters, 2010; Szeliski, 2010). SfM uses a 3D structure of the environment and camera motion within the scene to automate 3D modeling (Szeliski, 2010). Where 3D laser scanning technology is an expensive technology, image processing SfM and DMVR software is cost-effective and efficient (cf. Golparvar-Fard et al., 2011).

8.7 Conclusion

The digital construction era has ushered in an increasing need for optoelectronic devices paired with other solutions witnessed in manufacturing setting to improve the efficiency and performance of project management. When coupled with other augmentations, like BIM, machine vision, and photogrammetry, laser scanning provides an ideal solution to rapidly create and update as-built data. Experience of these new technologies has proven that while these changes have improved on the traditional means of reconstructing the built environment, challenges doggedly persist. These challenges stem from the incredible cost of point cloud processing time, which could be resolved by the process of automation. This augmentation could transform the raw data of point cloud data into useful quality or progress inspection information and eliminate the potential for human error. Readily available semantic information in an accessible format is crucial for Construction 4.0 professionals. This requirement is set to grow in prominence especially since the mandate to deliver projects to a BIM Level 2 standard in the UK.

To better resolve the issues arising in the AECO sector, looking toward the resolutions created in other industries, such as the aviation and automotive industries. This is compounded by the OEMs development of commercial products that are not only expensive but also are driven by consumers poorly define specifications and monetary gain. Over time, BIM platforms, 3D scanners, and standalone software packages have evolved to meet the needs of demand.

Perhaps the mindset of industry and academia should change to consider a cluster of multifaceted technologies under the singular converged umbrella-term of 'digital construction' vis-à-vis ever finer granulation of individual approaches.

8.8 Summary

- There are four fundamental applications of optoelectronics in construction, namely: progress tracking; quality control and assessment; structural health monitoring; and development of as-built data.
- The automation of scan data object recognition is propelled by technologies from more advanced industries such as machine vision and photogrammetry.
- BIM provides a much-needed reference point for a semantic link between photogrammetry and automated point cloud data recognition, thus enabling automatic creation of as-built data.
- Construction 4.0 will transform in the way in which optoelectronic devices will be utilized ostensibly to replicate the manufacturing *modus operandi* through improved progress monitoring and quality control.

References

Akanmu, A. and Anumba, C. J. (2015) Cyber-physical Systems Integration of Building Information Models and the Physical Construction. *Engineering, Construction and Architectural Management*, 22, No. 5, pp. 516–535.

Alvarez-Romero, S. O. (2014) Use of Building Information Modeling Technology in the Integration of the Handover Process and Facilities Management. *Worcester Polytechnic Institute*. Available via: https://web.wpi.edu/Pubs/ETD/Available/etd-090914-170046/unrestricted/Disertation_final_SA.pdf (Accessed: April 2019).

Andujar, D., Ribeiro, A., Fernandez-Quintanilla, C., and Dorado, J. (2011) Accuracy and Feasibility of Optoelectronic Sensors for Weed Mapping in Wide Row Crops. *Sensors*, 11, No. 3, pp. 2304–2318.

Barazzetti, L., Banfi, F., Brumana, R., Gusmeroli, G., Previtali, M., and Schiantarelli, G. (2015) Cloud-to-BIM-to-FEM: Structural Simulation with Accurate Historic BIM from Laser Scans. *Simulation Modelling Practice and Theory*, 57, pp. 71–87.

Beraldin, J. A., Blais, F., Rioux, M., Cournoyer, L., Laurin, D., and MacLean, S. G. (2000b) Eye-safe Digital 3-D Sensing for Space Applications. *Optical Engineering*, 39, No. 1, pp. 196–211.

Blais, F., Beraldin, J. A., and El-Hakim, S. F. (2000) Range Error Analysis of an Integrated Time-of-flight, Triangulation, and Photogrammetric 3D laser Scanning System. Proc. SPIE 4035, Laser Radar Technology and Applications V, (5 September 2000); DOI: https://doi.org/10.1117/12.397796.

Blais, F. O. (2004) Review of 20 Years of Range Sensor Development. *Journal of Electronic Imaging*, 13, No. 1, pp. 231–240.

Bosché, F. (2010) Automated Recognition of 3D CAD Model Objects in Laser Scans and Calculation of As-built Dimensions for Dimensional Compliance Control in Construction. *Advanced Engineering Informatics*, 24, No. 1, pp. 107–118.

Bosché, F., Ahmed, M., Turkan, Y., Haas, C. T., and Haas, R. (2015) The Value of Integrating Scan-to-BIM and Scan-vs-BIM Techniques for Construction Monitoring Using Laser Scanning and BIM: The Case of Cylindrical MEP Components. *Automation in Construction*, 49, pp. 201–213.

Breuckmann, B. (2014) 25 Years of High Definition 3D Scanning: History, State of the Art and Outlook. Proceedings of the EVA London, July 8–10, 2014, pp. 262–266.

Brilakis, I., Lourakis, M., Sacks, R., Savarese, S., Christodoulou, S., Teizer, J., and Makhmalbaf, A. (2010) Toward Automated Generation of Parametric BIMs Based on Hybrid Video and Laser Scanning Data. *Advanced Engineering Informatics*, 24, No. 4, pp. 456–465.

Chen, K., Lu, W., Peng, Y., Rowlinson, S., and Huang, G. Q. (2015) Bridging BIM and Building: From a Literature Review to an Integrated Conceptual Framework. *International Journal of Project Management*, 33, No. 6, pp. 1405–1416.

Cointault, F., Simeng Han, G. R., Sylvain Jay, D. R., Billiot, B., Simon, J.-C., and Salon, C. (2016) 3D Imaging Systems for Agricultural Applications: Characterization of Crop and Root Phenotyping in O. Sergiyenko and J. C. Rodriguez-Quinonez (eds), *Developing and Applying Optoelectronics in Machine Vision*, Hershey: IGI Global, pp. 236–272.

Creath, K., and Wyant, J. C. (1992) Moiré and Fringe Projection Techniques in D. Malacara (ed), *Optical Shop Testing*, Hoboken: Wiley & Sons, pp. 653–681.

Derriso, M. M., McCurry, C. D., and Schubert Kabban, C. M. (2016) *2 – A Novel Approach for Implementing Structural Health Monitoring Systems for Aerospace Structures A2 – Yuan, Fuh-Gwo Structural Health Monitoring (SHM) in Aerospace*. Duxford: Structures Woodhead Publishing, pp. 33–56.

El-Omari, S. and Moselhi, O. (2009) Data Acquisition from Construction Sites for Tracking Purposes. *Engineering, Construction and Architectural Management*, 16, No. 5, pp. 490–503.

Faro. (2004). FARO Laser Scanner Focus Series Technical Specification. Available via: www.faro.com/products/3d-surveying/laser-scanner-faro-focus-3d/overview/(Accessed: April, 2019).

Gao, T., Akinci, B., Ergan, S., and Garrett, J. (2015) An Approach to Combine Progressively Captured Point Clouds for BIM Update. *Advanced Engineering Informatics*, 29, No. 4, pp. 1001–1012.

Godin, G., Borgeat, L., Beraldin, J. A., and Blais, F. (2010) 17–19 March 2010 *Issues in Acquiring, Processing and Visualizing Large and Detailed 3D Models. Paper presented at the Information Sciences and Systems (CISS), 2010 44th Annual Conference on Information Sciences and Systems (CISS)*, 17–19 March 2010, Princeton, pp. 1–6.

Golparvar-Fard, M., Bohn, J., Teizer, J., Savarese, S., and Peña-Mora, F. (2011) Evaluation of Image-based Modeling and Laser Scanning Accuracy for Emerging Automated Performance Monitoring Techniques. *Automation in Construction*, 20, No. 8, pp. 1143–1155.

Gordon, S. J. and Lichti, D. D. (2007) Modeling Terrestrial Laser Scanner Data for Precise Structural Deformation Measurement. *Journal of Surveying Engineering*, 133, No. 2, pp. 72–80.

Hoffmeister, D. (2016) Laser Scanning Approaches for Crop Monitoring in G. R. F. A. Viviana Scognamiglio and P. Giuseppe (eds), *Comprehensive Analytical Chemistry*, Vol. 74, Amsterdam: Elsevier Science, pp. 343–361 Available via: www.sciencedirect.com/science/article/pii/S0166526X16300319 (Accessed: April, 2019).

Hornberg, P. D. A. (2006) *Handbook of Machine Vision*. P. D. A. Hornberg (ed). Weinheim: Wiley-VCH.

Hyojoo Son, J. N. and Kim, C. (2015) *Semantic As-built 3D Modeling of Buildings Under Construction from Laser Scan Data Based on Local Convexity without an As-planned Model. 32nd ISARC Conference, Oulu, Finland*. Available via: www.iaarc.org/publications/fulltext/FFACE-ISARC15-3002169.pdf (Accessed: April 2019).

Irizarry, J. and Johnson, E. N. (2014) *Feasibility Study to Determine the Economic and Operational Benefits of Utilizing Unmanned Aerial Vehicles (UAVs)*. Atlanta: Georgia Institute of Technology.

Kim, C., Son, H., and Kim, C. (2013) Automated Construction Progress Measurement Using a 4D Building Information Model and 3D Data. *Automation in Construction*, 31, pp. 75–82.

Kim, M.-K., Cheng, J. C. P., Sohn, H., and Chang, C.-C. (2015a) A Framework for Dimensional and Surface Quality Assessment of Precast Concrete Elements using BIM and 3D Laser Scanning. *Automation in Construction*, 49, pp. 225–238.

Kim, M.-K., Sohn, H., and Chang, C.-C. (2015b) Localization and Quantification of Concrete Spalling Defects Using Terrestrial Laser Scanning. *Journal of Computing in Civil Engineering*, 29, No. 6, p. 04014086.

Klein, L., Li, N., and Becerik-Gerber, B. (2012) Imaged-based Verification of As-built Documentation of Operational Buildings. *Automation in Construction*, 21, pp. 161–171.

Koutsoudis, A., Vidmar, B., Ioannakis, G., Arnaoutoglou, F., Pavlidis, G., and Chamzas, C. (2014) Multi-image 3D Reconstruction Data Evaluation. *Journal of Cultural Heritage*, 15, No. 1, pp. 73–79.

Lehtomäki, M., Jaakkola, A., Hyyppä, J., Lampinen, J., Kaartinen, H., Kukko, A., Puttonen, E., Hyyppä, H. (2015) Object Classification and Recognition From Mobile Laser Scanning Point Clouds in a Road Environment. *IEEE Transactions on Geoscience and Remote Sensing*, 54, No. 2, pp. 1226–1239.

Lindner, L. (2016) Laser Scanners in O. Sergiyenko and J. C. Rodriguez-Quiñonez (eds), *Developing and Applying Optoelectronics in Machine Vision*, Hershey: IGI Global, pp. 108–145.

Longstreet, B. (2010). Finding the Right Solution to Create an As-built BIM of Portland International Airport's Baggage Handling Facility. Available via: www.leica-geosystems.us/en/ScanningPDX.pdf (Accessed: April 2019).

Love, P. E., Matthews, J., Simpson, I., Hill, A., and Olatunji, O. A. (2014) A Benefits Realization Management Building Information Modeling Framework for Asset Owners. *Automation in Construction*, 37, pp. 1–10.

Markets (2016) 3D Scanner Market by Offering (Hardware, Aftermarket Service), Type (Laser, Structured Light), Range, Product (Tripod Mounted, Fixed CMM Based, Portable CMM Based, Desktop), Application, Industry, and Geography – Global Forecast to 2022. Available via: www.marketsandmarkets.com/Market-Reports/3d-scanner-market-119952472.html (Accessed: April 2019).

Matthews, J., Love, P. E., Heinemann, S., Chandler, R., Rumsey, C., and Olatunj, O. (2015) Real Time Progress Management: Re-engineering Processes for Cloud-based BIM in Construction. *Automation in Construction*, 58, pp. 38–47.

McArthur, J. (2015) A Building Information Management (BIM) Framework and Supporting Case Study for Existing Building Operations, Maintenance and Sustainability. *Procedia Engineering*, 118, pp. 1104–1111.

Meadati, P., Irizarry, J., and Akhnoukh, A. K. (2010) BIM and RFID Integration: A Pilot Study. Advancing and Integrating Construction Education Second International Conference on Construction in Developing Countries (ICCIDC–II), August 3–5, 2010, Cairo, pp. 570–578.

Monserrat, O. and Crosetto, M. (2008) Deformation Measurement Using Terrestrial Laser Scanning Data and Least Squares 3D Surface Matching. *ISPRS Journal of Photogrammetry and Remote Sensing*, 63, No. 1, pp. 142–154.

Park, H. S., Lee, H. M., Adeli, H., and Lee, I. (2007) A New Approach for Health Monitoring of Structures: Terrestrial Laser Scanning. *Computer-Aided Civil and Infrastructure Engineering*, 22, No. 1, pp. 19–30.

Patacas, J., Dawood, N., and Kassem, M. (2015) BIM for Facilities Management: Evaluating BIM Standards in Asset Register Creation and Service Life Planning. *Journal of Information Technology in Construction*, 20, No. 10, pp. 313–318.

Pătrăucean, V., Armeni, I., Nahangi, M., Yeung, J., Brilakis, I., and Haas, C. (2015) State of Research in Automatic As-built Modelling. *Advanced Engineering Informatics*, 29, No. 2, pp. 162–171.

Perez, L., Rodriguez, I., Rodriguez, N., Usamentiaga, R., and Garcia, D. F. (2016) Robot Guidance Using Machine Vision Techniques in Industrial Environments: A Comparative Review. *Sensors (Basel)*, 16, No. 3, p. 335.

Peters, K. H. M. G. (2010) The Structure-from-motion Reconstruction Pipeline – a Survey with Focus on Short Image Sequences. *Kybernetika*, 46, No. 5, pp. 926–937.

Riaz, Z., Edwards, D. J., and Thorpe, A. (2006) SightSafety: A Hybrid Information and Communication Technology System for Reducing Vehicle/Pedestrian Collisions. *Automation in Construction*, 15, No. 6, pp. 719–728.

Rushmeier, F. B. A. H. (2002) The 3D Model Acquisition Pipeline. *Computer Graphics Forum*, 21, No. 2, pp. 149–172.

Safa, M., Shahi, A., Nahangi, M., Haas, C., and Noori, H. (2015) Automating Measurement Process to Improve Quality Management for Piping Fabrication. *Structures*, 3, pp. 71–80.

Scaioni, M., Barazzetti, L., Giussani, A., Previtali, M., Roncoroni, F., and Alba, M. I. (2014) Photogrammetric Techniques for Monitoring Tunnel Deformation. *Earth Science Informatics*, 7, No. 2, pp. 83–95.

Shoji, K. (2013) *Time-of-Flight Techniques Handbook of 3D Machine Vision*. Boca Raton: Taylor & Francis, pp. 253–274.

Son, H. and Kim, C. (2016) Automatic Segmentation and 3D Modeling of Pipelines into Constituent Parts from Laser-scan Data of the Built Environment. *Automation in Construction*, 68, pp. 203–211.

Son, H., Kim, C., and Kim, C. (2015) 3D Reconstruction of As-built Industrial Instrumentation Models from Laser-scan Data and a 3D CAD Database Based on Prior Knowledge. *Automation in Construction*, 49, pp. 193–200.

Son, H., Kim, C., and Turkan, Y. (2015) Scan-to-BIM – An Overview of the Current State of the Art. Paper presented at the 32nd ISARC Conference, Oulu, Finland. Available via: www.iaarc.org/publications/2015_proceedings_of_the_32st_isarc_oulu_finland/scan_to_bim_an_overview_of_the_current_state_of_the_art_and_a_look_ahead.html (Accessed: April 2019).

Sturm, P. and Triggs, B. (1996) A Factorization Based Algorithm for Multi-image Projective Structure and Motion in B. Buxton and R. Cipolla (eds), *Computer Vision — ECCV '96: 4th European Conference on Computer Vision Cambridge*, UK, April 15–18, 1996 Proceedings Volume II Berlin, Heidelberg: Springer Berlin, pp. 709–720.

Szeliski, R. (2010) *Computer Vision: Algorithms and Applications*. London: Springer.

Tang, P., Huber, D., Akinci, B., Lipman, R., and Lytle, A. (2010) Automatic Reconstruction of As-built BIM from Laser-scanned Point Clouds: A Review of Related Techniques. *Automation in Construction*, 19, No. 7, pp. 829–843.

Thiel, K. H. and Wehr, A. (2004) Performance Capabilties of Laser Scanners – An Overview and Measurement Principle Analysis. International Archives of Photogrammetry. *Remote Sensing and Spatial Information Sciences*, 36, No. 8, pp. 14–18.

UK Industrial Vision Association [UKIVA]. (2007) *Machine Vision Handbook*. 3rd ed. Wallington, UK: UKIVA. Available via: www.ukiva.org/pdf/machine-vision-handbook.pdf (Accessed: April 2019).

Volk, R., Stengel, J., and Schultmann, F. (2014) BIM for Existing Buildings – Literature Review and Future Needs. *Automation in Construction*, 38, pp. 109–127.

Wang, C., and Cho, Y. K. (2015) Performance Evaluation of Automatically Generated BIM from Laser Scanner Data for Sustainability Analyses. *Procedia Engineering*, 118, pp. 918–925.

Wang, C., Cho, Y. K., and Kim, C. (2015) Automatic BIM Component Extraction from Point Clouds of Existing Buildings for Sustainability Applications. *Automation in Construction*, 56, pp. 1–13.

Wehr, A. and Lohr, U. (1999) Airborne Laser Scanning – An Introduction and Overview, ISPRS. *Journal of Photogrammetry and Remote Sensing*, 54, No. 3, July 1999 pp. 68–82.

Xiong, X., Adan, A., Akinci, B., and Huber, D. (2013) Automatic Creation of Semantically Rich 3D Building Models from Laser Scanner Data. *Automation in Construction*, 31, pp. 325–337.

Zeibak-Shini, R., Sacks, R., Ma, L., and Filin, S. (2016) Towards Generation of As-damaged BIM Models using Laser-scanning and As-built BIM: First Estimate of As-damaged Locations of Reinforced Concrete Frame Members in Masonry Infill Structures. *Advanced Engineering Informatics*, 30, No. 3, pp. 312–326.

Zhang, C. and Arditi, D. (2013) Automated Progress Control Using Laser Scanning Technology. *Automation in Construction*, 36, pp. 108–116.

Zou, Y., Kiviniemi, A., and Jones, S. W. (2017) A Review of Risk Management through BIM and BIM-related Technologies. *Safety Science*, 97, 88–98.

9

THE POTENTIAL FOR ADDITIVE MANUFACTURING TO TRANSFORM THE CONSTRUCTION INDUSTRY

Seyed Hamidreza Ghaffar, Jorge Corker, and Paul Mullett

9.1 Aims

- Holistic overview of different additive manufacturing processes for the construction industry, including some of the 3D-printed demonstration projects completed to date.
- Current status of technology, limitations, and advantages with reference to materials, systems, and construction processes.
- Identification of future areas of research and development.

9.2 Introduction

Construction is an important economic engine, but also one of the largest consumers of resources and energy. The modern construction industry is undergoing a period of dramatic policy shift and priority change from a profit-charged business machine to a socio-economic and environmentally driven sector. The manufacture of construction products and materials accounts for the largest amount of emissions throughout the construction process. Progress in the construction industry to implement Additive Manufacturing (AM) as an eco-innovative solution will be discussed in this chapter with reference to essential parameters and the practical changes needed to realize a holistic solution that will potentially bring many benefits to the construction industry. Additive manufacturing technology has the potential to help the construction industry to transition into a responsive and technically advanced sector, although, different grades of advanced printable feedstock needs to be formulated/developed to make this technology effective for the making of structural load bearing elements (Ghaffar, Corker, and Fan, 2018; Ghaffar and Mullett, 2018). When using AM to fabricate load bearing components, the material can be printed in the form of either mesh or truss-like systems, potentially eliminating the molding and de-molding process, e.g. for a new concrete construction project, approximately 60% of the total cost is spent for the formwork and labor (Paul et al., 2018).

With almost 40 years of development, a wide range of industries are using AM technology. AM, also known as 3D printing, is one of the emerging advanced technologies aiming to minimize the supply chain in the construction industry through autonomous production of building components directly from digital models; without human intervention and complicated formwork (Bos et al., 2016; Gosselin et al., 2016; Jeon et al., 2013; Lim et al., 2012; Valkenaers et al., 2014; Wolfs, Salet, and Hendriks, 2015). AM uses a combination of materials science, architecture and design, computation and robotics (Figure 9.1). Yet in some ways, it is not as futuristic as it sounds. The key for its successful implementation is not necessarily technological, rather practical; industry stakeholder collaboration to provide a complete approach for the effective application of the existing technology including materials science, architecture/design, computation and robotics along with the fundamentals of construction engineering. The areas for further developments for AM technology lie within scaling up of a holistic mobilized integrated system with the emphasis on materials formulations, which are tailored for the 3D printing system.

AM, automation and robotics are at the forefront of building innovation and have the potential to disrupt the way in which structures are built. The construction industry is not the first industry to seek the benefits of robotically controlled, freeform manufacture; the medical, automotive, and aerospace industries are examples where 3D-printed objects are offered as standard products/component parts (Bishop et al., 2017; Singh, Prakash, and Ramakrishna, 2019; Ye et al., 2019). Although, it is still early days for this progressive technology in a rapidly advancing digital age, AM, along with its many benefits, has the potential to be one of the main driving forces for the construction industry's advancement, particularly with how, together with off-site manufacture, it could change construction methodology paradigms at a large scale. The constant development and implementation of AM in other sectors is unfortunately in stark contrast with that in the construction industry. The construction industry is very slow to implement changes for anything requiring extensive and vigorous amounts of research and standardization to be changed and/or implemented. It is also hard to change an industry that has always functioned on one-off transactional relationships and requires a sizable amount of manual labor. The industry suffers from significant 'lock-in' relating to procurement, regulation, technology, and funding. Nevertheless, innovation and transformative technologies will play a key part in the changing the current industry paradigms, where taking advantage of the use of technologies from other industries will help to reinforce competitive advantages.

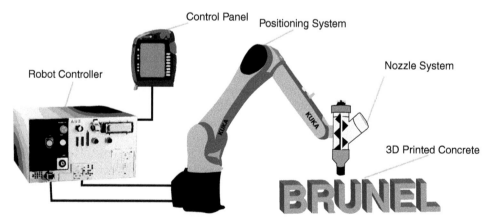

Figure 9.1   3D printing system mechanisms

In this chapter the fundamentals and latest processes along with printable raw materials for AM will be critically reviewed, identifying the future pathways for development. Other crucial dimensions of AM in construction will also be explored, such as the practical and commercial challenges for the successful implementation of this eco-innovative technology.

9.3 Additive manufacturing processes

From the extensive list of different conceptual AM processes that have arisen in the previous decades, many are steadily reaching the industrial and mainstream commercial market, sometimes driven by highly dynamic open-source communities.

Being already established commercially in many distinct sectors, AM is still taking its first steps within the construction industry and most commonly led by research programs' demonstration projects focused on piloting computer design-assisted extrusion of large prefabricated parts. Due to the multiple different terminologies currently being used to describe each specific AM process, some even sharing identical trademark names, recent authors have tried to provide a common classification and define standard terminology for AM within the scope of the construction industry. In this sense, one should highlight the works of Delgado Camacho et al. (Delgado Camacho et al., 2018), who have published a thorough overview based of the classification delivered by the American Society for Testing and Materials (ASTM) and made in collaboration with the International Organization for Standardization (ISO) (ISO/ASTM 52900:2015, 2015).

Amongst the many different AM technologies available up until the present day, the automation of material extrusion has been the most affordable and established one used along with the construction industry scope. Extrusion technologies manage to deliver continuous runs of layer-by-layer self-sustain printed bodies, in different desired shapes and properties, through synchronized material pumping systems and precise location deposition nozzles. Material extrusion systems for construction applications have been mostly addressing cementitious material formulations, and some of those have already accomplished the transition towards commercial businesses, with several printers currently available in the market.

Other AM techniques, such as Vat Photopolymerization 3D printing technology (which includes the common Stereolithography or SLA optical fabrication) and Material Jetting, are mainly associated with the usage of thermoset photopolymer resins. Whereas frequently chosen for other applications fields, these techniques are yet to be considered as feasible options for generating large-scale construction elements. The main reason for hindering its application relies on significant associated drawbacks such as the typical polymer degradation over time and the considerable amount of liquid light-activated resins that would be needed to manufacture bigger parts, along with the large size and costly equipment required. Altogether, these limitations turn these processes impracticable and expensive nowadays to be replicable for construction applications.

Metal AM for construction applications is still at its early stages of development and very much linked to non-commercial research and demonstration projects and associated with Powder Bed Fusion (PBF) and Directed Energy Deposition (DED) techniques. While both methods still pose budget, operational time, and maximum printing size restrictions, other solutions such as Wire and Arc Additive Manufacturing (WAAM), although having inferior dimensional accuracy and surface finishing capabilities, still offers quicker and more cost-effective operations, along with the absence of size limitations (Buchanan and Gardner, 2019).

Other potential advances in AM for the construction industry are quite noticeable, including a variety of solutions under rapid development such as the multi-material fabrication, the use of hybrid techniques pairing different AM processes and the association of both subtractive and additive capabilities. All these new AM trends are opening up a wide range of possibilities for both off-site element prefabrication and on-site manufacturing. Nonetheless, and with so many different AM technologies being engineered, the cementitious material extrusion automation is still the most affordable and common AM approach used today for construction applications. In this scope, 3D printers for construction can be roughly dived into two different groups: gantry systems or mechanical robotic arms.

9.3.1 Types of systems

9.3.1.1 Gantry

While able to secure the making of large individual components and structures or even complete buildings, the gantry system is currently by far the most established AM method used in the construction industry. Betabram, Concrete Printing, COBOD, 3D Concrete Printing (3DCP) and WinSun are just a few examples of companies working with gantry system types of equipment, most often using cementitious materials. Those are based on a movable setup having a three-axis Cartesian coordinate system with an extrusion nozzle and sometimes a building platform. Depending on the manufacturer, the moving and the static element may either be the platform or the nozzle system. The latter commonly has trowels attached on its side to level out and smooth the surface irregularities of the printed layers (Hwang and Khoshnevis, 2017). Typically addressing the continuous extrusion of thick multi-layer elements, the first gantry systems were envisaged following a rapid prototyping technology proposed by Pegna (1997) in the late 1990s, which was then further developed and patented (US Patent US7814937B2) as a single manufacturing process known as Contour Crafting (CC), by Behrokh Khoshnevis at the University of Southern California (USC). The overall goal was to be able to print entire houses following a single on-site and computer-assisted manufacturing process. Some of the most important advantages of CC in comparison with other layer-by-layer fabrication techniques are the faster fabrication speed and the broader choice of materials that can be used with (Khoshnevis, 2004). Although addressing on-site manufacturing capabilities, gantry systems main drawback relies on the need to use large size implementation areas for its printing equipment, which demands specific and careful transportation and labor intensive installations requirements. Moreover, this system also imposes significant design limitations, since it can mainly print perpendicular to the build surface, limiting curvature to the horizontal plane (Delgado Camacho et al., 2018 and Keating, 2015). Vertical curvatures can be achieved, but are limited and based on inter-layer stability, structural stability, and temporary works.

9.3.1.2 Robotic arms

By providing a six-axis, multi-directional motion that can be programmable for countless stepwise functions, mechanical robotic arms applied for construction applications are far more flexible and offer superior design freedom than the gantry counterparts. Another significant benefit of robotic arms is their greater on-site mobility, allowing working on narrow areas. Moreover, robotic arms can generally be effortlessly shipped and installed at different sites. The MIT Digital Construction Platform (DCP), CyBe, and C-Fab are just a few examples of AM technology based on robotic arms contenders operating in the construction market (Delgado Camacho et al., 2018). Recently, a Russian company called Apis Cor claimed to have on-site printed a full home using cementitious-based formulations and a proprietary robotic arm system (Apis Cor, 2019). Complementary, hybrid systems covering robotic

Table 9.1 13 house construction 3D printers

3D printer	Category	Type	Build size (m)	Country
BetAbram P1	Available	Gantry system	16 x 8.2 x 2.5	Slovenia
COBOD BOD2	Available	Gantry system	12 x 45 x 1.5	Denmark
Constructions-3D 3D Constructor	Available	Robotic arm	13 x 13 x 3.8	France
CyBe Construction CyBe RC 3Dp	Available	Robotic arm	2.75 x 2.75 x 2.75	Netherlands
ICON Vulcan II	Available	Gantry system	2.6 x 8.5 x ∞	United States
MudBots 3D Concrete Printer	Available	Gantry system	1.83 x 1.83 x 1.22	United States
Total Kustom StroyBot 6.2	Available	Gantry system	10 x 15 x 6	United States
WASP Crane WASP	Available	Delta system	Ø 6.3 x 3	Italy
Apis Cor	Project	Robotic arm	8.5 x 1.6 x 1.5	Russia
Batiprint3D 3D printer	Project	Robotic arm	Up to 7m high	France
S-Squared ARCS VVS NEPTUNE	Project	Gantry system	9.1 x 4.4 x ∞	United States
Contour Crafting	Service	Gantry system	-	United States
XtreeE	Service	Robotic arm	-	France

Source: (ANIWAA, 2019) (last updated on July 2019)

arms and movable frame gantry and crane assisted structures have been developed by WASP (WASP, 2019). Recently, a disruptive concept has been developed by the Spanish Institute for Advanced Architecture of Catalonia (IAAC), while creating a group of tiny and movable robotic printing devices called Minibuiders, that can simultaneously operate at location as a team of construction builders and generate building elements far bigger than the robots itself (IAAC, 2019).

A comprehensive list of the latest gantry and robotic arms systems available and specially engineered for construction applications is shown in Table 9.1 (ANIWAA, 2019).

9.4 Printable raw materials for construction

Material formulation and the science behind the interactions of different components certainly counts as a vital workforce that determines the successful implementation of AM technology in the construction industry. Many methods of 3D printing have been developed for several types of materials. The printable raw materials formulation are usually a combination of bulk materials (e.g. soil, sand, crushed stone, clay, recycled aggregates) mixed with a binder (e.g. Portland cement, fly ash, polymers) and workability additives, chemical agents, and viscosity modifying agents.

As a consequence of AM development and growth, a variety of 3D printing materials of different nature can be used, from filaments, wires, pastes to inks. The categorization provided by ISO/ASTM (ISO/ASTM 52900:2015, 2015) has a wide-ranging division of different AM material types, categorized from metallic, polymer, ceramics, and composites. To be aligned with the standard classification used in the construction field, authors like Delgado Camacho et al. (Delgado Camacho et al., 2018) have grouped AM materials into three different main sets: cementitious, polymeric, and metallic and alloys.

9.4.1 *Cementitious-based raw materials*

Cementitious-based materials are the most studied option for utilization in AM for the construction industry. This is because of their unique fresh and hardened characteristics and the extensive variety of possible raw materials to be generated (including rheology modifying agents) and admixtures available to tailor their performance (Ghaffar, Corker, and Fan, 2018). Printing a cementitious composite structure requires a mix formulation in which the setting time of cementitious paste, shape stability of the first few layers and interfacial bonding between the layers to be thoroughly controlled and investigated for optimization.

The layered buildability of printed objects, the lack of compaction, and the typical material compositions used in 3D printing of cementitious-based materials, makes this manufacturing technology different from others, especially concerning the structural properties of the manufactured product. The interfaces and/or interlayers that are established must create an intimate bonding and interlocking. The characteristics of adhesion are primarily determined by mechanical interaction, chemical bonding, and intermolecular and surface forces, therefore, an anisotropic dependency on 3D printing process parameters such as interlayer interval time, nozzle height, and surface moisture, should be expected (Wolfs, Bos, and Salet, 2019).

The fresh properties of printable cementitious-based raw materials and their subsequent hardened mechanical behavior have been investigated by many researchers as it is of great concern for current AM technologies (see: Al-Qutaifi, Nazari, and Bagheri, 2018; Feng et al., 2015; Kazemian et al., 2017; Le et al., 2012a; Ma, Li, and Wang, 2018; Marchment, Sanjayan, and Xia, 2019; Paul et al., 2018; Perrot, Rangeard, and Pierre, 2016; Wolfs, Bos, and Salet, 2018; Xia, Nematollahi, and Sanjayan, 2019). A comprehensive list of performance requirements and test methods for a cementitious-based raw material mixture has not yet been developed. Some previous research has focused on specific properties of printing mixtures such as shape stability, also known as shape retention and green strength (Le et al., 2012a; Perrot, Rangeard, and Pierre, 2016; Voigt et al., 2010). Creating acceptance standards and/or criteria for printable cementitious-based materials can be achieved only after various relevant studies with focus on different dimensions have been carried out culminating in a reasonable amount of data on the performance of different printable mix ratios for cementitious-based raw materials when implemented for construction projects.

The four key characteristics of the fresh concrete relevant to 3D printing identified by Le et al. (Le et al., 2012b) are: (1) Pumpability: the ease and reliability with which material is moved through the delivery system; (2) Printability: the ease and reliability of depositing material through a deposition device; (3) Buildability: the resistance of deposited wet material to deformation under load; and (4) Open time: the period where these properties are consistent within acceptable tolerances.

The rheology and flow-ability of the fresh cementitious-based material should enable the fluent extrusion and produce small filaments. Cementitious-based paste which has low fluidity and adhesion will most likely lead to the formation of voids between filaments and weak interlayer bonding, which subsequently reduces the overall mechanical properties of the printed structure (Ghaffar, Corker, and Fan, 2018; Le et al., 2012b).

The additive manufacturing technology in the Construction research group at Brunel University London has focused on two categories of printable formulations: A = inorganic binder (proprietary advanced and eco-sustainable geo-polymer) + fine aggregates; B = organic binder (thermal plastic polymers) + filler/fibers. All four groups of materials are mainly processed

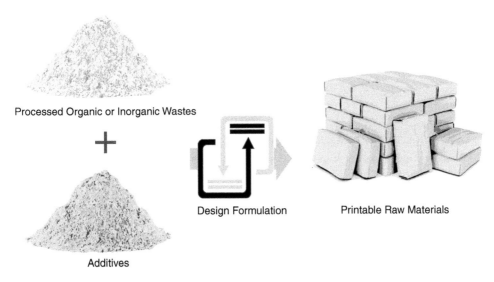

Processed Organic or Inorganic Wastes

+

Additives

Design Formulation

Printable Raw Materials

Figure 9.2   Printable feedstock (raw materials) formulation from waste using additives

from construction demolition waste. The notion and focused methodology is to design formu-lations that will lead to the realization of base printable feedstock being ready for printing i.e. printable feedstock (Figure 9.2).

The interlayer bonding quality and the cohesion mechanisms of cementitious-based mate-rials for AM have been the focus of some researchers (Al-Qutaifi, Nazari, and Bagheri, 2018; Marchment, Sanjayan, and Xia, 2019; Sanjayan et al., 2018; Zareiyan and Khoshnevis, 2017a, 2017b) and it is an important area which needs to be explored further.

Interlayer adhesion can be defined as the interaction of the materials in both the micro and macro scale. In the micro scale, there are many chemical reactions and in macro scale, the surface roughness plays a critical role for developing an intimate cohesion between layers. The interlayer bond strength of printed cementitious-based objects serves as a critical mechanical property that can create potential flaws between extruded layers and in turn, it can induce stress concentrations. The AM speed has to be carefully chosen to allow the layers to establish strong and cohesive properties that create enough shape stability to sustain their own weight and the weights of successive layers above them.

Zareiyan and Khoshnevis (2017b), analyzed the interlayer bond strength of a Contour Crafted structure, the results indicated a 16% improvement at interface by changing the fabri-cation parameter. They studied the influence of extrusion rate and layer thickness on interlayer adhesion, where it was revealed that thickness of the layer depends on the: (1) fresh properties of the concrete mixture, (2) the design of the nozzle, and (3) speed of the fabrication pro-cess. In another study, Zareiyan and Khoshnevis (2017a), tested the bond strength of speci-mens manufactured with interlocking at the interface. It was found that the bonding strength is increased by the interlocking layers (i.e. of 1.27 cm depth) due to the increased contact surface of layers. Panda et al. (2018), investigated the tensile bond strength of 3D-printed geo-polymer mortar in relation to the time interval between layers, nozzle speed, and stand-off distance. The results showed that the bond strength is a function of state of interface material between two nearby layers that can be affected by the rate of material strength devel-opment and 3D printing system parameters. Longer time intervals between the layers reduces the strength while lower printing speed and lower stand-off distance lead to better interlayer strength (Panda et al., 2018). Improvements in the interlayer adhesion can also be induced as a result of material formulation and mix designs, for example it was shown that low percentage calcium aluminate cement can lead to stronger interlayer adhesion (Van Zijl et al., 2016).

9.4.2 Non-cementitious raw materials

9.4.2.1 Polymeric

Due to their wide range of different compositional natures, polymers in their multiple forms (e.g. filaments, resins, and even powders) are still the most frequent type of materials to be used in AM, either at an industrial scale for automotive, aeronautics, and health care high-performance applications, or for personal DIY components and products.

Comprehensive reviews of the polymeric AM applications have been given recently by Delgado Camacho et al. (2018) and Tuan D. Ngo et al. (2018). Up to the present day, AM techniques operating with thermoplastics having low transition and melting points are amongst the most frequent methods for printing polymer and polymer composites. Materials such as environmentally friendly polylactic acid (PLA), up to the amorphous acrylonitrile-butadiene-styrene terpolymers (ABS) and polycarbonate (PC), are the three most common polymers being used in 3D printing, mostly processed by Fused Deposition Modeling (FDM) techniques. PLA-based composites having biocompatible or bioactive materials are becoming widely common to create porous scaffolds for medical applications and tissue engineering. Other materials like polystyrenes, polyamides and elastomers are among the most used by Selective Laser Sintering (SLS) (Ngo et al., 2018), while Stereolithography uses UV light photopolymerized resins in its process. Although offering superior levels of precision and accuracy, photo-solidification and resin printing techniques still lack further improvements, especially concerning the mechanical strength and temperature resistance of the final products, that are typically associated with the UV light intensity and with the homogenous light exposure of the resins during the printing operation.

Most of the times, advanced polymer composites having a fiber reinforcement (such as natural fibers or carbon fibers) are favored over polymer-only products which generally have a weaker mechanical and overall reduced performance. These composites typically allow generating more complex designs with improved structural integrity, superior strength, stiffness, wear toughness, and higher energy absorption capacity (Ngo et al., 2018). Also, the usage of nanocomposite materials (such as Nylon 6, nano SiO_2, silver (Ag) nanoparticles, carbon nanotubes, nanoclays, among others) is a fast-growing market for allowing generating lightweight products with many improved characteristics such as mechanical, thermal, and electrical properties, fire resistance and biocompatibility. One good example of composite AM for construction applications was provided by researchers at the Oak Ridge National Laboratory (ORNL) and Cincinnati Incorporated (CI) in the US, who have developed a particular gantry printer to (Big Area Additive Manufacturing – BAAM system) to print carbon-fiber-reinforced ABS compounds at large-scale geometries (Brenken et al., 2018). Although the use of polymer AM is much less common in the construction market, when compared to the one working with cementitious compositions, the potential offered by advanced composites can become an unmatchable solution to answer specific aesthetic or functional requirements. Probably the most solid example of the many possibilities that advanced polymer composites and AM technologies can deliver to the construction market segment has recently been given by Skanska. Using a Selective Laser Sintering machine, this construction company has effectively 3D printed several complex-shape designed Nylon PA12 parts, to be used as connectors between structural elements and as an alternative to conventional and costly cast steel nodes (CIOB, 2019).

9.4.2.2 Metallic and alloys

Metal AM has seen significant developments in the recent years and, contrary to the more traditional and reluctant construction industry, other sectors such as the aeronautics and automotive, defence, and medical industries are the ones that are most benefiting from additive technologies to produce highly innovative engineered computer-assisted parts. To understand better the many signs of progress being made in metal AM applications, a comprehensive review has been given by Tuan D. Ngo et al. (Ngo et al., 2018).

Powder Bed Fusion (PBF), Selective Laser Sintering (SLS), Selective Laser Melting (SLM), and Directed Energy Deposition (DED) are currently amongst the primary metallic 3D printing techniques used for industrial applications, all mainly based on the melting and sintering of metal powder or wire feedstock using a laser or electron beam. Once melted, the material is then converted into a solid part or object following a sequential layer-by-layer process (Ngo et al., 2018). However, the appropriate choice of metals and alloys for 3D printing is still restricted, and the search for a broader group of materials is one of the current drivers of metal AM ongoing research. Moreover, recent commercially driven projects developments, also covering the improvement of printing parameters and post-processing operations, are often including polymer-metal composites to generate more complex solutions (Ryder et al., 2018).

Additive metal manufacturing typically works with high performance and costly materials, like titanium alloys, which are mostly required by the aerospace, automotive, and medical industries. Other metal AM applications, either commercially or under research, are described by Tuan D. Ngo et al. (Ngo et al., 2018), including the usage of different steel alloys (covering from austenitic to stainless steels and tool steels), some aluminum alloys (most frequently $AlSi_{10}Mg$ and $AlSi_{12}$), a few nickel-based super alloys (e.g. Inconel alloys), a number of cobalt-based and magnesium alloys, up to pure substances such as tungsten, rhenium, gold, and copper. The same author also refers to recent research works including the usage of high-entropy alloys, magnetic alloys, bulk metallic glasses (amorphous metals), high-strength alloys, and functionally graded materials (FGM).

All the recent developments made in metal AM are becoming increasingly important, especially for allowing the reduction of time-consuming and costly post-processing machining, as opposed to other standard manufacturing techniques. The merits of metal AM also include increased design flexibility to create multiple part components or generate highly intricate and lightweight structures. However, the rise of metal AM solutions for the construction segment is still hindered by a natural market request for large full-scale components, that are still and most often far beyond the current small-scale parts production capacity. This is one of the main reasons why until now, metal AM in construction is still a small niche business and mainly associated with the making of small-scale metal parts designed for structural purposes. In this context, there are just a few companies using metal AM for building applications and those who are operating in this field primarily use metal Powder Bed Fusion (PBF) approaches for creating optimized lightweight parts and complicated geometry building elements under special high-priced requests. Even though metal AM can deliver comparable mechanical properties to the ones offered by conventional technologies, the price of the printed metallic parts is still regarded as a major commercial obstacle.

Recent reports, such as the one from Natasa Mrazovic (2018), have pointed out that metal AM will hardly be commercially competitive soon, mainly due to costs of the slow printing operation speeds and high initial costs. Nonetheless, new fast-paced research projects under development might change the current prospects. In this regard, recent examples are opening the pathway to future applications, such as the MX3D technology resulting from a research project started by the Joris Laarman Lab in collaboration with Acotech and supported by Autodesk. MX3D uses a gas-metal arc welding (GMAW) prototype machine connected to an industrial robot. Having Arup as lead engineering partner among a multidisciplinary consortium, MX3D has created a curved and complex 3D geometry metal footbridge by welding multiple small stainless steel segments (MX3D, 2019). This 12-meter long robot printed bridge is expected to be installed in Amsterdam in autumn 2019, after having been showcased at the Dutch Design Week, in October 2018. At the same time, a group led by Arup has developed a technique to design and 3D print complex-shaped steel node components for street lighting chandeliers in a project entitled the Grote Marktstraat project (The Hague, Netherlands). The AM solution has proven to be highly useful for allowing a significant reduction of time, material waste, and overall cost. According to the company, around 1,600 nodes were used in the project, of which 1,200 were of slightly different in terms of design to fit distinctive cable angles of the lighting structure (ARUP, 2019b). Thus, while still not very common, metal AM is regarded as one of the most promising manufacturing approaches for the construction industry of the future.

9.5 Practical and commercial challenges and opportunities for 3D printing construction

9.5.1 Additive manufacturing demonstration projects

Implementation of AM in the construction industry has resulted in different technical breakthroughs and improvements in output efficiency. There are some records claiming different levels of success in printing buildings (see Figure 9.4). For instance, WinSun, a Chinese construction company, claims to have 3D printed ten homes within just 24 hours in 2014. The company used a large extrusion-based printer to print the house components separately off-site before they were transported and assembled on site (3DPrint.com, 2019). In 2015, the company claims to have built a five-story apartment building with an area of about 1100 square meters, which is considered the tallest 3D-printed structure (Sevenson, 2019).

In 2016, WinSun 3D printed an office measuring 250 square meters in the United Arab Emirates (UAE). The building was printed using a factory-based gantry system. The model took 17 days to print and the company claimed that the labor cost was cut by more than 50%, in comparison to conventional buildings of similar size (Labonnote et al., 2016).

D-shape has developed a system by straining a binder on the material layer. D-shape is a gantry-based powder-bed 3D printer, which claims to have an effective printing method for large-scale objects (Cesaretti et al., 2014). The D-shape method uses a powder deposition process, which is similar to the inkjet powder printing process where a binder is used so that selective layers of printing materials are hardened. D-shape is mainly a dry process. The desired structure is erected in a single work session, starting from the bottom up.

Apis Cor, in 2016, printed a house in Russia using mobile 3D printing technology shown in Figure 9.3. The 3D printing of self-bearing walls, partitions, and building envelope was carried out in 24 hours, and totaled 38m² of printed building area (Apis Cor, 2019). This example attempts to demonstrate that an *in-situ* deposition approach can be more economical than off-site prefabrication of building blocks and the subsequent assembly on-site. Apart from the rotary gantry and its limitations concerning printing size, a disadvantage can also be the sensitivity of raw materials and the printing processes itself to site conditions, which can obstruct the on-site approach.

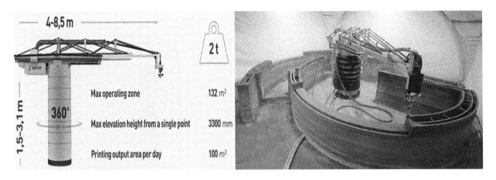

Figure 9.3   Apis Cor mobile 3D printer's specification and the 3D printer in action (Apis Cor, 2019)

The company claims that their second phase of development is to create equipment for use in high-rise construction and to enable the printing of foundations, floors, and roofing, which will allow entire buildings to be printed at once. They also plan to make their 3D printing equipment self-contained and autonomous, with the aim to explore the potential of constructing structures and buildings beyond this planet.

The Italian company World's Advanced Saving Project (WASP) is one of the pioneers in the field of 3D printing. Their latest project Gaia is an affordable home that's been printed out of mud using the modular printing system named the Crane Wasp (see Figure 9.4d). The Crane Wasp has been developed to print larger-scale structures. With a print diameter of about 6.6 meters in diameter by 3 meters in height, the Crane Wasp is easy to assemble and disassemble, and more than one can be set up in a modular way by adding more traverses and printer arms in order to print a larger structures or a whole village of structures, if need be (Mok, 2019). For this project, the printing materials used comprised of 10% silt, 7.5% clay, and 7.5% sand taken from the same site, 40% straw chopped rice, 25% rice husk, and 10% hydraulic lime. The mixture has been mixed through the use of a miller, which makes the mixture homogeneous and workable (Mok, 2019).

In 2017, Siam Cement Group (SCG) 3D printed a tall (i.e. 3 meter) cave structure called the 'Y-Box Pavilion, 21st-century Cave' in Thailand, where they used the 4-meter-high BIGDELTA WASP printer. They used a special type of cement with high compressive strength properties, with the inclusion of a binding agent and various fibers to support the tall concrete structure (Alec, 2019).

In 2017, XtreeE, a French company, used their *in-situ* robotic arm to 3D print a 4-meter high post which acts as supports for the playground roof of a school in Aix-en-Provence, France (XtreeE, 2019). The printing time is claimed to have been 15 hours and 30 minutes. The post is made of two parts: its envelope, used as lost formwork, and cast concrete inside the envelope.

In 2017, Construction Building On Demand (COBOD) was founded by '3D Printhuset' who made the first 3D-printed building in Europe, 'The Building On Demand' (The BOD) (Scott, 2019a). The building was constructed in Copenhagen and acts as an office space for the harbor industry surrounding it. Due to the speed, material, and low labor requirements, the cost of a 3D-printed building is relatively low compared to standard constructed buildings.

In 2018, engineering firm ARUP and architecture studio 'CLS Architetti' collaborated on 3D printing a 100 square meter house, which was exhibited in Milan, Italy. A small mobile

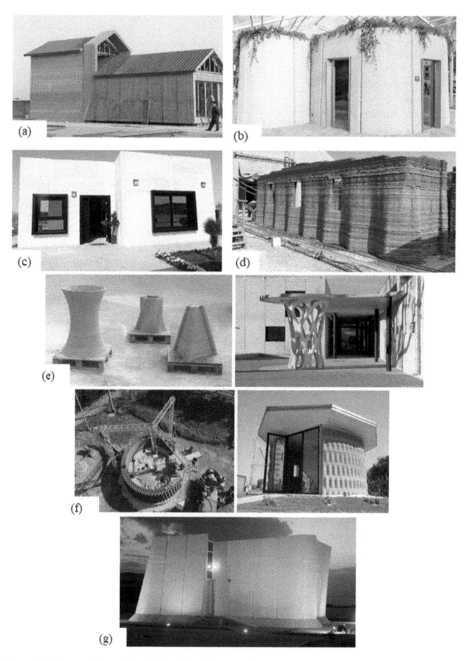

Figure 9.4 Some of the various AM project demonstrations around the world (a) WinSun in China; (b), (c), and (g) CyBe Construction B.V. in Italy, Saudi Arabia, and UAE respectively, (d) Marine Corps Systems Command in USA; (e) XtreeE project demonstrations in France; (f) WASP in Italy

robotic arm printed the house's curved walls, which was capable to print walls in 60 to 90 minutes. CyBe Construction from the Netherlands designed this printer. It is claimed that the walls were printed using a tailored mix of recycled concrete, from demolition waste, which had the capability to cure in five minutes. According to the architects at CLS, the cost of 3D printing a single square meter is currently €1,000, which is half the average price of traditional construction. Locatelli predicts that as the method becomes more advanced, this figure will drop to around €200 to €300 per square meter (Jordahn, 2019). CyBe continued their progress with another demonstration in 2018. They 3D printed a concrete house in Riyadh, Saudi Arabia (Cameron, 2019), which is intended to prove the feasibility of 3D printing homes and encourage more private industries to invest in the technology. CyBe also 3D printed the world's first drone laboratory in UAE. The 168 square meter building was 3D printed on-site in just three weeks with the CyBe RC 3DP machine. They 3D printed the walls and parapets for the R&Drone Laboratory, which consists of 27 separate printed elements (Saunders, 2019).

In 2018, the Marine Corps Systems Command (MCSC) of the US Armed Forces, 3D printed a 46 square meter concrete barracks in 40 hours. This was an on-site operation at the US Army Engineering Research and Development Center in Champaign, Illinois (Aouf, 2019). It is evident that mobile manufacture of various parts and components is on the horizon. The main goal is to send out army units with just the right amount of equipment to establish a mobile unit for on-demand 3D printing.

In 2018, ICON was the first in America to secure a building permit for a 3D-printed home (ICON, 2019). ICON developed the 3D printing robotics printer named Vulcan II, which they claim can 3D print entire communities with up to 186 square meter homes. The 33 square meters were printed in approximately 48 hours. The cost of the project is stated to be below $10,000 which is much cheaper than the average cost for a home of similar size and quality (Reggev, 2019).

The world's first 3D-printed bicycle bridge was made from reinforced and pre-stressed concrete in a collaborative project between Eindhoven University of Technology (TU/e) and BAM Infra in Gemert, Netherlands (Scott, 2019b). The six pieces of 3D-printed prefabricated blocks were reinforced with steel cable and assembled together to erect the bridge on-site. At about 800 layers of concrete, the nearly 8-meter-long bridge is placed above a channel.

In 2018, additive steel welding robotic arms completed 3D printing the structure of a steel pedestrian bridge devised by technology startup MX3D, which is set to span a canal in Amsterdam. With ARUP involved as lead structural engineer, MX3D created intelligent software that transforms welding machines into 3D printing robots to produce a fully functional steel bridge. This bridge weighs nearly 5 tonnes and is 12.5 meters long (ARUP, 2019a).

9.5.2 Benefits of additive manufacturing in construction

9.5.2.1 Market potential

With global challenges such as overpopulation and growing displacement due to environmental factors, it is clear that traditional construction will have difficulty keeping up with the demand for affordable housing. It is also imperative to build quickly, cheaply, and with high quality. In the UK alone it is estimated that the rate of home building needs to increase by 50% to keep up with current demand (Department for Communities and Local Government, 2018). The main drawbacks of traditional construction methods are the over-reliance on a skilled labor force, high material cost, long construction times, the risk of human error, cost of transport, construction waste, and the overuse of natural resources.

As discussed in Section 9.1, there is a growing anticipation that AM will have a significant role to play in the future development of an efficient, digitized construction industry, overcoming many of the current industry limitations. This anticipation has grown in a relatively short period of time from a few academic thought-leaders and optimists to a more widespread expectation amongst major industry stakeholders. This is witnessed by the increased collaboration and investment seen between contractors, research institutions, and smaller-scale AM companies. Contractors and suppliers such as Royal BAM Group, Besix, Skanska, and Peri Group have all made significant investments in AM technology. The Saudi Arabian contracting firm Elite for Construction and Development Company has recently invested in the world's largest 3D Printer (Global Construction Review, 2019).

However, given the current proof-of-concept status of AM, as demonstrated by the projects described in Section 9.4.1, the true potential of AM in construction has yet to be fully established or accurately predicted. Claims of the potential for widespread time and cost savings (Hager, Golonka, and Putanowicz, 2016; Sevenson, 2019) are often made by those within the niche industry however these predictions are not always based on a technically sound review of the opportunities offered and challenges faced by the technology in a real construction environment.

Based on the proof-of-concepts and a general rising of expectations, governments have also identified the opportunities offered by AM in construction and have, in some instances, brought in regulations and targets to encourage use and adoption of the technology. The Dubai Government for example has stipulated that 25% of buildings must be 3D printed by 2030 (Dubai Future Foundation, 2019a).

9.5.2.2 Potential benefits

When looking at the potential market opportunity for AM it is therefore important to take a view based on an exponential extrapolation of the technology as it currently stands, alongside some balanced assumptions regarding future development areas, commercial viability (for example by reviewing the track record of other technologies on the Gartner Hype Cycle (Gartner Hype Cycle, 2019) with an experienced understanding of how design and construction is actually implemented.

AM and its successful implementation as an eco-innovative practice in the construction industry may offer various benefits which are summarized in Figure 9.5 and include; reduced material waste (up to 30%) (Perkins and Skitmore, 2015), lower energy use, *in-situ* production, complex architectural and design freedom with lesser resource demands and related CO_2 emissions over the entire product life cycle (Gebler, Schoot Uiterkamp, and Visser, 2014). It also offers changes in labor structures, including gender equality, and safer working environments and both encourages and necessitates a shift towards more digital and localized supply chains.

Some of these opportunities are discussed further below:

- *Reduced wastage.* There is a significant amount of material waste generated in the process of traditional *in-situ* concrete. If AM can truly offer elimination, or significant reduction, in formwork then there are obvious benefits to be gained environmentally. Concrete over-supply, over-pouring, and construction errors all contribute to wastage (Sanjayan and Nematollahi, 2019). AM offers concrete delivered precisely only where it is required, and using only the specific amount of material required. It should be noted however that an assumption of a net environmental benefit only holds true however if the structure

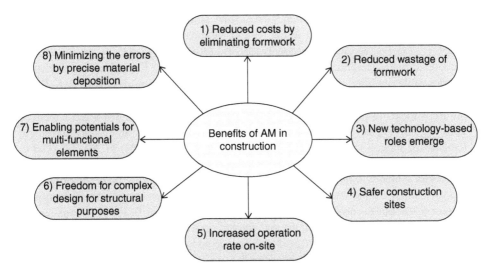

Figure 9.5 The opportunities presented by AM for the construction industry

itself is of comparable efficiency. It has yet to be seen if AM structures can rival traditional forms of construction for efficiency and practicality. Equally though, for certain applications AM could potentially offer a more efficient structure than traditional construction techniques can provide (based on similarly traditional design techniques).

- *Increased operation rate on site*. One of the biggest attractions of AM is the potential to build faster. This benefit is obvious for certain applications and not so obvious for others. Repetitive, low-rise construction that requires significant manual labor (e.g. single-story blockwork) can be implemented much faster at scale with the smart adoption of AM technology. This assumption is very dependent on the integration of AM into the design, the scale of the development and the availability of raw materials. Of course, it also depends on the benchmark. Other technologies are evolving that will rival AM in terms of speed (and other benefits) for certain forms of structure; however, the benefits for each type of technology are not yet clear. A study of robotic methods of construction (García de Soto et al., 2018) concluded that it only offered advantages for more complex building forms where labor and formwork costs were higher, however this is not universal. For example, systems such as Fast Brick Robotics (FBR, 2019) have the potential to be able to build a blockwork house much faster than traditional techniques.
- *Reduction of labor*. Along with the increased speed of construction comes an associated reduction in personnel on site. It is estimated that labor costs currently comprise between 15–50% of total construction costs (FBR, 2019), however labor costs vary significantly throughout the world. Over time, variances in labor costs will reduce due to international market forces. Even in places where labor is relatively cheap, the rise of (much-needed) regulation regarding worker welfare, will also serve to increase costs to employers whilst the issues of quality and consistency of production will remain. AM potentially offers to replace gangs of laborers with just one or two, technically proficient, supervisors and finishers. This argument for AM and robotics in general will only become more persuasive with time.
- *Safer construction sites*. Hand-in-hand with the expected reduction of labor, are the benefits to health and safety. With less labor on site, there is a proportional drop in risk

exposure and therefore less potential for accidents. However, besides a reduction in head-count, there will also be a shift in the type of roles carried out by staff that remain on site. Over-exertion historically accounts for 24% of workplace injury in construction in the United States (Everett, 2012) and UK Health and Safety Executive statistics show injuries while handling, lifting, or carrying account for 21% of injuries in 2017/18 (Dai et al., 2018). AM has the benefit of reducing manual handling hazards due to the elimination of formwork erection, and manual materials placement. In addition, AM offers a reduced need for working at height which, according to the UK Health and Safety Executive remains the biggest cause of construction workplace deaths (48%) in 2017/18 (Dai et al., 2018). The introduction of robotic techniques in construction does however present a new risk which will need to be effectively managed; namely the presence of humans and robots in the same working environment. There have been several reported cases of worker deaths due to robots (Dearden, 2015; *The Guardian*, 2019). The risk management approach traditionally adopted in the manufacturing industry, following the requirements of ISO 102,018 (ISO 10218-1:2018, 2019), is to keep humans and robots separate. There is however, an increasing need to consider the use of so-called collaborative robots that work side-by-side with humans. Requirements for such industrial robotic systems are now available such as ISO/TS 15,066 (ISO/TS 15066:2016, 2019), and on construction sites where there is inevitably an overlap in trades and processes requiring humans and robots to work side-by-side, similar guidance will need to be developed and integrated into site-specific risk assessments. Increasingly advanced technologies will also assist in this integration, offering smart robotic systems that can recognize humans and modify activities based on proximity and other risk factors.

- *Minimizing errors and improving consistency.* By replacing multiple manual tasks with those controlled and implemented by precision technology, it follows that human error is potentially reduced and the outcomes become more predictable and consistent. With suitable control, it is expected that AM will be able to produce the same, identical structure repeatedly with reliability, compared to traditional construction where it is inevitable that human error or variability will result in an inconsistent final deliverable. This follows the same logic for the introduction of robotics into the manufacturing industry; resulting in gains in both time and quality. At the present time, the AM process is currently still subject to a multitude of parameters; material constituents, environmental, technological, software, and human control which all influence the quality and repeatability of the final product (Ahmed et al., 2016). Of course, there are other factors to consider in assessing the impact of AM and robotics in construction quality. Many issues with construction quality do not reside wholly with operators or site staff. Aspects including suitable quality processes, staff training, supervision, supply chain management, relevant standards and specifications, design shortcomings, commercial, and time pressure all play a role in overall quality of construction. Whilst AM has potential to address human error aspects, the use of AM in construction is still in its infancy and standards and specifications for the application of AM, which would drive quality and consistency, do not yet exist. The successful application of AM is still dependent on its use within a functioning quality management system. Of course, as an automated system, the concept of 'rubbish in, rubbish out' still applies, so if the human operator makes an error which is not identified by quality procedures, then a robotic system has the potential to execute the error with even greater speed and efficiency.

- *Plant reduction and rationalization.* AM offers simplification of the construction process, and with it, reduction and rationalization of the plant required. Construction sites currently include a multitude of plant ranging from mobile plant such as crawler/mobile cranes, diggers, and trucks to fixed plant including tower cranes, concrete pumps, hoists, and jump forms systems. The potential elimination of formwork and blockwork means that less material has to be physically delivered and moved, either by crane or vehicle and the adoption of AM also requires less multiple handling; in other words it can be delivered directly to the point of application. The equipment components are also generic, and often scalable, meaning that it can be used repeatedly in the same fashion across sites and from site-to-site. For example, the COBOD gantry system is currently scalable to up to 45 meters in length, meaning that multiple, adjacent buildings could be constructed in parallel without repeated setup, and the material delivered to one single point for distribution and application where needed. As materials technology develops and our paradigms for providing reinforcement also change, then it is possible we could see many construction sites with no cranes at all, especially as heavy-lifting drones come of age.

- *Design freedom and flexibility.* An often lauded benefit of AM is its ability to produce free-flowing, unhindered geometrical forms which can give greater architectural freedom and expression (Gaudillière et al., 2019). Examining these claims more closely, and in the context of current technical constraints we can see that these benefits, whilst not necessarily untrue in terms of potential, are certainly tempered. As discussed in Section 9.1, we do see examples in manufacturing and biomechanical industries of AM being used to manufacture complex components that would be impossible or very difficult, to manufacture using subtractive methods or from casting in molds. The ability for AM to produce something geometrically complex of significant size is limited by both physical scale and material science (Ahmed et al., 2016). For example, at present, cementitious AM can produce in-plan curvature relatively easily. However, anything more than subtle non-verticality is still a significant challenge for many cementitious AM systems, in terms of both layer-to-layer stability and overall structural stability. Nevertheless, interesting geometrical forms can be created that would otherwise be impossible to create using traditional techniques. This can offer a structural advantage as well as an aesthetic one (Holt et al., 2019) as rectilinear forms are known to be structurally weaker than curvilinear forms.

- *Bespoke design.* In many respects AM is the antithesis of modularization. As the industry moves closer towards off-site fabrication and Design for Manufacture and Assembly (DfMA) there is a natural trend towards repetition, rationalization, simplification, and a 'copy-paste' mentality. In parallel with increased digitization, there is a natural place for this approach; however, it can lead to uninspiring architecture and ultimately materially inefficient construction. AM offers, in tandem with geometrical freedom, an opportunity for unique, efficient, bespoke design. Coupled with an ability to prepare AM designs that are, at a fundamental level, entirely integrated with made-to fit components and geometrical arrangements, then the attraction becomes clear. With a fragmented supply chain and an immature AM industry, the industry has yet to understand how the available spectrum of solutions can be blended to achieve the benefits of modularization with AM freedom of expression.

- *Integrated design.* Whilst AM offers different benefits and options compared to modular construction, one thing that the two approaches have in common are the benefits of an integrated design solution. Traditional construction requires the coordination of design from different disciplines, and this often causes great difficulty on projects mainly because the systems are designed separately and function in isolation (even when designed by a so-called multidisciplinary consultant). For example, beams have to be coordinated with air-conditioning ducts, floors have to consider the weights of mechanical plant, and core walls need to provide sufficient risers for services). Some attempts have been made to improve the integration of designed systems (e.g. HVAC systems integrated into columns and beams). However, in general, due to the design not being prepared holistically and the fact that the structure is traditionally built by a contractor after the design is completed, using a variety of sub-contractors that operate across different trades (that occupy areas of a construction site at different times), the opportunities for fully integrated solution are limited.

• As with modular solutions, an AM system, with vertical supply chain integration allows all stakeholders to engage in the design process upfront and ensure the design provides the best holistic solution. An AM solution offers an opportunity to prepare an integrated design that can provide for all the building needs in harmony. Based on a full Building Information Model (BIM) all the requirements and components can be 'designed-in' and the interfaces integrated into the AM solution. For example, HVAC ducts can be integrated into the natural curves of wall cavities, lighting, and electrical sockets can be positioned and constructed with upfront precision. Of course, an integrated AM solution also offers the ability for structure and architecture to become intimately intertwined, rather than just uncomfortable counterparts. Structure can become softer, more organic, and with a range of textures, allowing it to be expressed more readily.

9.5.2.3 Cost comparison

One of the fundamental drivers for the adoption of AM in the construction industry is cost. This aspect is multifaceted, complex, and regionally sensitive, but the above factors all play a part in the overall commercial viability and associated risk management.

The opportunities are best understood by way of an example. Table 9.2 presents some commercial data from a proposed confidential AM project in the Middle East. The project was a small, single-story residential building of 50 square meters. A large, competent contractor priced the project using typical market rates for both traditional construction methods and for a 3D-printed alternative design. An experienced 3D printing company provided the prices for the 3D-printed components.

Table 9.2 Cost comparison of AM versus traditional construction

Description of work	Estimated construction cost [US$]	
	3D printed	Traditional
Preliminaries	1,044,811	1,970,049
Contractor works	102,451	112,777
3D printing works	41,359	0
Other sub-contractors	406,750	397,000
Design costs	240,000	240,000
Other costs	200,830	271,586
Total	**2,036,201**	**2,991,410**
Difference		995,209

The numbers presented in Table 9.2 are quite startling, indicating the potential for one-third (33%) reduction in overall construction cost based on the application of AM on a single small-scale project. A more detailed breakdown of the figures indicates the following:

- A significant reduction in preliminaries including plant (refers to heavy machinery and equipment used during construction works), site supervision, and site establishment. Plant costs are reduced by 42%, site supervision reduced by 68% and site establishment by 20%.
- Contractor works show a reduction due to savings in formwork, labor, and materials. Labor costs are reduced by 34%, and formwork reduced by 72%.

These commercial figures appear to present a compelling case for AM in construction; however, there are many factors that make the cost comparison sensitive to regional market conditions and AM supply and demand.

The true viability can only be measured by scalability. The above example has a disproportionately high cost associated with preliminaries, whereas for a larger project this will be much lower proportionally thereby reducing the relative impact of AM. A larger project comparison study is required for an assessment. Preliminary costs are also potentially higher for large contractors who have more overheads than smaller, nimbler builders do.

Design costs are indicated as being equal for the two schemes; however, this is unlikely due to the lack of design standards, unfamiliar design methodologies, and difficulties in gaining approvals from the relevant local authorities. Initially design costs are expected to be approximately double that of traditional design, however this will reduce with time.

The cost of the 3D printing itself is highly sensitive to AM supply and demand and utilization. A lease model is assumed including the 3D printer, operatives, and consumables. As with the lease of other plant, the cost-rate is highly dependent on the capital cost of the printer, the assumed depreciation and utilization. These factors will change as the AM market matures; intuitively capital costs reduce and utilizations increase resulting in a reduced unit cost, however ultimately supply and demand will drive the pricing structures.

Nevertheless, the small case study demonstrates that adoption of developing technology can be cost-effective for the right application. Further work is obviously needed to understand the drivers and subtleties of pricing and delivering AM projects and to understand the best types of projects for the differing AM systems on the market.

9.5.2.4 Market opportunities

The benefits of AM in construction highlighted above are compelling; however, they are subject to regional variation and are limited in terms of technical complexity (discussed in Section 9.4.3).

It is apparent that many of the benefits and limitations align with the use of AM in technically simple, repetitive applications where architectural efficiency and deviation from traditional forms of construction may offer benefits in terms of cost, materials, and time.

An attractive solution is the use of AM in remote locations, often where low-cost or emergency housing is required and access for construction plant is difficult. As indicated in the cost comparison study above, a significant reduction in labor, materials, formwork, and plant would have tangible benefits in remote locations or where such resources are scarce and supply chains for modular or prefabricated solutions are not available. In contrast, AM, with locally sourced bulk materials and a supply of water, could in principle provide a means to construct low-cost housing or emergency shelter almost anywhere in the world. The only necessary imported materials would be a suitable binder, necessary additives and the printer itself.

This type of application is still subject to the development of an AM system that can cope with the variation in locally sourced materials and challenging environmental factors. It is however an area that has already been recognized and a low-cost housing solution is already being developed by ICON (2019).

Other related applications include the potential to use AM for large-scale one-to-two-story residential developments. The rise of modularization and other automated construction methods will make competing in such markets difficult; however, the ability for AM to provide freedom of architectural form may make it suitable for specific applications or development types.

Further applications of AM, for example in high-rise, cannot be discounted and also have significant potential but are likely to be dependent on further developments in materials technology to make solutions structurally efficient and economically viable.

9.5.3 Challenges in implementing additive manufacturing in construction

Having considered the potential benefits and opportunities presented by AM in the construction industry, for a balanced view it is necessary to also look at the challenges facing the adoption of the technology. The following sections look at both technical and practical challenges to the implementation of AM in construction.

9.5.3.1 Standards and certification

Certifying new building products and characterizing them is, and probably will remain, one of the biggest challenges for AM in the construction industry (Ghaffar, Corker, and Fan, 2018). The construction industry primarily depends on established standards for the products to ensure the consistency and quality, therefore, the lack of those related to AM will inhibit the adoption of this technology in the coming years.

In the absence of such standards, anything more than a 'proof-of-concept' (i.e. fully functional and compliant with building/municipal standards) requires demonstration using first principles and compliance testing. This is often prohibitively expensive and time-consuming for stakeholders wishing to apply AM to real projects. The industry and various standards and quality committees and agencies are however responding. For example, the American Concrete Institute has recently set up ACI Committee 564 to develop and publish design and specification guidance for concrete AM. Until such time as material standards and codified design approaches become available then AM will only take confident strides forward if the stakeholders take time to understand what is needed to bring a new technology to the market and give consultants, contractors, and AM companies the commercial space they need to deliver.

9.5.3.2 Skills development

Whilst there may be some concern that greater automation in the construction industry may result in increased unemployment there is also potential for many new jobs related to the implementation and operation of robotics in the industry. A discussion on the impact of automation on employment is outside the scope of this chapter, however there are many reports that discuss the key issues and predict the possible outcomes across different industries (McKinsey & Company et al., 2017; Reis and Durkin, 2018).

In basic terms, there will always be better economies resulting from advancement and utilization of technology and the use of AM in construction, which has demonstrated credible potential similar to other industries. Such changes will not happen overnight and there will be a period of transition where the construction workforce will need to be trained to learn new skills for the jobs that will emerge as a result of automation. The World Economic Forum Report 'The Future of Jobs' (WEF, 2018), states that 75 million jobs may be displaced worldwide by a shift in the division of labor between humans and machines, although 133 million new roles may emerge that are more compatible with the new division of labor between humans, machines, and algorithms (WEF, 2018). Investments that are more intensive will be dedicated by the industry to develop the skills necessary to implement roles such as robotics engineers, data analysts, software and applications developers and user-experience interaction designers. Governments and industry will need to work together to ensure that workers are given opportunities for upskilling and do not, as has happened in other disrupted industries such as manufacturing and retail, end up either unemployed or facing employment in less skilled (but less automated) jobs or tasks.

9.5.3.3 Lack of drivers for change in the construction industry

Despite encouraging proof-of-concept projects and an apparent acceleration in technological development, the belief that AM will transform the industry is not widespread. A survey summarized in the World Economic Forum Report 'Shaping the Future of Construction' (2019) indicates that leaders in the industry are still somewhat unconvinced. In this report, it is shown that many key figures in the industry still do not consider 3D printing of components will offer a significant impact on the industry, and even fewer consider Contour Crafting of whole buildings to provide a significant benefit.

This perspective is indicative of an industry that is highly risk-averse and resistant to change. It is notable that the technologies that are touted as having the most likely impact are those that are essentially already well established (i.e. BIM and prefabrication) and are therefore those that present less risk. Whilst such views are unlikely to prevent the eventual penetration of AM into the industry, they may well serve to stifle investment, limit opportunities for collaboration, and reduce the speed of adoption. However, this view may also be symptomatic of an industry that does not have an appetite for transformation. According to the KPMG 2017 Global Construction Survey whilst 60% of contractors admitted that in the last three years, adverse project performance had significantly impacted their business, more than 80% had confidence in delivering projects on time and on budget. Similarly, the UK Construction Industry Performance Report 2018 shows that whilst cost and time predictability of construction projects is at only 66% and 63% respectively, overall client satisfaction with the finished product is at 87%. It therefore appears that despite the time and cost risks, clients are satisfied with the product they are provided with. The impetus to make transformational change is therefore limited by a lack of market ambition, low expectations, and a false sense of security with the status quo.

9.5.3.4 Material rheology

Probably the greatest challenge for AM in the construction industry is scalability and the largest single factor in this respect is the material itself. The primary issue is the ability of the material to tread the fine line between extrudability (i.e. the ability to push the material through the nozzle) and strength and stiffness development; namely that sufficient to provide support to the next layer (shape retention). As discussed in Section 9.3.1 of this chapter, this is not a simple equation, but comprises a plethora of parameters that interact in a way that currently only a few experts in the industry truly understand. Nozzle design, layer width, layer thickness, print speed, length of continuous print, ambient temperature, water temperature, and mix design all play their part in this black art.

There are considerable rheological properties requirements for 3D printable cementitious materials, which are significantly affected by many factors, including sand gradation and packing fraction (Le et al., 2012a; Weng et al., 2018). Rheological properties are usually characterized by static/dynamic yield stress and plastic viscosity in Bingham Plastic model. Both pumpability and buildability are closely related to the rheology performance of materials, namely static/dynamic yield stress and plastic viscosity. All the rheological properties are attributed to the inter-particle force (Flatt and Bowen, 2006). Typically, higher static/dynamic yield stress and plastic viscosity would enhance buildability and diminish pumpability (Perrot, Rangeard, and Pierre, 2016). Hence, balancing the buildability and pumpability is essential in material rheology design for 3D printing.

The industry is a long way from providing standard mixes or guidance on these aspects and much of the current knowledge is locked behind patents, non-disclosure agreements, and intellectual property clauses in contracts.

9.5.3.5 Reinforcement

Reinforced concrete is a persistent, tried and tested construction paradigm, which has been gradually optimized over the last one-and-a-half centuries. Much of today's construction supply chain, plant, logistics, and labor force are based on the reinforced concrete paradigm. Yet the construction industry has failed to develop any viable alternatives and perhaps, because of deep-rooted industrial lock-in, has lacked either a desire or need to do so. However, the rise of construction automation suddenly brings the shortcomings of reinforced concrete into focus. Reinforced concrete requires the use of discrete reinforcement, arranged in accordance with specific design requirements, and cast within a molded concrete continuum. The bond between the hardened concrete matrix and the discrete reinforcement provides the composite behavior on which modern reinforced concrete design theory is based. Although there are some automated systems to generate pre-arranged reinforcement arrangements (e.g. piles, columns, or even slabs) these systems are limited. Future development in robotics and automation is unlikely to obviate the significant manual labor required to form and place discrete reinforcement in most normal scenarios.

Meanwhile, as we explore the practicalities of AM applied to structural concrete, it becomes apparent that the presence of discrete reinforcement presents substantial difficulties including clashes between the robotic arm/gantry or printer head nozzle and the pre-placed reinforcement, ensuring composite behavior between the reinforcement and the 3D-printed concrete. Not to mention the inability of discrete reinforcement to achieve the architectural forms AM has the potential to unlock. The current paradigm of composite, discrete reinforcement therefore lacks the efficiency, freedom, and flexibility of homogeneous materials typically found at a smaller scale necessary for AM.

There have been significant developments in the field of fiber-reinforced concrete, with some types of Ultra-High-Performance Fiber-Reinforced Concrete (UHPFRC) achieving significantly enhanced compressive and flexural strengths compared to unreinforced Ordinary Portland Cement (OPC) concretes. However, these specialist concretes still require very specific methods of preparation, placement, and curing and have been limited to special applications. Whilst it is quite common to add fibers to AM concrete mixes, this is typically for crack control, and materials technology has yet to provide a fiber-reinforced solution that can provide significant strength improvements that can be delivered via AM (Panda, Chandra Paul, and Jen Tan, 2017).

Whilst accelerating materials technology and nanotechnology in particular, may help drive the industry forward to a solution, existing AM designs have to consider how reinforcement can be incorporated to meet the necessary strength, stability, and serviceability performance requirements under the imposed temporary, permanent, and transient actions. Current proof-of-concept examples approach this in one of two ways:

1. They ignore it, i.e. the buildings are not reinforced. Whilst this may prove to be a reasonable, practical approach for the simplest forms of construction in some geographies, or for uninhabited demonstration projects, we do not yet have a particular understanding of thin-walled, layered construction, in the same way we do blockwork or brickwork, which has been used for hundreds, or even thousands, of years. Some research has been carried out in this area (Suiker, 2018), although, much more is required to develop industry guidance. We are therefore some time away from developing a rule-based, codified approach to the use of this form of construction to provide strength and stability under load.
2. Discrete reinforcement is included. The 3D-printed structure typically provides permanent formwork within which a traditional reinforced concrete structure can be created. This method has the benefit of potentially complying with current building standards but impairs the AM process (as discussed above) as it has to be coordinated with the placement of discrete reinforcement and the pouring of *in-situ* concrete. Additional formwork and temporary support must be provided in key areas such as lintels and ring-beams. In addition, unless the AM elements can be shown to be acting compositely with the *in-situ* structure then the outcome is inevitably less efficient than a traditionally formed equivalent.

Some systems have attempted to integrate discrete reinforcement into the 3D printing process (e.g. The Office of the Future, Dubai by WinSun), however these systems require extensive Finite Element modelling to consider the fundamental behaviors and extensive testing to predict the performance of the composite structure. It is debatable as to whether such systems will be able to satisfy building control engineers across the globe until internationally recognized standards and specifications are available.

9.5.3.6 *Stability in the temporary condition*

AM offers the ability to place layers of concrete in unique and interesting geometries, however if the vertical inclination exceeds the ability of one layer to overhang the next then the structure will become unstable whilst still gaining strength and the layers can 'collapse' during or after printing. Some systems include plastic or metal 'ties' between the inner and outer walls to mitigate stability issues of the individual walls, however this still does not prevent overall instability of the system where vertical inclinations are required. The degree to which vertical inclination can be accommodated therefore depends on the extruded properties of the layer being printed and the stiffened properties of the previous layers (all of which are tied to the rheology and mix parameters discussed above). Figure 9.6 shows the collapse of a 3D-printed section during printing under the weight of subsequent layers. In addition, obviously, an assessment of wall stability is required to ensure the hardened wall section is stable during construction and temporary works (e.g. structures) may be necessary to provide support

Figure 9.6 Collapse of 3D-printing under the weight of subsequent layers

Figure 9.7 Temporary shoring to provide stability

until any permanent structure proving support is completed – something that will detract from 3D printing's attractiveness. Figure 9.7 shows the implementation of an inclined 3D-printed wall and the temporary shoring required to ensure stability during construction.

Some systems, such as D-shape (D-shape, 2019) provide temporary stability through the use of a binder-jet applied to a bed of conglomerate material, thereby providing temporary stability to the permanent structure during the addition of each layer. These systems are able to provide greater freedom of architectural form, however have the same challenges regarding the introduction of reinforcement, and currently have their own specific issues relating to definition of form, finishes, and construction logistics.

9.5.3.7 Interfaces, joints, and connections

AM is conveniently imagined as a way to deliver a building in its entirety; a structure that grows before our eyes, eventually providing a finished article complete with roof, windows, doors, insulation, and finishes. This is obviously a simplistic view, and when considering the opportunities offered by AM it is important to consider that buildings are made from thousands of components, structural and non-structural, all of which must be coordinated, installed, and connected. At present, AM proof-of-concept structures are incomplete. They are part of the whole and must be constructed whilst interacting and connecting with these other components. This creates difficulty, as it requires the AM process, and built form, to accommodate these interfaces. Simple examples might be connections to roof elements, lintels, doorframes, electrical conduits, and of course foundations. Each of these interfaces requires new details, which will influence the continuity and complexity of the printing process.

It is therefore vitally important that the design team works closely with the contractor and AM specialist to deliver a design that considers this coordination and the interfaces from the outset. This is discussed further in Section 9.6.

9.5.3.8 Additive manufacture of horizontal elements

Perhaps one of the most significant challenges for AM is that it is naturally suited to the printing of vertical elements in layers that remain stable under their own weight under gravity. Printing of horizontal elements presents practical difficulties for the technology, for example printing a floor slab or a beam *in situ* is simply not currently practicable, due to both the formwork and shoring required and the fundamental limitations of extrusion-based methods. Alternative approaches exist, for example printing horizontal elements vertically in sections and then lifting and rotating them into place, as undertaken for the bicycle bridge in the Netherlands (Scott, 2019b), however, this approach limits the application of AM and necessitates the addition of other construction activities to form the final structure. In addition, as permanent works, horizontal elements generally work in flexure therefore reinforcement or pre-stressing is needed to overcome tension forces, which complicates the design and requires consideration of additional technical issues. Such issues can be overcome using solutions as employed on The Office of the Future (Dubai Future Foundation, 2019b), however, this significantly complicates the design and relies on the direct bond between discrete reinforcement and 3D-printed concrete. Materials technology may provide future solutions by providing a means to integrate tensile capacity into homogeneous feedstock. However, shape retention and temporary support will continue to remain a challenge.

The option of using geometry to provide structural strength and stiffness, through arching or dome structures, is a potential solution (e.g. Project Milestone in the Netherlands (3D Print House, 2019)) however this is limited by the ability of the AM system to provide vertical inclination during printing and may also require temporary support and formwork.

The challenge of printing horizontal elements could necessitate a rethink on the AM system employed. The use of a binder-jetting AM system similar to D-shape (D-shape, 2019) would permit the construction of freeform, temporarily supported vertical and horizontal structures. Whilst the challenges of implementation in the field, and the large quantities of temporary material would be a significant factor, coupled with development of materials technology it could revolutionize both architecture and construction through the integration of temporary and permanent works.

9.5.3.9 Environmental performance

AM is often hailed as less environmentally damaging that traditional construction (see earlier discussion on wastage), however one aspect that needs careful consideration is the energy and thermal performance of AM structures. The 'double-walled' approach favored by many of the AM building systems offers the potential for insulation to be included in between the cavity, however many systems actually connect the two thin walls together for stability, which creates thermal bridging issues. Of course, insulation can be provided on the exterior. Nevertheless, this hides the AM structure, which may be architecturally desirable or undesirable, and may present its own problems if the geometry is irregular. Some study has been undertaken on this subject (Gosselin et al., 2016), although, more is required to develop thermally efficient wall and roof constructions.

9.6 Future areas of research and development

The World Economic Forum (2018) proposed 70 plausible futuristic visions of construction, emphasizing prefabrication and modularization boosting construction cost-efficiency, as well as virtual reality, intelligent systems, and robots running the smart construction. Another increasing research area targets at exploiting the Internet of Things (IoT) and data science for collaborative planning and design, efficient building elements production, integrated supply chains, smart project delivery, effective maintenance, and improved energy efficiency (Hinkka and Tätilä, 2013; Zhong et al., 2015). In this section, several main areas for future research and developments will be discussed.

9.6.1 3D-printing robotics

One further aspect of robotics development for AM is making the systems more suitable for use in a construction environment. Construction is, by its very nature, transient, messy, and exposed to all sorts of weather and ground conditions. Most plant and construction equipment has evolved over many years to a point where the equipment is robust and reliable under transport, erection, and use, even in the harshest of environments and with the least sympathetic of operatives. Most AM systems however have evolved from laboratory prototypes or have been developed based on existing robotics systems that are intended for internal usage in a controlled environment. The systems are not therefore particularly suitable for repeat or prolonged use on construction sites.

The physical and digital operational reliability and robustness of AM systems must evolve to a level where the systems can be employed on construction sites, moved and erected repeatedly and operated reliably with confidence by non-expert operatives. Furthermore, the systems must have the flexibility to be employed for a range of applications; e.g.:

- Mobile, scalable gantry solutions for repeat building construction (e.g. villas).
- Specialist gantry solutions on slip-form style systems for multi-story buildings.
- Small mobile robotic arm solutions for complex small-scale applications.
- Foldable or flat-packed solutions for quick out the back-of-a truck set-up in difficult to reach locations.

For this reason, it is likely that the future of AM in construction may include collaborations between existing construction plant and equipment manufacturers and AM systems manufacturers to drive development in a direction that will consider the needs and preferences of contractors.

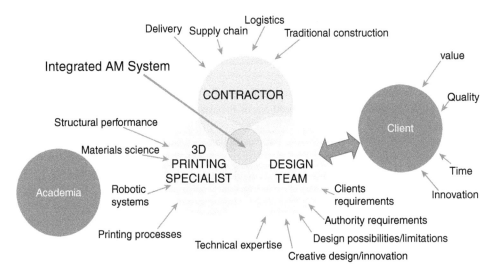

Figure 9.8   Integrated delivery of an AM system for construction (Robert Bird Group, 2018)

9.6.2 *Integrated delivery*

As referenced previously, there are multiple reports in the market that discuss the reasons for the lack of productivity in the construction industry and explore the reasons for the slow uptake in new technology. The fractured nature of the industry, both within the supply chain and between projects, presents a real challenge to investment in new areas, and AM is no exception. However, to fully realize the benefits of a paradigm such as AM in the industry a more holistic long-term view is needed; one that encompasses research, feasibility, conceptual studies, multidisciplinary design, full BIM integration, and application of IoT along with consideration of construction logistics, materials, supply chain management, and constructability. Figure 9.8 shows the required interrelationship of the traditional stakeholders in order to deliver such a holistic AM construction solution.

To achieve an integrated design and delivery approach requires involvement from a range of stakeholders, supply chain partners, and a change to the current project-based transactional methods of operation. It is necessary to move away from a project-based mindset to one that is based on AM system development and implementation to deliver unique integrated solutions that can be applied on projects. If the industry is to realize the same progress as seen in other industries, such as the automotive industry, then it must move towards full engagement of supply chain partners and a long-term approach.

There have been moves in some parts of the world to trial enterprise-based approaches to construction, however, the success of these schemes has been limited to large infrastructure projects or clients with a large asset portfolio. It remains to be seen if the approach is scalable or attractive for clients who are currently still intent on competitive tender for individual projects. A detailed discussion of this subject is beyond the scope of this chapter; however, it is paramount to the successful adoption of technology in the industry, not only AM.

9.6.3 Constructability

An AM system suitable for construction must provide benefits to contractors and other stake-holders in order for it to be of interest. Current construction paradigms, despite their draw-backs, are well understood and, with good planning and sufficient budget, can be used to deliver a very broad range of engineering solutions. As projects increase in complexity the degree of planning and engineering forethought increases significantly to deliver the final permanent works safely, on time and on budget. Complex mega-projects or iconic projects unfortunately often encounter problems and are delivered late as a result. The adoption of a Construction Methodology and Erection Sequence (CMES) ahead of actual construction using Virtual Design and Construction (VDC) techniques ultimately pays dividends as it helps mitigate risks ahead of construction where they can become extremely expensive to manage. Whilst AM projects are not likely to be mega-projects (at least not for some time) the approach necessitates a similar methodology. Construction should be visualized using VDC techniques based on a BIM environment, exploring the various stages of construction; assessing the sequence of construction and the mitigating any risks, understanding logistics issues and temporary works required, and managing interfaces such as precast or other build-ing components.

As pointed out in Section 9.6.3 the constructability of the solution must be part of the integrated design process, involving the contractor and other stakeholders in the review and assessment of risk during construction and mitigating any such risks through design decisions or supply chain involvement in solution development.

As discussed in Section 9.5.3.3, until contractors understand the benefits of AM and the risks of implementation are explained and mitigated, adoption will remain low.

9.7 Conclusion

Although the introduction of AM will be beneficial for construction businesses, to reap signif-icant rewards from the technology the industry needs to adopt an innovative mindset, whereby the benefits of AM can be realized in combination for the benefit of all stakeholders. At pres-ent, despite the exciting prospects offered by AM, both the positive and negative aspects need to be understood and managed for it to be truly transformative.

From a materials perspective, currently, the materials applicable for AM of construction components are limited and require further development to overcome the shortcomings espe-cially with respect to structural strength and stiffness, and also the constraints imposed by the extrudability of the material vs rapid strength gain. Further research is ongoing and a further understanding of rheology and different studies into fiber technology may be significant for the future direction of AM materials for scalable applications.

From a cost perspective, whilst there have been no large-scale studies undertaken, it would appear that AM can be competitive for specific projects in particular markets. However, the range of potential applications of AM means that it is unlikely that there will be a 'one-size-fits-all' solution. As with the current range of materials, products and methods, AM will intro-duce many more (and significantly more varied) ways of achieving a solution. For example, the use of AM for construction components may offer an interesting dynamic with modular construction, which is also seeing a rise across the industry. Capital costs of 3D printers for construction (e.g. robotic arm or gantry systems) are still finding a suitable price point, how-ever, as such technology moves from bespoke laboratory equipment into a necessary phase of offering more efficient, mass-produced, construction-ready, and site-capable equipment, the cost of the technology will reduce and for those who invest and apply the technology effectively it will rapidly pay for itself. Therefore, AM can be, and will continue to become, an economical and more reliable method of component fabrication and building construction.

From a people perspective, whilst there will inevitably be a reduction in semi-skilled labor, the introduction of AM gives an opportunity (if managed correctly) for staff to be upskilled and for a significant reduction in health and safety risk. The innovative nature of AM technology provides a new platform for builders to progress in the market, expanding into sectors that previously were not possible, something that may become increasing important with the rise of off-site manufacture and modular construction.

As the AM of construction components arrives to the market, it will require a period of time before the positive impacts will be felt by the industry, e.g. it will require builders to learn new techniques and develop new skills, which may prove to be temporarily disruptive, give rise to new risks and detract from business-as-usual.

The industry must be collectively ready for this. The stakeholders in the industry must adopt more collaborative approach to business, one that facilitates research and development with shared risk and reward, to provide the funding and space required for AM to take the next step into providing value as a mainstream construction option. As stated in the World Economic Forum Report 'Shaping the Future of Construction' (WEF, 2019):

> Other industries, such as the automotive industry, have already undergone radical and disruptive changes, and their digital transformation is now well under way. Construction companies need to act quickly and decisively: lucrative rewards await nimble companies, while the risks are serious for hesitant companies.

The adoption of AM in construction is a perfect case study.

9.8 Summary

- Additive manufacturing is a groundbreaking technological advancement that is increasingly surpassing different methods of manufacturing.
- Comprehensive review on the implementation of additive manufacturing in the construction industry as an eco-innovative solution.
- The benefits and limitation of additive manufacturing for the construction industry.

References

3D Print House (2019) *3dprintedhouse: Project-project milestone*. Available at: https://3dprintedhouse. nl/en/project-info/project-milestone/ (Accessed: 23 May 2019).

3DPrint.com (2019) *Chinese construction company 3D Prints an entire two-story house on-site in 45 Days. 3DPrint.com The Voice of 3D Printing/Additive Manufacturing* Available at: https://3dprint. com/138664/huashang-tengda-3d-print-house/ (Accessed: 30 August 2017).

Ahmed, Z. Y. et al. (2016) 'Design considerations due to scale effects in 3d concrete printing', in *8th International Conference of the Arab Society for Computer Aided Architectural Design*, pp. 115–124.

Alec (2019) *3ders.org – Thai cement maker SCG develops an elegant 3m-tall 3D printed "pavilion" home, 21st C. Cave, 3D Printer News & 3D Printing News*. Available at: http://3ders.org/articles/20160427-thai-cement-maker-scg-develops-an-elegant-3m-tall-3d-printed-pavilion-home-21st-c-cave.html (Accessed: 13 May 2019).

Al-Qutaifi, S., Nazari, A., and Bagheri, A. (2018) 'Mechanical properties of layered geopolymer structures applicable in concrete 3D-printing', *Construction and Building Materials*, 176, pp. 690–699. doi: 10.1016/j.conbuildmat.2018.04.195.

ANIWAA (2019) *The 13 best construction 3D printers in 2019 (May 2019)*. Available at: https://aniwaa. com/house-3d-printer-construction/ (Accessed: 21 August 2019).

Aouf, R. M. (2019) *US military 3D prints concrete barracks on site*. Available at: https://dezeen. com/2018/09/05/us-military-3d-prints-concrete-barracks-on-site-technology/ (Accessed: 14 May 2019).

Apis Cor (2019) *The first on-site house has been printed in Russia Apis Cor. We print buildings*. Available at: http://apis-cor.com/en/about/news/first-house (Accessed: 21 April 2017).

ARUP (2019a) *Amsterdam 3D printed steel bridge – Arup*. Available at: https://arup.com/projects/mx3d-bridge (Accessed: 23 May 2019).

ARUP (2019b) *Additive manufacturing: Bringing 3D printing to hard-hat construction global research*. Available at: https://research.arup.com/projects/additive-manufacturing-bringing-3d-printing-to-hard-hat-construction/ (Accessed: 21 August 2019).

Bishop, E. S. et al. (2017) '3-D bioprinting technologies in tissue engineering and regenerative medicine: Current and future trends', *Genes and Diseases*, pp. 185–195. doi: 10.1016/j.gendis.2017.10.002.

Bos, F. et al. (2016) 'Additive manufacturing of concrete in construction: Potentials and challenges of 3D concrete printing', *Virtual and Physical Prototyping*, 2759(October), pp. 1–17. doi: 10.1080/17452759.2016.1209867.

Brenken, B. et al. (2018) 'Fused filament fabrication of fiber-reinforced polymers: A review', *Additive Manufacturing*, doi: 10.1016/j.addma.2018.01.002.

Buchanan, C. and Gardner, L. (2019) 'Metal 3D printing in construction: A review of methods, research, applications, opportunities and challenges', *Engineering Structures*, doi: 10.1016/j. engstruct.2018.11.045.

Cameron (2019) *3ders.org – Saudi Arabia 3D prints a house in two days 3D Printer News & 3D printing news*. Available at: https://3ders.org/articles/20181231-saudi-arabia-3d-prints-a-house-in-two-days. html (Accessed: 14 May 2019).

Cesaretti, G. et al. (2014) 'Building components for an outpost on the lunar soil by means of a novel 3D printing technology', *Acta Astronautica*, 93, pp. 430–450. doi: 10.1016/j.actaastro.2013.07.034.

CIOB (2019) *Skanska claims first with 3D printed cladding. Construction Manager – News* Available at: http://constructionmanagermagazine.com/news/skanska-claims-industry-first-3d-printed-cladding/ (Accessed: 21 August 2019).

D-shape (2019) *D-shape*. Available at: https://d-shape.com/ (Accessed: 23 May 2019).

Dai, F. et al. (2018). 'Health and safety statistics for the construction sector in Great Britain', *Visualization in Engineering*, 39(6), https://gartner.com/smarterwithgartner/confron. doi: 10.1111/j.1743-6109.2008. 01122.x Endothelial.

Dearden, L. (2015) 'Worker killed by robot in welding accident at car parts factory in India', *The Independent*. Available at: https://independent.co.uk/news/world/asia/worker-killed-by-robot-in-welding-accident-at-car-parts-factory-in-india-10453887.html (Accessed: 21 August 2019).

Delgado Camacho, D. et al. (2018) 'Applications of additive manufacturing in the construction industry – A forward-looking review', *Automation in Construction*, doi: 10.1016/j.autcon.2017.12.031.

Department for Communities and Local Government (2018) 'House building; new build dwellings, England: December Quarter 2017', *Housing Statistical Release*, 350 August p. 26. doi: 10.1021/ jo0162078.

Dubai Future Foundation (2019a) *Dubai 3D printing strategy – Dubai Future Foundation*. Available at: https://dubaifuture.gov.ae/our-initiatives/dubai-3d-printing-strategy/ (Accessed: 22 May 2019).

Dubai Future Foundation (2019b) *3D printed office – Dubai Future Foundation*. Available at: http:// officeofthefuture.ae/ (Accessed: 23 May 2019).

Everett, J. G. (2012) 'Ergonomics, Health, and Safety in Construction: Opportunities for Automation and Robotics', *Automation and Robotics in Construction Xi*, pp. 19–26. doi: 10.1016/ b978-0-444-82044-0.50007-2.

FBR (2019) *FBR (Fastbrick Robotics) Industrial automation technology*. Available at: https://fbr.com. au/ (Accessed: 22 May 2019).

Feng, P. et al. (2015) 'Mechanical properties of structures 3D printed with cementitious powders', *Construction and Building Materials*, 93, pp. 486–497. doi: 10.1016/j.conbuildmat.2015.05.132.

Flatt, R. J. and Bowen, P. (2006) 'Yodel: A yield stress model for suspensions', *Journal of the American Ceramic Society*, doi: 10.1111/j.1551-2916.2005.00888.x.

García de Soto, B. et al. (2018) 'Productivity of digital fabrication in construction: Cost and time analysis of a robotically built wall', *Automation in Construction*, 92, pp. 297–311. doi: 10.1016/j. autcon.2018.04.004.

Gartner Hype Cycle (2019) *Hype cycle research methodology*. Available at: https://gartner.com/en/ research/methodologies/gartner-hype-cycle (Accessed: 22 May 2019).

Gaudillière, N. et al. (2019) 'Building applications using lost formworks obtained through large-scale additive manufacturing of ultra-High-Performance concrete', *3D Concrete Printing Technology*, doi: 10.1016/B978-0-12-815481-6.00003-8.

Gebler, M., Schoot Uiterkamp, A. J. M., and Visser, C. (2014) 'A global sustainability perspective on 3D printing technologies', *Energy Policy*, 74(C), pp. 158–167.doi: 10.1016/j.enpol.2014.08.033.

Ghaffar, S. and Mullett, P. (2018) 'Commentary: 3D printing set to transform the construction industry', *Proceedings of the Institution of Civil Engineers – Structures and Buildings*, 171(10), pp. 737–738. doi: 10.1680/jstbu.18.00136.

Ghaffar, S. H., Corker, J., and Fan, M. (2018) 'Additive manufacturing technology and its implementation in construction as an eco-innovative solution', *Automation in Construction*, pp. 1–11. doi: 10.1016/j.autcon.2018.05.005.

Global Construction Review (2019) *Saudi firm buys "world's largest 3D printer" for housing – News – GCR*. Available at: http://globalconstructionreview.com/news/saudi-firm-buys-worlds-largest-3d-printer-housing/(Accessed: 22 May 2019).

Gosselin, C. et al. (2016) 'Large-scale 3D printing of ultra-high performance concrete – a new processing route for architects and builders', *Materials and Design*, 100, pp. 102–109. doi: 10.1016/j.matdes.2016.03.097.

The Guardian (2019) 'Robot kills worker at Volkswagen plant in Germany', *The Guardian*. Available at: https://theguardian.com/world/2015/jul/02/robot-kills-worker-at-volkswagen-plant-in-germany (Accessed: 23 May 2019).

Hager, I., Golonka, A., and Putanowicz, R. (2016) '3D printing of buildings and building components as the future of sustainable construction?', *Procedia Engineering*, pp. 292–299. doi: 10.1016/j.proeng.2016.07.357.

Hinkka, V. and Tätilä, J. (2013) 'RFID tracking implementation model for the technical trade and construction supply chains', *Automation in Construction*, doi: 10.1016/j.autcon.2013.05.024.

Holt, C. et al. (2019) 'Construction 3D printing', *3D Concrete Printing Technology*, pp. 349–370. doi: 10.1016/B978-0-12-815481-6.00017-8.

Hwang, D. and Khoshnevis, B. (2017) 'An innovative construction process-contour crafting (CC)', in *Proceedings of the 22nd International Symposium on Automation and Robotics in Construction*. doi: 10.22260/isarc2005/0004.

IAAC (2019) *Minibuilders*. Available at: https://robots.iaac.net/(Accessed: 21 August 2019).

ICON (2019) *Our Story | ICON*. Available at: https://iconbuild.com/about (Accessed: 22 May 2019).

ISO/ASTM 52900:2015 (2015) "Standard Terminology for Additive Manufacturing-General Principles-Terminology"; ICS Number Code 01.040.25, 25.030.

ISO 10218-1:2018 (2019) *ISO 10218-1:2011 – Robots and robotic devices – Safety requirements for industrial robots – Part 1: Robots*. Available at: https://iso.org/standard/51330.html (Accessed: 23 May 2019).

ISO/TS 15066:2016 (2019) *ISO/TS 15066:2016 – Robots and robotic devices – Collaborative robots*. Available at: https://iso.org/standard/62996.html (Accessed: 23 May 2019).

Independent (2019) *Worker killed by robot in welding accident at car parts factory in India | The Independent*. Available at: https://independent.co.uk/news/world/asia/worker-killed-by-robot-inwelding- accident-at-car-parts-factory-in-india-10453887.html (Accessed: 21 August 2019).

Jeon, K.-H. et al. (2013) 'Development of an automated freeform construction system and its construction materials', *Proceedings of the 30th International Symposium on Automation and Robotics in Construction and Mining*, 33(4), pp. 1359–1365.

Jordahn, S. (2019) *Arup and CLS Architetti's 3D-printed house was built in a week*. Available at: https://dezeen.com/2018/11/19/video-mini-living-3d-printing-cls-architetti-arup-movie/?li_source=LI&li_medium=bottom_block_1 (Accessed: 13 May 2019).

Kazemian, A. et al. (2017) 'Cementitious materials for construction-scale 3D printing: Laboratory testing of fresh printing mixture', *Construction and Building Materials*, 145, pp. 639–647. doi: 10.1016/j.conbuildmat.2017.04.015.

Keating, S. (2015) 'Beyond 3D printing: The new dimensions of additive fabrication', in J. Follett (ed.), *Designing for Emerging Technologies: UX for Genomics, Robotics, and the Internet of Things*, Sebastopol, CA: O'Reilly Media, pp. 379–405.

Khoshnevis, B. (2004) 'Automated construction by contour crafting – Related robotics and information technologies', *Automation in Construction*, doi: 10.1016/j.autcon.2003.08.012.

Labonnote, N. et al. (2016) 'Additive construction: State-of-the-art, challenges and opportunities', *Automation in Construction*, 72, pp. 347–366. doi: 10.1016/j.autcon.2016.08.026.

Le, T. T. et al. (2012a) 'Mix design and fresh properties for high-performance printing concrete', *Materials and Structures*, doi: 10.1617/s11527-012-9828-z.

Le, T. T. et al. (2012b) 'Mix design and fresh properties for high-performance printing concrete', *Materials and Structures*, 45, pp. 1221–1232. doi: 10.1617/s11527-012-9828-z.

Lim, S. et al. (2012) 'Developments in construction-scale additive manufacturing processes', *Automation in Construction*, 21(1), pp. 262–268. doi: 10.1016/j.autcon.2011.06.010.

Ma, G., Li, Z., and Wang, L. (2018) 'Printable properties of cementitious material containing copper tailings for extrusion based 3D printing', *Construction and Building Materials*, 162, pp. 613–627. doi: 10.1016/j.conbuildmat.2017.12.051.

Marchment, T., Sanjayan, J., and Xia, M. (2019) 'Method of enhancing interlayer bond strength in construction scale 3D printing with mortar by effective bond area amplification', *Materials and Design*, 169, doi: 10.1016/j.matdes.2019.107684.

McKinsey & Company et al. (2017) 'A future that works: Automation, employment, and productivity', *McKinsey Global Institute* (January), p. 148. Available at: http://njit2.mrooms.net/pluginfile.php/688844/mod_resource/content/1/ExecutiveSummaryofMcKinseyReportonAutomation.pdf.

Mok, K. (2019) *Huge modular 3D printer creates $1,000 tiny house out of mud (Video), TreeHugger*. Available at: https://treehugger.com/green-architecture/gaia-house-3d-printed-out-mud-wasp.html (Accessed: 13 May 2019).

Mrazovic, N. (2018) *Metal 3D print search engine series metal 3D printing tech feasibility study to 3D print a full scale curtain wall frame as a single element 001*. Available at: http://laserstatics.com/publications/metal-3d-printing-tech-02/pages/feasibility-study-to-3d-print-full-scale-curtain-wall-frame-as-single-element-001.htm (Accessed: 21 August 2019).

MX3D (2019) *MX3D Bridge*. Available at: https://mx3d.com/projects/bridge-2/(Accessed: 21 August 2019).

Ngo, T. D. et al. (2018) 'Additive manufacturing (3D printing): A review of materials, methods, applications and challenges', *Composites Part B: Engineering. Elsevier*, 143, pp. 172–196. doi: 10.1016/J.COMPOSITESB.2018.02.012.

Panda, B. et al. (2018) 'Measurement of tensile bond strength of 3D printed geopolymer mortar', *Measurement: Journal of the International Measurement Confederation*, 113, pp. 108–116. doi: 10.1016/j.measurement.2017.08.051.

Panda, B., Chandra Paul, S., and Jen Tan, M. (2017) 'Anisotropic mechanical performance of 3D printed fiber reinforced sustainable construction material', *Materials Letters*, 209, pp. 146–149. doi: 10.1016/j.matlet.2017.07.123.

Paul, S. C. et al. (2018) 'Fresh and hardened properties of 3D printable cementitious materials for building and construction', *Archives of Civil and Mechanical Engineering*, 18(1), pp. 311–319. doi: 10.1016/j.acme.2017.02.008.

Pegna, J. (1997) 'Exploratory investigation of solid freeform construction', *Automation in Construction*, doi: 10.1016/S0926-5805(96)00166-5.

Perkins, I. and Skitmore, M. (2015) 'Three-dimensional printing in the construction industry: A review', *International Journal of Construction Management*, pp. 1–9. doi: 10.1080/15623599.2015.1012136.

Perrot, A., Rangeard, D., and Pierre, A. (2016) 'Structural built-up of cement-based materials used for 3D-printing extrusion techniques', *Materials and Structures*, pp. 1213–1220. doi: 10.1617/s11527-015-0571-0.

Reggev, K. (2019) *How ICON is building the $4,000 3D-printed homes of the future – Dwell*. Available at: https://dwell.com/article/icon-3d-printed-homes-for-4000-dollars-23d715bf (Accessed: 23 May 2019).

Reis, C. and Durkin, S. (2018) *Workforce of the future: The competing forces shaping 2030*, PricewaterhouseCoopers GmbH.

Ryder, M. A. et al. (2018) 'Fabrication and properties of novel polymer-metal composites using fused deposition modeling', *Composites Science and Technology*, doi: 10.1016/j.compscitech.2018.01.049.

Sanjayan, J. G. et al. (2018) 'Effect of surface moisture on inter-layer strength of 3D printed concrete', *Construction and Building Materials*, 172, pp. 468–475. doi: 10.1016/j.conbuildmat.2018.03.232.

Sanjayan, J. G. and Nematollahi, B. (2019) '3D concrete printing for construction applications', *3D Concrete Printing Technology*, pp. 1–11. doi: 10.1016/B978-0-12-815481-6.00001-4.

Saunders, Sarah (2019) *CyBe Construction announces that 3D printing is complete for Dubai's R&Drone laboratory*. Available at: https://3dprint.com/176561/cybe-3d-printed-dubai-laboratory/(Accessed: 22 May 2019).

Scott, Clare (2019a) *3D printhuset announces new company dedicated to construction 3D printing*. Available at: https://3dprint.com/225945/3d-printhuset-new-company/(Accessed: 22 May 2019).

Scott, Clare (2019b) *TU/e and BAM infra get to work on 3D printed concrete bicycle bridge*. Available at: https://3dprint.com/178462/eindhoven-3d-printed-bridge/(Accessed: 22 May 2019).

Sevenson, B. (2019) *Shanghai-based WinSun 3D prints 6-story apartment building and an incredible home 3DPrint.com The Voice of 3D Printing/Additive Manufacturing*. Available at: https://3dprint.com/38144/3d-printed-apartment-building/(Accessed: 13 May 2019).

Singh, S., Prakash, C., and Ramakrishna, S. (2019) '3D printing of polyether-ether-ketone for biomedical applications', *European Polymer Journal*, pp. 234–248. doi: 10.1016/j.eurpolymj.2019.02.035.

Suiker, A. S. J. (2018) 'Mechanical performance of wall structures in 3D printing processes: Theory, design tools and experiments', *International Journal of Mechanical Sciences*, 137, pp. 145–170. doi: 10.1016/j.ijmecsci.2018.01.010.

Valkenaers, H. et al. (2014) 'Additive manufacturing for concrete: A 3D printing principle', *Conference Proceedings – 14th International Conference of the European Society for Precision Engineering and Nanotechnology, EUSPEN 2014*, 1, pp. 139–142. Available at: https://engineeringvillage.com/share/document.url?mid=cpx_M8c435c614c32ee2a18M7bcd1017816338&database=cpx. (Accessed: 13 May 2019).

Van Zijl, A. et al. (2016) 'Properties of 3D printable concrete', *2nd International Conference on Progress in Additive Manufacturing*, (May), pp. 421–426. Available at: http://hdl.handle.net/10220/41820. (Accessed: 13 May 2019).

Voigt, T. et al. (2010) 'Using fly ash, clay, and fibers for simultaneous improvement of concrete green strength and consolidatability for slip-form pavement', *Journal of Materials in Civil Engineering*, 22(2), pp. 196–206. doi: 10.1061/(asce)0899-1561(2010)22:2(196).

WASP (2019) *3d printer house Crane WASP 3D Printers. WASP* Available at: https://3dwasp.com/en/3d-printer-house-crane-wasp/(Accessed: 21 August 2019).

WEF (2018) *The future of jobs report 2018, Centre for the New Economy and Society*. doi: 10.1177/1946756712473437.

WEF (2019) *Shaping the future of construction: A breakthrough in mindset and technology. World Economic Forum* Available at: https://weforum.org/reports/shaping-the-future-of-construction-a-breakthrough-in-mindset-and-technology (Accessed: 23 May 2019).

Weng, Y. et al. (2018) 'Design 3D printing cementitious materials via Fuller Thompson theory and Marson-Percy model', *Construction and Building Materials*, doi: 10.1016/j.conbuildmat.2017.12.112.

Wolfs, R., Salet, T., and Hendriks, B. (2015) '3D printing of sustainable concrete structures', *Proceedings of the International Association for Shell and Spatial Structures (IASS) Symposium 2015* (August), (pp. 1–8).

Wolfs, R. J. M., Bos, F. P., and Salet, T. A. M. (2018) 'Early age mechanical behaviour of 3D printed concrete: Numerical modelling and experimental testing', *Cement and Concrete Research*, 106, pp. 103–116. doi: 10.1016/j.cemconres.2018.02.001.

Wolfs, R. J. M., Bos, F. P., and Salet, T. A. M. (2019) 'Hardened properties of 3D printed concrete: The influence of process parameters on interlayer adhesion', *Cement and Concrete Research*, doi: 10.1016/j.cemconres.2019.02.017.

Xia, M., Nematollahi, B., and Sanjayan, J. (2019) 'Printability, accuracy and strength of geopolymer made using powder-based 3D printing for construction applications', *Automation in Construction*, 101, pp. 179–189. doi: 10.1016/j.autcon.2019.01.013.

XtreeE (2019) *Post in Aix-en-Provence | XtreeE*. Available at: http://xtreee.eu/post-in-aix-en-provence/(Accessed: 22 May 2019).

Ye, W. et al. (2019) 'Separated 3D printing of continuous carbon fiber reinforced thermoplastic polyimide', *Composites Part A: Applied Science and Manufacturing*, 121, pp. 457–464. doi: 10.1016/j.compositesa.2019.04.002.

Zareiyan, B. and Khoshnevis, B. (2017a) 'Effects of interlocking on interlayer adhesion and strength of structures in 3D printing of concrete', *Automation in Construction*, Elsevier 83, pp. 212–221. doi: 10.1016/J.AUTCON.2017.08.019.

Zareiyan, B. and Khoshnevis, B. (2017b) 'Interlayer adhesion and strength of structures in Contour Crafting – Effects of aggregate size, extrusion rate, and layer thickness', *Automation in Construction*, 81, pp. 112–121. doi: 10.1016/j.autcon.2017.06.013.

Zhong, R. Y. et al. (2015) 'A big data approach for logistics trajectory discovery from RFID-enabled production data', *International Journal of Production Economics*, doi: 10.1016/j.ijpe.2015.02.014.

10

DIGITAL FABRICATION IN THE CONSTRUCTION SECTOR

Keith Kaseman and Konrad Graser

10.1 Aims

- Provide context and key references relating to the development, proliferation and adoption of digital fabrication principles and practices in Architecture, Engineering and Construction (AEC).
- Survey recent and current academic research projects that delineate and demonstrate various facets of the state of the art in digital fabrication.
- Describe the expansive and collaborative nature of digital fabrication at the cutting edge of AEC today through a case-study of a touchstone architectural practice.
- Identify established and emerging technologies, companies and approaches imbued with potential to significantly impact future AEC practices with respect to digital fabrication in Construction 4.0.

10.2 Introduction

10.2.1 Overview

Digital fabrication is currently evolving from a status of relative anomaly to practical ubiquity in today's AEC practices (Bock, 2016; Marble, 2012). This chapter sets out to survey its preliminary roots, examine the state of the art in both academia and professional practice, and show current trends in its development. While far from conclusive or comprehensive, the primary aim of this chapter is to provide an overview of digital fabrication in architecture and construction, and to illustrate the expansive nature of the many roles it currently plays in AEC.

Digital fabrication in construction has grown into a multifaceted array of systems and processes at various levels of complexity, technological maturity, and scale. Amplified levels of digital agility in the profession are prerequisite to effectively operate within this territory. It requires the capacity to collaborate across discipline boundaries in a fluid environment spanning computation, digitally driven physical building processes, and new organizational models. The adoption of Industry 4.0 principles in AEC is ongoing with the rise of new models of industrialized construction and the use of design platforms and configurators. This will increasingly lead to the types of integrated project structures and supply chains required for the successful diffusion of digital fabrication technologies.

10.2.2 General definition

Digital fabrication in the AEC industry is commonly defined as a fabrication or building process driven by the seamless conversion of design and engineering data into fabrication code for digitally controlled tools (Gershenfeld, 2012). We define a digital fabrication process in construction as a construction method or system which relies on digital fabrication entirely or to a significant degree, either in prefabrication or on-site construction. Examples of digital fabrication processes in construction include robotic fabrication and assembly, large scale additive manufacturing, and the use of specialized automation systems for material processing in areas ranging from advanced fabrication of metal or timber assemblies to various forms of concrete processing to the fabrication of multi-material composites.

Common tools utilized in digital fabrication production practices include computer numerically controlled (CNC) routers, multi-axis industrial robotic arms, laser-cutters, plasma cutters, water-jet cutters and hot-wire foam cutters. Recent advancements in mobile robotics, large-scale 3D printing, cable robots, unmanned aerial systems (UAS) along with a wide array of customized tools, robotic end-effectors and other machinery continue to widen the field of achievable possibilities for digital fabrication workflows in AEC.

10.2.3 Benefits of digital fabrication in AEC

Digital fabrication in architecture and construction does not always aim at the highest possible degree of automation. Instead, many recent initiatives in both research and practice have pushed for exploring the potential of data driven fabrication to achieve higher geometrical complexity, precise material processing, and improved time and cost control. This can lead to new architectural possibilities, e.g. by allowing economic serial fabrication of customized parts to overcome the high degree of standardization and repetition still dominating today's construction processes (Hack et al., 2013).

At the same time, digital fabrication practices can contribute to lowering the environmental footprint of construction (McKinsey & Company, 2016; World Economic Forum in collaboration with Boston Consulting Group, 2016), for example through reduced use of material, structurally effective shapes and graded assemblies, and functional integration such as thermal activation, acoustics, electrical or HVAC systems. While affording the potential to integrate such varied and complex systems, digital fabrication also requires advanced digital planning tools and processes and integrated planning processes. Recent advances in digital planning methods such as Building Information and Modeling (BIM) and Virtual Design and Construction (VDC) therefore lay the groundwork for the advancement of digital fabrication approaches that could transform the AEC industry.

10.3 Digital fabrication in architecture and construction

While the history of digital fabrication stands upon decades of key breakthroughs in computer science, engineering, design and manufacturing sectors (Carpo, 2017), the early 2000s demarcate the beginning of a significantly transformative era within the field of architecture and construction. Many highly publicized projects laid substantial groundwork for the propagation of digital fabrication in architecture through construction achievements that were only attainable through the utilization of digital tools. Facilitated by advancements in modeling and parametric software applications, the capacity for designers and architects to operate within this territory naturally expanded, and many architects directed their attention towards expanding the scope of digital fabrication more directly into practice and ultimately, construction. Early innovation in extending digital design tools directly into fabrication developed through the work of emerging practices from this time, many of which were small offices primarily composed of young, digitally astute architects and designers. Generally, early pursuits in digital fabrication tended to either center around the promise of mass customization as it may relate to architecture and construction or bespoke material assemblies.

10.3.1 Digital fabrication applications in architecture

As built examples and associated protocols were being developed and deployed at accelerated rates during this period, a significant shift in digital fabrication practice and utilization began to permeate the field. Whereas many initial applications were geared towards the rationalization and digitally driven production of complex *geometry*, architecture practices began to develop approaches that expanded the scope of digital fabrication into the realm of project *logistics*. Architects began to expand operative principles into approaches and practices in which digital models served as the direct source of fabrication data, production processes, assembly orchestrations and construction sequences.

For example, SHoP Architects exploited opportunities and protocols enabled by digital fabrication practices for all phases of work, from inception to construction, for their Camera Obscura at Mitchell Park in Greenpoint, NY (2002–2005). Notably, in much the same way that assembly information is conveyed in an instruction set for model airplane kits, both bid and construction documents illustrated digitally produced components and assemblies through descriptive catalog sheets and sequential three-dimensional drawings (Allen and Iano, 2004) (Figure 10.1). Laser-cut steel base components were etched with labels and marks to guide varying weld conditions throughout the process, facilitating the fabrication of complex structural parts without reliance upon traditional drawings. Likewise, laser-cut and anodized aluminum alignment shelves were fabricated to serve as permanent jigs used to guide the variably pitched wood members with which a CNC produced composite stressed-skin was assembled in place. (Sharples Holden Pasquarelli, 2002)

At a much larger scale and embedded with amplified complexity, the Mercedes Benz Museum by UN Studio in Stuttgart, DE (2001–2006) exemplifies this shift in the scope of construction methods attainable through digital fabrication. Predating many of today's BIM operational standards (Marble, 2012), custom software applications and digital file sharing protocols were developed by UN Studio for a working approach referred to as "digital sustainability", in which a "mother model" was developed and collaboratively maintained, providing digital production data through all phases of design, planning, fabrication and construction administration (van Berkel, 2012). The net result of this approach is perhaps most readily apparent in the remarkably complex concrete structure of the building, the formwork of which was digitally fabricated and coordinated. Many facets of this project proved to be compellingly influential throughout the field of architecture at the time of its completion, as the magnitude with which geometric data drove crucial operations and logistics provided a fresh and early vision for what the deployment of digital fabrication principles may look like at the scale of a construction site.

As demand for the utilization of increasingly sophisticated digital fabrication principles propagated AEC, specialty fabricators began to play important roles as primary consultants to architects. Firms such as custom metalwork fabricator Zahner (www.azahner.com), composite material fabricator Kreysler and Associates (www.kreysler.com) and CW Keller, a firm specialized in CNC fabrication of custom components with various materials (www.cwkeller.com)

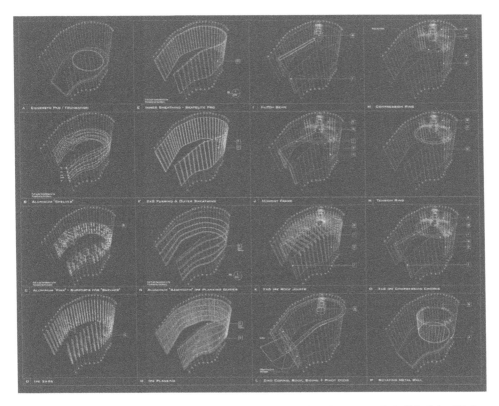

Figure 10.1 Component catalog from bid documents for Camera Obscura at Mitchell Park by SHoP Architects, ca 2003

and many others have contributed their expertise to projects with service arrangements ranging from fabrication and delivery to design-assist and design-build roles, significantly shaping the execution and craft of many illustrative projects by establishing fabrication-aware design approaches in early project development phases. A small sample of work where the effectiveness of such collaborations in building envelopes readily apparent includes the De Young Museum by Herzog de Meuron in San Francisco, CA (2002–2005), 41 Cooper Square by Morphosis in New York, NY (2004–2009), Broad Museum by Diller + Scofidio Renfro in Los Angeles, CA (completed 2015) and SF MoMA by Snøhetta in San Francisco, CA (2010–2016).

10.3.2 *Construction industry applications of digital fabrication*

Digital fabrication in the construction industry is not an entirely new development. An early boom in the development of specialized robotic systems for building construction began in Japan in the 1980s with systems largely aimed at the automation of standardized building construction (Bechthold, 2010). However, a broader application of these systems never occurred, which is often ascribed to the high degree of standardization and design restrictions they imposed (Bock, 2004; Bonwetsch, 2012; Gramazio, Kohler, and Willmann, 2014). Today, industry adoption of digital fabrication is still very limited, and digital fabrication is not yet deployed at scale in the industry (Graser et al., 2019). However, we can observe several emerging approaches to commercial implementation of digital fabrication in the AEC sector.

Japan led the way in automation in construction in the 1980s and the 1990s. Early applications ranged from tele-operated and pre-programmed robotic material manipulators to large-scale integrated construction automation systems. One example is the Obayashi ABCS (Automated Building Construction System), an integrated, on-site "factory" style automation system. It consisted of a large-scale climbing platform extending over the entire building footprint on which guided cranes and industrial robots were deployed. Automated procedures included the erection and welding of steel elements, the installation of prefabricated floor panels, fitting of curtain walls and the jacking of the construction operation platform (Taylor, Wamiziri, and Smith, 2003). These automation-geared systems were developed in part to respond to skilled labor shortages and an ageing workforce (Taylor, Wamiziri, and Smith, 2003), problems which are reported to be on the rise in economies like the US today as well (Garcia de Soto, 2019).

Today we can see two trends at large for commercial implementation of digital fabrication in the AEC sector. First, product innovations on physical implementation systems such as construction robots or large-scale 3D printers. Second, specialized consultants administering custom design to fabrication workflows on a project basis.

Main proponents of the first trend are on-site robotic systems or 3D extrusion printers and off-site robotic prefabrication systems. Examples for on-site systems are SAM (Semi Automated Mason) by US-based Construction Robotics (www.construction-robotics.com), a site-deployed robotic arm mounted on a linear track for automated bricklaying. A similar approach is taken by Australian-based Fastbrick Robotics's Hadrian X system (www.fbr.com.au), a truck-mounted telescopic boom equipped with a gripper and an integrated mortar or adhesive dispenser for masonry work. A pioneering example for the on-site additive manufacturing approach is D-Shape, a large-scale adaptation of the binder-jet principle for building construction which has been under development since the mid-2000s. On-site 3D extrusion printing systems include US-based CC Corp's Contour Crafting technology (Contourcrafting.com) and Chinese Winsun Building Technology (www.winsun3d.com). In the US, ICON BUILD (www.iconbuild.com) has recently released Vulcan II, a commercially available site-deployed gantry robot for homebuilding using cementitious material extrusion. Its first permitted application was the 2018 Chicon House, a single-story prototype for a basic home to address homelessness in developing countries in collaboration with the NGO New Story.

Robotic applications have also entered commercial use in the prefabrication domain. Swiss ROBMade Facades (keller-systeme.ch/en/robmade-facades) is a commercially available system for robotically assembled non-standard brick facades. Its first application was the 2006 Gantenbein Winery project by Bearth Deplazes Architects and Gamazio Kohler (Bonwetsch, Gramazio, and Kohler, 2007), and the system has been applied on multiple projects since. Other enterprises use robotic prefabrication as part of a vertically integrated industrialized construction approach: The Swiss timber prefabrication contractor ERNE Holzbau operates a digital production line equipped with versatile gantry robot with an automated tool changer. The system was used to fabricate the Sequential Roof at the ETH Zurich ArcTecLab building (Willmann et al., 2016), a custom prefabricated system, but is also in use to fabricate standard modular timber structures. US contractor Digital Building Components has recently commercialized robotic assembly of welded metal stud wall and facade panels (www.digitalbuilding.com). An example of additive manufacturing applied in prefabrication is Branch Technology, a US-based startup using a polymer extrusion technology to create free-form architectural elements currently expanding its capacity beyond the production of non-loadbearing interior elements and open pavilion structures (www.branch.technology).

The second trend is the emergence of specialized consultants streamlining digital fabrication applications in collaboration with digitally advanced contractors within the framework of construction projects. The collaboration of contractors or design practices with an external specialized technology integrator offers a more flexible approach to the development of innovative digital fabrication workflows. The involvement of a new type of consultant administering design to fabrication workflows and producing fabrication data from design documents and models can expand design, engineering and fabrication capabilities simultaneously over the span of multiple projects. One example of this trend is Design-to-Production, a Swiss-based consultancy that specializes in workflow development linking design activities to digital manufacturing and assembly by means of parametric CAD modelling, software design optimization and constructability analysis (designtoproduction.com). Among many other projects, this strategy has helped the realization of a series of noteworthy free-form timber structures such as the Nine Bridges Golf Club (Scheurer, 2019) in Seoul, South Korea (2008) and the Centre Pompidou (Figure 10.2) in Metz, France (2008), both by Shigeru Ban Architects, the Cambridge Mosque (2017) by Marks Barfield Architekten, and the Omega Swatch HQ in Bienne, Switzerland (under construction) by Shigeru Ban Architects.

The adoption of digital fabrication strategies in AEC is in its early stages, in product innovation in digital fabrication systems as well as in process innovation. Merging these two yet separate trends in the future should lead to exciting new possibilities in the ways digital fabrication will shape design and construction.

Figure 10.2 Roof module, Heasley Nine Bridges Golf Resort by Shigeru Ban Architects in Seoul, South Korea (2008)

10.4 State of research in digital fabrication

A wide array of tools, techniques and approaches comprise the current state of research in digital fabrication. In addition to the common CNC toolset that has traditionally signified work in the field (Carpo, 2017), there are now robots that fly, weave, climb, rove, and perform intricate collaborative work with other humans and other robots. Countless types of end-effectors for robotic arms have been configured and utilized, ranging from highly customized mechanisms to novel incorporations of standard industrial tools such as chainsaws, bandsaws and welders.

Much of today's research focuses on physical fabrication principles, such as: large scale robotic fabrication or 3D printing (Hack et al., 2017; Jipa, Bernhard, Meibodi, and Dillenburger, 2016; Kaseman and Anderson, 2016; McGee, Velikov, Thun, and Tish, 2018; Parascho et al., 2017); the control and optimization of complex structures and material systems (Fleischmann and Menges, 2011; Rippmann et al., 2016); digitally controlled material processing methods (Lloret-Fritschi et al., 2018; Wangler et al., 2016); or on automation and control of construction tools (Helm, Ercan, Gramazio, and Kohler, 2012; Dörfler et al., 2016; Giftthaler et al., 2017, Hutter et al. 2016; Shahmiri and Gentry, 2016).

Advancements in *on-site robotic fabrication* have been demonstrated through work performed by MIT Digital Construction Platform (Keating et al., 2014), ICD/ITKE Research Pavilion 2014–2015 in which an integral pneumatic form was internally lined with robotically distributed carbon fiber thus achieving structural integrity through a composite skin (Knippers and Menges, 2018) (Figure 10.3), and Rock Print Pavilion in Winterthur, CH (2018), which advanced ETH research into Jammed Architectural Structures (JAS) composed of crushed stone and string (Aejmelaeus-Lindström et al., 2018).

Unmanned aerial systems (UAS) are now also intricately woven into digital production workflows, as has been recently demonstrated through work including: ICD/ITKE Research

Figure 10.3 ICD/ITKE Research Pavilion 2014–2015, a composite structure utilizing robotically woven carbon fibers within a pneumatic shell

Pavilion 2016–2017 (Felbrich et al., 2017), in which a drone was utilized in conjunction with two robotic arms placed relatively far apart, cyclically transferring a strand of carbon fiber between each through a weaving operation that produced a long-span cantilevered structure; and "On-Site Robotics for Sustainable Construction" (Dubor et al., 2018), which utilized UAS to capture thermal data of ceramic extruded by a large cable robot, provide feedback to the system and thus regulate the printing operation accordingly.

As is evident in the sample outlined above, the current state of digital fabrication research involves intricate configurations of digital tools and operational protocols. Demonstrations performed through such *multisystem platforms* provide new conceptions of the field itself, further expanding the definition and projected scopes of digital fabrication.

10.5 Research demonstrator case study: DFAB HOUSE (ETH, 2017–2019)

10.5.1 Introduction and background

DFAB HOUSE is a building-scale demonstrator of digital fabrication in architecture completed in 2019 (Figure 10.4). It was initiated the Swiss National Centre of Competence in Research (NCCR) Digital Fabrication, a leading long-term research program in the field of digital fabrication in architecture and construction established in 2014. The NCCR's research explores the potential of a seamless combination of digital technologies and physical building

Figure 10.4   DFAB HOUSE by ETH as completed in February 2019.

processes to transform today's design and building practice. Digital fabrication at the building scale involves new technologies that cut across traditional discipline and supply chain boundaries and require new ways of system integration. To implement it in practice will require systemic, multidisciplinary innovation (Graser et al., 2019). To answer this challenge, the NCCR is set up as an interdisciplinary initiative combining research expertise from multiple fields, including architecture; civil, structural, and industrial engineering; robotics; computer science; material sciences; sustainability studies; and social sciences.

The site of DFAB HOUSE is NEST, a construction research and demonstration facility near Zurich, Switzerland, designed to help accelerate innovation in the construction sector. The initiator and operator of NEST is Empa, the Swiss Federal Laboratories for Material Science and Technology. NEST is a "living lab" with special emphasis on open innovation, technology transfer and user involvement (Richner et al., 2017). Its physical structure consists of a "backbone" infrastructure offering empty floor slabs on three levels serving as building sites for up to 15 independent modular research modules called Units. Its service core contains circulation, technical services, and a multi-modal central utility hub (Richner et al., 2017). DFAB HOUSE constitutes one Unit within NEST.

10.5.2 *Project objective*

The project objective of DFAB HOUSE was to maximize the use of novel digital fabrication technologies for the construction of all major constituting building parts. The intention of combining multiple novel processes in a single project was to rethink the overall planning and construction process and to study the effects of closing the digital thread from design to fabrication on the following aspects: design flexibility; material economy; time and cost efficiency; and improved quality control (www.dfabhouse.ch). In addition, upscaling and integration of research projects had to be achieved within a prescribed project time schedule and budget. This way, the projects seek to demonstrate that emerging digital fabrication technologies can be applied at full scale and meet the high demands of building codes, budget and time constraints, and strict performance requirements (Graser et al., 2020).

The design and construction of DFAB HOUSE also required the crossing of boundaries between research and industry. It was executed together with a network of more than 40 industry partners, including Swiss and international market and technology leaders (dfabhouse.ch). Industry partners contributed resources and expertise on several levels: at the single research project level; as partners in R&D, upscaling and execution of the technology applications for DFAB HOUSE; as specialist planners and consultants; and as execution contractors (Graser et al., 2019).

DFAB HOUSE underwent a standard permitting process to obtain its building permit and certificate of occupancy. It therefore required full compliance to Swiss building codes and regulations to ensure structural and operational safety and meet energy code standards. Beyond code compliance, NEST has an additional set of requirements pertaining to operational energy balance, on-site renewable energy generation, connectivity to the existing energy systems provided by NEST, and user comfort (Richner et al., 2018).

As a multisystem demonstrator project, DFAB HOUSE intends to prove feasibility, boost TRL, expand interdisciplinary R&D partnerships between academia and industry, and expose new ideas to industry and the public. In addition to technical aspects, it reflects upon the impact of an emerging digital practice on architectural design, as well as on the proposition of a future "digital building culture" (Gramazio Kohler, 2018).

10.5.3 Description of featured technologies

The bulk of DFAB HOUSE, including its main structural components, was digitally fabricated using six novel technologies based on NCCR Digital Fabrication research, labeled "Innovation Objects" (IOs). DFAB HOUSE IOs address both on-site digital fabrication and custom digital prefabrication. This section provides an overview of the digital fabrication technologies applied in DFAB HOUSE and references for further reading.

On-site digital fabrication

DFAB HOUSE demonstrates on-site digital fabrication in a joint application of two Innovation Objects, In situ Fabricator (IF) and Mesh Mould.

The *In situ Fabricator (IF)* is a mobile, versatile autonomous on-site construction robot developed at ETH Zurich. A further development of earlier robotic applications by Gramazio Kohler Research, such as dimRob (Helm, Ercan, Gramazio, and Kohler, 2012) and the mobile robotic production facility R-O-B (2008), the IF deployment on the DFAB HOUSE construction site constitutes the first construction application of an autonomous robot for on-site construction (www.dfabhouse.ch, Sandy et al., 2016). IF is equipped with sensing and feedback systems specifically designed to allow it to operate autonomously in unstructured environments such as construction sites, as opposed to the controlled environment of industrial production lines (Lussi et al., 2018).

Mesh Mould is a fabrication system for free-form cast-in-place concrete structures (Figure 10.5). In this system, a steel rebar mesh fabricated by a robot acts as structural reinforcement and as stay-in-place formwork, unifying the reinforcement and formwork

Figure 10.5 In situ Fabricator (IF) deploying the Mesh Mould, a fabrication system for free-form cast-in-place concrete structures, at the DFAB House

production into a single robotically controlled fabrication system (Hack et al., 2017). A customized special purpose end-effector bends, welds, and cuts steel reinforcement to additively manufacture a digitally designed mesh on site. The process is inherently waste free. In contrast to most other digital fabrication approaches in the sector, such as 3D extrusion printing, it intrinsically incorporates continuous steel reinforcement (Hack et al., 2017). This enables the construction of high performance non-standard building parts at competitive cost (García de Soto et al., 2018), material savings through structural optimization, and the reduction of construction waste reduction compared to conventional construction of similarly complex structures (Agustí-Juan et al., 2017).

Digital prefabrication

Smart Dynamic Casting (SDC) is an automated small-scale slip-forming process for concrete prefabrication. The process is based on a reusable, actuated formwork with a variable cross-section. Like Mesh Mould, SDC is a zero-waste process and allows the systematic integration of reinforcement (Lloret-Fritschi et al., 2018). The SDC process is highly automated. The setup is controlled by a custom software that synchronizes the process of material filling, cross section change of the formwork, and slipping speed. In addition, highly controlled concrete processing, using retardation and acceleration through admixtures and real-time feedback from inline friction measurement, helps produce high surface quality and size accuracy without post-processing (Lloret-Fritschi et al., 2018). This allows the fabrication of customized, non-standard elements without the use of disposable formwork. SDC was applied in DFAB HOUSE to produce non-standard prefabricated concrete facade mullions with structurally optimized geometry.

Smart Slab is a multifunctional building component fabricated with 3D printed formwork. Large-scale industrial 3D sand print technology was adapted to the context of construction. A customized generative design software took advantage of the high resolution, sub-millimeter tolerances and formal freedom of additive manufacturing to broaden the design possibilities of prefabricated concrete in architecture. The slab design took into consideration structural optimization to reduce material volume and the integration of building systems, resulting in a structure about 70% lighter than a comparable conventional concrete slab (Aghaei Meibodi et al., 2018).

Spatial Timber Assemblies is a collaborative robotic prefabrication process for non-standard modular timber prefabrication (Figure 10.6). It was the first full-scale project assembled at the Robotic Fabrication Lab at ETH Zurich, a facility with four gantry-mounted versatile industrial robots for building-scale prefabrication. A two-robot setup with an integrated CNC controlled saw was controlled by fabrication code derived from a fabrication-aware algorithmic design model. The robotic setup was equipped with an automated tool changer to switch from grippers to milling/drilling end-effector. This allowed cutting and placement of timber beams as well as high precision milling of pre-defined screw channels. Automating the complete fabrication and assembly sequence from the standard timber beam to the three-dimensional assembly allowed for bespoke spatial geometry, structural optimization and waste reduction. For example, using custom diagonal studs at no extra cost helped eliminate structural paneling and replace it with truss action, achieving material savings through geometry. The system was used to fabricate a two-story modular residential portion of the DFAB HOUSE (Thoma et al., 2019).

With the *Lightweight Translucent Façade* system, DFAB HOUSE features a high performance, aerogel insulated membrane based facade system for complex building geometries.

Figure 10.6   Spatial Timber Assemblies in production in the Robotic Fabrication Lab at ETH Zurich

The system was custom-engineered for integration with the ruled, truss-like external wall geometries of Spatial Timber Assemblies. A fabrication workflow derived from the integrated 3D model ensured precise manufacturing of eight custom-shaped, non-planar envelope panels. Building upon the properties of its digitally fabricated timber substructure, it reduced weight and footprint or the building envelope while maximizing daylighting in the building, high-lighting possible effects of digital fabrication on a building's energy performance and indoor environmental quality (dfabhouse.ch).

10.5.4 DFAB HOUSE: key lessons

DFAB HOUSE is remarkable with respect to principles of Construction 4.0, as there is no distinction between digital fabrication and construction in its conception, design or realiza-tion. DFAB HOUSE is a multi-technology demonstrator realized with academia and indus-try involvement (Figure 10.7). Its technological diversity makes DFAB HOUSE a uniquely comprehensive building-scale demonstrator of the possibilities of digital fabrication in the AEC sector today. Interdisciplinary collaboration lies at the heart of this approach. Research disciplines involved in DFAB HOUSE include robotics, structural and industrial engineer-ing, computer science, material science, digital building technologies, architecture and digital design, building physics, and sustainability science. Construction trades involve in-situ con-crete construction, concrete prefabrication, large-scale industrial 3D printing, modular tim-ber prefabrication and tensile facade manufacturing (Graser et al., 2019). Further, the DFAB HOUSE case study allows early stage observations on organizational models Industry 4.0 can offer to foster diffusion of systemic innovations such as digital fabrication technologies in the AEC industries (Graser et al., 2019).

Beyond the advances in fabrication technology and insight into innovation processes, DFAB HOUSE has been a source for further studies of the environmental assessment of dig-ital fabrication (Agustí-Juan et al., 2017), productivity of digital fabrication (García de Soto et al., 2018), and rethinking the roles in the AEC industry to accommodate digital fabrication

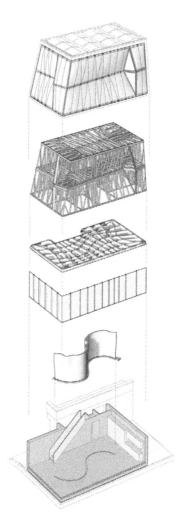

Figure 10.7 DFAB House Innovation Objects (IOs): Lightweight Translucent Façade, Spatial Timber Assemblies, Smart Slab, Smart Dynamic Casting (SDC), Mesh Mould (listed from top to bottom)

(Garcia de Soto et al., 2019). Being the first of its kind in expanding the scope of digital fabrication to a full-featured building construction, DFAB HOUSE has become a cornerstone case study for understanding the emergent potential of digital fabrication and its implications for the AEC domain.

10.6 Practice case study: SHoP architects

SHoP has uniquely operated at the forefront of digital fabrication in architecture over the past 20 years, consistently developing advanced modes of professional operation throughout its evolution as a practice. With notable projects including PS1 Dunescape in Queens, NY (2000), the aforementioned Camera Obscura at Mitchell Park in Greenpoint, NY (2002–2006), Barclays Center in Brooklyn, NY (completed 2012) (Figures 10.8 and 10.9), Botswana Innovation

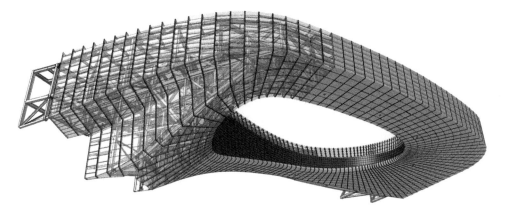

Figure 10.8   Structure and envelope model for Barclays Center in Brooklyn, NY by SHoP Architects (completed 2012)

Figure 10.9   Installation of digitally fabricated facade component assemblies at Barclays Center

Hub in Gaborone, BW (2011–present), the supertall tower currently in construction at 111 West 57th Street in Manhattan, NY, among many others, their body of work and demonstrated methods of practice have exerted significant impact throughout many portions of the AEC community. (Rice, 2014)

This section aims to synopsize key principles and excerpts from extended interviews between the authors of this chapter, William Sharples (Founding Partner, SHoP) and John Cerone (Associate Principal, SHoP) in early 2019, which focused on the current state of digital fabrication and its role as a prime mode of professional practice.

10.6.1 Digital fabrication: operational mindset

SHoP has extended its command of digital design, fabrication and delivery into sophisticated modes of operation that permeates all aspects of their practice. Positioning its work in relation to the tradition of craft management across a long historical arc, SHoP proclaims that it is essential to first understand material, fabrication and construction methods in order to develop and coordinate advanced design and fabrication workflows, schedules and costs. Enmeshing this mindset with a robust digital toolset allows SHoP to push certain boundaries in relation to both design and performance goals.

By integrating digital fabrication protocols through all phases of work, SHoP is also able to utilize project data to inform many other critical professional aspects, including such practicalities as building relationships and maintaining trust with both clients and communities, meeting timelines and performing within budget constraints. Additionally, this digitally fluid approach affords the anticipation and avoidance of many problems that typically arise when deploying digitally fabricated components into numerous scales and scopes of construction.

Such a methodology proved to be effective in facilitating the design, production and installation of the digitally fabricated facade panels for Barclays Arena, for example, in which laser-cut steel components were accurately bent with a CNC press brake and then fabricated into larger assemblies (Holden et al., 2012). These facade assemblies were then pre-weathered through a specialized factory process for a period of time before installation on site. A dynamic digital platform built for the specific manufacturing and delivery chain was utilized to manage all aspects of design and delivery, from automating printed production tickets for factory workers from updated parametric design models, to itemizing and tracking the production, delivery and installation of every assembly facilitated through customized web-based interfaces and applications.

10.6.2 Digital fabrication and human interaction

SHoP has thrived in its ability to cultivate sincere engagement in the digital production of projects, operating under the premise that when everyone understands the system and how their specific work fits within it, the channels of communication become especially fluid and productive. For example, with Botswana Innovation Hub (2011–present), SHoP knew every parameter tied to all of the Computer-Aided Manufacturing (CAM) tools that were acquired for the project in Cape Town, South Africa. More importantly, however, their team made it imperative to get to know everyone involved with all aspects of digital fabrication and delivery.

Striking productive relationships with everyone running the CNC machines, knowing specifically who will work on which parts or steps in the process, along with those who are delivering components or managing databases proved to be critical in collaboratively designing and maintaining the entire production chain from scratch. Remarkably, the factory floor and data flows were organized such that no drawings or fabrication tickets were produced for the fabrication, assembly or installation of digitally produced components, and the primary coordination channel for the digitally fabricated facade components was between updated lidar scans of the structure (Figure 10.10) and automated adjustments to the singular-source digital model, which in turn fed all CNC operations on the factory floor for the digitally fabricated facade panels. This process relied upon the infusion of human interaction into direct model-to-manufacturing.

Figure 10.10   Lidar scan of Botswana Innovation Hub under construction.

10.6.3 Digital design capacity

SHoP focuses significant attention on problem solving and developing efficient ways of tackling new challenges, citing that out-of-box software solutions are often not adequate to achieve the levels of resolution and logistics typically embedded within their work. As such, project teams customize digital tools and workflows as necessary. Relying upon a digital fabrication design fluency that is shared throughout the entire office, SHoP augments project teams as necessary with internal groups who specialize in advanced computation, database management, customized AR and VR applications along with input from diverse fields of engineering. Further, while many projects call for high-fidelity deliverables, SHoP has a core team that is well-versed with Direct-to-Manufacturing practices; this group's cumulative knowledge is actively disseminated to the whole office as the work develops through any given project.

10.6.4 Execution vs. experimentation

SHoP does not have a dedicated research and development team. Instead, customized workflows, tools and protocols are developed for real project challenges, with all efforts typically directed towards an active project or what they see as potential fabrication and construction challenges for upcoming work. Noting that their entire operation is fundamentally geared for execution, a key challenge they have found in pursuing experimentation and raw research in practice lies in the difficulty of projecting whether or not the investment in time and resources will actually pay-off and when that might happen.

In this light, SHoP continues to seek opportunities to perform research in advanced digital fabrication techniques and procedures through experimental commissions, such Flotsam and Jetsam for Design Miami and WAVE/CAVE for Milan Design Week, both designed and fabricated in 2017. For Flotsam and Jetsam, SHoP worked with Branch Technology and Oak Ridge National Laboratory to develop customized large-scale 3D-printing protocols utilizing various forms of embedded analysis through the entire design and delivery process (Daas and Wit, 2018). The focus for WAVE/CAVE centered on generating a customized material process that combined advanced digital design with the craftsmanship built into traditional Italian ceramic extrusion techniques (Overall et al., 2018).

10.6.5 Digital fabrication platform

While the field of digital fabrication has evolved from how material and geometry come together to its expanded and integral role in the design of total project team structure and logistics, many firms have sought new ways to improve upon their design, production and management processes. In recent years, SHoP has embraced Project Lifecycle Management (PLM) practices as an effective way to conceive of, design and manage projects. Operating within such principles allows for the conflation of digital fabrication, material properties, logistics and project management, while providing access to various types of live simulation, analysis, approvals, administrative tools and assembly instruction to everyone involved in the project. Custom-built data connectors and agnostic databases are utilized as platforms through which analysis and feedback are as close to real-time as possible during all phases of work. As such, criteria for stakeholder decisions is consolidated through a singular, digitally dynamic platform that entails all aspects of design and coordination from fabrication to construction. In this sense, SHoP holds the operational view that the platform *is* the project.

10.7 Conclusion

This chapter attempts to provide an overview of the emerging potential of digital fabrication in the AEC sector. In closing the digital thread from early design stages throughout the planning process through manufacturing an assembly of the built environment, it constitutes an essential step in the progressing digitalization of the AEC sector. Many factors will play a role in its progress and ultimate success, and the definition of digital fabrication will continue to evolve and expand as new technologies and methods, workflows, protocols and practices arise. New organizational models and forms of collaboration will emerge to support this development. Platforms and integrators will do their part in enabling innovation in the building process and linking digital fabrication to the growing field of industrialized construction. Building systems integration, multi-constraint optimization, and life cycle management will pave the way to more sustainable design and building practice. And lastly, new forms of man-machine collaboration and the merging of fabrication with virtual and augmented reality will open up a vast space of opportunity.

However, today the development of digital fabrication in AEC is still in its early stages. While it holds enormous potential, its first application represents a tiny fraction of the USD 1.1 trillion spent annually in the global construction industry (McKinsey & Company, 2016). Digital fabrication technologies are not yet fully commercialized or suitable for the vast majority of construction projects. We expect that future research, demonstrator projects, and innovations driven by the rapidly changing industry will ultimately transform the way we design and build. The years to come hold great promise but equally great challenges academia and practice must embrace to open new doors to the digital future of construction.

10.8 Summary

- Digital fabrication has evolved into tools, techniques and approaches in Architecture, Engineering and Construction (AEC).
- Digital fabrication has advanced into a multi-machine and multi-process environment and is expanding in scope and sophistication.
- It offers heightened capacity to deliver a fluid operational continuum spanning computation, organizational models and innovative physical building processes.
- Understanding of digital fabrication is becoming essential for all professions involved within design and construction.
- The DFAB House (ETH, 2017–2019), is a building-scale demonstrator that incorporates six novel technologies based on research initiated and performed by the Swiss National Centre of Competence in Research (NCCR) Digital Fabrication.
- The current tendencies in architectural practice are illustrated, with focus on SHoP Architects (New York), who actively participate in the evolution of the multifaceted field.

References

Aejmelaeus-Lindström, P., Rusenova, G., Mirjan, A., Gramazio, F. and Kohler, M. (2018). "Direct Deposition of Jammed Architectural Structures." In J. Willmann, P. Block, M. Hutter, K. Byrne and T. Schork, eds, *Robotic Fabrication in Architecture, Art and Design 2018*, 1st Cham: Springer Nature Switzerland AG. pp. 270–281.

Aghaei Meibodi, M., Jipa, A., Giesecke, R., Shammas, D., Bernhard, M., Leschok, M., Graser, K. and Dillenburger, B. (2018). "Smart Slab: Computational Design and Digital Fabrication of a Lightweight Concrete Slab." In P. Anzalone, M. Marcella del Signore and A. J. Andrew John Wit (eds.,) *Re/Calibration: On Imprecision and Infidelity*, Mexico City: Acadia Publishing Company. pp. 320–327.

Agustí-Juan, I., Müller, F., Hack, N., Wangler, T. and Habert, G. (2017). "Potential Benefits of Digital Fabrication for Complex Structures: Environmental Assessment of a Robotically Fabricated Concrete Wall." *Journal of Cleaner Production or International Journal of Life Cycle Assessment*, 154, pp. 330–340.

Allen, E. and Iano, J. (2004). *Fundamentals of Building Construction: Materials and Methods*. 4th Hoboken, NJ: John Wiley & Sons.

Bechthold, M. (2010). "The Return of the Future." *Architectural Design* 80(4), pp. 116–121.

Bock, T. (2004). "Construction Robotics and Automation: Past-Present-Future." *Robotics and Applications; WAC 2004; ISORA 2004; TSI Press SERIES* 15, pp. 287–294.

Bock, Thomas. (2016). "The Future of Construction Automation: Technological Disruption and the Upcoming Ubiquity of Robotics." *In Automation in Construction* 59, pp. 113–121.

Bonwetsch, T. (2012). "Robotic Assembly Processes as a Driver in Architectural Design." *Nexus Network Journal* 14(3), pp. 483–494.

Bonwetsch, T., Gramazio, F. and Kohler, M. (2007). "Digitally Fabricating Non-Standardised Brick Walls." In *ManuBuild, conference proceedings*, 191–196. Rotterdam: D. M. Sharp.

Carpo, M. (2017). *The Second Digital Turn: Design Beyond Intelligence*. Cambridge: MIT Press.

Cerone, J. and Sharples, W. (2019), Interviewed by Graser, K. and Kaseman, K., February 1, 2019, March 19, 2019 and February 19, 2019.

Daas, M. and Wit, A. eds (2018) *Towards a Robotic Architecture*. Novato: Applied Research and Design Publishing, pp. 52–54.

Dörfler, K., Sandy, T., Gifthaler, M., Gramazio, F., Kohler, M. and Buchli, J. (2016). "Mobile Robotic Brickwork – Automation of a Discrete Robotic Fabrication Process Using an Autonomous Mobile

Robot." In D. Reinhardt, R. Saunders and J. Burry, eds, *Robotic Fabrication in Architecture, Art and Design 2016*, Cham: Springer International Publishing, pp. 204–217.

Dubor, A., Izard, J., Cabay, E., Sollazzo, A., Markopoulou, A. and Rodriguez, M. (2018). "On-Site Robotics for Sustainable Construction." In J. Willmann, P. Block, M. Hutter, K. Byrne and T. Schork, eds, *Robotic Fabrication in Architecture, Art and Design 2018*, 1st Cham: Springer Nature Switzerland AG. pp. 390–401.

Felbrich, B., Früh, N., Prado, M., Saffarian, S., Solly, J., Vasey, L., Knippers, J. and Menges, A. (2017). "Multi-Machine Fabrication: An Integrative Design Process Utilising an Autonomous UAV and Industrial Robots for the Fabrication of Long-Span Composite Structures." In T. Nagakura, S. Tibbits, M. Ibañez, and C. Mueller, eds, *ACADIA 2017 Discipline & Disruption: Projects Catalog of the 37th Annual Conference of the Association for Computer Aided Design in Architecture*, Cambridge, MA: ACADIA, pp. 248–259.

Fleischmann, M. and Menges, A. (2011). "ICD/ITKE Research Pavilion: A Case Study of Multi-disciplinary Collaborative Computational Design." In C. Gengnagel, A. Kilian, N. Palz and F. Scheurer, eds, *Computational Design Modelling*, Berlin: Springer International Publishing, pp. 239–248.

Garcia de Soto, B. (2019). "The Construction Industry Needs a Robot Revolution," *Wired Magazine*, published online April 6, accessed April 12, 2019 under www.wired.com/story/the-construction-industry-needs-a-robot-revolution/.

García de Soto, B., Agustí-Juan, I., Hunhevicz, J., Joss, S., Graser, K., Habert, G. and Adey, B. (2018). "Productivity of Digital Fabrication in Construction: Cost and Time Analysis of a Robotically Built Wall." *Automation in Construction* 92, pp. 297–311.

Garcia de Soto, B., Agustí-Juan, I., Joss, S. and Hunhevicz, J. (2019). "Implications of Construction 4.0 to the Workforce and Organizational Structures." *International Journal of Construction Management* January. Published online May 20, 2019. https://doi.org/10.1080/15623599.2019.1616414.

Gershenfeld, N. (2012). "How to Make Almost Anything: The Digital Fabrication Revolution." *Foreign Affairs* 91(6), pp. 43–57.

Giftthaler, M., Farshidian, F., Sandy, T., Stadelmann, L. and Buchli, J. (2017). Efficient Kinematic Planning for Mobile Manipulators with Non-Holonomic Constraints using Optimal Control. In: *IEEE International Conference on Robotics and Automation.*

Gramazio, F., Kohler, M. and Willmann, J. (2014). *The Robotic Touch: How Robots Change Architecture*. Zurich: Park Books.

Graser, K., Wang, Y., Hoffman, M., Bonanomi, M., Kohler, M. and Hall, D. (2019). "Social Network Analysis of DFAB HOUSE: A Demonstrator of Digital Fabrication in Construction," *Proceedings of Engineering Project Organization Conference (EPOC)*, Vail, CO.

Graser, K., Baur, M., Apolinarska, A., Dörfler, K., Hack, N., Jipa, A., Lloret-Fritschi, E., Sandy. T., Sanz Pont, D. and Kohler, M. (2020). "DFAB HOUSE – A Comprehensive Demonstrator of Digital Fabrication in Architecture." In Fabricate 2020 Conference Publication, Design and Making, London (accepted).

Hack, N., Lauer, W., Langenberg, S., Gramazio, F. and Kohler, M. (2013). "Overcoming Repetition: Robotic Fabrication Processes at a Large Scale." *International Journal of Architectural Computing* 11(3), pp. 285–299.

Hack, N., Wangler, T., Mata-Falcón, J., Dörfler, K., Kumar, N., Walzera, A., Graser, K., Reiter, L., Richner, H., Buchli, J. and Kaufmann, W. (2017). Mesh Mould: An On Site, Robotically Fabricated, Functional Formwork. In: *Second Concrete Innovation Conference (2nd CIC).*

Helm, V., Ercan, S., Gramazio, F. and Kohler, M. (2012). "Mobile Robotic Fabrication on Construction Sites: Dim-Rob." In *IEEE/RSJ International Conference on Intelligent Robots and Systems*, 43, 35–41.

Holden, K., Pasquarelli, G., Sharples, C., Sharples, C. and Sharples, W. (2012). *SHoP: Out of Practice.* 1st New York: Monacelli Press, pp. 396–481.

Hutter, M., Leemann, P., Hottiger, G., Figi, R., Tagmann, S., Rey, G. and Small, G. (2017). "Force Control for Active Chassis Balancing." In *IEEE/ASME Transactions on Mechatronics*, 22(2), pp. 613–622, April. doi: 10.1109/TMECH.2016.2612722.

Jipa, A., Bernhard, M., Meibodi, M. and Dillenburger, B. (2016). 3D-Printed Stay-in-Place Formwork for Topologically Optimized. In: *TxA Emerging Design + Technology Conference 2016*. San Antonio.

Kaseman, K. and Anderson, J. (2016). dFOIL: Drone Deployment Station and Augmented Reality Application. In: K. Velikov, S. Manninger, M. Campo, S. Ahlquist and G. Thun, ed., *Acadia 2016*

Posthuman Frontiers: Data, Designers, and Cognitive Machines Projects Catalog of the 36th Annual Conference of the Association for Computer Aided Design in Architecture. Wolfville: Acadia University, pp. 152–157.

Keating, S., Spielberg, N., Klein, J. and Oxman, N. (2014). "A Compound Arm Approach to Digital Construction." In W. McGee and M. Ponce de Leon, eds, *Robotic Fabrication in Architecture, Art and Design 2014*, Cham: Springer, pp. 99–110.

Knippers, J. and Menges, A. (2018). "Fibrous Tectonics." In M. Daas and A. Wit, eds, *Towards a Robotic Architecture*, Novato: Applied Research and Design Publishing, pp. 64–75.

Kohler, Gramazio. (2018). "Are We Approaching a Digital Building Culture?." In *Royal Academy of Arts Talk*, London: Royal Academy of Arts. Accessible online at www.royalacademy.org.uk/event/gramazio-kohler, accessed on October 1, 2019.

Lloret-Fritschi, E., Scotto, F., Gramazio, F., Kohler, M., Graser, K., Wangler, T., Reiter, L., Flatt, R. and Mata-Falcón, J. (2018). "Challenges of Real-Scale Production with Smart Dynamic Casting." In T. Wangler and R. Flatt, eds, *First RILEM International Conference on Concrete and Digital Fabrication – Digital Concrete 2018*, 1st Springer International Publishing.

Lussi, M., Sandy, T., Dorfler, K., Hack, N., Gramazio, F., Kohler, M. and Buchli, J. (2018). Accurate and Adaptive in Situ Fabrication of an Undulated Wall Using an On-Board Visual Sensing System. *2018 IEEE International Conference on Robotics and Automation (ICRA)*, pp. 1–8.

Marble, S. (ed) (2012). *Digital Workflows in Architecture*, Basel: Birkhäuser.

McGee, W., Velikov, K., Thun, G. and Tish, D. (2018). "Prologue for a Robotic Architecture." In.

McKinsey & Company (2016). Imagining Construction's Digital Future. Technical Report, June.

Overall, S., Rysavy, J., Miller, C., Sharples, W., Sharples, C., Kumar, S., Vittadini, A. and Saby, V. (2018). "Made-to-Measure: Automated Drawing and Material Craft." *Technology\Architecture + Design* 2(2), pp. 172–185.

Parascho, S., Gandia, A., Mirjan, A., Gramazio, F. and Kohler, M. (2017). "Cooperative Fabrication of Spatial Metal Structures." In A. Menges, B. Sheil, R. Glynn and M. Skavara, eds, *Fabricate: Rethinking Design and Construction*. London: UCL Press.

Rice, A. (2014). From Barclays Center to Modular High Rises, SHoP Architects is Changing the Way We Build Buildings. *Fast Company*, [online] pp. 114–130. Available at: www.fastcompany.com/3025601/shop-architects-the-new-skyline [Accessed July 30, 2019].

Richner, P. Heer, Ph., Largo, R., Marchesi, E. and Zimmermann, E. (2017). "NEST – A Platform for the Acceleration of Innovation in Buildings." *Informes de la Construcción*, 69(548), pp. 222ff.

Rippmann, M., Mele, T., Popescu, M., Augustynowicz, E., Echenagucia, T., Barentin, C., Frick, U. and Block, P. (2016). "The Armadillo Vault: Computational Design and Digital Fabrication of a Freeform Stone Shell." In S. Adriaenssens, F. Gramazio, M. Kohler, A. Menges and M. Pauly, eds, *In Advances in Architectural Geometry 2016*, Zurich: vdf Hochschulverlag AG, pp. 344–363.

Sandy, T., Giftthaler, M., Dorfler, K., Kohler, M. and Buchli, J. (2016). "Autonomous Repositioning and Localization of an In Situ Fabricator." *2016 IEEE International Conference on Robotics and Automation, ICRA 2016, Stockholm, Sweden, May 16–21, 2016*. IEEE.

Scheurer, F. (2019). "Digital Craftsmanship: From Thinking to Modeling to Building." In S. Marble, ed, *Digital Workflows in Architecture: Design-Assembly-Industry*, Basel: Birkhäuser, pp. 118–123.

Shahmiri, F. and Gentry, R. (2016) A Survey of Cable-Suspended Parallel Robots and their Applications in Architecture and Construction, in *Blucher Design Proceedings*, SIGraDi 2016, pp. 914–920.

Sharples Holden Pasquarelli/SHoP. (2002). "Eroding the Barriers." *Architectural Design: Versioning: Evolutionary Techniques in Architecture* 72(5), pp. 90–100.

Taylor, M., Wamiziri, S. and Smith, I. (2003). "Automated Construction in Japan," in *Proceedings of the Institution of Civil Engineers, Civil Engineering* 156, No.12562, pp. 34–41.

Thoma, A., Adel, A., Helmreich, M., Wehrle, T., Gramazio, F. and Kohler, M. (2019). "Robotic Fabrication of Bespoke Timber Frame Modules." In J. Willmann, P. Block, M. Hutter, K. Byrne and T. Schork, eds, *Robotic Fabrication in Architecture, Art and Design 2018*, 1st Cham: Springer Nature Switzerland AG. pp. 447–458.

van Berkel, B. (2012). "Diagrams, Design Models and Mother Models." In S. Marble, ed, *Digital Workflows in Architecture: Design-Assembly-Industry*, Basel: Birkhäuser, pp. 74–89.

Wangler, T., Lloret, E., Reiter, L., Hack, N., Gramazio, F., Kohler, M., Bernhard, M., Dillenburger, B., Buchli, J., Roussel, N. and Flatt, R. (2016). "Digital Concrete: Opportunities and Challenges." *RILEM Technical Letters* 1, p. 67.

Willmann, J., Knauss, M., Bonwetsch, T., Apolinarska, A., Gramazio, F. and Kohler, M. (2016). "Robotic Timber Construction – Expanding Additive Fabrication to New Dimensions." *Automation in Construction* 61, pp. 16–23.

World Economic Forum in collaboration with Boston Consulting Group (2016). *Shaping the Future of Construction, A Breakthrough in Mindset and Technology.* [online] www3.weforum.org. Available at: www3.weforum.org/docs/WEF_Shaping_the_Future_of_Construction_full_report__.pdf [Accessed September 9, 2017].

11

USING BIM FOR MULTI-TRADE PREFABRICATION IN CONSTRUCTION

Mehrdad Arashpour and Ron Wakefield

11.1 Aims

- Evaluate the performance of prefabrication as a core component of Construction 4.0.
- Explore the interface between multi-trade prefabrication (MTP) and building information modeling (BIM).
- Analyze the process of BIM implementation for MTP and optimize relevant decisions.

11.2 Introduction – prefabrication as a core component of Construction 4.0

The construction industry accounts for a large share of global GDP (Global Construction Perspectives and Oxford Economics, 2011). However, evidence of below par performance in construction projects has been recognized by government and industry bodies around the world (Cole and Hudson, 2003; Salidjanova, 2011). This problem has been partially caused by variability in resource availability and shortage of specialized contractors (Thomas and Flynn, 2011; Gurevich and Sacks, 2014). Current methods of assigning processes to individually specialized contractors are resulting in excessive fragmentation and inflexibility in the construction industry (Kumaraswamy and Matthews, 2000; Polat, 2015). In such settings, trades with the greatest work content limit the progress rate and productivity of the whole project (Loosemore and Andonakis, 2007).

Offsite prefabrication has significant potential to improve productivity and performance of the construction industry. Prefabrication is a unique hybrid of manufacturing and construction in which structural and nonstructural elements are manufactured in controlled factory environments (Goulding, Pour Rahimian, Arif, and Sharp, 2014). However, at present, offsite prefabrication is being criticized for replicating the traditional subcontracting approach and therefore the fragmented practice in the construction industry (Global_Construction, 2015). Operations in this environment are often undertaken without the necessary coordination to prevent work blockages in the production network (Goulding et al., 2012). Therefore, there is currently not much difference between on-site and offsite construction processes where defragmentation initiatives such as multi-trade prefabrication are yet to be adopted.

Flexibility in prefabrication networks is increased by MTP since production networks are enabled to dynamically address variability in demand and resource availability (Bartholdi III

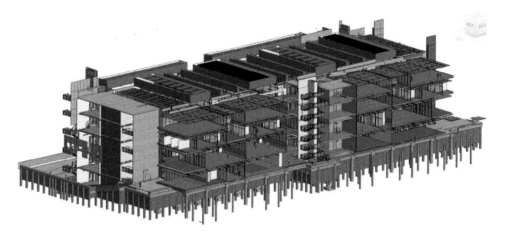

Figure 11.1 Case project – an educational complex that deploys MTP and involves various prefabricated components and multiple trades

et al., 2006, Alvanchi et al., 2012). The focus of previous research has been on designing various work integration architectures that can be used for MTP (Arashpour et al., 2018b; Nasirian et al., 2019). This is due to the fact that coordination of MTP is notoriously difficult because of various decision criteria and parameters involved (Seresht and Fayek, 2015).

The current chapter analyzes the use of BIM for MTP as a core component of Construction 4.0, where BIM is capable of improving constructability, safety, knowledge transfer and coordination. The structure of this chapter is as follows. First, the most important applications of building information modeling for multi-trade prefabrication are identified and explained. Then, the implementation process of BIM for MTP is modelled using system dynamics methods in a case project (Figure 11.1). Finally, conclusions are drawn at the end of the chapter.

11.3 Background – applications of BIM for multi-trade prefabrication

BIM has become pivotal to the prefabrication industry as it has increased collaboration between project trades, thereby improving the processes that are sequentially done in traditional construction (Eadie et al., 2013; Evelyn, George, Imelda, and Hanjoon, 2017). Decisions made in the early design stages has a significant influence on the actual construction (Arashpour et al., 2018a). Traditional design methods are limited in terms of continually analyzing constructability during the design process due to fragmented information (Merschbrock et al., 2018). BIM models are used as databases for data exchange and integration for MTP. Regarding the design phase, BIM allows information required by trades to be superimposed on one model, which creates an opportunity for constructability measures to be incorporated throughout design processes (Hosseini et al., 2018; Kim et al., 2016). With the aid of BIM applications, trades can more effectively access information related to integrated processes. BIM can also help designers to optimize configurations for MTP during early phases of new design (Chien et al., 2014; Oraee et al., 2017).

Another important application of BIM for MTP is related to the control of variability in offsite manufacture and assembly. To control geometric variability in prefabrication of building elements, all discrepancies should be quantified precisely. This is possible by comparing as-designed and as-built states and measuring deviation between the two (Sharma, Sawhney, and Arif, 2017; Wee et al., 2017). When the as-designed status is presented by BIM models of different elements and assemblies, there are different methods to capture the as-built status such as registration of three-dimensional cloud points generated by laser scanners (Balado et al., 2017; Bueno et al., 2018; Bueno et al., 2017). After registration of 3D point clouds, they are compared with BIM models, representing the as-designed status. This comparison quantifies existing deviations at critical interface points and informs decisions on best corrective actions.

BIM for MTP can also facilitate safety management (Sun et al., 2019; Tomek and Matějka, 2014). Construction projects are unique, fragmented and complex, requiring temporary and multidisciplinary teams. Therefore, effective knowledge transfer would prevent similar safety breaches, reduce risks of safety hazards, and enhance safe work practices (Ni et al., 2018; Sun. et al., 2015). In addition to the safety knowledge in the form of personal experience, and project-specific information, numerous safety rules and regulations, best practices, and accident records exist in diverse and fragmented formats (Hsu et al., 2015; Mignone et al., 2016). Moreover, emerging sensor-based safety management systems will create enormous data, which needs to be processed, analyzed, and transformed into knowledge. Bilal et al. (2016) identified several potential opportunities for utilizing Big Data within BIM, such as linking BIM with external sources, using BIM for the Internet of Things (IoT) applications, and enabling Big Data storage systems that are compliant with Industry Foundation Classes (IFC). The research surrounding knowledge management can be categorized into three groups: (1) knowledge capture and reuse; (2) knowledge sharing; and (3) knowledge storage and retrieval (Abdulaal et al., 2017; Aram et al., 2014; Wang and Meng, 2019).

BIM platforms are developed using an object-oriented structure to enable utilization of customized parameters in the models for capturing additional information related to a specific building element. In addition, BIM as the central platform increases reusability of the captured knowledge. For instance, information about a particular solution that has been proved effective in previous projects can be easily retrieved when encountering a similar problem. Similarly, the information concerning safety as related to the construction of a building element can be stored as a parameter of the corresponding building element in the BIM model, thereby facilitating capture and reuse of the relevant knowledge.

The most effective safety risk control for MTP is elimination of risks as a significant portion of risks arising during construction are inherent to design and can be eliminated (Behm, 2005; Zou et al., 2016). Conditions leading to construction accidents are often identifiable during the design phase (Xiahou et al., 2018; Zhang et al., 2016). Identification of design choices with inherent safety problems has been significantly facilitated by the use of BIM, as it enables analysis of all building elements and their relations in a single platform (Malekitabar et al., 2016; Xu et al., 2014). A number of studies have focused on development of BIM-based tools to identify violations of safety standards using rule-checking methods and to provide decision support in the early stages of construction (Poghosyan et al., 2018). The process requires safety rules to be translated into machine-interpretable language, and correlations between various building elements and safety rules need to be specified (Hongling, Yantao, Weisheng, and Yan, 2016). For instance, Cooke et al. (2008) developed a prototype for assisting design engineers in making safety-conscious decisions, utilizing argument trees to estimate the likelihood of a hazard based on certain design choices. Mirahadi et al. (2019) developed a framework for evaluating evacuation performance of a building using two risk indices, crowd simulation, and BIM in the format of IFC to assist designers in optimizing the building layout for better safety performance in case of an emergency. Yuan et al. (2019) created a Prevention through Design (PtD) knowledge base using safety design regulations, documents, and best practices.

The emergence of new innovative technologies and techniques is transforming the process of safety management for MTP (Edirisinghe, 2018). Many of the developed frameworks, however, are standalone applications. In recent years, BIM has significantly enhanced the process of project delivery (Tang et al., 2019). BIM provides geometrical information at the component level, as well as enabling effective management of metadata about various aspects of the project. In addition to spatial information, it allows incorporation of project schedule, which provides numerous benefits to project management. BIM, therefore, is a highly suitable platform for the integration of sensor-based safety management systems. The information obtained from various sensor-based monitoring systems can be incorporated with BIM for real-time monitoring of the construction site. The visualization capability of BIM renders it as an effective tool for efficient communication of real-time information with site managers.

In addition to automated safety monitoring capabilities that this approach provides, in case of an arising unsafe and hazardous situation, safety managers can be quickly notified of the exact location, and the current state of the site, and therefore act in a timely manner to address the hazard and take appropriate measures to control the risks. A number of studies have explored using wireless network systems in conjunction with BIM for enhanced real-time safety monitoring. In these studies, environmental factors such as temperature, humidity, and oxygen levels are monitored and visualized within a BIM platform (Cheung et al., 2018; Parn et al., 2019; Riaz et al., 2014).

Dong et al. (2018) developed a framework, integrating real-time location systems with pressure sensors embedded in personal protective equipment (PPE) to identify and monitor the misuse of PPE. Using a BIM environment accessible through a web application, the managers can specify danger zones, and the system can automatically identify misuse of PPE in hazardous areas. In another study, Park et al. (2017) used Bluetooth Low Energy (BLE) mobile tracking sensors to track workers' locations in real-time. Hazardous areas are specified in the BIM platform by the managers, and workers' presence in unsafe zones are automatically detected.

Although previous studies have highlighted benefits of multi-trade prefabrication (MTP) based on building information modeling (BIM), the implementation process has been challenging. The next section of this chapter investigates the causes for such challenge including lack of proper BIM training.

11.4 Decision making on BIM for multi-trade prefabrication

11.4.1 Dynamics of innovation diffusion

In conventional construction settings, head contractors often engage several trades including tilers, plasterers, carpenters, plumbers, and electricians. Multi-trade prefabrication, however, coordinates trades to achieve a synchronized delivery and significantly lowers construction costs (Hamdan et al., 2016). Using BIM for MTP facilitates coordination and communication amongst client, head contractor, trades and supply chain members (Jang, 2018). In the long term, there will be time and cost reductions for labor and logistics along with higher quality and lower defect rates.

To analyze the implementation process of BIM for MTP, a case project (Figure 11.1 and 11.2) was selected. The project is an educational complex that deploys MTP and involves various prefabricated components and multiple trades. The selected building has a steel structure and a total construction period of 24 months.

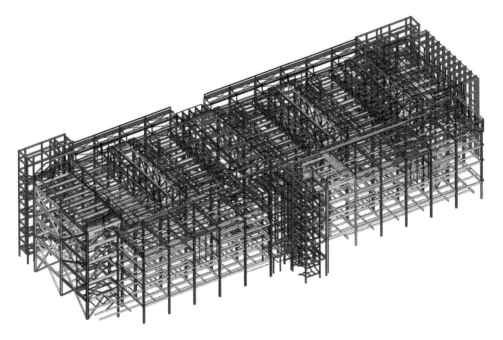

Figure 11.2   Structural BIM model of the case project

From a system point of view, the process of BIM adoption as a result of interactions amongst trades is not trivial to model. Decision making on the optimum level of client investment to optimize the use of BIM for MTP is critical to prefabrication projects. Although previous research shows that tier 2 and 3 contractors are slow to adopt BIM, interactions within the MTP environment can accelerate the adoption process (Ghaffarianhoseini et al., 2016). As modeled in Figure 11.3, trades can be divided into two groups of BIM users (innovators) and potential BIM adopters (imitators) in prefabrication projects. Mathematically, the model of MTP in Figure 11.3 is similar to a Riccati equation with constant coefficients (Bittanti et al., 2012).

Dynamic process simulation and multi-objective optimization can be used to model the use of BIM for MTP. From a system dynamics perspective, BIM adoption acts as an inflow to the BIM user population over the time. There are other parameters affecting trades' decision to adopt the technology including peer influence and the extent to which BIM is found useful by trades (Oraee et al., 2017). As can be seen in Figure 11.4, the adoption process in a medium-size project including 40 trade groups will reach equilibrium in around 20 months.

Results of system dynamics simulations show that population of BIM adopters increases with a non-linear trend over the project duration. In the analyzed case, the first 50% of population (innovators) have adopted BIM quickly (requiring less than 40% of the project duration). The second half of trades (imitators) required significantly longer times to start using BIM for MTP. Another important observation is related to the BIM adoption rate. As can be seen in Figure 11.5, maximum rate of BIM adoption belongs to months seven and eight. After reaching the peak, the adoption rates decrease again that is related imitator attitudes in slow BIM adoption.

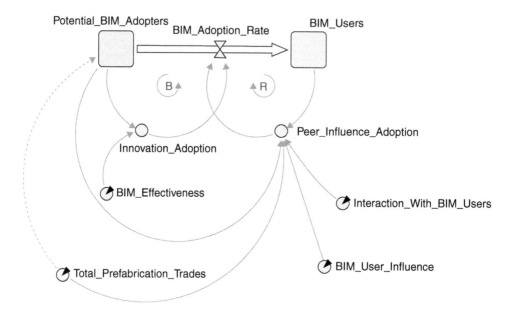

Figure 11.3   BIM for MTP – diffusion of innovations within the Construction 4.0 context

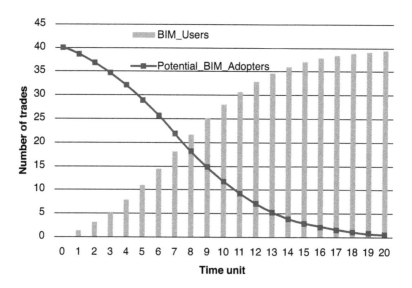

Figure 11.4   Number of BIM users (innovators) vs. potential BIM adopters (imitators)

11.4.2 *Optimum training decisions on BIM for multi-trade prefabrication*

Previous research has shown the effectiveness of training and relevant investments on BIM adoption (Merschbrock et al., 2018). Training can provide knowledge support for trades to overcome challenges of BIM implementation (Bradley et al., 2016). With increasing pressure

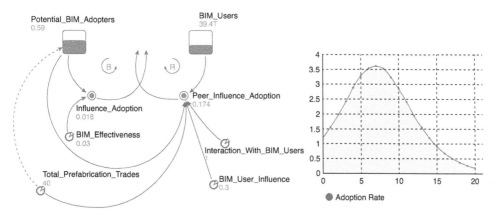

Figure 11.5   Simulation results – number of BIM adopters influenced by dynamic variables

to adopt BIM in publicly funded projects, it is critical to invest in training trade contractors (Lam et al., 2017). Prefabrication provides an ideal setting for BIM training where interaction of multi trades supports peer learning processes (J. and David, 2017). This results in optimization of investment in training so that faster BIM implementation does not come at excessive training costs (Arashpour and Aranda-Mena, 2017).

In the first system model (Figure 11.3), training effectiveness for multi-trade prefabrication is assumed to be constant. To better reflect real-world conditions, the system model in Figure 11.6 is improved to be able to manage BIM training expenditures. Changing the monthly promotion expenditures we will affect the advertising effectiveness. The dynamic variable of BIM effectiveness is now influenced by training expenditures.

Simulation results of the first system model in Figure 11.4 show the fast BIM implementation for MTP in early project times (first 40% of the total project duration). Consequently, it will be optimum to only continue BIM training and relevant expenditures to this point of time. Understandably the BIM implementation for MTP will continue after that as a result of peer learning and influence by earlier BIM users. The state chart added to the bottom of the model in Figure 11.4 defines the model behavior and causes the transition from project period with BIM training to the next.

Results of system dynamics simulations in Figure 11.7 show the effectiveness of BIM training for MTP in increasing the population of BIM adopters immediately after starting the project. As can be seen, after stopping the investments on BIM training in the fifth month, the adoption rates decreases by continues steadily until the very end of project. Implementation of BIM after the training period is driven by peer learning and influence by earlier BIM users.

Optimum training decision on BIM for multi-trade prefabrication requires the best training plan to reach the maximum BIM implementation at the specified moment of time with minimal expenditures. The objective in this study is to reach 80% BIM implementation at the end of the eighth month. To achieve this objective, selected model parameters including monthly expenditure on BIM training and desirable duration for training are systematically adjusted. As can be seen in Figure 11.8a, the total cost of training is systematically decreased in optimization experiments. The optimized values for monthly expenditure and training duration are fed into the dynamic simulation model (Figure 11.6) to yield results. As can be seen in Figure 11.8b, the optimized training strategy results in significant increase in BIM implementation as compared to the base case.

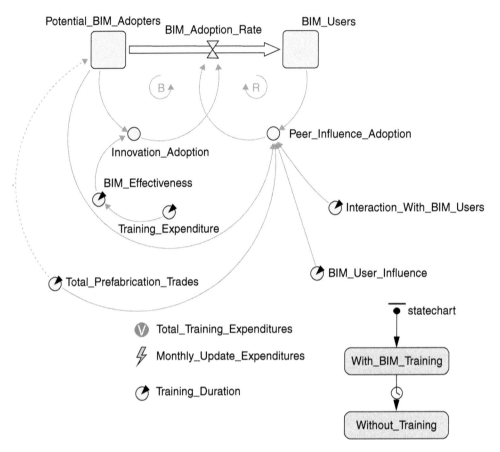

Figure 11.6 Training effect on BIM for MTP implementation

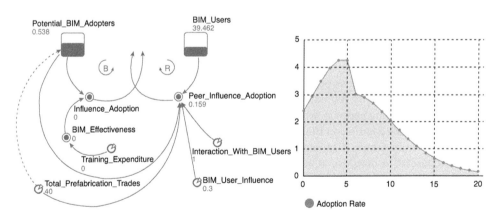

Figure 11.7 Importance of BIM training for MTP in increasing early implementation

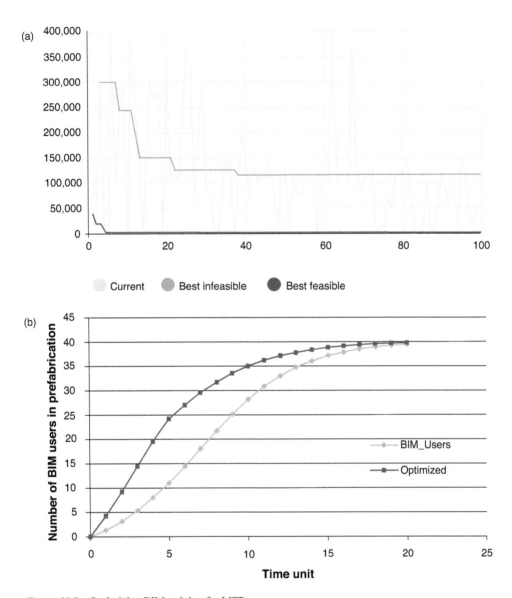

Figure 11.8 Optimizing BIM training for MTP

11.5 Conclusion

This chapter examined the implementation of BIM for multi-trade prefabrication in construction. Offsite prefabrication has significant potential to improve productivity and performance of the construction industry. However, operations in this environment are often undertaken without the necessary coordination to prevent work blockages in the production network. Flexibility in prefabrication networks is increased by MTP since production networks are enabled to dynamically address variability in demand and resource availability.

BIM is pivotal to the prefabrication industry as it improves collaborations between project trades, knowledge transfer, quality management, and safety monitoring. BIM models are used as databases for data exchange and integration for MTP. In the design phase, BIM allows information required by trades to be superimposed on one model, which creates an opportunity for constructability measures to be incorporated throughout design processes. BIM for MTP can also facilitate safety management. Construction projects are unique, fragmented, and complex, requiring temporary and multidisciplinary teams. Therefore, effective knowledge transfer through BIM can prevent similar safety breaches, reduce risks of safety hazards, and enhance safe work practices. BIM for MTP is also used to control geometric variability in prefabrication of building elements. This is possible by comparing as-designed and as-built states and measuring deviation between the two.

For successful implementation of BIM for MTP, an optimized training program is required to provide knowledge support for trades to overcome challenges. With increasing pressure to adopt BIM in publicly funded projects, it is critical to invest in training trade contractors. Prefabrication provides an ideal setting for BIM training where interaction of multi trades supports peer learning processes. The modeling framework provided in Section 11.4 of this chapter results in optimization of investment in training so that faster BIM implementation does not come at excessive training costs.

11.6 Summary

- Although offsite prefabrication can improve productivity and performance of the construction industry, operations can be effected without sufficient coordination to prevent work blockages.
- Production networks that dynamically address variability in demand and resource availability through MTP increase flexibility in prefabrication networks.
- BIM improves collaboration between project trades, knowledge transfer, quality management and safety monitoring to enable and enhance offsite production.
- BIM models are used as databases for data exchange and integration for MTP.
- Information for trades is superimposed on one BIM model so constructability measures can be incorporated throughout design processes and facilitates safety management.
- Effective use of BIM supports safe operations in construction projects, which are unique, fragmented and complex, requiring temporary and multidisciplinary teams.
- BIM for MTP controls geometric variability in prefabrication of building elements by comparing as-designed and as-built states, measuring variance.
- Successful implementation of BIM for MTP requires optimized training programs to provide knowledge support for trades – investment here is critical.
- Prefabrication is an ideal setting for BIM training where interaction of multi trades supports peer learning processes.

References

Abdulaal, B., A. Bouferguene, and M. Al-Hussein (2017). "Benchmark Alberta's architectural, engineering, and construction industry knowledge of building information modelling (BIM)." *Canadian Journal of Civil Engineering* **44**(1): 59–67.

Alvanchi, A., S. Lee, and S. M. Abourizk (2012). "Dynamics of workforce skill evolution in construction projects." *Canadian Journal of Civil Engineering* **39**(9): 1005–1017.

Aram, S., C. Eastman, and R. Sacks (2014). A knowledge-based framework for quantity takeoff and cost estimation in the AEC industry using BIM. 31st International Symposium on Automation and Robotics in Construction and Mining, ISARC 2014, University of Technology Sydney.

Arashpour, M. and G. Aranda-Mena (2017). Curriculum renewal in architecture, engineering, and construction education: Visualizing building information modeling via augmented reality. 9th

International Structural Engineering and Construction Conference: Resilient Structures and Sustainable Construction, ISEC 2017, ISEC Press.

Arashpour, M., Y. Bai, V. Kamat, R. Hosseini, and I. Martek (2018a). Project production flows in off-site prefabrication: BIM-enabled railway infrastructure. 35th International Symposium on Automation and Robotics in Construction and International AEC/FM Hackathon: The Future of Building Things, ISARC 2018, International Association for Automation and Robotics in Construction I.A.A.R.C).

Arashpour, M., R. Wakefield, B. Abbasi, M. Arashpour, and R. Hosseini (2018b). "Optimal process integration architectures in off-site construction: Theorizing the use of multi-skilled resources." *Architectural Engineering and Design Management* **14**(1–2): 46–59.

Balado, J., L. Díaz-Vilariño, P. Arias, and M. Soilán (2017). "Automatic building accessibility diagnosis from point clouds." *Automation in Construction* **82**: 103–111.

Bartholdi III, J. J., D. D. Eisenstein, and Y. F. Lim (2006). "Bucket brigades on in-tree assembly networks." *European Journal of Operational Research* **168**(3): 870–879.

Behm, M. (2005). "Linking construction fatalities to the design for construction safety concept." *Safety Science* **43**(8): 589–611.

Bilal, M., L. O. Oyedele, J. Qadir, K. Munir, S. O. Ajayi, O. O. Akinade, H. A. Owolabi, H. A. Alaka, and M. Pasha (2016). "Big Data in the construction industry: A review of present status, opportunities, and future trends." *Advanced Engineering Informatics* **30**(3): 500–521.

Bittanti, S., A. J. Laub, and J. C. Willems (2012). *The Riccati Equation*, Springer Science & Business Media.

Bradley, A., H. Li, R. Lark, and S. Dunn (2016). "BIM for infrastructure: An overall review and constructor perspective." *Automation in Construction* **71**: **Part 2** 139–152.

Bueno, M., F. Bosché, H. González-Jorge, J. Martínez-Sánchez, and P. Arias (2018). "4-Plane congruent sets for automatic registration of as-is 3D point clouds with 3D BIM models." *Automation in Construction* **89**: 120–134.

Bueno, M., H. González-Jorge, J. Martínez-Sánchez, and H. Lorenzo (2017). "Automatic point cloud coarse registration using geometric keypoint descriptors for indoor scenes." *Automation in Construction* **81**: 134–148.

Cheung, W. F. F., T. H. H. Lin, and Y. C. C. Lin (2018). "A real-time construction safety monitoring system for hazardous gas integrating wireless sensor network and building information modeling technologies." *Sensors (Switzerland)* **18**(2).

Chien, K. F., Z. H. Wu, and S. C. Huang (2014). "Identifying and assessing critical risk factors for BIM projects: Empirical study." *Automation in Construction* **45**: 1–15.

Cole, T. and R. Hudson (2003). Final report of the royal commission into the Australian building and construction industry. Melbourne.

Cooke, T., H. Lingard, N. Blismas, A. Stranieri, A. S. H. L. T. Cooke and N. Blismas (2008). "ToolSHeDTM: The development and evaluation of a decision support tool for health and safety in construction design." *Engineering, Construction and Architectural Management* **15**(4): 336–351.

Dong, S., H. Li, and Q. Yin (2018). "Building information modeling in combination with real time location systems and sensors for safety performance enhancement." *Safety Science* **102**: October, 226–237.

Eadie, R., M. Browne, H. Odeyinka, C. McKeown, and S. McNiff (2013). "BIM implementation throughout the UK construction project lifecycle: An analysis." *Automation in Construction* **36**: 145–151.

Edirisinghe, R. (2018). "Digital skin of the construction site: Smart sensor technologies towards the future smart construction site." *Engineering, Construction and Architectural Management* **26**(2): 184–223.

Evelyn, T. A. L., O. George, T. Imelda, and K. Hanjoon (2017). "Framework for productivity and safety enhancement system using BIM in Singapore." *Engineering, Construction and Architectural Management* **24**(6): 1350–1371.

Ghaffarianhoseini, A., J. Tookey, A. Ghaffarianhoseini, N. Naismith, S. Azhar, O. Efimova, and K. Raahemifar (2016). "Building Information Modelling (BIM) uptake: Clear benefits, understanding its implementation, risks and challenges." *Renewable and Sustainable Energy Reviews*, 75: 1046–1053.

Global_Construction (2015). *A Global Forecast for the Construction Industry to 2030*, London: Oxford Economics and Global Construction Perspectives.

Global Construction Perspectives and Oxford Economics (2011). "Global construction 2020: A global forecast for the construction industry over the next decade to 2020." Final report 3.

Goulding, J., W. Nadim, P. Petridis, and M. Alshawi (2012). "Construction industry offsite production: A virtual reality interactive training environment prototype." *Advanced Engineering Informatics* **26**(1): 103–116.

Goulding, J. S., F. Pour Rahimian, M. Arif, and M. D. Sharp (2014). "New offsite production and business models in construction: priorities for the future research agenda." *Architectural Engineering and Design Management* **11**(3): 163–184.

Gurevich, U. and R. Sacks (2014). "Examination of the effects of a KanBIM production control system on subcontractors' task selections in interior works." *Automation in Construction* **37**: 81–87.

Hamdan, S. B., A. Alwisy, M. Al-Hussein, S. Abourizk, and Z. Ajweh (2016). Simulation based multi-objective cost-time trade-off for multi-family residential off-site construction. Winter Simulation Conference, WSC 2015, Institute of Electrical and Electronics Engineers Inc.

Hongling, G., Y. Yantao, Z. Weisheng, and L. Yan (2016). *BIM and Safety Rules Based Automated Identification of Unsafe Design Factors in Construction*, Tsinghua University, Beijing 100084. Beijing: Elsevier Ltd.

Hosseini, M. R., I. Martek, E. Papadonikolaki, M. Sheikhkhoshkar, S. Banihashemi, and M. Arashpour (2018). "Viability of the BIM Manager Enduring as a Distinct Role: Association Rule Mining of Job Advertisements." *Journal of Construction Engineering and Management* **144**(9): 04018085.

Hsu, K. M., T. Y. Hsieh, and J. H. Chen (2015). "Legal risks incurred under the application of BIM in Taiwan." *Proceedings of the Institution of Civil Engineers: Forensic Engineering* **168**(3): 127–133.

J., G. B. and G. David (2017). "The adoption of 4D BIM in the UK construction industry: An innovation diffusion approach." *Engineering, Construction and Architectural Management* **24**(6): 950–967.

Jang, S. (2018). "Comparative analysis of multi-trade prefabrication construction methods." *Journal of Asian Architecture and Building Engineering* **17**(3): 503–509.

Kim, K., Y. Cho, and S. Zhang (2016). "Integrating work sequences and temporary structures into safety planning: Automated scaffolding-related safety hazard identification and prevention in BIM." *Automation in Construction* **70**: 128–142.

Kumaraswamy, M. M. and J. D. Matthews (2000). "Improved subcontractor selection employing partnering principles." *Journal of Management in Engineering* **16**(3): 47–57.

Lam, T. T., L. Mahdjoubi, and J. Mason (2017). "A framework to assist in the analysis of risks and rewards of adopting BIM for SMEs in the UK." *Journal of Civil Engineering and Management* **23**(6): 740–752.

Loosemore, M. and N. Andonakis (2007). "Barriers to implementing OHS reforms – The experiences of small subcontractors in the Australian Construction Industry." *International Journal of Project Management* **25**(6): 579–588.

Malekitabar, H., A. Ardeshir, M. H. Sebt, and R. Stouffs (2016). "Construction safety risk drivers: A BIM approach." *Safety Science* **82**: 445–455.

Merschbrock, C., M. R. Hosseini, I. Martek, M. Arashpour, and G. Mignone (2018). "Collaborative role of sociotechnical components in BIM-based construction networks in two hospitals." *Journal of Management in Engineering* **34**(4): 05018006.

Mignone, G., M. R. Hosseini, N. Chileshe, and M. Arashpour (2016). "Enhancing collaboration in BIM-based construction networks through organisational discontinuity theory: A case study of the new Royal Adelaide Hospital." *Architectural Engineering and Design Management* 1–20.

Mirahadi, F., B. McCabe, and A. Shahi (2019). "IFC-centric performance-based evaluation of building evacuations using fire dynamics simulation and agent-based modeling." *Automation in Construction* **101**: 1–16.

Nasirian, A., M. Arashpour, B. Abbasi, and A. Akbarnezhad (2019). "Optimal work assignment to multiskilled resources in prefabricated construction." *Journal of Construction Engineering and Management* **145**(4): 04019011.

Ni, G., Q. Cui, L. Sang, W. Wang, and D. Xia (2018). "Knowledge-sharing culture, project-team interaction, and knowledge-sharing performance among project members." *Journal of Management in Engineering* **34**(2): 04017065.

Oraee, M., M. R. Hosseini, E. Papadonikolaki, R. Palliyaguru, and M. Arashpour (2017). "Collaboration in BIM-based construction networks: A bibliometric-qualitative literature review." *International Journal of Project Management* **35**(7): 1288–1301.

Park, J., K. Kim, and Y. K. Cho (2017). "Framework of automated construction-safety monitoring using cloud-enabled BIM and BLE mobile tracking sensors." *Journal of Construction Engineering and Management* **143**(2): 05016019.

Parn, E. A., D. Edwards, Z. Riaz, F. Mehmood, and J. Lai (2019). "Engineering-out hazards: Digitising the management working safety in confined spaces." *Facilities* 37(3–4): 196–215.

Poghosyan, A., P. Manu, L. Mahdjoubi, A. G. F. Gibb, M. Behm, and A. M. Mahamadu (2018). "Design for safety implementation factors: A literature review." *Journal of Engineering, Design and Technology* 16(5): 783–797.

Polat, G. (2015). "Subcontractor selection using the integration of the AHP and PROMETHEE methods." *Journal of Civil Engineering and Management* 22: 1–13.

Riaz, Z., M. Arslan, A. K. Kiani, S. Azhar, S. A. M. A. Zainab Riaz and Adnan K. Kiani (2014). "CoSMoS: A BIM and wireless sensor based integrated solution for worker safety in confined spaces." *Automation in Construction* 45: 96–106.

Salidjanova, N. (2011). *Going Out: An Overview of China's Outward Foreign Direct Investment*, US-China Economic and Security Review Commission, Washington DC, United States-China Economic and Security Review Commission.

Seresht, N. G. and A. R. Fayek (2015). "Career paths of tradespeople in the construction industry." *Canadian Journal of Civil Engineering* 42(1): 44–56.

Sharma, S., A. Sawhney, and M. Arif (2017). Parametric Modelling for Designing Offsite Construction. Creative Construction Conference, CCC 2017, Elsevier Ltd.

Sun, C., Q. Man, and Y. Wang (2015). "Study on BIM-based construction project cost and schedule risk early warning." *Journal of Intelligent and Fuzzy Systems* 29(2): 469–477.

Sun, J., X. Ren, and C. J. Anumba (2019). "Analysis of knowledge-transfer mechanisms in construction project cooperation networks." *Journal of Management in Engineering* 35(2): 04018061.

Tang, S., D. R. Shelden, C. M. Eastman, P. Pishdad-Bozorgi, and X. Gao (2019). "A review of building information modeling (BIM) and the Internet of Things (IoT) devices integration: Present status and future trends." *Automation in Construction* 101: 127–139.

Thomas, H. R. and C. J. Flynn (2011). "Fundamental principles of subcontractor management." *Practice Periodical on Structural Design and Construction* 16(3): 106–111.

Tomek, A. and P. Matějka (2014). "The impact of BIM on risk management As an argument for its implementation in a construction company." 2014 Creative Construction Conference, CCC 2014 85: 501–509.

Wang, H. and X. Meng (2019). "Transformation from IT-based knowledge management into BIM-supported knowledge management: A literature review." *Expert Systems with Applications* 121: 170–187.

Wee, T. P. Y., M. Aurisicchio, and I. Starzyk (2017). Evaluating modularisation tools in construction. 34th International Symposium on Automation and Robotics in Construction, ISARC 2017, International Association for Automation and Robotics in Construction I.A.A.R.C.

Xiahou, X., J. Yuan, Q. Li, and M. J. Skibniewski (2018). "Validating DFS concept in lifecycle subway projects in china based on incident case analysis and network analysis." *Journal of Civil Engineering and Management* 24(1): 53–66.

Xu, J., A. H. Li, H. Q. Liu, M. Z. Ye, and J. R. Zhang (2014). "Application and risk analysis of BIM in railway systems." *Journal of Railway Engineering Society* 31(3): 129–133.

Yuan, J., X. Li, X. Xiahou, N. Tymvios, Z. Zhou, and Q. Li (2019). "Accident prevention through design (PtD): Integration of building information modeling and PtD knowledge base." *Automation in Construction* 102: 86–104.

Zhang, L., X. Wu, L. Ding, M. J. Skibniewski, and Y. Lu (2016). "BIM-based risk identification system in tunnel construction." *Journal of Civil Engineering and Management* 22(4): 529–539.

Zou, Y., A. Kiviniemi, and S. W. Jones (2016). "A review of risk management through BIM and BIM-related technologies." *Safety Science* 97: 88–98.

12

DATA STANDARDS AND DATA EXCHANGE FOR CONSTRUCTION 4.0

Dennis R. Shelden, Pieter Pauwels, Pardis Pishdad-Bozorgi, and Shu Tang

12.1 Aims

The objective of this chapter includes:

- Provide an overview of building information exchange standards.
- Examine the current building standard's application and limitation.
- Explore alternative data standards for Construction 4.0.
- Discuss motivation and potential for the development of next generation open standards.

12.2 Introduction

Data standards and principles of information exchange have been central to the success of computing. The "stack" of open information protocols from Transmission Control Protocol/Internet Protocol (TCP/IP) to Hyper Text Markup Language (HTML) has been one of the main drivers for the proliferation of computing in its current connected form. In contrast, building information standards have had limited impact on the industry, with the most successful building information exchange standards being proprietary. The ambitious Industry Foundation Classes (IFC) project to develop an open building object schema and exchange standard has not achieved the sort of transformative or disruptive impacts of the internet in adoption, richness of application and business value creation.

Today, a new generation of digital technologies is rapidly advancing capabilities in other industry sectors, from health care to transportation. Software is moving to the cloud, involving continuous transactions between remote systems, server-based data, and mobile communications. New digitally driven industrial paradigms including web services, the Internet of Things (IoT) and Blockchain are changing the ways business is being conducted, promising the next generation of digitally driven efficiency (Tang et al., 2019), and are central to the industry advances of Construction 4.0 discussed elsewhere in this book.

These advances are beginning to appear in the Architecture, Engineering and Construction (AEC) industries. Software companies are migrating their technologies to the cloud and providing new ways for interfacing with the information generated by their tools. A network of new innovators within professional design and construction companies and at startup companies are developing tools to support aspects of integrated project workflow. Promising applications of building information are repurposing Building Information Modeling (BIM) data into smart building and smart commerce applications.

Today's AEC interoperability standards – developed to support file-based exchange of whole building models – have not kept up with this evolution in technology. In response, a new generation of data exchange standards is currently being designed – one that draws from the architecture and supporting protocols of web- and cloud-services to rethink the base assumptions underlying today's building oriented open standards.

This chapter will provide an overview of building information exchange standards, including motivations and potential, current state and next generation developments. The basic principles of IFC and its place in the broader stack of open information exchange standards will be provided, along with a discussion of IFC's current applications and limitations. Alternative data standards and associated information architectures relevant to aspects of Construction 4.0 such as IoT and geospatial systems will be discussed. An overview of modern web services oriented schema, encoding standards and web services communications protocols from eXtensible Markup Language (XML) and JavaScript Object Notation (JSON) to Representational State Transfer (REST), Web Ontology Language (OWL), and GraphQL will be discussed, with specific focus on current and potential construction applications. Finally, the motivations and potential for a redeveloped open standard for next generation information exchanges will be presented including both the technical underpinnings and the industry economic and collaboration context impacting adoption.

12.3 Elements of a data standard

Data exchange standards imply a set of commonly held assumptions between a transmitter and recipient of information. The success of the internet and web have been founded on the incremental development and evolution of a layered set of standards and protocols, that span from low level hardware and signal transmission to high level information structuring and communication. Abstractions between these layers allow multiple parties to tie into standards at their particular level of interest in the stack while assuming that other layers will be taken care of by others.

The internet itself is such a set of standard protocols and serves as a model that defines many of both the lower and higher level constructs that define the building industry data standards. The internet stack is defined by five layers (as shown in Figure 12.1).

The Link, Internet and Transport are responsible for the inter-device communications and information routing. The application and data layers are typically considered when defining a domain specific information exchange. In particular the application layer is of importance to data standardization in the construction industry. It consists of the following parts:

- *Data model* – At the core of standards stand premises on how information is structured, and the relationships among discrete pieces of information. (Enhanced) Entity – Relation (EER) diagrams are key tools in defining the organization of information. The Unified Modeling Language (UML) allows the same thing. The resulting EER and/or UML diagrams can be used to structure data in diverse data models. Such data models have diverse structures and could be hierarchical (e.g. XML), name value pairs (e.g. JSON) or graphs (e.g. Recourse Description Framework (RDF)).

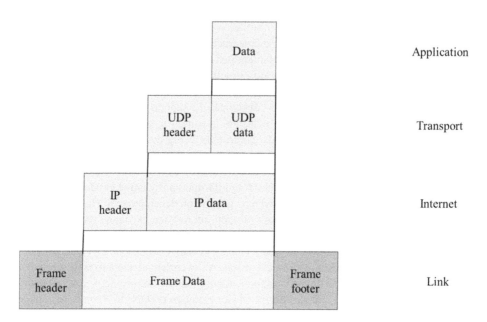

Figure 12.1 Layers of internet stack

- *Data schema and object model.* Data standards – particularly in the AEC industry – seek to encode the definition of entities and their relationships in a standardized manner that supports replicability, machine validation, and potentially translation to other application support contexts. Therefore, data schemas are devised in information modeling and/ or schema languages. Example schema languages are XML schema definition (XSD), EXPRESS, and OWL.
- *Serialization of instance data.* This is the "payload" of information that describes the objects or phenomena of interest. Such instance data can be serialized in diverse formats. For example, diverse XML serializations can be generated from the same XML data model. In the case of RDF graphs, serializations can be made in Turtle, N-Triples, RDF/XML, JSON-LD, etc. Even though these serializations have a different syntax and look, the data model stays identical. Well-structured instance data follows an agreed data schema (e.g. XSD) within a certain data model (e.g. hierarchical XML).

12.4 Industry Foundation Classes: overview, application, and limitation

The IFC standard is one of the most important data standards for building information description and exchange. IFC has been developed over the past 20 years as an attempt to create a public, open framework for the representation, encoding, and exchange of BIM-based building descriptions, with potential extensions to supporting the exchange of broader sets of data related to buildings, products, components, and infrastructure. IFC itself represents an extension of the broader Standard for the Exchange of Product (STEP) model data standard. This section will provide the basic principles of IFC and its place in the broader stack of open information exchange standards, along with a discussion of IFC's current applications and limitations.

12.4.1 IFC overview

12.4.1.1 Technical terms, history versions, basic architecture of IFC

IFC is an industry developed data model to represent building life cycle data. The IFC standard, developed by BuildingSMART, supports the aim to represent, semantically describe, and exchange BIM data between software applications, various stakeholders and throughout construction stages.

The IFC schema defines an extensible and consistent data representation with the ISO-STEP EXPRESS language (Sacks et al., 2018). The EXPRESS data specification language is defined in the 10,303–11:1994 International Organization for Standardization (ISO) standard (International Organization for Standardization, 2005). The EXPRESS language uses types, entities, properties, and rules to construct a specific EXPRESS schema (.exp) (Pauwels and Terkaj, 2016a).

The IFC standard has several successively developed versions; the current IFC4_ADD2 version of the IFC schema contains 766 entities, 413 property sets, and 130 defined data types. These schemas represent construction related information in objects with interrelated meanings and purpose, guaranteeing that descriptions in different BIM environments can be generally mapped to the IFC format. The complexity of the current IFC schema provides semantic richness of building information, supports different BIM environments, and addresses various application needs like cost estimation, energy simulation, and construction logistic, therefore improves the information interoperability. The community of data services provided by the IFC community have expanded to serve some of these needs; others are in development or under consideration. For example, there is a significant new effort to expand the IFC to support infrastructure applications such as roads, highway, bridges, and railways (buildingSMART International, 2019c). Figure 12.2 illustrates the IFC Infra project schema organization.

IFC is conceptually structured as a collection of objects and their relationships. The extensions of these objects and relations for specific use are domain-specific extensions

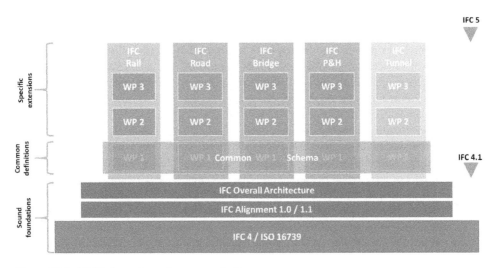

Figure 12.2   IFC Infra project schema organization

that comprise the top level of the IFC data models (Sacks et al., 2018). The IFC specification contains four conceptual layers namely Resource Layer, Core Layer, Interoperability Layer, and Domain Layer (see below). Each layer contains a group of individuals without duplication.

- *The Resource Layer* is the lowest layer which contains all the base entities. These base entities allow to define common AEC objects such as building elements, building service elements, process elements, management elements, and generic features.
- *The Core Layer* contains kernel entity definitions which define core concepts like actors, groups, processes, products, and relationships.
- *The Interoperability Layer*, also named shared element data schema, provides entity definitions that are subject to several domains for inter-domain data exchange.
- *The Domain Layer* is the highest layer and contains entity definitions that are specific to individual domains like architecture, structural analysis, and facility management. These entities are categorized according to industry disciplines and are used for intra-domain data exchange.

In addition to the IFC EXPRESS schema, BuildingSMART has developed the Property Set Definition (PSD) schema as part of the IFC specification (part of IFC2X3). The PSD schema defines how custom property sets and properties can be defined (buildingSMART, 2008). In addition to this PSD schema, which is defined using XSD, a number of PSET XML files are made available, which follow this PSD schema. These PSET XML files provide a number of standard property sets and properties, which are directly linked to the corresponding IFC EXPRESS schema. As of IFC4, property sets and properties are now included directly in the IFC EXPRESS schema.

12.4.1.2 Geometry

Geometric constructs are central to building design and construction information. The central principles of modern two- and three-dimensional coordinate geometry have persisted since their development in the 16th-century Cartesian geometric system. In the most basic form, geometries are based on two- and three-dimensional points expressed as ordered sets of real numbers. Higher level constructs are generated from organized sets of these points. A bounded region in the plan may for example be constructed as a series of points that describe the sequence of vertices comprising the boundary segments. The constellation of geometric objects is perhaps surprisingly diverse. Within the general set of building descriptions, the following taxonomy might be considered:

- *Primitives.* Simple geometries are described by collections of points and simple Euclidean expressions. Point clouds – collections of discrete points – have emerged recently as a surprisingly robust means for capturing project geometry. This methodology has emerged as a product of laser based point scanners that can capture large volumes of points from the physical environment. These points clouds may include additional semantic information such as detected color and sample intensity as well as scan identity and potentially identifications on specific points identified during data capture. High volume data storage capabilities and capabilities for performing operations on point clouds have assisted making these relatively low level data constructs appropriate for performing relatively robust operations.

- *Linear geometries.* Points may be assembled through higher level data structures to provide linearized representations of project geometry – either in two or three dimensions. In the above example, a site boundary is represented as a sequence of linear segments implicitly defined by a sequence of line segment vertices. By extension, triangulated meshes or polygons in two and three dimensions may be constructed to support the delineation of surfaces; volumes may be constructed as assemblages of tetrahedral, cubes, closed surfaces. Much of today's 3D environmental applications such as Google Earth rely on triangulated meshes generated from captured images using photogrammetry methods.
- *Boundary representations.* Modern advanced geometric modelers have established NURBS surface geometries as a base convention for the description of non-linear/non-planar geometries. Briefly, these modeling systems rely on continuous functions over regularized organizations of one, two, or higher dimensional arrays of control points to describe continuously curved geometries of curves, surfaces, and even volumes. Lower dimensional boundary representations define the edge geometries of these shapes. Bounded regions may be connected by establishing connectivity between their boundaries, resulting in the ability to define assemblages of open or closed primitives.
- *Parametric geometry descriptions.* At the highest level, geometries may be defined by "recipes" based on lower level primitives. For example, a beam may be described as an extrusion of a profile swept along a line or curve with the profile having a certain orientation in space (assumed to be vertical in the Z direction unless otherwise specified). The profile may be described in terms of linear or curved segments, or alternatively may simply be referenced by the name of the profile, with the assumption that the profile geometry associated with the profile identifier may be commonly determined from some shared data set. By extension, parametric variables and functions on these variables may be used as part of the geometric descriptions. Parametric geometric representations have the benefit of extremely compact descriptions of geometry relative to some of the other conventions (Pauwels et al., 2017). However, these parametric descriptions are frequently the origin of geometric discrepancies between reading and writing applications, and ambiguities or errors in interpretation of the geometric recipes are applied at each end of the transmission.

12.4.1.3 Related Standards

IFC is not an isolated standard. In addition, BuildingSMART developed several other standards around IFC to support functionality and extensibility for the schema development. The most important standards are organized in a triangle of data (as shown in Figure 12.3), process, and terms (buildingSMART International, 2019d).

In total, there are five connected standards associated with IFC:

- *Industry Foundation Classes (IFC).* The main "BuildingSMART data model" standard is an ISO standard (ISO 16,739). It is designed based on ISO-STEP technologies. Each version of the IFC standard is published with a number of serializations, all of which are available via one specifications page (buildingSMART International, 2019a).
- *Information Delivery Manual (IDM).* The main "BuildingSMART process definition" standard is an ISO standard (ISO 29,841). As data exchange in practice does not always require the full data model or schema, IDM is designed to provide a standardized methodology to capture process, information flow and exchange requirements for specific scenarios. This methodology facilitates documenting processes, and associated information

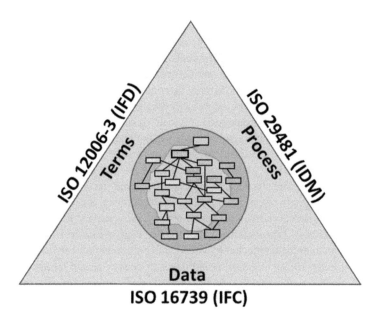

Figure 12.3 IFC standards triangle

that need to be exchanged between parties, software, or activities. The output of IDM contains a target work process like a process map and a specification of information required by the target process (Exchange Requirement (ER)). Further development of the IDM output also assists the formation of a software development process (building-SMART, 2011).

- *International Framework for Dictionaries (IFD)*. The main standard for "Building-SMART dictionary terms" is also defined as the ISO 12,006–3 data dictionary standard. This standard sets out an international framework to develop data dictionaries: collection of terms and definitions used to develop a data model. The common framework enables internal references between classification system, information models, object models, and process models (International Organization for Standardization, 2007). The Build-ingSMART Data Dictionary is one example implementation of the IFD standard.
- *BIM Collaboration Format (BCF)*. A candidate standard has been proposed for the rep-resentation of collaboration messages and issues (issue management). This standard is developed externally in the realm of BIM for workflow communication and collabo-ration. It supports BIM process workflow communication in an open file XML format "bcfXML" and enables BCF data exchange using RESTFUL web service "bcfAPI" (buildingSMART, 2014). This standard has been adopted by many BIM software inde-pendent of the other standards as a means of sharing views and capturing issues across disparate products and their models.
- *Model View Definition (MVD)*. The MVD standard is closely related to both IFC (data) and IDM (process). Namely, data exchange in practice (the IDM process) uses only a small subset of a data model (IFC). Such subsets can be defined as MVDs, with each of those subsets responding to one of the exchange requirements represented in the IDM process. MVD-defined subsets aim to satisfy one or many exchange requirements of the

AEC industry. The exchange requirement can be defined following the IDM methodology. The MVD can be encoded in mvdXML format. The official tool for developing an MVD is IfcDoc (buildingSMART International, 2019b). The combined use of IDM and MVD facilitates defining subsets of IFC for certain exchange scenarios which can be used by software developers as a guidance for importing/exporting exchange scenarios. The currently approved MVD are IFC 4 Design Transfer View, IFC 4 Reference View and Trust in BIM deliverables (mvdXML 1.1). MVDs for previous versions of IFC include the Coordination View, Space boundary add-on view, basic FM handover view, and structural analysis view (Pinheiro et al., 2018).

12.5 Uses and applications

As an extension of the STEP initiative, IFC was developed as a way of encoding generalized data and relationships for two main purposes:

- As a means of exchanging data across applications.
- As a way of archiving data outside of proprietary systems.

Original ambitions to achieve "round robin" interoperability of "live" native geometry with full operational capacity on re-import have been largely ineffective and the ambition of supporting these capabilities are a primary motivation for the open standards community are diminishing.

12.5.1 Current limitations

One of the key reasons why the construction industry data standards have lagged behind other industries is the continued reliance on STEP. Even though IFC sets an excellent basis for data exchange in the construction industry, it has a number of limitations, most of which arise from the strong dependency on the EXPRESS information modeling language, which is seldom used in any other industry. Relatively newer data models such as XML, JSON, and RDF are much more widely used and known to software developers. The following limitations can typically be recognized in relation to IFC:

- Complexity: The EXPRESS information modeling language is very expressive and allows the definition of almost anything, including the possibility to link anything to anything. WHERE rules can be defined, as well as procedures, and complex constructs like arrays of lists of lists of arrays are possible. As much of this functionality is used in IFC, the resulting schema is particularly complex, difficult to use, and variations in its interpretation can result in discrepancies and errors between applications.
- Extensibility: It is recognized to be very difficult to extend IFC, e.g. for new domains. Further additions are made over the course of time to one and same schema, and it is a big challenge to avoid inconsistencies in the schema. Many current data modeling languages (e.g. XSD, RDF, etc.) can be extended and implemented much more easily.
- Modularity: Even though EXPRESS allows the definition of subschemas, this functionality is not used; and the entire IFC schema is a single file download. This makes it difficult to use only a part of the schema (e.g. bridge) in a specific application. The MVD functionality can be used for this purpose, yet implementation of an MVD again takes a lot of time.

12.6 Evolutions of building data standards

Emerging scenarios for industry interoperability represent a marked departure from this current state. Anticipated scenarios will remove the reliance on human consensus documentation, manual compliance checking, and vendor development resources in order to implement new versions of the data schema. Instead, modern communications, distributed development and version control, and cloud data sources will automate many of today's manual processes.

Web and cloud data sources are expected to occupy a significantly larger aspects of these data strategies. These resources will allow immediate access to evolutions of the data standard. Top tier AEC software vendors are increasingly moving their products to the cloud, and are in need of interoperability data tools to connect information generated by a suite of different technologies targeted at various constituents across the building supply chain.

A much larger set of constituencies beyond end user communities and AEC software vendors will need to be served. There is an emerging community of innovators in design and engineering firms who are developing automated design and engineering tools that incorporate building data sets as a core aspect of their functionality – and potential interchanges with other innovators. Integrated building delivery companies are developing automated ways of connecting design – fabrication – install – operate aspects of the building systems they are developing and have internal interoperability needs. Building operations strategies are integrating building information models with real time operations – not simply at the building operating system level but in order to improve occupant and business organizational performance. Building product manufacturers are seeking ways to present their products into BIM modeling environments at the point of design.

In short, a large number of high value use cases are appearing in the market, each with significant interoperability needs that share common characteristics, including:

- *Data validation.* This ambition would support automated validation of requirements for the structure as well as information content of a given standard or view (Solihin, Eastman, and Lee, 2015). Aspects of validation might include:
  - That a declared encoding (well formedness) is adhered to in the transmission.
  - That the set of elements, relations and attributes in a transmission represents a restricted set of the possible such data elements in the standard (i.e. ONLY a certain set of data elements is contained).
  - Reciprocally, that a required set of such data elements is provided for defined elements in the standard (i.e. that the transmission is complete according to some definition).
  - That the values contained in the transmission adhere to a given set of rules.
- *Encoding interoperability.* The potential for a standard to be based on a singular serialization convention has increasingly come into question. As a consequence of the rapid development of new software languages, encoding formats and data conventions, new data standards are emerging with increasingly frequency. The possibility of automated machine translation from one encoding to another is an emerging goal of data standards.
- Fortunately, the translation of machine encodings is a relatively mature field. Compiler and machine translation capabilities – originally developed to translate between higher level programming languages and lower level machine and assembly code – have evolved to serve the translation between object descriptions. The existence of schema definitions – encodings of the rules for expressing information in a given data convention – are key to the translation across standards.

- *Federated information standards.* There is currently a rapidly expanding set of disconnected data standards supporting distinct domains of interest. Convergence of technologies and use cases is resulting in overlaps and collisions among these standards. For example, GIS and BIM use cases for building and city level data are increasingly converging, and information associated each of these are of increasing relevance to each other. Similarly, data associated with building level concerns (BIM) is relevant to the building control systems information data maintained in standards such as Building Automation and Control networks (BACnet), HAYSTACK and others. These domain specific standards may overlap in that they share certain entity definitions, parameters that may be of relevance to each other.

- *Data storage and operations.* While the original ambitions of IFC and other interoperability data protocols concern the transmission/serialization of information for exchange among systems, new "NoSQL" data storage systems are increasingly available for direct storage of data in data storage mechanisms closely associated with certain data encodings as "native" information. For example, the MongoDB system is closely associated with JSON structured data and can be operate without heavy data conversion (Bilal et al., 2016). Similarly, there are increasing capabilities and use cases for data to be operated on directly – outside of the contexts of proprietary software applications. The ability to extract, display, process, and manipulate data in the open standard directly is an area of increasing interest. A wealth of opensource and proprietary systems for working with data in open formats has emerged along with use cases of interest. The Construction Operations Building Information Exchange (COBie) standard is an example of an open data standard encapsulating proprietary data generated by BIM systems has been expressed in an EXCEL based format of tables that allows both human and machine based validation, editing, and visualization.

- *Cloud and web services applications.* As web databases and web accessible systems have proliferated, data standards that can support communications between these systems is an increasing need. AEC software is increasingly is becoming cloud enabled – moving away from desktop software to cloud applications and micro-services. Data exchange is no longer just file based – where a "complete" building model is exported from one application and imported into another. Rather, subsets of the overall building model are exchanged through REST APIs and other transactional exchanges. The open standards of the future should support partial model and partial data exchanges, where limited amounts of IFC formatted data are requested and delivered around a specific need (Curry et al., 2013). These use cases present new demands on data standards including partial model exchange and update. The emergence of standard web transactions through REST and related client server transactions is creating an emerging need for standards to evolve beyond the "data payload" level to potentially support standard querying and response conventions (Isikdag, 2015).

12.6.1 Hierarchical data standards: XML and JSON

A number of initiatives have emerged over the coming years to respond to these emerging use cases. Most of efforts to expand and evolve the existing EXPRESS based data encodings to support more recent data standards and protocols. One approach to improving IFC standard is by making it available in XML (Nisbet and Liebich, 2007). The IfcXML is derived from IFC EXPRESS model as an XSD. The IfcXML schema is automatically created from the EXPRESS schema by through "XML representation of EXPRESS schemas and data" language binding.[1]

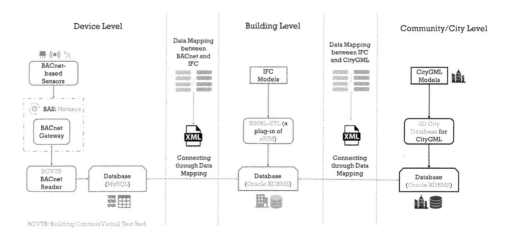

Figure 12.4 The experiment framework: connecting BACnet, IFC, and CityGML based databases (Gao et al., 2018)

The mapping between EXPRESS to XML schema is guided by a standardized configuration file ensuring the bi-directional conversion between ".ifc" and ".ifcXML". The latest release of IfcXML 4 contains IfcXML XSD schema model, configuration setting of the data translation process, documentation of configuration setting, and configuration schema definition (Afsari, Eastman, and Castro-Lacouture, 2017). With the newer data model, it is possible to integrate IFC with other domain at different levels such as device level, building level, and city level for information exchange.

An example is designed as demonstration for a federated data framework. In this example, BACnet (American Society of Heating Refrigerating and Air-Conditioning Engineers, 2016), IFC, and CityGML (Open Geospatial Consortium, no date) are adopted as the data protocol of device level, building level, and city level, respectively.

Figure 12.4 shows the framework. At device level, the data generated by the Building Automation Servers' sensor network are collected and stored in a MySQL database conforming to the BACnet protocol. An open-source tool named BCVTB (Building Controls Virtual Test Bed) is used to read the data generated by BACnet devices and write them to the database.

IFC models are transformed into a simplified RDBMS (Relational Database Management System) data format and stored in an Oracle Database Server (Oracle RDBMS). The tool we used in this step is called BIMRL-ETL (BIM Rule Language – Extract, Transform, and Load), which is a plug-in of xBIM.

At the community/city level, a CityGML city model is also transformed and stored in an Oracle Database Server. A tool named 3D City Database (3DCityDB) is used to automate this process. The 3DCityDB is an open-source package consisting of a database schema and a set of software tools to import, manage, analyze, visualize, and export CityGML city models. "The database schema results from a mapping of the object-oriented data model of CityGML 2.0 to the relational structure of a spatially enhanced relational database management system" (SRDBMS) (Technical University of Munich, 2015).

To federate the data in each level's database by using the common data fields, we have identified the overlaps (not exhaustively) between BACnet XML and ifcXML (IFC in the XML form), and that between ifcXML and CityGML XML. The common data fields used for database federation are shown in Tables 12.1 and 12.2.

Table 12.1 The common data fields between BACnet XML and IFC XML (Gao et al., 2018)

BACnet XML (DR-034A-28)	IFC XML (IFC4 ADD2)
▼ Object	▼ ifcBuilding
propertyIdentifier	Id
Name	Name
Type	ObjectType
Description	Description
contextTag	Tag

Table 12.2 The common data fields between IFC XML and CityGML (Gao et al., 2018)

IFC XML (IFC4 ADD2)	CityGML 2.0 XML
▼ ifcBuilding	▼ Building
Id	Gmlid
Name	gmlname
Description	gmldescription
ElevationOfRefHeight	measuredHeight
ifcBuildingAddresss	address

This example demonstrates the framework for (1) studying different IoT data schemas, (2) identifying common data fields between these schemas and BIM data schemas, (3) exchange data between different database using XML data model, and thus (4) establishing the federated dataset for IoT-enabled Smart City applications (Gao et al., 2018).

12.6.2 Linked data and web ontologies

IfcOWL is another recent effort to expand IFC's capabilities, which makes IFC data available in RDF graphs (Pauwels and Terkaj, 2016b). This initiative aspires to make the IFC data fully web-based and to leverage the expanding capabilities for connecting both data and knowledge over the web. The ifcOWL effort started as early as 2005 and culminated in 2016 with the emergence of the ifcOWL draft international BuildingSMART standard. One of the main criteria in preparing the IFC schema in OWL (as shown in Figure 12.5) has been to maintain backwards compatibility with the EXPRESS IFC schema. The resulting OWL ontology for IFC can hence be considered almost identical to the EXPRESS and XSD schemas of IFC, encoded in RDF (Open Geospatial Consortium, no date).

The RDF data model inherently functions as a graph (as shown in Figure 12.6). Everything that is described with the RDF data model, follows a directed labeled graph structure. Every statement in RDF consists of three elements, namely subject, predicate, and object. These statements are called triples. The predicate labels a directed arrow between subject and object. When one keeps linking different items and concepts, a potentially endless network of linked data results, describing potentially any domain of discourse.

This RDF data model is obviously very powerful as it potentially has an ever-increasing size, which is ideal to allow a web-based data infrastructure. This Linked (Open) Data (LOD)

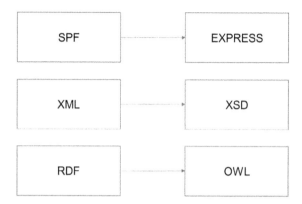

Figure 12.5 IFC data and schema (Pauwels et al., 2017)

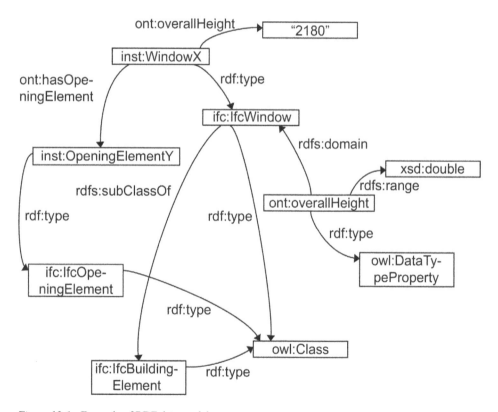

Figure 12.6 Example of RDF data model

cloud or Web of Data has by now covered various domains, including health care, biology, media, and geography – domains which can be readily connected to building design, construction, and operations data and use cases.

Diverse partial clouds of data are made available on diverse online servers. The linking mechanism allows for direct and explicit linking across servers. It hereby relies on the Uniform

Resource Identifier (URI) mechanism, which is a foundational building block for anything in the global web. To be precise, every concept in the LOD cloud has a URI, which typically follows the structure of an HTTP address. These URIs serve as unique identifiers. As much as any document or website on the web is unique through its URI, which directly mapped to IP addresses using a Domain Name Server (DNS), any concept defined in an RDF graph is by default unique because of its reliance on this URI infrastructure. This feature is crucial to allow global representing and interlinking of data and is therefore potentially a major improvement over the use of file-based GUIDs.

Similar to other data models, also the RDF data model has a data schema language, namely the OWL. Each IFC EXPRESS schema is transformed into an OWL ontology, with the namespace following the structure: https://standards.BuildingSMART.org/IFC/DEV/IFC4/ADD2/OWL/. Any concept in IFC (e.g. IfcWall, IfcWindow, …) thus has a unique URI identifier within this web address.

With this ontology, it is possible to use modern RDF-based technologies to access the IFC data. For example, the ifcOWL ontology can simply be loaded in a state-of-the-art ontology editor, such as Protégé, and it can be extended using standard and widely known techniques. Furthermore, diverse well-known software libraries are available in each programming language (Java, C#, Python, JavaScript, etc.) that enable the creation and use of an ifcOWL-compliant RDF graph. Furthermore, diverse commercial and open-source RDF triple stores are available, which can be used to store the IFC-based RDF graphs. As a result, graphs as displayed in Figure 12.7, become available, almost for free, in diverse servers and repositories.

With this basic transformation to the RDF data model, extensibility of the data model using state of the art tools is within reach out of the box. However, merely transforming the EXPRESS IFC schema to an IFC ontology does not solve the issues of complexity and lack of modularity. Indeed, transforming the monolithic single IFC schema into one OWL ontology still results in a monolithic single OWL ontology for IFC, with equal complexity and lack of modularity.

Therefore, the most recent research and industry initiatives around the IFC standard target the reduction of the complexity and the increase of modularity. These have been among the core goals of the W3C Linked Building Data community group. In inspiration of the IFC standard, this group decided to make a number of ontologies available, each of them covering a certain domain. The most core ontology in this regard is the building topology ontology (BOT) (Rasmussen et al., 2019). This ontology only covers the basic structure of a building, including zone, site, building, building storey, space, and element classes and a number of object and datatype properties (not more). With this ontology, one can easily represent the most basic structure of a building and publish it as linked data. Second, this group looked at product data, thereby distinguishing product taxonomies (window, wall, beam, flowterminal, etc.) and corresponding properties (fire rating, thermal transmittance, etc.). The representation of products relies on the existing GoodRelations (Hepp, 2016) and Schema.org ontologies yet recommends building a number of product ontologies with corresponding properties. These product and properties ontologies form separate modules and aim to not be as complex as the original IFC EXPRESS versions. As a result, they serve as excellent candidates for a linked data-based product data repository. 3D geometry is to a large extent kept out of the RDF data model, either simplified into "data blobs" or externally linked to. Largely inspired by the approach in the geospatial domain, simplification consists of including a simple string snippet with geometric data for a particular part or object in the IFC data.

As a result, this W3C LBD group made available a number of ontologies (modularity) that are much simpler in kind, compared to the full IFC data standard (complexity). With

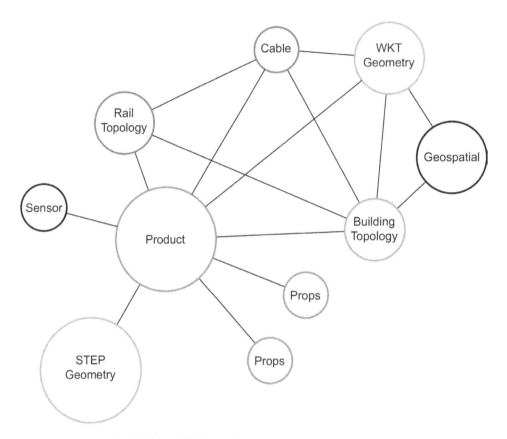

Figure 12.7 Example of IFC-based RDF graph

these ontologies, one can represent any sort of building data in a web-compliant manner, thereby limiting to the desired scope (e.g. only key product data and building topology). The results are Linked Building Data graphs in the RDF graph format, as displayed in the Figure 12.8. Data about a single building can be spread out over multiple servers, providing access to users in an equal modular granularity. Many current use cases start with these graphs to connect the data to sensor data or geospatial data, which has become trivial, simply by the shift in data model.

In addition to this shift of data representation, it is now possible to publish building data and deploy also the other languages and state of the art techniques within this realm of linked data. For example, it is perfectly feasible to rely on query languages (e.g. SPARQL (Prud'Hommeaux et al., 2008)) to retrieve, and even slightly reformat, subsets of data. These query languages have an enormous potential in improving the MVD workflows previously defined in BuildingSMART. Because standard query languages and query engines are available, vendors don't need to "implement certain MVDs", instead they only need to provide a query interface based on the existing technologies.

Further potential improvements lie in (1) the use of rule languages and reasoning engines for building code compliance checking, and (2) the use of web services which are able to be triggered on demand to enable small RDF-based functionalities (Pauwels et al., 2018).

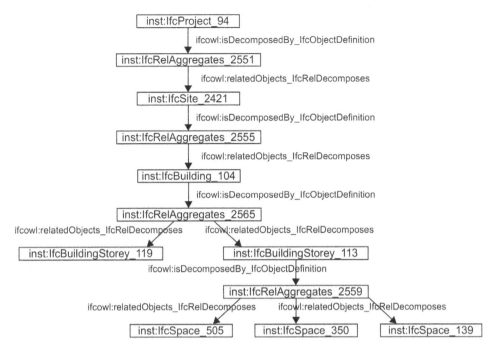

Figure 12.8 Linked Building Data graph

12.7 Conclusion

Advances in digital technologies that are established in many industry sectors such as IOT and Blockchain are beginning to appear in the AEC industries. The readily available access to software and data through the cloud provide new ways for operators to engage with data, information, and tools. There is a proliferation of innovative tools, developed by a growing sector of professionals and companies to support integrated project workflow. BIM data is being transitioned into smart building and smart commerce applications. Current AEC interoperability standards have not kept pace with this evolution in technology. As such, a new generation of data exchange standards is now being designed, drawing from the architecture and supporting protocols of web and cloud services to reconsider the basic assumptions that underlie today's building oriented open standards.

12.8 Summary

- Motivations and current state of the ongoing open building information exchange standards development efforts.
- Discussed increasing semantic and connectivity demands of Construction 4.0 applications.
- Communication of both core geometric and higher-level semantic information among an expanding set of distributed, connected applications and devices.
- Core object modeling and semantic descriptions of existing building data exchanges provide an excellent basis for this work.
- Evolving the data modeling paradigms of these existing building data standards into today's web connected data models supported by the broader technology community, can be a key strategy for supporting the broad and expanding set of emerging Construction 4.0 applications.

Note

1   ISO 10,303 part 28 edition 2

References

Afsari, K., Eastman, C. M. and Castro-Lacouture, D. (2017) 'JavaScript Object Notation (JSON) data serialization for IFC schema in web-based BIM data exchange', *Automation in Construction*, Elsevier, 77, pp. 24–51. doi: 10.1016/j.autcon.2017.01.011.

American Society of Heating Refrigerating and Air-Conditioning Engineers (2016) ANSI/ASHRAE Standard 135-2016: BACnet A data communivation protocol for building automation and control networks. Atlanta, GA. Available at: www.ashrae.org/technical-resources/bookstore/bacnet (Accessed: 22 March 2018).

Bilal, M. et al. (2016) 'Big Data in the construction industry: A review of present status, opportunities, and future trends', *Advanced Engineering Informatics* pp. 500–521. doi: 10.1016/j.aei.2016.07.001.

buildingSMART (2008) Summary of property set releases – Welcome to buildingSMART-Tech.org. Available at: www.buildingsmart-tech.org/specifications/pset-releases (Accessed: 2 March 2019).

buildingSMART (2011) Information delivery manuals – buildingSMART international user group. Available at: http://iug.buildingsmart.org/idms (Accessed: 30 January 2019).

buildingSMART (2014) BCF intro – Welcome to buildingSMART-Tech.org. Available at: www. buildingsmart-tech.org/specifications/bcf-releases (Accessed: 2 March 2019).

buildingSMART International (2019a) IFC Schema Specifications - buildingSMART Technical. Available at: https://technical.buildingsmart.org/standards/ifc/ifc-schema-specifications/(Accessed: 23 April 2019).

buildingSMART International (2019b) IfcDoc – buildingSMART Technical. Available at: https:// technical.buildingsmart.org/resources/ifcdoc/(Accessed: 23 April 2019).

buildingSMART International (2019c) Infrastructure room – buildingSMART. Available at: www. buildingsmart.org/standards/rooms-and-groups/infrastructure-room/ (Accessed: 23 April 2019).

buildingSMART International (2019d) Open Standards – the basics – buildingSMART. Available at: www.buildingsmart.org/standards/technical-vision/open-standards/(Accessed: 23 April 2019).

Curry, E. et al. (2013) 'Linking building data in the cloud: Integrating cross-domain building data using linked data', *Advanced Engineering Informatics*, 27 (2), pp. 206–219. doi: 10.1016/j.aei.2012.10.003.

Gao, X. et al. (2018) Foundational research in integrated building Internet of Things (IoT) data standards research report. Atlanta, GA. doi: 10.13140/RG.2.2.17378.99525.

Hepp, M. (2016) Goodrelations: The web vocabulary for E-Commerce, *GoodRelations homepage*. Available at: www.heppnetz.de/projects/goodrelations/ (Accessed: 28 June 2019).

International Organization for Standardization (2005) ISO 10303-14:2005 industrial automation systems and integration – Product data representation and exchange – Part 14: description methods: The EXPRESS-X language reference manual. Available at: www.iso.org/standard/38047.html (Accessed: 2 March 2019).

International Organization for Standardization (2007) ISO 12006-3:2007 Building construction – Organization of information about construction works – Part 3: framework for object-oriented information. Available at: www.iso.org/standard/38706.html (Accessed: 30 January 2019).

Isikdag, U. (2015) *Enhanced Building Information Models Using IoT Services and Integration Patterns*. *SpringerBriefs in Computer Science*, ed S. Zdonik et al. Springer International Publishing. doi: 10.1007/978-3-319-21825-0.

Nisbet, N. and Liebich, T. (2007) ifcXML Implementation Guide. Available at: www.buildingsmart-tech.org/downloads/accompanying-documents/guidelines/ifcXML Implementation Guide v2-0.pdf (Accessed: 23 April 2019).

Open Geospatial Consortium (no date) CityGML. Available at: www.opengeospatial.org/standards/citygml (Accessed: 23 April 2019).

Pauwels, P. et al. (2017) 'Enhancing the ifcOWL ontology with an alternative representation for geometric data', *Automation in Construction*, Elsevier, 80, pp. 77–94. doi: 10.1016/j.autcon.2017.03.001.

Pauwels, P. et al. (2018) 'Linked Data', in A. Borrmann, M. König, C. Koch and J. Beetz eds, *Building Information Modeling*. Cham: Springer International Publishing, pp. 181–197. doi: 10.1007/978-3-319-92862-3_10.

Pauwels, P. and Terkaj, W. (2016a) 'EXPRESS to OWL for construction industry: Towards a recommendable and usable ifcOWL ontology', *Automation in Construction*, Elsevier, 63, pp. 100–133. doi: 10.1016/j.autcon.2015.12.003.

Pauwels, P. and Terkaj, W. (2016b) ifcOWL ontology (IFC4). Available at: https://standards.buildingsmart.org/IFC/DEV/IFC4/ADD2/OWL/index.html (Accessed: 23 April 2019).

Pinheiro, S. et al. (2018) 'MVD based information exchange between BIM and building energy performance simulation', *Automation in Construction*, 90, pp. 91–103. doi: 10.1016/j.autcon.2018.02.009.

Prud'Hommeaux, E. et al. (2008) 'SPARQL Query Language for RDF'. Available at: www.w3.org/TR/rdf-sparql-query/ (Accessed: 28 June 2019).

Rasmussen, M. et al. (2019) Building topology ontology. Available at: https://w3c-lbd-cg.github.io/bot/ (Accessed: 28 June 2019).

Sacks, R. et al. (2018) *BIM Handbook: A Guide to Building Information Modeling For Owners, Designers, Engineers, Contractors and Facility Managers*. Hoboken, NJ: John Wiley & Sons, Inc, doi: 10.1002/9781119287568.

Solihin, W., Eastman, C. and Lee, Y. C. (2015) 'Toward robust and quantifiable automated IFC quality validation', *Advanced Engineering Informatics*, 29 (3), pp. 739–756. doi: 10.1016/j.aei.2015.07.006.

Tang, S. et al. (2019) 'A review of building information modeling (BIM) and the Internet of Things (IoT) devices integration: present status and future trends', *Automation in Construction*, Elsevier, 101, pp. 127–139. doi: 10.1016/j.autcon.2019.01.020.

Technical University of Munich (2015) 3D. Available at: www.lrg.tum.de/en/gis/projects/3dcitydb/ (Accessed: 23 Sep 2018).

13

VISUAL AND VIRTUAL PROGRESS MONITORING IN CONSTRUCTION 4.0

Jacob J. Lin and Mani Golparvar-Fard

13.1 Aims

- Investigate comprehensively the emerging computer vision and BIM-driven technologies that facilitate physical-to-digital transformation and digital-to-physical transformation in the Construction 4.0 framework.
- Explore thoroughly computer vision and BIM-driven data analytics for construction performance monitoring at project and operation level.
- Discuss the challenges of current state-of-the-art methods and commercially available solutions for monitoring construction and operation in built environment.
- Present potential solutions for using computer vision techniques on project control problems.
- Present multiple case studies to validate practical significance and opportunities.

13.2 Computer vision for monitoring construction – an overview

The exponential increase in the volume of images and videos captured in the built environment with smartphones, fixed cameras, unmanned aerial and ground vehicles, and laser scanning devices, provides a unique opportunity to digitally record and analyze the entire construction life cycle. To make these ambient big visual data actionable, research has developed and applied state-of-the-art computer vision techniques to produce 3D reality models of ongoing operations and automatically organize and manage them over project timelines. The integration of these models with Building Information Modeling (BIM) and project schedules has enabled development and use of many new solutions for project control, safety inspection, quality assessment/control, productivity analysis, operation maintenance, and building energy performance. This process of capturing the Reality, comparing to Planned and then analyzing through emerging computer vision technologies illustrate the built environment assets' transformation from physical to digital and back to physical in the Construction 4.0 framework.

These transformation process are not only studied in research but also started to get implemented in real-world construction site. Construction technology startup companies and construction companies of all sizes are taking advantages of their collective power to implement these technologies to benefit their customers and business. With the ever-increasing technology investment and the high expectation of technology transformation from the executive level in construction companies (Armstrong and Gilge, 2017), the construction industry seems to start catching up with other industry in terms of technology adoption for performance improvement. Nowadays, construction project control is not recording progress and safety issues on paper during job walks; progress and safety information are analyzed through photos taken from drones, robots, smartphones, 360 cameras and any kind of platform that captures images. Construction workers are used to seeing drones taking pictures for progress tracking and project engineers always carry commodity smartphones and tablets to document valuable information. Currently, owners and contractors with strategic vision are starting to adopt new technologies in construction, but there is still a lack of understanding of solving the problems with the right technologies. To find a holistic solution for construction, a comprehensive understanding of the potential of the technology and the problems in construction is required.

Computer vision is an interdisciplinary field that trains a computer to interpret and understand visual data. Computer vision research started in the early 1970s with digital image processing, computing 3D models from partially overlapping images, until today's advancement on object recognition, tremendous works have been achieved to make the computer perceive the same three-dimensional structure of the world as human (Szeliski, 2010). On top of the scientific contribution, application of computer vision in the industry ranges from automotive driver assistance, sports analysis, film and audio, safety monitoring, security, web and cloud to medical purposes (Lowe, 2015) where it shares the nature of automating the human analysis from visual data. Computer vision has attracted growing research interest in construction in recent years on automating, monitoring and improving labor intensive, time-consuming, repetitive tasks. The researcher applies computer vision techniques on several domains such as automated progress monitoring by detecting progress deviation through comparing operational level 4D BIM to as-built model (Golparvar-Fard et al., 2011; Han and Golparvar-Fard, 2014, Turkan et al., 2012b), scalable workface assessment by deriving visual construction activities from videos (Khosrowpour et al., 2014, Peddi et al., 2009, Rezazadeh Azar and McCabe, 2012, Yang et al., 2015), safety inspection by identifying violations and potential safety hazards (Q. Fang et al., 2018; Han and Lee, 2013, Seo et al., 2015, Yu et al., 2017), quality assessment/control by inspecting defects and the spacing of rebar cage (Akinci et al., 2006, Koch et al., 2015). The recent development on deep learning which allows the machine to automatically discover and formulate the features needed for classification (LeCun et al., 2015), also provides a great opportunity in construction to scale the current solutions to accommodate different environmental settings. It seems to be an easy decision to apply computer vision technologies in construction by looking at the potential and the benefits that come with it. However, it is still important to analyze the current problems in construction to better understand the opportunity of using visual data as a source of capture, monitor and analysis construction performance.

Construction still suffers from project delays and cost over budget. Recent reports indicate that more than 35% of all construction projects will incur a major change which leads cost overruns and schedule delays. The numbers become worse when we look into mega projects, 98% of these projects suffer from cost overrun and schedule delay (Changali et al., 2015). While many factors are contributing to the cost overruns, low productivity is one of the major issues. When advancement in computer vision, machine learning, and visual data analytics are emerging, construction has the opportunity for immense progress.

In the following section, the opportunity of using visual data as a source for capture, analytics and representation of Reality and Planned data on construction projects are discussed. Next, we present how the cutting-edge computer vision and machine learning techniques such as deep learning, object detection, and Structure from Motion (SfM) together with BIM and schedule are integrated and applied in the context of built environments to enable performance monitoring at both project task and operations level. The comparison of reality vs. plan will be discussed in detail for progress monitoring. For each use case, we will provide a concise literature review and assessment of current state-of-the-art solutions in the market, and will discuss the key underlying methods and recent solutions in detail. We will demonstrate their real-world performance by using several building and infrastructure project case studies. We will also discuss the tangible benefits in forms of Return on Investment and will share lessons learned across these projects. The challenges of applying these techniques in typical project workflows and open areas for research and development are also discussed in detail.

13.2.1 *The unprecedented growth of visual data*

Today, commodity hardware usually comes with built-in cameras such as smartphones, tablets, wearables, and camera-equipped unmanned ground/aerial vehicles. The easy access of cameras and the wide range of technologies boosted the exponential growth in the volume of images and videos that are being recorded on construction sites on a daily basis (Ham et al., 2016). Besides the daily photo logs taken by the construction team, professional photographers are also hired to systematically document and collect images and videos. One of the popular construction documentation service providers has claimed that about 325,000 images are taken by professional photographers, 95,400 images by webcams, and 2000 images by construction project team members at a typical commercial building project (~750,000 sf). On top of these 400,000 images in total, with 18% of construction companies using drones for photogrammetry and mapping (Budiac, 2018), this number can be much higher (Figure 13.1). This trend of visual documentation provides a unique opportunity to digitally record and analyze the entire construction life cycle in a fraction of cost comparing to the current solution such as laser scanning.

13.2.2 *Why visual data and what problems they can solve*

Visual data plays an important role from the pre-construction design development, the project execution to the later stage after the project handover. It visually presents the conditions of the construction site, building, and structure which delivers the best interface for communication, coordination, and planning. In the early construction phase, one of the most important processes is to accurately understand the current site conditions for site planning and evaluation. Visual data help engineers to develop different plans and simulate the logistics with the consideration of current site conditions. Project engineers collect visual data for progress documentation, quality assessment and safety inspection, visual data enhance the transparency and efficiency for communication. In the post-construction phase, these documented visual data can be used for facility management (Taneja et al., 2011) and legal proof for litigation to visually verify the quality and execution. The applications of visual data are not only seen in vertical construction, but has also been widely used for transportation and infrastructure

Figure 13.1 Various forms of visual data and their frequency of capture

projects such as bridge inspection, crack detection, mining material recognition, and excavation volume measurement.

With the broad implementation of n-dimensional BIM (i.e. 3D models enriched with information such as time, cost, safety, and productivity), enhanced 3D visualization with semantic building information improved communication, coordination, and planning. Different use cases have shown the value-added by utilizing BIM from early design phase to facility management, such as Lu et al. (2014) report 6.92% cost saving by using BIM, Staub-French and Khanzode (2007) report 25–30% productivity improvement by using BIM for coordination and constructability reviews to identify design conflicts. The integration of Reality (visual data collected onsite) with Plan (nD BIM) can efficiently communicate the necessary information for successful project control.

In summary, visual data brings four different aspects to improve construction: (1) transparency: visual data provides a graphical picture that people can see; (2) efficiency: visual data can deliver clear and complete information for communication; (3) accessibility: visual data can be easily shared and accessed through different interfaces; (4) durability: visual data can be stored digitally for a long period of time and can be accessed when needed easily. Today, visual data are widely used for solving different problems because of the benefits discussed above, and can be categorized into the following groups: (1) documentation: progress photos, safety issues, quality problems are visually documented to better communicate with the responsible person; (2) analytics: the documented visual data are analyzed through algorithms to make it actionable, this ranges from basic volume measurement for excavation to complicated safety violation detection; (3) inspection: with limited human resources and accessibility for inspection, visual data provides a flexible way to examine the site; (4) maintenance: facility management is often challenging because of the discrepancies between the drawings and the actual conditions, visual data provides the exact conditions as it shows. Figure 13.2 shows an

Figure 13.2 The top image shows MEP and structural system overlay on top of a 360 image, the bottom left shows a 2D drawing overlay on orthographic photo, the bottom right shows volumetric measurement on a point cloud model

orthographic photo overlay on 2D drawings for site layout and logistics planning, Mechanical Electrical and Plumbing (MEP) system maintenance using visual data and volume measurement on excavation projects.

13.3 Review on the current state-of-the-art for computer vision applications in the industry and research

Over the past few years, researchers have been developing and validating computer vision analytics in the construction domain, and many of these applications has been extended and used in the industry. The following sections introduce the state-of-the-art computer vision applications in research and industry in the terms of project control, we will discuss the fundamental theory behind the applications, and the implementation of case studies.

13.3.1 Reality capture techniques

Reality capture is the process of generating a digital model representation of real-world subjects such as construction examples: buildings, site conditions, mechanical rooms, dams, etc. The result is 3D point clouds or mesh models that are based on millions of data points mapping the entire subject. This process provides site engineers quick and accurate access to the current site conditions, where it has the potential to replace the traditional time-consuming site survey and daily job walks. It has also been widely used in construction to document the progress and issues by leveraging the visual data collected on the construction site. Laser scanning and image-based 3D reconstruction are two different types of reality capture method that are implemented onsite depending on the purpose and requirement. Laser scanning is mainly used for interior and high accuracy requirement capture due to its time-consuming and labor-intensive process. Image-based 3D reconstruction provides a flexible solution that accepts images taken from all sources such as smartphone, digital camera, 360 camera, site surveillance camera, drones. SLAM (Simultaneous Localization and Mapping) is the method that creates a map and localizes itself simultaneously which are used for automated data collection. While most of them share the same purpose of capturing existing condition, typically the results are shown as a point cloud. In this section, we will introduce the typical workflow of generating the point cloud as well as different techniques that are used to automate and improve the data capture process. Here we will focus on image-based 3D reconstruction and the applications of reality capture, the technical details of laser scanning is not in the scope of this chapter.

13.3.1.1 Computer vision techniques for monitoring project at the schedule task level

Schedule task-level progress monitoring focuses on using computer vision techniques to analyze the physical occupancy of the corresponded task location, which is usually represented as an area in 2D image or volume in a 3D point cloud. Task location can be extracted from the WBS (Work Breakdown Structure) of the schedule, rule-based naming conventions from the task name, task ID coding system or 2D drawing grid system. For example, to examine the progress for task "Area A Column", volumetric measurement in a 3D point cloud of the pre-defined Area A can be used for geometry analysis to check the occupancy.

Today, there are two dominant practices for leveraging images for tracking work in progress:

1. Generating large panoramic images of the site and superimposing these large-scale high-resolution images over existing maps – While these images provide excellent visuals to ongoing operation, they lack 3D information to assist with area-based and volumetric-based measurements necessary for progress monitoring. Also, none of the current commercially available platforms provide a mechanism to communicate who is working on which tasks at what location and they mainly deliver high-quality maps of construction sites.
2. Producing 3D point cloud models – The state-of-the-art in image-based 3D modeling methods from computer vision domain has significantly advanced significantly over the past decade. These developments have led to several commercially available platforms that can automatically produce 3D point cloud models from collections of overlapping images.

Several AEC/FM firms have started to leverage these platforms to produce as-built 3D point cloud models of their project sites via images taken from camera-equipped Unmanned Aerial Vehicles (UAVs). Nevertheless, today's practices are mainly limited to measuring excavation work and stockpiles (e.g. efforts led by SkyCatch, Kespy, DroneDeploy, and Propeller). This is because highly overlaying images taken with a top-down view can produce high-quality 3D models of these operations. However, creating complete 3D point cloud models for building and infrastructure systems also requires the UAVs to fly around the structure to capture work in progress. Because there is no automated mechanism for identifying most informative viewpoints, often the produced 3D point cloud models are incomplete. Also the state-of-the-art Structure from Motion (SfM) and Multi-View Stereo (MVS) techniques for image-based 3D reconstruction – as used in (Golparvar-Fard et al., 2010; 2012) – may distort angles and distances in generated point cloud models. Figure 13.3 shows several examples of a point cloud model that was generated via

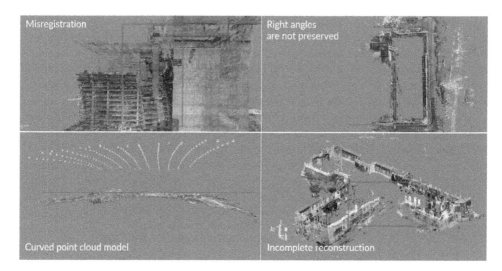

Figure 13.3 Typical challenges in using Standard SfM techniques for image-based 3D reconstruction: camera misregistration due to repetitive structure and lack of long range shots (top left); distortions in angle and distance (top right); curved point cloud due to camera calibration failure and lack of oblique photo capture (bottom left), and incomplete point cloud reconstruction due to fail loop closure

highly overlapping images taken from construction sites. The reconstructed 3D point clouds are up-to-scale and still exhibit problems in completeness and accuracy.

Recent research addresses the challenges mentioned above by developing comprehensive capture plans which include different camera angles and complete coverage of the whole target building. While the current commercially available capture plan only supports two-dimensional top-down capture plans.

In the following sections, three different techniques to generate point cloud are introduced with a detail discussion with its underlying theory.

IMAGE-BASED 3D RECONSTRUCTION

Over the past decade, there has been significant progress on research related to image-based 3D reconstruction techniques. These techniques can be divided into two groups: (1) variants of SfM-MVS pipeline such as (Golparvar-Fard et al., 2012, Lin et al., 2015) (Figure 13.4) and (2) leverage BIM as a priori and adopt a constrained-based procedure for image-based 3D reconstruction (Karsch et al., 2014).

In the SfM process, visual features are extracted from each image using the GPU to accelerate the algorithm (Wu, 2015). Researchers use Scale Invariant Feature Transforms (SIFT) (Lowe, 2004) and a variant of SIFT with Hessian Matrix as our visual features. Matched feature pairs are detected rapidly in terms of location, rotation, scale, and scene illumination during the multi-core CPU matching over a span of sequential images or among all the images. The 3D location is then calculated based on the Nister's five-point algorithm and camera matrix is also extracted using the Direct Linear Transform by Hartley and Zisserman (2004 in a Random Sample Consensus (RANSAC) algorithm loop which helps to remove the false matches. Then initial pair for reconstruction needs to be selected for the 3D reconstruction process. Typical heuristics (Snavely et al., 2008) choose the pair that have a maximum percentage

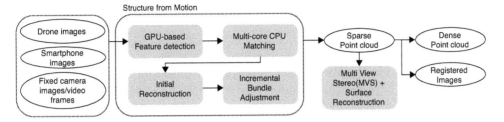

Figure 13.4 Image-based 3D reconstruction workflow

of matching features inliers before fitting the Fundamental Matrix in the RANSAC loop and the minimum percentage of after fitting the Homography matrix in the RANSAC loop. The images are incrementally added in the 3D reconstruction process until cameras cannot observe any 3D points. The bundle adjustment process (see Equation 13.1) minimizes the distance between the visual features and their re-projected 3D points iteratively throughout the process.

$$\underset{\mathbb{P}\backslash\mathbb{P}_a,\,t}{\mathrm{argmin}} \sum_{i=1}^{N}\left[\sum_{u\,\in\,\mathrm{tracks}_a} \mathrm{project}\left(\mathbb{P}_i, X_u\left(t_u\right)\right)-u\right] \tag{13.1}$$

The sparse point cloud is then generated with measured camera matrix that can be used to input for the MVS dense point cloud generation and images localization (see Figure 13.5).

The accuracy and completeness of image-based reconstruction can be enhanced by adding BIM as a constraint. (Karsch et al., 2014) presents a new BIM-assisted SfM procedure together with Multi-View Stereo that improves completeness and accuracy in 3D reconstruction. Leveraging all the camera parameters calculated in the previous step, MVS algorithm aims to improve the density of the point cloud by finding visual correspondence in the images. These visual correspondences are triangulated to fill in the sparse point cloud. Various MVS algorithms are used in different purposes, PMVS (Furukawa and Ponce, 2010), MVE (Fuhrmann

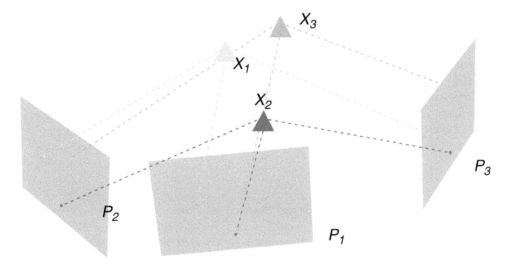

Figure 13.5 Image-based 3D reconstruction where X1, X2, X3 are three back-projected 3D points from the visual features of three Images P1, P2, and P3

et al., 2014), and COLMAP (Schönberger and Frahm, 2016, Schönberger et al., 2016) are introduced in the following section.

The PMVS (Furukawa et al., 2009) algorithm consists of three steps: match, expand, and filter. Utilizing the camera information from the SfM process, a set of new features are detected and matched among all the images yielding 3D information of a sparse set of patches with important regions of the images. It expands the initial matches to the nearby pixels to generate denser patches around the initial patch and then filter out the incorrect matches to reduce the noises. Although PMVS is widely used, MVE (Fuhrmann et al., 2014) aims to generate a denser point cloud for multi-scale geometry mesh reconstruction to preserve more details in the result. It uses the depth map from every view to reconstruct the dense point cloud, even though the redundancy of many views are overlapped with similar parts, this helps to only reconstruct the other small parts of the scene that are not visible. Also, the depth map can be directly embedded with all the parameterized information such as color. After the MVS from MVE, to merge the depth map into a globally consistent geometry, it uses hierarchical Signed Distance Fields (SDF) to interpolate and approximate points based on sample scale and redundancy. The final result produces a globally consistent surface mesh. COLMAP (Schönberger and Frahm, 2016, Schönberger et al., 2016) further improved the MVS process through embedding pixelwise depth and normal estimation with a photometric and geometric prior restricted view selection. The PatchMatch stereo is performed in four different directions and optimized through generating normal and depth candidates from the previous iteration to perform the patch based matching. With a multi-view geometric consistency term developed, the integration can simultaneously refine and fuse the image depth and normal. After the final step to filter remaining outliers that are not photogrammetrically and geometrically stable with support from multiple views, the dense point cloud is colored and generate with normal. The MVS process results in a dense point cloud that is necessary for construction purposes such as occupancy analysis, material recognition, and safety inspection.

This typical SfM-MVS pipeline is used for most commercially available reality capture solutions and extensively applied in construction sites. Although it generally can produce an accurate result, SfM can still suffer from curvature and distortion due to camera calibration and capture strategies. For example, captures with only top-down images at the same elevation generate ambiguity and accumulate error that results in a distorted and curved point cloud. To ensure accuracy and completeness, researchers developed process that leverage fiducial markers, Ground Control Points (GCP) and BIM to improve the results for construction, we will introduce related works in the next section. Also, with the increasing demands on generating as-built models for interior construction, 360 images and videos provide a quicker and easier approach to completely capture the interior space. However, the result of the MVS process is often restricted by the illumination and motion blur of the capture which highly depends on the capture strategies. To overcome these limitations of computer vision techniques, construction research has also focused on developing automated data capture system that is discussed in the next section.

SIMULTANEOUS LOCALIZATION AND MAPPING (SLAM)

In robotics, Simultaneous Localization and Mapping (SLAM) is the method to create and update the map of an unknown environment while simultaneously identifying the location of itself in the map. This computational method is widely used in robotics and employed in self-driving cars, drones, rovers, and robots. In construction, SLAM has received attraction because of its efficiency and low cost to quickly localize images taken in the building and also the potential to automate the visual data collection process. Hector SLAM (Kohlbrecher et al., 2014, 2011) uses 2D LiDAR sensor to build a navigation map and Iterative Closest Point (ICP) algorithm (Censi, 2008) to detect the location of the LiDAR. ORB SLAM (Mur-Artal et al., 2015) uses a monocular camera to detect 3D locations of visual features and localize the camera with respect to the features through Visual Odometry. SLAM technology is well suited for indoors mapping however it is associated with drift errors leading to misalignment of local maps. Wheels Odometry and Internal Measurement Unit (IMU) data are usually fused via Extended Kalman Filtering (EKF) to improve the accuracy of localization and reduce the mapping errors (Einicke and White, 1999).

SLAM-related research in construction has focused on generating and registering point clouds in an efficient and inexpensive way (Amer and Golparvar-Fard, 2018, Brilakis et al., 2011a, Jog et al., 2011), and the potential to automate the visual data collection and analytics (Asadi et al., 2018, Jin et al., 2018; P. Kim et al., 2018b). SLAM provides a tangible 3D point cloud model in a short amount of time for large areas and also has the potential to increase the automation and frequency of data collection. However, limited research investigates the implementation in dynamic construction environment, integration of hardware and software for construction purposes and the accuracy and completeness of the resulted point cloud. BIM offers geometry knowledge that could be used as a priori to improve the performance of SLAM and reality capture techniques in general. The following section discusses the role of BIM in computer vision techniques.

13.3.1.2 The role of BIM in computer vision monitoring techniques

BIM and 4D BIM are often served as a baseline for planned progress at task-level and operational-level monitoring. To enable progress deviation analysis, the BIM and point cloud models are necessary to be aligned in the same coordinate system. We discuss the underlying theory, state-of-the-art research that automates this process and the limitations of BIM for computer vision monitoring in the following section.

REALITY CAPTURE INTEGRATION WITH BUILDING INFORMATION MODELING

Point clouds generated from the SfM-MVS pipeline have arbitrary coordinate systems. They are also up-to-scale and thus their pixel units do not directly translate to real-world Cartesian coordinates. To register these point clouds into the site coordinate system, a minimum of three correspondences is needed. These corresponded could be based on (1) setting visual survey-ing benchmarks with known real-world coordinate systems such that the user (manually or through an automated detection procedure) can establish their correspondence with site coor-dinates, or (2) manually finding correspondence between up-to-scale point clouds and BIM. At least three points or correspondences with BIM are required to solve for the similarity transfor-mation between the two coordinate systems (Golparvar-Fard et al., 2009, 2012). Figure 13.6 shows an example of the markers that were used at McCormick Place project in Chicago, IL. These markers can be automatically or manually detected and their coordinates from the point cloud data can be matched to their equivalent from 4D BIM.

Several researchers have also focused on automating the process of alignment between BIM and point clouds without markers or GCPs. This especially becomes a difficult problem in built environments where structures and elements usually share similar geometry shape with symmetric characteristic. Previous works achieved limited automated registration with pre-defined constraints (Nahangi et al., 2015), semi-automated approaches (Bosché, 2012), limited symmetric geometry identification or partial or pre-processed data (Son et al., 2015) and prior

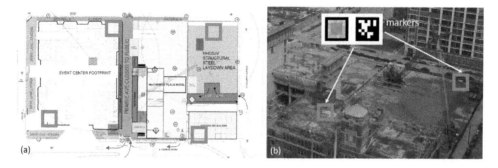

Figure 13.6   Surveying targets used for registration of point clouds and point cloud to BIM

information assisted system (Bueno et al., 2018). While this research area remains open, general purposes such as progress monitoring can be satisfied with manual registration discussed above. With having the BIM model registered to point cloud model, progress information extraction and comparison of reality vs. plan are discussed in section 13.3.2. The discussion so far primarily focuses on project- and task-level monitoring that examines the performance output such as work amount of concrete placement, but does not capture the detail performance of the input resources such as equipment active time. In the following section, we will discuss computer vision techniques that are used to analyze progress information at the operation level.

13.3.1.3 Computer vision techniques for monitoring project at the operation level

Operation-level monitoring is commonly conducted through manual examination from onsite fixed cameras' video footage to better understand productivity and safety. However, manual human-based monitoring is labor intensive and expensive. Recent research focuses on vision-based automated visual data analysis to facilitate productivity and safety monitoring. The following section discusses the latest research on applying computer vision techniques to track resources and activities through object detection, tracking and pose estimation.

OBJECT DETECTION, TRACKING, AND POSE ESTIMATION FOR TRACKING
RESOURCES AND THEIR ACTIVITIES

Object detection is one of the most challenging tasks for a computer to perform (Szeliski, 2011), objects could be occluded, appear in different poses and look differently depending on the environmental settings. All these varieties make it unlikely to successfully perform a brute force matching against a large database. Early research used techniques that rely on the presence of features (Bag of Words), relative positions and segmenting the image into semantical meaningful patches, these usually depend heavily on the context of the surrounding objects and scene. Because of the variabilities and the high dependency of the surrounding environment, the machine learning approach gains more popularity in the recent development of object detection. As early as using boosting, neural networks, support vector machines to the recent deep learning approach, object detection is still one of the fundamental problem researchers trying to solve in computer vision. While tracking an object in a video could be also seen as object detection in sequential images, it also involves object prediction and association between frames.

Computer vision based operation-level monitoring focuses on tracking the construction resources (workers, equipment, and materials) and analyzing the interaction between each other via visual data collected on the construction site. Object detection and activity recognition techniques are applied on construction equipment and workers to track the trajectory and motion for measuring the input resources in each activity. For equipment productivity analysis, single and multiple equipment activity recognition methods are developed to examine the earthmoving and dump truck efficiency (Golparvar-Fard et al., 2013; J. Kim et al., 2018b; H. Kim et al., 2019), then dirt loading cycling time is evaluated through the identified activity to improve productivity (Rezazadeh Azar et al., 2013). Point cloud volumetric measurement with video analysis for finer time scales productivity estimation are fused to analyze the productivity onsite to the schedule task level (Bügler et al., 2017). These productivity data are also inputted into simulation models to better estimate task completion and project duration (H. Kim et al., 2018a). Besides equipment productivity analysis, pose estimation and worker detection method are also developed for work sampling automation and productivity assessment. To be able to train a machine learning method to detect worker's activity, researchers have developed a crowdsourcing web-based annotation tool to gather ground truth data efficiently (Liu and Golparvar-Fard, 2015). Ironworker and carpenter activity are classified into 16 different types of individual activity through analyzing surveillance videos (X. Luo et al., 2018b). Through the spatial and temporal relevance between workers and objects 17 types of construction activities are recognized (H. Luo et al., 2018; X. Luo et al., 2018a). The majority of these works only track the location of the workers. However, without interpreting activities and purely based on location information, deriving meaningful workface data is challenging (Khosrowpour et al., 2014, Yang et al., 2015). For example, for drywall activities, distinguishing between idling, picking up gypsum boards, and cutting purely based on location is difficult, as the location of a worker would not necessarily change during these tasks.

Construction health and safety is another area that computer vision techniques are heavily applied. Commercial solutions such as SmartVid generates safety reports and identify top risks through detecting objects such as hard hats, safety vest, boots, ladder, man lift, crane, etc. from site images. To make detected objects actionable and be able to assess safety issues, researchers have developed language-image framework to transfer domain specific safety rules into computer vision paradigm (Tang and Golparvar-Fard, 2017). Automatic safety assessment system for surface mining activities are developed based on the speed of equipment movement and distance (Chi and Caldas, 2012). Research in safety management also developed object detection techniques with a more complex environment such as detecting safety harnesses when workers are working at heights (W. Fang et al., 2018b). Methods that detect unsafe worker behavior or action that could lead to potential safety hazard are also developed to improve safety performance, such as posture recognition from ordinary 2D cameras (Yan et al., 2017), real-time image-skeleton-based action (climb, dump, lean) analysis (Yu et al., 2017), and ladder climbing detection (Ding et al., 2018, Han and Lee, 2013, Han et al., 2013).

However, computer vision methods are also not advanced enough to conduct detailed assessments from videos or RGB-D data because methods for fully automated detection and tracking (Brilakis et al., 2011b, Escorcia et al., 2012, Memarzadeh et al., 2013), and deriving activities from long sequences automatically (especially when workers interact with tools) are not mature (M Golparvar-Fard et al., 2012, Gong et al., 2011, Khosrowpour et al., 2014). The current taxonomy of construction activities also does not enable "visual activity recognition" at a task level to be meaningful for workface assessment (Liu and Golparvar-Fard, 2015). While full automation is appealing, training machine learning methods require very large amount of empirical data which is not yet available to the construction informatics community (Liu and Golparvar-Fard, 2015).

The following section discussed methods comparing reality vs. plan at both schedule task and operation levels.

13.3.2 Methods for comparing reality vs. plan at both schedule task and operation levels

13.3.2.1 State-of-the-art in computing in civil engineering

The state-of-the-art methods of automated comparison are still in its infancy. Largely because these methods leverage geometry of the 3D reconstructed scenes to reason about the presence of elements on the construction sites. As such, they are unable to differentiate operations details such as finished concrete surfaces vs. forming stage and cannot accurately report on the state of work-in-progress. On the other hand, methods that detect and classify construction material from 2D images have primarily been challenged in their performance due to their inability to reason about geometrical characteristics of their detected components.

OCCUPANCY BASED PROGRESS MONITORING

Golparvar-Fard et al. (2010, 2012) utilized image-based 3D point clouds and BIM model to reason about the occupancy and visibility of the elements. They proposed a supervised machine learning method to infer the construction process. Bosché et al. (2014) compare laser scanning point cloud models against BIM to monitor the progress of Mechanical Electrical Plumbing (MEP) systems. Turkan et al. (2012b) further introduce a method to differentiate different operational details of concrete construction objects such as formwork and rebar (Turkan et al., 2012b). These methods are limited in their ability to detect operational details and the occlusion and visibility of elements in the point clouds (Figure 13.7).

APPEARANCE-BASED PROGRESS MONITORING

To address the limitation of occupancy-based progress monitoring, K. Han and Golparvar-Fard (2015) introduced a computer vision method to back-project the elements in a point cloud to the corresponding images and extract image patches to detect the material of the elements. This reasoning enables progress tracking of operational details and linking to the correct task level. They further leverage the geometry feature of the image patches to enhance the accuracy of material recognition (Han et al., 2018). However, these methods are unable to utilize the geometrical characteristics of their detected components (Figure 13.8).

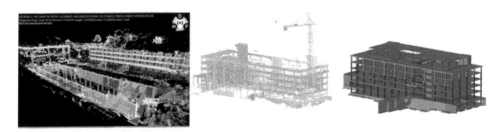

Figure 13.7 Progress is shown in as-built and 4D BIM models with color-coded status superimposed together (left) (M. Golparvar-Fard et al., 2012), laser scanned as-built (middle) and 4D BIM model (right) (Turkan et al., 2012a)

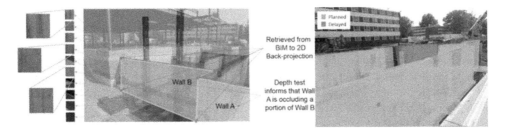

Figure 13.8 Using patches retrieved from BIM to 2D back-projection to classify material and performing depth test to exclude occluding area (K. Han and Golparvar-Fard, 2015) (left); progress status is extracted by comparing as-built and as-planned after occupancy detection and material classification (right)

While significant advancement in research has been made, still applying these methods to a full-scale projects requires (1) accounting for the lack of details in 4D BIM, (2) addressing as-built visibility issues, (3) creating large-scale libraries of construction materials that could be used for appearance-based monitoring purposes; and (4) methods that can jointly leverage geometry, appearance, and interdependency information in BIM for monitoring purposes.

13.3.2.2 Machine learning and deep learning

Machine learning techniques are widely used in recognition algorithms and almost become the fundamental element for recognition tasks, as it has been proven that only using basic patterns without learning to build a recognition system will lead to failure (Szeliski, 2011). Researchers develop algorithms from principal component, subspace, discriminant analysis, and more sophisticated discriminated classification algorithms such as support vector machines, boosting and deep to analyze images. Conventional learning techniques rely on careful engineering and domain expertise to design feasible feature extractors to transform the images to computer understandable format, then separating the features through a hyperplane in n-dimensional space. Deep learning automatically discovers and formulates the features needed for classification with multiple levels of representation retrieved by generating non-linear modules that transform representation to a higher abstract format at each level. Complex functions are then learned in the process of the transformation to generate the classifier with the last layer representation amplify important features and discriminate irrelevant components. In the past decade, very large databases that contain tens of thousands of training images and with the advancement on computational resources opened up the fast development of learning techniques. A lot of new semantic segmentation models now can achieve region-based object detection such as Fast R-CNN, Faster R-CNN, R-FCN, and FPN, they could reach to pixel-wise segmentation through Conditional Random Fields (CRF) model or models such as MASK R-CNN.

Vision-based learning applications in construction focus on the aforementioned subject of monitoring at project and operation level and utilizes the most feasible model to address the problems. Support Vector Machine (SVM) has been widely used as a classifier for recognition tasks in construction. Rezazadeh Azar and McCabe (2011) used SVM with Haar-like and HOG shape feature to detect excavators; Golparvar-Fard et al. (2013) classify equipment action such as digging, hauling, dumping, and swinging by SVM classifier with a set of spatio-temporal patterns of descriptors; Dimitrov and Golparvar-Fard (2014) developed an SVM classifier that can robustly learn and infer construction material categories using features from a Bag of Words pipeline of filter responses and HSV color values; Han and Golparvar-Fard

(2015) developed a similar approach with input of back-projection patches from as-planned BIM model to corresponded with registered image patches. Other conventional machine learning models such as k-Nearest Neighbors (k-NN) and Adaptive Boosting (AdaBoost) have also been used to detect construction workers and equipment from images (Park and Brilakis, 2012, Rezazadeh Azar and McCabe, 2011).

Deep learning models such as Convolutional Neural Network (CNN) are used for guardrail detection (W. Fang et al., 2018; Kolar et al., 2018) use Faster R-CNN to detect workers without hardhats from surveillance camera and verify if workers are wearing harnesses when working at heights. Kim et al. (2019) developed a framework that uses R-FCN to track truck license plate for productivity simulation and a similar framework with CNNs and Long Short-Term Memory (LSTM) to analyze earthmoving processes (H. Kim et al., 2018a). H. Luo et al. (2018) developed a deep stream CNN that can simultaneously capture static spatial features, short-term motion and long-term motion in a video. While tracking and prediction remain as open problems in computer vision, research has developed models with LSTM encoder-decoder networks to forecast future locations and Mixture Density Network (MDN) to model uncertainty in predictions (Tang and Golparvar-Fard, 2017).

While these models have shown promising results, the training data are usually limited and applied in a controlled environment. The following section specifically discussed the current research opportunities, challenges, and limitations for monitoring ongoing construction at both project and operation levels.

13.3.3 Opportunities, challenges, and limitations for monitoring ongoing construction at both project and operation levels

With the exponential growth of visual data in construction and the fast pace advancement in computer vision, it provides a unique opportunity to investigate construction problems in an interdisciplinary lens. In the previous sections, we can see that significant achievements have been made over the past decades, yet a large number of problems still remain as open research challenges. In the following section, potential problems that hinder improvement for projects and operation-level monitoring are discussed and possible resolutions are introduced.

Research and commercial solutions using visual data to monitor construction progress still have to address several critical problems to potentially automate the process. Although recent algorithm improvement on image-based and video-based 3D reconstruction has shown promising results, the accuracy and completeness of the point cloud can still be improved on generating consistent good results in different environmental settings. For example, the reconstruction tends to produce poor quality results on reflective surfaces and thinner structures which are commonly seen in construction as curtain walls and steel components. As automated data collection can better assure the quality of the final reconstruction, little research has investigated possible solutions for complex interior construction sites. In addition, even though the interior reconstruction from 360 images has shown very promising results, the dense point cloud quality still needs to be improved to enable occupancy-based progress monitoring. Currently, the point cloud generation often resulted in curvature or misregistration due to large area coverage that causes error accumulation or low light environment that causes motion blur. On the other hand, occupancy-based monitoring only provides binary results of the observed object which does not necessarily support project control decisions. Occupancy-based monitoring can leverage the recent development on point cloud semantic segmentation that utilizes the scene context or the integration with appearance-based methods that extract colors, textures, shapes, and semantic information from the 2D image to streamline the automated progress monitoring process. While object recognition and material classification on 2D images could potentially improve the process, it requires a large collection of dataset specifically for construction material to train the machine learning model and evaluate the result. As synthetic data in computer vision has successfully gained performance for object detection, pose estimation, etc., it could also provide an inexpensive and efficient way to generate a large database. Having a feasible LOD as-planned model remains challenging, automated as-planned 4D model generation could not only offer value for progress monitoring research but also resolve the most time-consuming part in virtual design and construction practices.

For research in monitoring operation-level, classification subjects are still restricted to several objects such as dump trucks, cranes, loaders, excavators, and workers. It is clear that to fully monitor operation-level tasks, we still need to expand the collection of objects for detection. It is also a common issue that research only tested on their own dataset which restricted the possibility to generalize the methodology to be applied on dynamic and cluttered environments such as construction. With assumptions on highly controlled environments and limited variables, the results are usually showing very impressive numbers but still remain doubtful when testing in the wild. Construction activities are also unlike regular daily life activities, hence, it is not always applicable to use existing pre-trained computer vision databases for construction activities. Assembling a complete database of equipment and worker activity is required to develop a practically feasible methodology. Group activity recognition that has not been well studied in the computer vision domain could also be useful for analyzing the different crew and trade resource deployments. Prediction and tracking of workers' behavior are also important to proactively manage safety issues. While prediction and tracking still remain challenging problems in computer vision, construction could use a priori information such as schedule, BIM, and documents to enhance the results.

The following section introduces the latest application on using state-of-the-art computer vision techniques in construction; implementation details and practical feedback from construction practitioners are also discussed.

13.3.4 Case studies

With exciting research advancement on applying computer vision techniques in construction, real-world construction sites have also started to embrace the latest technology. The following case studies are observations of state-of-the-art research implementation and interviews with site engineers.

13.3.4.1 Progress monitoring

To leverage reality capture models to communicate progress efficiently for project control decisions, researchers have developed a visual production management system to enhance planning and coordination (Lin and Golparvar-Fard, 2018). The visual production management system takes in images captured at different times and locations to automatically generate 4D as-built point clouds and also localize unordered images in the same environment. By linking look-ahead schedules and BIM via Work Breakdown Structure locations, location-based 4D models are created and merged with the visual asset management platform to benchmark and monitor "who does what work in what location". By accessing the resulting visual production models through smartphones, worker hours are documented per task, per organization, and per location. The reliability of the look-ahead schedule is measured based on actual production and productivity rates and top locations at risk for potential delays are highlighted on a weekly basis. Subcontractors can understand the current progress through reality capture and the weekly work plan through color-coded 4D BIM model (Figure 13.9, Figure 13.10). Through this case study, it shows that progress tracking via reality capture can enhance awareness of ongoing work and better flow of information. The visual production management system

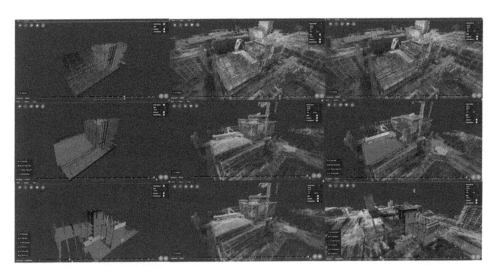

Figure 13.9　The first column shows progress via 4D point cloud and BIM; the second column shows that location-based 4D BIM model and work-in-progress tracking integrated with point clouds; the third column shows the 4D BIM with subcontractor responsible tasks color-coded to communicate who does what work in what location

Figure 13.10　The web-based system has been used on different construction site during the coordination meetings, it has been proved that it can efficiently enhance accountability and traceability, and predictive analytics improve reliability in short-term planning

allows project participants to directly commit to work tasks without the need for addressing difference in the level of detail between 4D BIM and actual work in progress, communicate performance deviations based on commitments from coordination sessions, measure readiness of the upcoming tasks based on interdependencies of current schedule tasks and their constraints, and highlight locations that are at-risk of potential performance problems.

13.3.4.2 Quality assessment/control

Reality capture through laser scan and image-based reconstruction are essential these days for quality assessment/control on construction sites. The Virtual Design and Construction (VDC) department usually are equipped with laser scan devices or 360 cameras that could be used onsite consistently. Currently, floor flatness, wall plumbness, and opening analysis are the most common application of reality capture. A $130 million 36 floor residential building project utilizes laser scan point cloud right after the concrete placement to measure floor flatness (Tang et al., 2011), slab edges alignment, embeds positions, and opening analysis. The unique shape of the slab edges differs from floor to floor, to ensure the constructability of the upcoming curtain wall installation, slab edges are required to perfectly align with the design drawings. On top of laser scan point clouds, crane camera collects high-quality top-down images to generate point clouds for the same purpose to support incomplete laser scan. Embeds and openings position are also verified by comparing reality capture to BIM models. Another $99 million 10 story education building performs before and after concrete placement scans to ensure the quality. Reality capture provides accurate quality assessment/control to keep transparency and accountability for construction activities.

13.3.4.3 Safety inspection

Researchers have successfully implemented human object interaction models with safety rules to identify and predict unsafe behavior (Tang and Golparvar-Fard, 2017). While previous research has tested on thousands of internet collected images, commercial solutions such as SmartVid and indus.ai are also implemented on construction sites with multiple companies. Even though these solutions can only reliably provide object detection on Personal Protection Equipment (PPE) for compliance checking, it still provides safety reports and identifies top risks based on daily jobsite photos. Another platform indus.ai claims to provide real-time tracking of equipment and workers, the actual proactive safety management application is limited. Despite the limitation of this existing application, it does provide a consistent and higher frequency inspection from all visual data sources. According to interaction with several general contractors' safety departments, safety managers typically have to manage multiple jobsites at the same time, as a result, they can only maintain weekly to bi-weekly jobsite visits. With the help of automatic safety analytics of jobsite photos and weekly safety reports of risk trends and the top risk items, safety inspection becomes more efficient. However, proactive safety inspection still remains an open challenge.

13.4 Conclusions

The application of computer vision techniques with BIM-driven data analytics is one of the major components that could drive the transformation process in the Construction 4.0 framework. This chapter provides a detailed review of the research and current practice of computer vision methods for monitoring construction performance both at project and operation level which are representing the process of physical and digital transformation. A detailed assessment of current state-of-the-art research and commercial solutions in the market with the recent development and future opportunities are discussed. While most of the research is still in early stages of development, this chapter has introduced several case studies in practice to evaluate the applicability and potential of the techniques. The performance and accuracy are compared and discussed for specific monitoring problems. Research opportunities and future research on advanced computer vision and machine learning techniques for construction are also proposed.

13.4.1 Lesson-learned from the state of the applications

There is still a gap between research and industrial application. Reports shows an impressive increase of 324% to nearly $3.1 billion in 2018 compared with $731 million in 2017 for funding in construction technology startups (Azevedo, 2019), more and more effort is working on transferring research and development projects into commercial solutions. However, research results do not necessarily translate to actual performance gain in practical project settings. Research outcomes should accommodate different configurations such as realistic training data and user interfaces for a variety of personas. To embrace technology in construction, research has to consider the practicality on a much larger scale to stimulate the adaption.

13.4.2 Discussion on tangible return of investments (ROI)

The ROI of applying computer vision technology in construction for monitoring can be boiled down into time and cost. In terms of time, several case studies have shown that with the implementation of vision-based monitoring platform can save time on planning, coordination, and communication. For example, to update the master schedule according to the actual progress involved the participation of a scheduler, executive level project coordinator, project manager, and superintendents, and took hours. With the vision-based platform assistance, the superintendent can update the progress by himself and can only discuss contractual date requirements with the executive level. The time and cost of man hours from different people could represent a large amount of savings for the project. In terms of cost, while reports show 4–6% of the total project cost is the median cost of rework (Armstrong and Gilge, 2017) and most of this is due to miscommunication, the current vision-based system has significantly improved communication through effective visuals and models. Several other metrics are also used to measure the ROI of vision-based systems, such as Earned Value Analysis, Percent Plan Complete (PPC), and delayed tasks.

13.4.3 Open opportunities for research

A comprehensive comparison using a large construction database is still required to objectively validate different approaches. However, to collect and label the data takes a lot of effort. The computer vision community has advanced significantly after major datasets such as ImageNet, AlexNet, CoCo, and KITTI, there is a need of a large construction data repository for researchers to share and contribute to. Furthermore, to integrate and link the process from the input of construction resources to the output of built infrastructure in the monitoring perspective could enhance the understanding of the whole construction process. With the SLAM approach fusing different information collected from sensors, the concept of generating a living digital twin of the construction product can help all stakeholders in the project transparently monitor the productivity, quality, and progress in an efficient manner.

13.5 Summary

- With full understanding of construction project control problems, computer vision, and BIM-driven data analytics for monitoring construction and operation has shown promising results in both research and commercial solution.
- While project- and task-level monitoring has shown successful implementation in research and industry, most of the research on operation-level monitoring is still in early development; to reduce the gap between research and industrial application, research must accommodate various settings and flexible user interface.
- Several case studies show tangible return of investment in terms of time and cost.

References

Akinci, B., Boukamp, F., Gordon, C., Huber, D., Lyons, C., & Park, K., 2006. A formalism for utilization of sensor systems and integrated project models for active construction quality control. *Automation in Construction* 15, 124–138. doi: https://doi.org/10.1016/J.AUTCON.2005.01.008.

Amer, F. & Golparvar-Fard, M., 2018. Decentralized visual 3d mapping of scattered work locations for high-frequency tracking of indoor construction activities, in: Construction Research Congress 2018. American Society of Civil Engineers, Reston, VA, pp. 491–500. https://doi.org/10.1061/9780784481264.048.

Armstrong, G. & Gilge, C., 2017. Global construction survey: make it, or break it – Reimagining governance, people and technology in the construction industry. URL https://assets.kpmg/content/dam/kpmg/xx/pdf/2017/10/global-construction-survey-make-it-or-break-it.pdf.

Asadi, K., Ramshankar, H., Pullagurla, H., Bhandare, A., Shanbhag, S., Mehta, P., Kundu, S., Han, K., Lobaton, E., & Wu, T., 2018. Vision-based integrated mobile robotic system for real-time applications in construction. *Automation in Construction* 96, 470–482. doi: https://doi.org/10.1016/J.AUTCON.2018.10.009.

Azevedo, M. A., 2019. Investor momentum builds for construction tech [WWW document]. URL https://news.crunchbase.com/news/investor-momentum-builds-for-construction-tech/ (accessed 4.19.19).

Bosché, F., 2012. Plane-based registration of construction laser scans with 3D/4D building models. *Advanced Engineering Informatics* 26, 90–102. doi: https://doi.org/10.1016/J.AEI.2011.08.009.

Bosché, F., Guillemet, A., Turkan, Y., Haas, C., & Haas, R., 2014. Tracking the built status of MEP works: assessing the value of a scan-vs-BIM system. *Journal of Computing in Civil Engineering* 28.

Brilakis, I., Fathi, H., & Rashidi, A., 2011a. Progressive 3D reconstruction of infrastructure with videogrammetry. *Automation in Construction* 20, 884–895. doi: https://doi.org/10.1016/j.autcon.2011.03.005.

Brilakis, I., Park, M.-W.-W., & Jog, G., 2011b. Automated vision tracking of project related entities. *Advanced Engineering Informatics* 25, 713–724. doi: https://doi.org/10.1016/j.aei.2011.01.003.

Budiac, D., 2018. Construction technology trends – 2018 Report [WWW document]. URL https://softwareconnect.com/construction/technology-trends-2018-report/ (accessed 4.21.19).

Bueno, M., Bosché, F., González-Jorge, H., Martínez-Sánchez, J., & Arias, P., 2018. 4-Plane congruent sets for automatic registration of as-is 3D point clouds with 3D BIM models. *Automation in Construction* 89, 120–134. doi: https://doi.org/10.1016/J.AUTCON.2018.01.014.

Bügler, M., Borrmann, A., Ogunmakin, G., Vela, P. A., & Teizer, J., 2017. Fusion of photogrammetry and video analysis for productivity assessment of earthwork processes. *Computer-Aided Civil and Infrastructure Engineering* 32, 107–123. doi: https://doi.org/10.1111/mice.12235.

Censi, A., 2008. An ICP variant using a point-to-line metric, in: 2008 IEEE International Conference on Robotics and Automation, pp. 19–25. https://doi.org/10.1109/ROBOT.2008.4543181.

Changali, S., Azam, M., & van Nieuwland, M., 2015. The construction productivity imperative. [WWW Document]. URL www.mckinsey.com/industries/capital-projects-and-infrastructure/our-insights/the-construction-productivity-imperative (accessed 11.11.19).

Chi, S. & Caldas, C. H., 2012. Image-based safety assessment: automated spatial safety risk identification of earthmoving and surface mining activities. *Journal of Construction Engineering and Management* 138, 341–351. doi: https://doi.org/10.1061/(ASCE)CO.1943-7862.0000438.

Dimitrov, A. & Golparvar-Fard, M., 2014. Vision-based material recognition for automated monitoring of construction progress and generating building information modeling from unordered site image collections. *Advanced Engineering Informatics* 28, 37–49.

Ding, L., Fang, W., Luo, H., Love, P. E. D., Zhong, B., & Ouyang, X., 2018. A deep hybrid learning model to detect unsafe behavior: integrating convolution neural networks and long short-term memory. *Automation in Construction* 86, 118–124. doi: https://doi.org/10.1016/J.AUTCON.2017.11.002.

Einicke, G. A. & White, L. B., 1999. Robust Extended Kalman Filtering. *Transactions on Signal Processing* 47, 2596–2599. doi: https://doi.org/10.1109/78.782219.

Escorcia, V., Dávila, M. A., Golparvar-Fard, M., & Niebles, J. C., 2012. Automated vision-based recognition of construction worker actions for building interior construction operations using RGBD cameras, in: Proceedings of the Construction Research Congress, May 21–23, 2012 West Lafayette, Indiana, United States.

Fang, Q., Li, H., Luo, X., Ding, L., Luo, H., Rose, T. M., & An, W., 2018. Detecting non-hardhat-use by a deep learning method from far-field surveillance videos. *Automation in Construction* 85, 1–9. doi: https://doi.org/10.1016/J.AUTCON.2017.09.018.

Fang, W., Ding, L., Luo, H., & Love, P. E. D., 2018. Falls from heights: a computer vision-based approach for safety harness detection. *Automation in Construction* 91, 53–61. doi: https://doi.org/10.1016/J.AUTCON.2018.02.018.

Fuhrmann, S., Langguth, F., & Goesele, M., 2014. MVE: A multi-view reconstruction environment, in: Proceedings of the Eurographics Workshop on Graphics and Cultural Heritage, GCH '14. Eurographics Association, Aire-la-Ville, Switzerland, Switzerland, pp. 11–18. https://doi.org/10.2312/gch.20141299.

Furukawa, Y., Curless, B., Seitz, S. M., & Szeliski, R., 2009. Reconstructing building interiors from images, in: 2009 IEEE 12th International Conference on Computer Vision, pp. 80–87.

Furukawa, Y. & Ponce, J., 2010. Accurate, dense, and robust multiview stereopsis. *IEEE Transactions on Pattern Analysis and Machine Intelligence* 32, 1362–1376.

Golparvar-Fard, M., Heydarian, A., & Niebles, J. C., 2013. Vision-based action recognition of earthmoving equipment using spatio-temporal features and support vector machine classifiers. *Advanced Engineering Informatics* 27, 652–663.

Golparvar-Fard, M., Peña-Mora, F., Arboleda, C. A., & Lee, S., 2009. Visualization of construction progress monitoring with 4d simulation model overlaid on time-lapsed photographs. *Journal of Computing in Civil Engineering* 23, 391–404.

Golparvar-Fard, M., Peña-Mora, F., & Savarese, S., 2010. D4AR–4 Dimensional augmented reality-tools for automated remote progress tracking and support of decision-enabling tasks in the AEC/FM industry, in: Proceedings of the 6th International Conference on Innovations in AEC.

Golparvar-Fard, M., Peña-Mora, F., & Savarese, S., 2011. Integrated sequential as-built and as-planned representation with tools in support of decision-making tasks in the AEC/FM industry. *Journal of Construction Engineering and Management* 137, 1099–1116. doi: https://doi.org/10.1061/(ASCE)CO.1943-7862.0000371.

Golparvar-Fard, M., Peña-Mora, F., & Savarese, S., 2012. Automated progress monitoring using unordered daily construction photographs and IFC-based building information models. *Journal of Computing in Civil Engineering*, 147–165. doi: https://doi.org/10.1061/(ASCE)CPPPP.1943-5487.0000205.

Gong, J., Caldas, C. H., & Gordon, C., 2011. Learning and classifying actions of construction workers and equipment using bag-of-video-feature-words and Bayesian network models. *Advanced Engineering Informatics* 25, 771–782.

Ham, Y., Han, K. K., Lin, J. J., & Golparvar-Fard, M., 2016. Visual monitoring of civil infrastructure systems via camera-equipped unmanned aerial vehicles (UAVs): a review of related works. *Visualization in Engineering* 4, 1. doi: https://doi.org/10.1186/s40327-015-0029-z.

Han, K., Degol, J., & Golparvar-Fard, M., 2018. Geometry- and appearance-based reasoning of construction progress monitoring. *Journal of Construction Engineering and Management* 144, 4017110. doi: https://doi.org/10.1061/(ASCE)CO.1943-7862.0001428.

Han, K. K. & Golparvar-Fard, M., 2014. Multi-sample image-based material recognition and formalized sequencing knowledge for operation-level construction progress monitoring, in: *Computing in Civil and Building Engineering*, pp. 364–372.

Han, K. K. & Golparvar-Fard, M., 2015. Appearance-based material classification for monitoring of operation-level construction progress using 4D BIM and site photologs. *Automation in Construction* 53, 44–57.

Han, S. & Lee, S., 2013. A vision-based motion capture and recognition framework for behavior-based safety management. *Automation in Construction* 35, 131–141.

Han, S., Lee, S., & Peña-Mora, F., 2013. Vision-based detection of unsafe actions of a construction worker: case study of ladder climbing. *Journal of Computing in Civil Engineering* 27, 635–644. doi: https://doi.org/10.1061/(ASCE)CP.1943-5487.0000279.

Hartley, R. & Zisserman, A., 2004. *Multiple View Geometry in Computer Vision*, 2nd ed. Cambridge: Cambridge University Press.

Jin, M., Liu, S., Schiavon, S., & Spanos, C., 2018. Automated mobile sensing: towards high-granularity agile indoor environmental quality monitoring. *Building and Environment* 127, 268–276. doi: https://doi.org/10.1016/J.BUILDENV.2017.11.003.

Jog, G. M., Fathi, H., & Brilakis, I., 2011. Automated computation of the fundamental matrix for vision based construction site applications. *Advanced Engineering Informatics* 25, 725–735. doi: https://doi.org/10.1016/j.aei.2011.03.005.

Karsch, K., Golparvar-Fard, M., & Forsyth, D., 2014. ConstructAide: analyzing and visualizing construction sites through photographs and building models. *ACM Transactions on Graphics* 33, 176.

Khosrowpour, A., Niebles, J. C., & Golparvar-Fard, M., 2014. Vision-based workface assessment using depth images for activity analysis of interior construction operations. *Automation in Construction* 48, 74–87. doi: https://doi.org/http://dx.doi.org/10.1016/j.autcon.2014.08.003.

Kim, H., Bang, S., Jeong, H., Ham, Y., & Kim, H., 2018a. Analyzing context and productivity of tunnel earthmoving processes using imaging and simulation. *Automation in Construction* 92, 188–198. doi: https://doi.org/10.1016/J.AUTCON.2018.04.002.

Kim, H., Ham, Y., Kim, W., Park, S., & Kim, H., 2019. Vision-based nonintrusive context documentation for earthmoving productivity simulation. *Automation in Construction* 102, 135–147. doi: https://doi.org/10.1016/J.AUTCON.2019.02.006.

Kim, J., Chi, S., & Seo, J., 2018b. Interaction analysis for vision-based activity identification of earthmoving excavators and dump trucks. *Automation in Construction* 87, 297–308. doi: https://doi.org/10.1016/J.AUTCON.2017.12.016.

Koch, C., Georgieva, K., Kasireddy, V., Akinci, B., & Fieguth, P., 2015. A review on computer vision based defect detection and condition assessment of concrete and asphalt civil infrastructure. *Advanced Engineering Informatics* 29, 196–210. doi: https://doi.org/10.1016/J.AEI.2015.01.008.

Kohlbrecher, S., Meyer, J., Graber, T., Petersen, K., Klingauf, U., & von Stryk, O., 2014. Hector open source modules for autonomous mapping and navigation with rescue robots BT RoboCup 2013: Robot World Cup XVII, in: S. Behnke, M. Veloso, A. Visser, and R. Xiong (eds.), Berlin, Heidelberg: Springer, pp. 624–631.

Kohlbrecher, S., von Stryk, O., Meyer, J., & Klingauf, U., 2011. A flexible and scalable SLAM system with full 3D motion estimation, in: 2011 IEEE International Symposium on Safety, Security, and Rescue Robotics, pp. 155–160. https://doi.org/10.1109/SSRR.2011.6106777.

Kolar, Z., Chen, H., & Luo, X., 2018. Transfer learning and deep convolutional neural networks for safety guardrail detection in 2D images. *Automation in Construction* 89, 58–70. doi: https://doi.org/10.1016/J.AUTCON.2018.01.003.

LeCun, Y., Bengio, Y., & Hinton, G., 2015. Deep learning. *Nature.* 521, 436.

Lin, J., Han, K., & Golparvar-Fard, M., 2015. Model-driven collection of visual data using UAVs for automated construction progress monitoring, in: International Conference for Computing in Civil and Building Engineering 2015. Austin, TX.

Lin, J. J. & Golparvar-Fard, M., 2018. Visual data and predictive analytics for proactive project controls on construction sites BT – Advanced Computing Strategies for Engineering, in: I.F.C. Smith and B. Domer (eds.), Cham: Springer International Publishing, pp. 412–430.

Liu, K. & Golparvar-Fard, M., 2015. Crowdsourcing construction activity analysis from jobsite video streams. *Journal of Construction Engineering and Management* 4015035. doi: https://doi.org/10.1061/(ASCE)CO.1943-7862.0001010.

Lowe, D., 2004. Distinctive image features from scale-invariant keypoints. *International Journal of Computer Vision* 60, 91–110.

Lowe, D., 2015. The computer vision industry. www.cs.ubc.ca/~lowe/vision.html (accessed 11.12.19).

Lu, W., Fung, A., Peng, Y., Liang, C., & Rowlinson, S., 2014. Cost-benefit analysis of building information modeling implementation in building projects through demystification of time-effort distribution curves. *Building and Environment* 82, 317–327. doi: https://doi.org/10.1016/J.BUILDENV.2014.08.030.

Luo, H., Xiong, C., Fang, W., Love, P. E. D., Zhang, B., & Ouyang, X., 2018. Convolutional neural networks: computer vision-based workforce activity assessment in construction. *Automation in Construction* 94, 282–289. doi: https://doi.org/10.1016/J.AUTCON.2018.06.007.

Luo, X., Li, H., Cao, D., Dai, F., Seo, J., & Lee, S., 2018a. Recognizing diverse construction activities in site images via relevance networks of construction-related objects detected by convolutional neural networks. *Journal of Computing in Civil Engineering* 32, 04018012. doi: https://doi.org/10.1061/(ASCE)CP.1943-5487.0000756.

Luo, X., Li, H., Cao, D., Yu, Y., Yang, X., & Huang, T., 2018b. Towards efficient and objective work sampling: recognizing workers' activities in site surveillance videos with two-stream convolutional networks. *Automation in Construction* 94, 360–370. doi: https://doi.org/10.1016/J.AUTCON.2018.07.011.

Memarzadeh, M., Golparvar-Fard, M., & Niebles, J. C., 2013. Automated 2D detection of construction equipment and workers from site video streams using histograms of oriented gradients and colors. *Automation in Construction* 32, 24–37.

Mur-Artal, R., Montiel, J. M. M., & Tardós, J. D., 2015. ORB-SLAM: A versatile and accurate monocular SLAM system. *IEEE Transactions on Robotics* 31, 1147–1163. doi: https://doi.org/10.1109/TRO.2015.2463671.

Nahangi, M., Yeung, J., Haas, C. T., Walbridge, S., & West, J., 2015. Automated assembly discrepancy feedback using 3D imaging and forward kinematics. *Automation in Construction* 56, 36–46. doi: https://doi.org/10.1016/J.AUTCON.2015.04.005.

Park, M.-W.-W. & Brilakis, I., 2012. Construction worker detection in video frames for initializing vision trackers. *Automation in Construction* 28, 15–25.

Peddi, A., Huan, L., Bai, Y., & Kim, S., 2009. Development of human pose analyzing algorithms for the determination of construction productivity in real-time. *Construction Research Congress* ASCE, Seattle, WA, 11–20.

Rezazadeh Azar, E., Dickinson, S., & McCabe, B., 2013. Server-customer interaction tracker: computer vision-based system to estimate dirt-loading cycles. *Journal of Construction Engineering and Management* 139, 785–794. doi: https://doi.org/10.1061/(ASCE)CO.1943-7862.0000652.

Rezazadeh Azar, E. & McCabe, B., 2011. Automated visual recognition of dump trucks in construction videos. *Journal of Computing in Civil Engineering* 6, 769–781.

Rezazadeh Azar, E. & McCabe, B., 2012. Part based model and spatial-temporal reasoning to recognize hydraulic excavators in construction images and videos. *Automation in Construction* 24, 194–202. doi: https://doi.org/10.1016/j.autcon.2012.03.003.

Schönberger, J. L. & Frahm, J.-M., 2016. Structure-from-motion revisited, in: 2016 IEEE conference on computer vision and pattern recognition (CVPR). IEEE, pp. 4104–4113. https://doi.org/10.1109/CVPR.2016.445.

Schönberger, J. L., Zheng, E., Frahm, J.-M., & Pollefeys, M., 2016. *Pixelwise View Selection for Unstructured Multi-View Stereo*. Cham: Springer, pp. 501–518. doi: https://doi.org/10.1007/978-3-319-46487-9_31.

Seo, J., Han, S., Lee, S., & Kim, H., 2015. Computer vision techniques for construction safety and health monitoring. *Advanced Engineering Informatics* 29, 239–251. doi: https://doi.org/10.1016/J.AEI.2015.02.001.

Snavely, N., Garg, R., Seitz, S. M., & Szeliski, R., 2008. Finding paths through the world's photos, in: Proceedings of SIGGRAPH.

Son, H., Bosché, F., & Kim, C., 2015. As-built data acquisition and its use in production monitoring and automated layout of civil infrastructure: a survey. *Advanced Engineering Informatics* 29, 172–183. doi: https://doi.org/10.1016/j.aei.2015.01.009.

Staub-French, S. & Khanzode, A., 2007. 3D and 4D modeling for design and construction coordination: issues and lessons learned. *Journal of Information Technology in Construction* 12, 381–407.

Szeliski, R., 2010. *Computer Vision: Algorithms and Applications*, 1st ed. Berlin, Heidelberg: Springer-Verlag.

Szeliski, R., 2011. *Computer Vision, Texts in Computer Science*. London: Springer. doi: https://doi.org/10.1007/978-1-84882-935-0.

Taneja, S., Akinci, B., Garrett, J. H., Soibelman, L., Ergen, E., Pradhan, A., Tang, P., Berges, M., Atasoy, G., Liu, X., Shahandashti, S. M., & Anil, E. B., 2011. Sensing and field data capture for construction and facility operations. *Journal of Construction Engineering and Management* 137, 870–881. doi: https://doi.org/10.1061/(ASCE)CO.1943-7862.0000332.

Tang, P., Huber, D., & Akinci, B., 2011. Characterization of laser scanners and algorithms for detecting flatness defects on concrete surfaces. *Journal of Computing in Civil Engineering* 25, 31–42. doi: https://doi.org/10.1061/(ASCE)CP.1943-5487.0000073.

Tang, S. & Golparvar-Fard, M., 2017. Joint reasoning of visual and text data for safety hazard recognition, in: Computing in Civil Engineering 2017. American Society of Civil Engineers, Reston, VA, pp. 450–457. https://doi.org/10.1061/9780784480847.056.

Turkan, Y., Bosché, F., Haas, C., & Haas, R., 2012a. Automated progress tracking using 4D schedule and 3D sensing technologies. *Automation in Construction* 22, 414–421.

Turkan, Y., Bosché, F., Haas, C., & Haas, R., 2012b. Toward automated earned value tracking using 3d imaging tools. *Journal of Construction Engineering and Management* 139, 423–433. doi: https://doi.org/10.1061/(ASCE)CO.1943-7862.0000629.

Wu, C., 2015. VisualSFM: A visual structure from motion system [WWW document] http://ccwu.me/vsfm/.

Yan, X., Li, H., Wang, C., Seo, J., Zhang, H., & Wang, H., 2017. Development of ergonomic posture recognition technique based on 2D ordinary camera for construction hazard prevention through view-invariant features in 2D skeleton motion. *Advanced Engineering Informatics* 34, 152–163. doi: https://doi.org/10.1016/J.AEI.2017.11.001.

Yang, J., Park, M.-W., Vela, P. A., & Golparvar-Fard, M., 2015. Construction performance monitoring via still images, time-lapse photos, and video streams: now, tomorrow, and the future. *Advanced Engineering Informatics* 29, 211–224. doi: https://doi.org/10.1016/J.AEI.2015.01.011.

Yu, Y., Guo, H., Ding, Q., Li, H., & Skitmore, M., 2017. An experimental study of real-time identification of construction workers' unsafe behaviors. *Automation in Construction* 82, 193–206. doi: https://doi.org/10.1016/J.AUTCON.2017.05.002.

14

UNMANNED AERIAL SYSTEM APPLICATIONS IN CONSTRUCTION

Masoud Gheisari, Dayana Bastos Costa, and Javier Irizarry

14.1 Aims

- Providing a comprehensive review of Unmanned Aerial System (UAS) technology application in construction.
- Discussing UAS definition and its components and technological features.
- Discussing UAS application in various pre-, during-, and post-construction phases.
- Reviewing UAS implementation challenges.

14.2 Introduction

The rise of the Industry 4.0 concept created a new industrial paradigm that embraces a set of future industrial developments regarding Cyber-Physical Systems (CPS), Internet of Things (IoT), Internet of Services (IoS), Robotics, Big Data, Cloud Manufacturing, and Augmented Reality (Pereira and Romero, 2017).

In Cyber-Physical Systems (CPS), all objects and physical processes are digitized, but it is a bilateral interaction, where the real provides the data for the virtual, and the latter provides decisions for the real. Integration through a CPS system aims to reduce or eliminate data collection manually in the field and provide real-time reports (Correa, 2018). Thus, it is possible to reduce the time spent by professionals in collecting tasks, leaving them with more time for value-added tasks and decision making (Correa, 2018). Among Cyber-Physical Systems (CPS), the Unmanned Aerial Systems (UASs) can be integrated in the construction processes as a tool for monitoring and digitizing construction projects from the early design to construction and post-construction phases.

Unmanned Aerial Systems (UASs) are aircraft that do not carry human pilots (Liu et al., 2014). UASs have generally been constructed for the use of the military and have a history that can be traced back over 150 years; unmanned balloons were employed in Venice in 1849 at a time of civil strife, and the British Royal Navy created a UAS named "Queen Bee" in the 1930s for training in anti-aircraft gunnery techniques (Howard et al., 2017). UAS technology has improved massively over time, and these aircraft are now extremely popular for a number of peaceful uses, e.g. monitoring the environment (Capolupo et al., 2015), monitoring wildlife (Martin et al., 2012), traffic surveillance (Liu et al., 2014), inspecting borders

(Rios-Morentin, 2011), imaging and mapping (Bendig et al., 2012) and fighting forest fires (Merino et al., 2012).

A report from the Association for Unmanned Vehicle Systems International (AUVSI) estimated that in the first three years of UAS being integrated into the National Airspace System (NAS) they will generate $13.6 billion of business, with exponential growth meaning they will generate over $82 billion in the decade from 2015 to 2025 (Jenkins and Vasigh, 2013). The swift expansion of the UAS market has created significant demands for their use in several industrial sectors, including the construction industry. In construction, these aircraft are most frequently used for safety inspections, damage assessments, monitoring progress, site inspections, and building maintenance tasks.

The utility of UAS in the sector is attributable to a number of factors, including how easy they are to use, how mobile they are, their agility, low costs, ability to capture data at a number of different heights, angles, and areas that are hard to access, and also the fact that numerous devices may be attached to them, including infrared sensors, lasers, radars, thermal sensors, cameras, gyroscopes, Global Positioning Systems (GPS) and others (Puri, 2005; Eisenbeiß, 2009; Siebert and Teizer, 2014).

The latest UASs have a significant degree of autonomy, which can lower threats to human health and safety on construction sites. Commercial UASs can use a number of different sensors for autonomous navigation and data collection, feeding data back to a control station instantly. Recent improvements in batteries, autonomous controls, and sensor technology have made these aircraft considerably more reliable for use in the construction sector.

In this chapter a comprehensive overview of UAS technology will be provided, its applicability to the construction industry, what challenges implementation presents and what benefits are accrued, and how it might be used in the future, especially in Construction 4.0. To begin with, what UASs are, how they are built, and what technology they use, what forms they take, the hardware, software and so forth employed, what they cost, how they are regulated in different countries, and other technical elements will be examined. There will follow a discussion on how UASs can be deployed before, during, and after building construction, with a focus on the reasoning behind deploying UASs in each of these phases, and then possible future applications of the technology in construction will be discussed. Lastly, a review will be undertaken of challenges to UAS implementation, including potential legal difficulties, liabilities, safety issues, and provision of the requisite training.

14.3 Unmanned Aerial Systems (UASs)

14.3.1 Definition

UASs are platforms for aerial hardware that do not carry a pilot, can collect and process data, and are capable of operating autonomously or with direct human control. Frequently referred to as drones, the definition of such systems is "a powered aerial vehicle that does not carry a human operator, uses aerodynamic forces to provide vehicle lift, can fly autonomously or be piloted remotely" (Newcome, 2004). The US Federal Aviation Administration (FAA) officially uses the term "Unmanned Aircraft System (UAS)" rather than "Unmanned Aerial Vehicle (UAV)" – frequently referred to as "Drone" – because Unmanned Aircraft (UA) do not carry crews on board (FAA, 2017). UAS generally consist of a portable controller for human operation and one, or more than one, UAV. The UAVs employed may be fitted with a variety of sensors, e.g. still/video cameras (including ones with the ability to see near and far infrared), specialized communications systems, radar, or laser-based rangefinders, amongst many others.

The majority of UASs can transfer data in real-time between the UAV(s) and the controller; some types also carry data storage facilities on board to collect data. UASs can undertake many assignments that can also be undertaken by manned vehicles, but they can frequently perform more quickly, with greater safety, and more economically (Puri, 2005).

14.3.2 Types of UASs

The two most common forms of UAS are fixed-wing and rotary-wing. Fixed-wing UASs employ aerodynamic wings to create lift, and rotary-wing aircraft gain lift from engine thrust. While fixed-wing UASs are capable of high-speed flight over a long duration and have more simple structures, they always need a runway or launcher for takeoff and landings and do not have hover capabilities. Rotary-wing UAS are capable of hovering and vertical takeoff and landing, but they cannot fly as fast, and do not have the flight range, of a fixed-wing machine (Pereira and Pereira, 2015). Figure 14.1 illustrates several examples of fixed-wing and rotary-wing UASs used for commercial purposes. There are two other forms of UAS, VTOLs, and blimps, but these are not frequently employed in the construction sector. Vertical Takeoff and Landing (VTOL) machines are fixed-wing types that can take off and land vertically; blimps, sometimes called aerostats, are lighter-than-air craft that gain lift via the gas they carry within them, which gives them longer flying times than other UAVs (Brooks et al., 2017).

Rotary-wing UAS can be helicopters or multicopters (e.g. quadcopter, hexacopter), according to the number of propellers they carry. Gheisari et al. (2015) assert that quadrotors are less mechanically complex than standard helicopters, which means that UASs have the potential to be more robust. Ellenberg et al. (2014) suggest that helicopters provide extended flying times, but no motor redundancy. Dobson et al. (2013) state that multi-rotor UASs provide greater stability for taking photographs, but they cannot carry as heavy a payload, or offer the same flight times, as a helicopter. Eschmann et al. (2012) and Wefelscheid et al. (2011) experimented with octocopters and suggested that they would be better able for retaining control of the vehicle in instances of in-flight engine failure. Roca et al. (2013) also suggested that octocopters were more robust and maneuverable.

In the construction sector, rotary-wing UASs are more frequently employed than fixed-wing models (Zhou and Gheisari, 2018). The unique advantage of a rotary-wing aircraft is that it can take off and land vertically and also hover over any chosen area, something that fixed-wing vehicles cannot do, requiring extensive runways for takeoff and landings, which can be difficult to find on many construction sites. However, when circumstances allow the aircraft to

(a) DJI Parrot DISCO FPV (fixed-wing)   (b) DJI Parrot ANAFI (quadcopter)   (c) Yuneec - Typhoon H (hexacopter)

Figure 14.1   UAS types and examples

fly at greater altitude and over larger areas (e.g. extensive construction sites, highways), fixed-wing UASs could be a preferable solution.

14.3.3 UAS forms of control

The ways in which UASs are controlled are different depending on the type of aircraft and how they are used: they can be controlled manually, be semi-autonomous, or completely autonomous. Figure 14.2 shows a brief illustration of the three types of UAS control available.

With manual operation, UAS are under control of operators on the ground via telemetry (Wen and Kang, 2014). Although this method is frequently used, it imposes limitations on maneuverability because the aircraft must always be within the operator's line of sight. Autonomous aircraft can be used for greater maneuverability, better economy, and reduced risk. There are two forms of autonomous flight frequently employed:

a) UAS following a fixed course set through preprogrammed GPS coordinates.

b) UAS employing sensor techniques and closed-loop feedback.

When operating using GPS, a flight can be planned and put into action using integrated wayfinding computer programs and GPS. One example of this was used by Siebert and Teizer (2014), employing waypoints generated by computer to establish a flight path for a UAS to follow using internal GPS; Freimuth et al. (2017) planned a path through definition of object boundaries in geo-referenced BIM and finding the UAS using the Global Navigation Satellite System (GNSS). The problem with planning flights via waypoints is a lack of accuracy. If aircraft cannot be located precisely near a building facade during inspection, or if shadowing effects are created by nearby buildings, GPS navigation can become unsuitable (Eschmann et al., 2012). GPS systems have a further drawback in that they do not permit accuracy regarding altitude (ibid.), and that when undertaking lengthy tasks, for example inspecting roads or pipelines, widely distanced waypoints or inaccurate GPS data could cause the aircraft to miss required sections of video (Rathinam et al., 2008). Because of this, onboard sensors have been integrated to enhance accuracy; these are used together with the GPS waypoints. In order to prevent collision with obstacles, UASs can be supplied with visual sensors (ibid.) or light detection and ranging (LiDAR) (Merz and Kendoul, 2011) to allow independent navigation. By employing a combination of active sensors and waypoint navigation, UASs can be employed to allow operation out of line of sight, in proximity to obstacles, and at low altitude (ibid.).

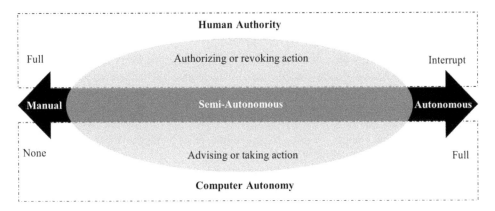

Figure 14.2 UAS degree of autonomy continuum

Enhanced mobility is one of the chief advantages of autonomous flight control. Bolourian et al. (2017) also contend that automated flight paths are more likely to be optimal. UASs that fly autonomously can access areas of roofs or tall structures more economically (Freimuth et al., 2017). In certain situations, autonomous flight should be included in the operation as deficiencies in line-of-site vision would prevent the UAS being fully manually controlled by the pilot (Michael et al., 2012); this also has safety advantages as it can prevent crashes when visual systems are outdistanced (Tomita et al., 2017). Regulatory restrictions may impose limits on UAS operating fully autonomously around construction sites (Wefelscheid et al., 2011).

Semi-autonomous UAS employ autonomous control at some points during a flight, or may only have certain functions operating autonomously, e.g. takeoff, landings, or return to base. When faced with challenges in terms of environmental complexity or obstacles to line-of-sight control that militates against full manual control of the aircraft, allowing for semi-autonomous flight means the operators can switch easily between fully autonomous control and semi-autonomous control, which allows the aircraft to inspect a location more closely, and its actions may be overridden as necessary (Michael et al., 2012).

14.3.4 Human-UAS interaction and UAS flight team

Human-robot interactions are simply defined as one or more humans interacting with one or more robots (Goodrich and Schultz, 2007). Robot systems, such as UASs, generally need an element of direct control, either through telemetry or another form of supervision depending on how autonomous they are. Telemetry implies that the robot is directly and entirely under human control; fully automated systems have robots with superior cognitive abilities that are able to interact with humans (see Figure 14.3). It is possible to study and analyze UAS operations and systems in theory and also technically using the same concepts of human-robot interactions. In UAS deployments, a human operator (pilot-in-command) has to work with members of their flight team (e.g. vision observers) and other stakeholders (e.g. construction workers, project managers, or safety engineers) to establish full control of the UAV, directing its flight by employing communication sensors (e.g. Wi-Fi signals, GPS, cameras). Best practice is to have a flight team consisting of at least three persons (de Melo et al., 2017), comprising the pilot-in-command, one observer guiding the pilot to the correct areas for data collection, and another observer checking that the flight remains safe (watching the aircraft and its environs). A flight crew of three was also proposed by Murphy et al. (2008), comprising pilot, flight director, and mission specialist.

14.3.5 UAS data collection sensors

A variety of different cameras are the most frequently employed sensors when UASs are employed in the construction sector. Cameras can take aerial photographs (digital/time lapse) or videos, which can then be processed and either inspected or used for the generation of 3D project models. In the construction sector, various forms of camera may be added to the UAS, for example:

- RGB: cameras that employ the same red/green/blue color bands as human eyes, producing an almost exact simulation of human vision.
- Thermal: cameras that are capable of detecting the heat given off by virtually every object or material and turning the received data into videos or images.
- Multispectral: cameras that employ Near Infrared, Red-Edge, Red and Green wave bands to record images of objects both visible and invisible to the human eye.

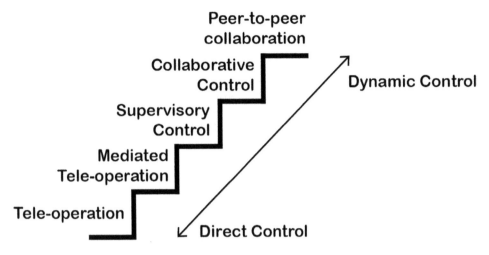

Figure 14.3 Level of autonomy

When a building or site is visually inspected by employing UAS equipment cameras there are generally two essential stages: acquiring the data and analyzing the data. For the initial stage, the UAS is employed to take aerial videos and photographs and in the second stage the data collected is analyzed either by using sophisticated image processing technology (Sankar-asrinivasana et al., 2015) or by having decision-makers directly observe the photographs or videos taken (Irizarry and Costa, 2016).

Employing technology that captures digital images is economic, practical, and simple to operate (Golparvar-Fard et al., 2009b). There are, however, some drawbacks to this more simple technology, e.g. the operators can only see what is revealed within the camera's fixed field of vision; images are slightly distorted, making them difficult to overlap; and they can be adversely affected by changes in lighting, weather, or site conditions (Álvares et al., 2018).

These limitations notwithstanding, Wang et al. (2014) employed a UAS using thermal cameras and visual RGB cameras for the diagnosis of visual and thermal issues of build-ings. Other researchers have employed photogrammetry techniques to create point cloud data allowing for three-dimensional models to be created or to measure distances when, for example, examining structural issues or detecting cracks in buildings (Ellenberg et al., 2014). Photogrammetry can also be employed to monitor the progress of construction, allowing the construction team to use the images taken for visualization, recognition, and location of the built environment and undertake analysis of their characteristics; this is seen as a useful way of measuring and communicating the physical progression of construction and to use for record-keeping and generating progress reports (Golparvar-Fard et al., 2009a, 2009b; Teizer, 2015; Álvares et al., 2018). UASs are not limited to carrying visual imaging sensors to provide project teams with site data: laser scanners can be employed for direct capture of services' 3D geometry, with point clouds generated by scanning the scene and assigning 3D coordinates to each point; laser pulse systems that measure transmission and return times are particularly effective for this (Eisenbeiß, 2009; Remondino, 2011). Eisenbeiß (2009) explains that laser pulses are directed at the surface that requires mapping and the scanner will register when the pulse is returned; the sensor/surface distance can then be derived from the operation time (transmission/return) and the speed of the pulse. With aerial laser scanners, further information is imported using the spatial coordinates offered by GPS and flight altitude (Eisenbeiß, 2009).

The chief advantages of employing laser scanning systems that they offer high levels of precision, even when surveying irregular complex shapes rather than smooth surfaces, they can capture large amounts of data very quickly, and they create the point cloud directly without data having to be processed after the operation (Groetelaars and Amorim, 2012). However, such sensors are not suited to all situations as they are costly, heavy, less portable, and do not provide a good representation of texture, which means that photography is required to be used in conjunction with the scanned points compiled to reconstruct textures (Remondino, 2011; Bhatla et al., 2012).

Light detection and ranging (LiDAR) has also been employed with UASs; this is a form of laser scanning that creates closely populated and accurate point counts (Merz and Kendoul, 2011). However, initial prototypes of such technology were heavy and expensive in comparison to high-resolution photography. Furthermore, LiDAR scanners need expert operators (Wefelscheid et al., 2011). Merz and Kendoul (Merz and Kendoul, 2011) used a COTS two-dimensional LiDAR with UAS to mitigate the expense and weight limitations of 3D equipment. The technology has now progressed to a point that it can be more broadly applied.

Explorations have also been undertaken into alternatives to traditional sensors. Roca et al. (2013) employed an Xbox Kinect sensor, an inexpensive low weight RGB-D camera that offered sufficient visual and depth information for the generation of three-dimensional point clouds. Although these point clouds were nonuniform and noisy, they propose that these problems can be overcome through mounting a variety of sensors on the UAS (ibid.). Ellenberg et al. (2014) employed photogrammetry techniques for processing images taken by a camera and compared their findings with the results from an independent Xbox Kinect device.

As well as using imaging sensors to collect visual data, a variety of other sensors have been used with UAS, for example, Radio Frequency Identification (RFID). RFID is a type of voice communication that employs electrostatic or electromagnetic coupling in the radio frequency segment of the electromagnetic spectrum for unique identification of objects, animals, or people. Hubbard et al. (2015) used a UAS fitted with an RFID unit to follow materials around a construction site. This research involved installing an RFID reader on the UAS and then putting tags with a variety of reading ranges along the route that the UAS would follow and observe the acquired system readings (Hubbard et al., 2015). This research demonstrated that integrating such technology can enhance the management of construction projects, and that information acquired by RFID units can be combined with BIM and software for managing project supply chains (Hubbard et al., 2015).

14.3.6 UAS costs

UAS technology has developed rapidly from the 1990s onwards. Estimations are that the $2.6 billion spent on this technology in 2016 will rise to $10.9 billion by 2025 (Teal Group, 2016). New advances in the design of vehicles, battery life, GPS navigation, and reliability of control have allowed many new economical and lightweight aircraft to be designed. In the current market, there are hundreds of manufacturers of UASs offering many different aircraft of different weights, payload capacities, and battery life, all factors that influence vehicle cost. With such a wide variety available, construction professionals need assistance in deciding what they should buy, how it should be used, and how they should maintain their systems in a sustainable manner. The cost of acquiring UASs are closely linked to a collection of general parameters that assist in determining the full expense of any particular system (Eiris and Gheisari, 2017):

- Weight: The full mass of the aircraft, to include the battery, propellers, motors, controllers and frame, i.e. all elements required for the aircraft to take off.
- Payload capacity: The greater the payload capacity of an aircraft, the more energy it will consume and the greater the power it requires, which will increase its costs. Payload capacity frequently comprises gimbles, cameras, and other sensors.
- Total takeoff weight: This is a combination of the weight of the aircraft and its maximum payload capacity. This represents the highest mass the aircraft's motors can get into the air.
- Endurance: How long the aircraft battery can keep the aircraft in flight. This metric varies according to the weight of payload, how the aircraft is configured, and the conditions in which it is flying. Manufacturer figures indicate how long the battery will last in the worst possible situation, i.e. carrying a full payload and in non-ideal weather conditions.

Other elements which influence UAS costs are whether or not the vehicle's flight path is automated and what form of navigation is employed. The avionics of the aircraft are driven by software; these include navigation, monitoring, controls, and communication. Additionally, operations may require mission-critical software for both mobile and fixed sites on the ground. There are various software packages (e.g. Agisoft, Pix4D) and cloud-based platforms (e.g. DroneDeploy or Skyward) that can undertake different tasks for AUS flights, ranging from planning flights and coordinating them to collecting, processing, and sharing data. Such software should be regarded as one of the costs of operating and managing a UAS and as being separate from the cost of acquisition.

14.3.7 Regulations and flight safety requirements

In terms of operating UASs in different countries, each country has its own local regulations regarding civilian use of the vehicles. A number of examples of such regulations are offered here to offer a broad overview of the ways in which UASs are regulated in various regions. In the USA, commercial use of UASs is subject to regulation and monitoring by the Federal Aviation Administration (FAA). The FAA introduced regulation for commercial employment of small UASs in August 2016 (see Table 14.1). The FAA regulations for the deployment of small UASs for the purposes of work or business (Title 14 of the Code of Federal Regulations Part 107), mandate that all flights should be controlled by a pilot with the correct certification for such tasks. Aircraft weighing over 25 kg have to be registered with the FAA (U.S. Department of Transportation, 2016). These regulations were altered in 2019, when the FAA ruled that every UAS, whether being used personally or commercially, must be registered with the FAA and the UAS must display a visible registration number (FAA, 2017). The operating regulations of the FAA also mandate that small UASs may only fly in daytime within line-of-sight of the operator, must not fly above 120 m or faster than 160 km/h (FAA, 2017). These regulations permit operators to receive commercial pilot certification for all forms of small UAS aerial operations, which includes those undertaken in construction. The FAA regulations as they currently stand have made purchasing or leasing small UASs an attractive proposition for construction companies, and it is predicted that increasing numbers of such companies will be investigating the potential of UAS technology for their operations (Eiris and Gheisari, 2017).

In the UK, the Civil Aviation Authority–United Kingdom (CAA UK) regulates UAS deployment under regulation CAP722 (CAA, 2015). The chief safety requirements of this regulation are that using a drone must not represent a risk to persons or property; that a pilot may only fly their vehicle if they meet all safety requirements and conditions; and that the aircraft must remain within visual line of sight of the pilots, who must follow its trajectory and avoid potential

Table 14.1 Main requirements of 14 CFR, Part 107 (FAA, 2017)

	Work/business purpose flights
Remote Pilot Certification	• Must be at least 16 years of age
	• Must pass an initial aeronautical knowledge test at an FAA-approved testing center
	• Must undergo Transportation Safety Administration (TSA) security screening
	• Must pass a recurrent aeronautical knowledge test every 24 months
Aircraft requirements	• Must weigh less than 25 kg, including payload, at takeoff
	• Must be registered if over 0.25 kg.
	• Must be registered under Part 107 if unmanned aircraft not flown under section 336
	• Registration identification number must be visibly displayed regardless of use
	• Must undergo pre-flight check to ensure that UAS is in condition for safe operations
Location requirements	• Fly in Class G airspace*
Operating rules	• Must keep the aircraft within visual line-of-sight (VLOS)*
	• Must fly under 120 m*
	• Must fly during the day or civil twilight*
	• Must fly at or below 160 km/h*
	• Must yield the right of way to manned aircraft*
	• Must NOT fly directly over people*
	• Must NOT fly from a moving vehicle, unless in a sparsely populated area*

* Part 107 Sections Subject to waiver: Operation from a moving vehicle or aircraft (§ 107.25), Daylight operation (§ 107.29), Visual line of sight aircraft operation (§ 107.31), Visual observer (§ 107.33), Operation of multiple small unmanned aircraft systems (§ 107.35), Yielding the right of way (§ 107.37(a)), Operation over people (§ 107.39), Operation in certain airspace (§ 107.41), Operating limitations for small unmanned aircraft (§ 107.51)

collisions. Under these regulations UASs are placed in three categories depending on weight: small (sUASs) under 20 kg, light UASs weighing between 20 kg and 150 kg, and those weighing in excess of 150 kg. Different regulations apply according to which category a vehicle comes into, with the greatest variation being for the heaviest aircraft. However, for all of these categories, a UAS must be registered, operations must be authorized, and pilots must be qualified (CAA, 2015).

In Europe, the European Aviation Safety Agency (EASA) has responsibility for the regulation of airspace, and a new regulatory framework for operating UAS safely is currently being developed. The EASA Committee voted unanimously, on February 28, 2019, to approve the proposal of the European Commission that an Implementing Act should be passed for regulating the operation of UAS (EASA, 2019). This Implementing Act has an adjoining Delegated Act that sets out the technical specifications of UASs. This was adopted by the European Commission on March 12, 2019, and forwarded to the EU Parliament and Council for a mandatory two-month period of scrutiny (EASA, 2019). The proposed EASA regulations define three forms of UAS operations, ranging from low risk to high risk (EASA, 2019). In the lower risk category, only VLOS flights are permitted below 150 m, and the UAS must have a maximum takeoff mass of 25 kg or less (ibid.). The new regulation will also address issues around insurance, liabilities, data protection, privacy, and safety.

For Brazil, the commercial deployment of UAS is subject to regulation and monitoring by ANAC (Agência Nacional de Aviação Civil), the national agency for civil aviation. Experimental operations employing UASs require authorization from ANAC and the regulations vary depending on how the Remote Pilot Aircraft (RPA) is classified. Aircraft are classified on the basis of their Maximum Takeoff Weight (MTOW), what it is being used for (experimental purposes, commercial, or corporate use), and whether it will remain within the pilot's line of sight (ANAC, 2017). For the purposes of experiment, any aircraft weighing less than 25 kg just has to be registered with ANATEL (Agência Nacional de Telecomunicações), the national agency of telecommunication to be legal to operate (ibid.). No pilot's license is needed, and it is not necessary to obtain authorization for flight. The flight criteria do demand that the aircraft remains within the pilot's visual line of sight and does not exceed an altitude of 120 m, or 60 m in urban areas (ANAC, 2017).

In Chile, operating UASs comes under the regulation of the *Direccion General de Aeronautica Civil* (DGAC – General Direction for Civil Aviation) and the regulation DAN 151 (DGAC, 2015). Under this regulation, any person or institution wishing to operate any form of UAS must register their aircraft with, and obtain authorization from, the DGAC. Additionally, the aircraft must be operated in VLOS mode, only be operated in good weather and in daylight (except with specific authorization), must not operate within 2 km of any airport, and must not operate over 120 m (DGAC, 2015). Table 14.2 illustrates the primary requirements for small

Table 14.2 Main requirements for small UASs (20–25 kg) according to international regulations

REQUIREMENTS	USA (FAA 2016, 2019)	UK (CAA, 2015)	EUROPE (EASA, 2019)[1]	BRASIL (ANAC, 2017)	CHILE (DGAC, 2015)
Aircraft approval and registration	Online registration, markings required	Requires authorization from CAA	Requires registration (three-year validity)	Online registration	Requires authorization from DGAC
Remote Pilot Certification	Yes	Yes	Yes	No	Yes
Autonomous operation	Allowed in VLOS	Forbidden	Forbidden	Allowed in VLOS	Forbidden
Minimum distance to operate nearby airports	8 km	5 km	-	5 km	2km
Minimum distance to operate from people and buildings	-	50 m	10 m	30 m	30 m
Urban area restriction	Restricted	Forbidden	-	Below 60 m	Below 120 m
Operation in BVLOS	Requires authorization	Requires authorization	Requires authorization	Requires authorization	Requires authorization
Night operation	Requires authorization	Forbidden	Forbidden	Forbidden	Forbidden

1 Approved by the European Commission's proposal. Requires EU Parliament and EU Council approval

UASs (i.e. aircraft weighing less than 20–24 kg for the majority of international regulations referred to in this section).

14.4 UAS applications

There has been a steady increase in interest from both researchers and practitioners regarding the potential of using UAS for managing construction projects in terms of undertaking safety inspections (Irizarry et al., 2012; Gheisari et al., 2014; de Melo et al., 2017), measuring earthworks (Siebert and Teizer, 2014), monitoring construction (Álvares et al., 2018), and aerial construction (Willmann et al., 2012; Augugliaro et al., 2015). Nevertheless, the widespread deployment of UAS in the construction industry is still in its infancy. One reason for this is that there is still a great deal to be learned regarding the capability and potential of UAS deployment in the construction sector. Given that contemporary modern projects are highly expensive and technologically complex, the deployment of advanced technology may assist in reaching the high productivity and performance targets demanded by the contemporary construction industry. This chapter will examine contemporary and possible future uses for UAS before construction, in the course of construction, and after construction.

14.4.1 Pre-construction stage

14.4.1.1 Project evaluation

Flexible and economic UASs are capable of efficient recording of spatial information regarding objects of interest (Liu et al., 2014). Still images or videos taken using UASs offer better resolution in both spatial and temporal dimensions in comparison to images taken with traditional satellite or other aerial platforms (Turner et al., 2014). When initiating projects, surveillance and/or sponsors may employ UASs to undertake aerial inspections for evaluation and determination of project feasibility. Deploying UASs is especially useful in urban districts (see Figure 14.4) where space is generally constricted. Spatial photographs can be obtained and measured with tools like Geographic Information Systems (GIS) to obtain estimates of the dimensions of a site, access paths, and height restrictions. Deploying UASs in this way is economical and saves time and other resources for project managers and companies (Zhou et al., 2018). Relevant spatial and attribute data can be collected more economically, faster, and more safely than previously, allowing project sponsors and investors to benefit from improved decision-making due to the greater accuracy of available data (Zhou et al., 2018).

14.4.1.2 Site planning

Standard management practices are not always the most efficient way of planning and scheduling activities on construction sites (Arkady and Aviad, 1999). Efficiently managing on-site activities with detailed plans and schedules for construction projects demands clear spatial analysis (Wen and Kang, 2014). Given that traditional ways of investigating sites and defining layouts do not generally offer a great deal of accuracy in terms of visualization, employing UAS offers project managers both physical and virtual visualizations, allowing them to visualize the layout of the site and plan and organize accordingly (Zhou et al., 2018). For example, logistical planning can be proposed and developed on the basis of aerial information regarding the surface and man-made features of the site and its environment. One example of

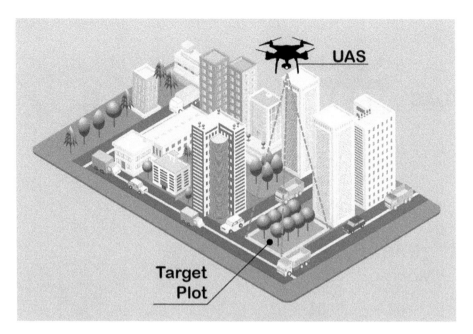

Figure 14.4 Applying a UAS to project evaluation in an urban area

technology being deployed for such purposes was an empirical study estimating how effective a combination of UAS and LiDAR technology could be for site planning (Cozart, 2016). UAS technology is also useful for the acquisition of reliable information regarding sources of water and electricity and the routes of conduits for same, which assists efficient planning regarding water and electricity supplies on site (Zhou et al., 2018).

14.4.2 Construction stage

14.4.2.1 Earth moving

A UAS can be deployed as an assistive tool by programming it to overfly construction sites requiring earth movement, transmitting images to computers that automatically construct 3D models of the environment (Siebert and Teizer, 2014). Unmanned excavators with autonomous intelligent controllers can then use this 3D model to undertake design plans through the excavation and movement of soil (Zhou et al., 2018). As the excavations and leveling of the site takes place, sensors on board the machine will gather data regarding the alterations to the site layout. This information will then be fed back into the site plan and inform further operations. By integrating UASs and unmanned excavators with autonomous intelligent controllers, earthwork can be carried out with greater precision, productivity, and economy without having to take breaks in operations for assessment. When earthwork operations are completed, a UAS can be deployed to take images of the site and match them against the designs for the earthworks to check the planning and reality are consistent (Zhou et al., 2018). Figure 14.5 demonstrates how UAS technology can be employed for earthworks. One example of the deployment of such processes is a collaboration between Komatsu, a manufacturer of construction machinery, and Skycatch, a UAS manufacturer, to design an excavator working with a UAS to excavate earthworks to specified elevations (Nicas, 2015).

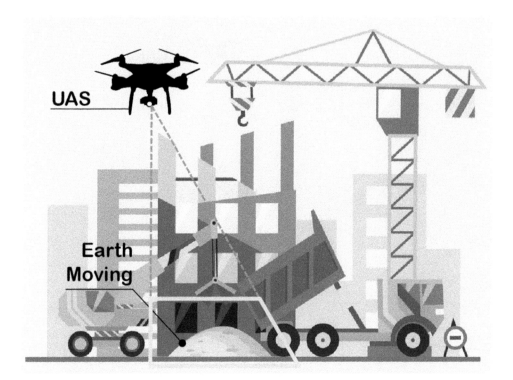

Figure 14.5   Applying a UAS to earth moving

14.4.2.2 Site communication

Fast and efficient communication between workers on a construction site and managers is vital for efficient project management (Zhou and Mi, 2017). A UAS may be used for the provision of instant communications on construction sites (Zhou et al., 2018). Using a UAS fitted with video cameras and voice transmitters, a project manager will be enabled to have direct interaction with their workers. This would offer great benefits, allowing managers to be present virtually at any point in any area of the construction site they chose to offer workers instant feedback and advice. This should be effective in eliminating many of the costs and errors associated with miscommunication between managers and workers.

14.4.2.3 Security and surveillance

In the construction industry, deploying UAS technologies for security purposes will be extremely beneficial (Zhou et al., 2018). UASs can access all areas and are far more economical than the human security guards, CCTV, and static video cameras that are currently deployed to protect construction sites. A UAS with long battery life or connected to a power source can offer continuous surveillance from previously inaccessible viewpoints. Security companies can deploy these vehicles in the hours of darkness to prevent theft and trespass, particularly on large construction sites. By giving security teams an "eye-in-the-sky", a UAS can positively enhance site security operations. UASs can use motion detectors to inform security officials of incursions (Zhou et al., 2018). A patrolling UAS is extremely visible and acts as a highly effective deterrent to theft.

14.4.2.4 Site transportation

It is impossible to avoid the requirement on construction sites to shift materials around the site using heavy lifting machinery. This machinery is generally cumbersome, slow-moving, and requires numerous workers for loading and moving. In comparison, small maneuverable UASs work perfectly as an enhancement to transportation strategies on construction sites (Zhou et al., 2018). In order to make multiple deliveries around a construction site, all a UAS requires is one worker to load and one to operate the vehicle. Although a typical UAS can support a weight of up to 5 lbs (2.3 kg), some variations are able to transport higher weights, making them perfect for the transport of small but vital tools and equipment. As UASs can fly over streets and walls, they are not constricted by any boundaries, meaning that they can achieve much greater efficiency and faster delivery times than standard vehicles that are handicapped by traffic and other physical limitations. The potential for UAS transportation is enormous (Opfer and Shields, 2014), particularly in rough terrain and inaccessible site locations. An example would be of a worker in the cabin of a tower crane who requires a certain small hand tool. An assistant on the ground could attach the tool to a UAS using a lanyard (Figure 14.6) and fly it up to the cab, saving the considerable time it would have taken the crane worker to climb down and back up again.

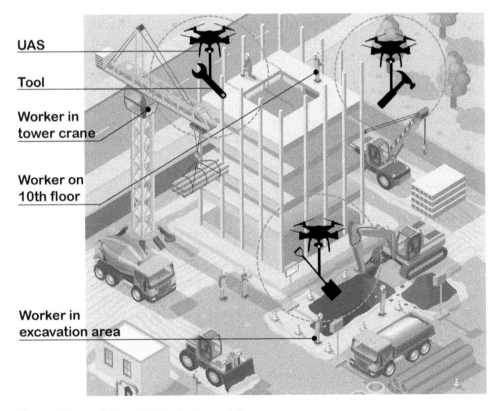

Figure 14.6 Applying a UAS to site transportation

14.4.2.5 Safety management

Construction companies are constantly striving to improve their safety record, but accidents are still a blight on the industry. The construction industry remains one of the most precarious industrial sectors (Perttula et al., 2006; Pinto et al., 2011). Construction site safety management generally entails planning for safety and training workers in safe practice before work commences and carrying out safety inspections on a regular basis. Safety officers are required to undertake continuous monitoring of the whole construction site to seek out and modify unsafe practices on the part of workers and any potential hazards involving equipment or materials. This can be a severely limited way of managing safety, particularly if there are insufficient safety officers on a large site. There is no way that there will ever be enough safety officers to pick up on all unsafe behavior and hazards in person as they arise.

It is also the case that in some locations on construction sites there are inherent safety risks that make it unsafe for such officers to enter the locations. By employing UASs, this problem can be resolved (Zhou et al., 2018). Safety officers can collect live data regarding unsafe practices or workplace hazards that could have a negative influence on construction or safety. Safety officers can also employ UASs to interact with construction workers, providing them with real-time feedback on safety issues. The data gathered by UASs, including images, videos, and feedback, may be retained to inform future safety plans for construction sites or to train workers in safety issues.

The feasibility of deploying UAS for safety inspections was investigated by Irizarry and Costa (2016) with a series of interviews with project engineers and project managers. de Melo et al. (2017) created safety checklists related to two Brazilian residential construction sites and undertook assessment of how feasible undertaking safety inspections with UAS would be. It was demonstrated that UAS could offer a wealth of detail regarding construction site safety conditions, and so enhance the quality of safety inspections. Similar research was undertaken by Irizarry et al. (2012), although only in a limited way, checking that all construction workers were wearing the requisite head protection. This research did demonstrate that UASs carrying video cameras and transmitters provided a way of directly inspecting safety issues more frequently and creating a means of instant interaction between workers and safety officers. In a recent study, Gheisari and Esmaeili (2019) investigated safety managers' perceptions on using UASs for safety monitoring purposes and found that the most important safety activities that can be improved using UASs were (1) monitoring cranes or boom vehicles in the proximity of overhead power lines, (2) monitoring activities in the proximity of cranes or boom vehicles, and (3) monitoring unprotected openings or edges.

One potential problem in deploying UASs in safety inspections, found by Irizarry et al. (2012), was that UAS flights could distract construction crews and also that UAS overflights could, in fact, create potential new safety problems. Usability assessments have estimated that a safety manager could become 50% more efficient by deploying UAS technology (Irizarry et al., 2012; Harris, 2015). Although it is necessary to recognize that the deployment of UASs for safety management has its limitations, using this technology is already showing promise in tests in its ability to enhance safety for inspections before construction commences (Goodman, 2017).

14.4.2.6 Construction monitoring

On sizable construction sites, construction activities involve the coordination of many hundreds of workers and items of equipment, frequently within confined spaces. Construction projects are often dynamic and complex, and sites need constant supervision. Traditionally, project managers and/or supervisors would walk around a site for inspections, having to climb from ground level to the highest level of construction. Monitoring in this way is ineffective and takes up considerable time. Employing UASs allows staff on site to undertake effective monitoring of every stage of construction activity with the technology removing the restrictions incumbent on human observers (Zhou et al., 2018). This concept was trialed in a project constructing a sports stadium with UAS monitoring construction and feeding back to software applications that automatically flagged areas where construction was too slow (Knight, 2015).

UASs offer global perspectives of construction sites, which are more useful than individual local perspectives for managers who wish to be able to see the whole project holistically (Liu et al., 2014). Global perspectives of construction sites permit managers/supervisors to have better levels of information regarding how the project is progressing and whether quality standards are maintained. With small UASs supplying detailed accurate images, the construction achieved can be compared with plans and every element of the process can be checked for quality standards and adherence to schedule. These managers can set up specific paths for the UAS to travel so that they can get a view of the most important and inaccessible areas.

One way in which the requisite automation levels for construction monitoring could be achieved has been suggested by Freimuth et al. (2017), a team which facilitated autonomous UAS monitoring by setting object boundaries in a geo-referenced BIM with flight coordination being achieved through the Global Navigation Satellite System (GNSS). If there is a desire to implement this means of site monitoring widely then determining what benefits it offers is important. Irizarry and Costa (2016) employed surveys to assess how effective UAS-assisted progress monitoring was. The main advantages they found for this form of progress monitoring were concerned with accessibility, time savings, and economy. Lin et al. (2015) noted that UAS monitoring provided economies, enhanced accessibility, and wider perspectives. Karan et al. (2014) contended that UAS monitoring was more time effective, cheaper, and safer.

14.4.2.7 Aerial construction

Compared to the majority of standard construction equipment, the flying robots that are UASs have certain distinct advantages, i.e. they can travel to any spatial point and fly around or even into existing objects (Zhou et al., 2018). With the guidance of mathematical algorithms, a fleet of UASs should be able to build tall structures without the assistance of cranes or scaffolding. In Figure 14.7, we see an example where two UASs work together to build an extremely tall chimney. Tall structures constructed using UASs might be able to implement new forms of architecture that were previously impossible using standard construction methodology, equipment, and material. To achieve this a multidisciplinary approach will be necessary, with new materials being developed, new construction processes, and advanced levels of additional design, and UASs would have to be programmed with adaptive strategies allowing them to fit with their environment and cooperate with each other in the assembly process. Research from the Institute for Dynamic Systems and Control at ETH Zurich related how four small UASs, driven by preprogrammed parameters and operating semi-autonomously, constructed a 6 m tall tower from 1500 foam blocks. The tower is a scale model of a planned future development, intended to reach to 600 m in height and housed 30,000 people (Stamp, 2013).

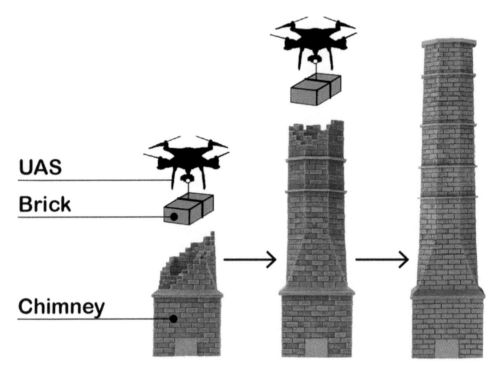

Figure 14.7   Applying UASs to chimney construction

14.4.2.8   General applications

Assisting with lighting and monitoring the dust on the jobsite are two general applications for which UASs can be used (Zhou et al., 2018). Construction workers are frequently required to work in challenging environments and projects are frequently racing against deadlines. On construction sites where heavy work schedules are in place, it is essential that adequate construction lighting is available to protect workers and keep projects on schedule. A key problem for construction sites is managing lighting and particularly nighttime light pollution. UASs may be employed during the hours of darkness to monitor all lighting on site, checking it has even coverage and is not breaching agreed boundaries without reason. Even when lighting schemes have been carefully planned, there may be variations between expectations and reality; UAS technology can assist in management and modification of lighting schemes, overcoming the problems that are inherent to current equipment such as fluorescent lighting, portable lighting, and incandescent lamps.

The production of dust is inherent to construction projects and UASs might be used for its monitoring. The production of dust is more than just irritating; it is one of the prime concerns regarding occupational health and construction workers. While there are mandatory regulations regarding the introduction of dust control schemes for construction activities, e.g. minimizing surface disturbance, roughening surfaces, and water spraying, not every construction site abides by the regulations or recommendations due to the financial costs of implementation. To check for compliance, environmental protection authorities may fly a suitable UAS over a site where significant dust levels are present and it is suspected that regulations have been breached. Real-time photography or videography from a UAS is extremely useful evidence that can help environmental protection authorities in dealing with dust issues when specific laws or regulations are being breached. Using UASs in this manner can make administration and enforcement of regulations more efficient and effective.

14.4.3 Post-construction stage

14.4.3.1 Manipulating simple items

When construction work has been finished on a project, it moves on to operational stages. At this point, it is inevitable that all buildings and infrastructure will require some minor modifications, repairs or maintenance. It is possible that UASs could be scaled down and produced to be extremely cost effective or even disposable, able to undertake simple tasks in inaccessible or dangerous spaces, e.g. cleaning the facade of skyscrapers, finding problematic corrosion points on the slope of a dam, or even undertaking "soft" testing on bridge deck undersides (Zhou et al., 2018).

14.4.3.2 Inspection tools

Photogrammetry based methodologies employing images collected by UASs could potentially measure minor deformities in ground objects more accurately than contact measuring methods. This means that UASs can offer a means of monitoring deformities in large infrastructures (Zhou et al., 2018). UASs can be employed for reaching inaccessible locations or ones that would usually demand complex and costly equipment to access. These methods could be employed for inspection of aging bridges and other architectural structures that have been constructed over large expanses of swiftly flowing water that make other inspections problematic (Metni and Hamel, 2007). One example was that of the USA's Delaware River & Bay Authority (DRBA), which in 2016 undertook trials on employing UAS (Maverik X8 and Inspire Pro 1) for the inspection of the New Jersey shore section of the Delaware Memorial Bridge

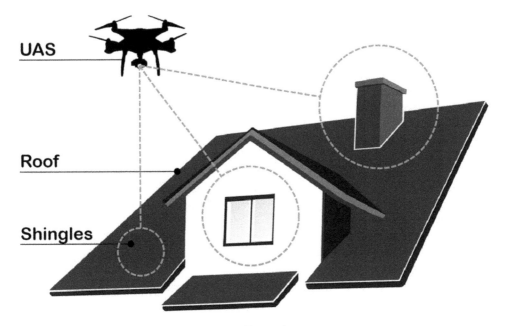

Figure 14.8   Applying a UAS to steep slope roof inspection

faster, more efficiently, and with much less necessity for lane closures than traditional methods, for example inspectors climbing over the edge of the bridge on ropes. Another possible use is for inspecting roofs. Roofs constructed with clay tiles or slates (see Figure 14.8) are difficult to walk on safely and often require scaffolding or ladders for inspection. There are also safety problems when roofs have a steep pitch. Employing UAS technology is both safer and faster for inspecting roofs and estimating work needed (Opfer and Shields, 2014). This form of inspection would greatly reduce the necessity for workers to spend time on roofs, lower the probability of errors in assessment, and make roof maintenance procedures more streamlined.

14.4.3.3 Energy analysis

When fitted with thermographic cameras, UASs can find roof leaks or electrical hotspots related to transformers that cannot be seen from ground level. Using aerial and inventory data gathered by UASs in combination with image processing software allows for the visualization of energy loss for large neighborhoods (Zhou et al., 2018). This data can then be rendered as a thermal map, which simplifies the identification of those buildings requiring renovation to become more energy efficient. In Austria, the Aspern, Vienna Urban Lakeside project has demonstrated how UASs technology can be employed for promoting energy-efficient and environmentally balanced standards (Gallagher, 2017).

14.4.3.4 Marketing tools

Marketing departments are always keen to exploit new technologies in order to show potential buyers what they have to offer. A number of companies are endeavoring to integrate aerial photography into their marketing offer (Zhou et al., 2018). Being able to show a project from an aerial perspective offers a unique view that cannot be achieved from ground level. In terms of selling or renting real estate, UASs can offer potential clients views of buildings as photorealistic 3D models. Integrated with virtual reality technology, this mapping will allow clients to undertake virtual walk-throughs of the interiors of buildings.

14.4.3.5 Disaster management

During natural disasters, e.g. landslides, hurricanes, and earthquakes, nearby buildings and other infrastructure are often unavoidably damaged. Being able to undertake quick reconnaissance of the structures is crucial for effective rescue missions. Standard techniques involve reconnaissance undertaken by humans which requires a great deal of labor. Natural disasters often take place in hard-to-reach areas. With careful coordination and planning, reconnaissance after disasters can be undertaken more effectively by using UASs fitted with appropriate sensing technology (Zhou et al., 2018). UASs help rescuers to estimate which buildings and other infrastructure have sustained too much damage to make it safe for rescuers to enter. UASs can also help with rescuing victims and cleaning and restoring buildings.

One example of this was the magnitude 9 earthquake that hit the northeast coast of Japan on March 11, 2011. The International Atomic Energy Agency deployed UASs for the measurement and calculation of radioactive emissions from the Fukushima nuclear power facility. The UASs deployed could get closer to the facility than piloted aircraft and did not expose any humans to radiation risks (Pamintuan-Lamorena, 2014). There are a number of other instances where UASs have undertaken reconnaissance after disasters. Pratt et al. (2008) used a tethered UAS for the evaluation of the damage in a parking garage collapse, as conventional helicopters were not able to get into the best position to review the damage and create dust clouds which could cover up important evidence. Michael et al. (2012) employed a UAS and ground robots for mapping the interior of a building damaged by an earthquake; this research demonstrated that flying robots can enter areas and collect data when ground robots are unable to access an area; the ability to hover is a notable benefit when undertaking observations and inspections. Yamamoto et al. (2014) employed UASs for assessing collapsed sections of slopes and river revetments; the technology was employed because close-photogrammetric software becomes less accurate when photographs are taken in too close juxtaposition, which is frequently the case with ground photography techniques. Kerle et al. (2014) employed UASs for collecting information regarding damage to facades and roofs of a number of buildings; they found that surveying using UASs offered more control and flexibility than Pictometry-style systems and were more economical. All these examples demonstrate that UASs could potentially offer significant enhancements to reconnaissance of natural disasters in many different situations.

14.5 UAS implementation challenges

One of the biggest challenges to those involved in construction projects who have an interest in deploying UAS technology is the regulatory issues they face. UASs are regarded as a form of aircraft that have to comply with civil aviation regulations. Legislation controlling aviation is different around the world as applied to UAS flights and must be adhered to if UASs are to be successfully deployed for use in commercial construction.

Another serious problem with the full deployment of UASs in construction is the issues surrounding legality and liability. A number of concerns have been expressed ranging from the possibilities of UASs damaging property or injuring bystanders to problems with insurance, trespass, invasion of privacy, and violation of property rights (Gheisari and Esmaeili, 2016, 2019). As UAS are deployed more frequently, concerns regarding personal privacy are growing. Many people fear that these aircraft could be used as a flying "peeping Tom", invading their private spaces (Bird, 2014). UAS deployment should not involve flying over pedestrians (14 CFR 107.39a), traffic (14 CFR 107.39b), or private property, both on safety grounds and in order to preserve privacy. Potential safety issues have been raised regarding the deployment of aircraft around a construction site, particularly in terms of collision or malfunction and also the potential for distracting those working on the site (Gheisari and Esmaeili, 2019). It is important that further research is ongoing regarding the deployment of UASs, particularly with reference to the ways in which construction workers and managers regard the issue (Costa and Mendes, 2016; de Melo and Costa, 2019). As UASs are frequently deployed at low altitudes of below 120 meters, the safety of the public, as well as construction site workers, is a matter of concern. Without a pilot to make checks, UASs may not be as well-maintained and so reliable as piloted aircraft. Operator distraction or error, misuse or failures can all be responsible for injuries caused by UASs (Finn and Wright, 2012). The fact that much legislation requires UASs to fly at low altitude means that there are noise issues that can impact on the welfare of workers and nearby residents have to be considered. If a UAS crashes on a construction site, the site may be damaged, the UAS can be damaged, and personal injury may be caused to site workers or third parties. As this is a relatively new area of operation, the majority of businesses and organizations do not have policies for dealing with accidents caused by the deployment of UASs. It is vital that operators should hold sufficient insurance to cover any claims arising from UASs accidents. While the insurance industry is in the process of developing a number of policies in this area, the issues are still being worked through, with the full range of problems related to risks in deploying such aircraft yet to be fully appreciated.

Technical problems may create issues that add to the complexity of deploying UASs on construction sites, including battery life, payload capabilities, interference from radio waves, and types of sensors deployed (Gheisari and Esmaeili, 2019). For UASs to be successfully deployed, training of pilots and other personnel must also be considered. Numerous national and international aviation bodies, including the FAA in the USA, mandate that for commercial purposes certified pilots must fly UASs. Appropriate workers on construction sites who could pilot UASs may have to be extensively trained in order to be certificated for the task. Other issues that must be taken into account when considering the use of UASs on construction projects are nighttime and beyond line of sight flying, weather conditions, limitations of space and congestion, and restrictions imposed by the size or height of the project itself.

14.6 Conclusion

This chapter has offered a review of the currently available UAS technology, including challenges of implementation, uses in construction, regulatory regimes, financial implications, deployment of sensors, human involvement, forms of control, and machine types, which is part of the new production paradigm, called Construction 4.0.

The construction industry can become more efficient and transparent in terms of real-time monitoring and inspections at all stages of construction projects by employing UAS technology, facilitating bi-directional coordination between virtual models and physical construction, as well.

The ability to acquire a sizable database of visual images, either regular photographs or more advanced data that can be processed with sophisticated software, is hugely beneficial. The data gathered via UAS assists managers in making better and faster decisions, deploying fewer resources, saving time and money, and using personnel and equipment more efficiently.

There are still issues to be resolved regarding safety risks, privacy, and adherence to regulations when using UASs safely and appropriately in the construction industry, especially as in most instances there will be large numbers of people living and working in the area of deployment. It is recommended that regular training and risk assessments should be a part of any UAS deployment to mitigate legal and safety issues.

In a number of countries, studies are being undertaken to improve the understanding of what a UAS can do and their potential to be of assistance in the construction industry, spurred by the need for the construction industry to fully embrace Construction 4.0 and associated new digital technologies. There is no doubt that once sufficient education and standardization has taken place, UASs have the potential to become a major element of new construction projects and facilitate the digital transformation of the construction industry.

14.7 Summary

- Reviewed UAS definition and its technical aspects such as vehicle types, forms of control, human-UAS interaction and UAS flight team, data collection sensors, cost, and flight regulations around the world.
- Discussed several UAS applications focusing on the logic behind using UASs in each of those applications and its benefits.
- Discussed UAS implementation challenges such as liability and legal concerns, safety challenges, and training requirements.

References

Agência Nacional de Aviação Civil (ANAC). (2017). "Regulamento Brasileiro Da Aviação Civil Especial (RBAC-E N°94)", Brasília, Brasil.

Álvares, J. S., Costa, D. B., and de Melo, R. R. S. (2018). "Exploratory Study of Using Unmanned Aerial System Imagery for Construction Site 3D Mapping", *Construction Innovation*, https://doi.org/10.1108/CI-05-2017-0049.

Arkady, R., and Aviad, S. (1999). "VR-Based Planning of Construction Site Activities", *Automation in Construction*, 8(6), 671–680.

Augugliaro, F., Zarfati, E., Mirjan, A., and D'Andrea, R. (2015). "Knot-Tying with Flying Machines for Aerial Construction", in *IEEE/RSJ International Conference on Intelligent Robots and Systems (IROS)*, Hamburg, Germany.

Bendig, J., Bolten, A., and Bareth, G. (2012). "Introducing a Low-Cost Mini-UAS for Thermal- and Multispectral-Imaging", International Archives of the Photogrammetry, Remote Sensing and Spatial Information Sciences – ISPRS Archives.

Bhatla, A., Choe, S. Y., Fierro, O., and Leite, F. (2012). "Evaluation of Accuracy of As-Built 3D Modeling from Photos Taken by Handheld Digital Cameras", *Automation in Construction*, 28, 116–127.

Bird, S. (2014). "What's that Buzzing Noise? Oh, It's Just a Drone Spying on You at the Beach", Available from: www.care2.com/causes/whats-that-buzzing-noise-oh-its-just-a-drone-spying-on-you-at-the-beach.html (accessed 06/2019).

Bolourian, N., Soltani, M. M., Albahria, A. H., and Hammad, A. (2017). "High Level Framework for Bridge Inspection Using LiDAR-Equipped UAV", in *34th International Symposium on Automation and Robotics in Construction (ISARC), 2017*, Vilnius Gediminas Technical University, Department of Construction Economics & Property, Vilnius, Vol. 34, pp. 1–6.

Brooks, C., Dobson, R., Banach, D., and Cook, S. J. (2017). *Transportation Infrastructure Assessment through the Use of Unmanned Aerial Vehicles*, Transportation Research Board 96th Annual Meeting, Washington DC, 2017-1-8 to 2017-1-12.

Capolupo, A., Pindozzi, S., Okello, C., Fiorentino, N., and Boccia, L. (2015). "Photogrammetry for Environmental Monitoring: The Use of Drones and Hydrological Models for Detection of Soil Contaminated by Copper", *Science of the Total Environment*, 514, 298–306.

Civil Aviation Authority. (2015). "Unmanned Aircraft System Operations in UK Airspace – Guidance, CAP 722", London.

Correa, F. R. (2018). "Cyber-Physical Systems for Construction Industry", in *2018 IEEE Industrial Cyber-Physical Systems (ICPS), May, 2018. Proceedings. IEEE Industrial Cyber-Physical Systems (ICPS)*, pp. 392–397.

Cozart, J. (2016). "The Effectiveness of Drone Based LiDAR", Available from: www.linkedin.com/pulse/effectiveness-drone-based-lidar-jeff-cozart (accessed 09/24/2016).

de Melo, R. R. S. and Costa, D. B. (2019), "Integrating Resilience Engineering and UAS Technology into Construction Safety Planning and Control", *Engineering, Construction and Architectural Management* (in press).

de Melo, R. R. S., Costa, D. B., Álvares, J. S., and Irizarry, J. (2017). "Applicability of Unmanned Aerial System (UAS) for Safety Inspection on Construction Sites", *Safety Science*, 98, 174–185, https://doi.org/10.1016/j.ssci.2017.06.008.

Dirección General De Aeronáutica Civil (DGAC) (2015). *Norma Técnica Aeronáutica, DAN 151- Operaciones de Aeronaves Pilotadas a Distancia (RPAS)*, de 02 de Abril de 2015. 1ª ed. C. Santiago, Chile.

Dobson, R. J., Brooks, C., Roussi, C., and Colling, T. (2013), "Developing an Unpaved Road Assessment System for Practical Deployment with High-Resolution Optical Data Collection Using a Helicopter UAV", in *Unmanned Aircraft Systems (ICUAS), 2013 International Conference in Atlanta, GA, 2013*, IEEE, pp. 235–243, https://doi.org/10.1109/ICUAS.2013.6564695.

Eiris, R. and Gheisari, M. (2017). "Evaluation of Small UAS Acquisition Costs for Construction Applications", in *Proceedings of the Joint Conference on Computing in Construction (JC3)*, I, 931–938.

Eisenbeiß, H. (2009). *UAV Photogrammetry*. Zurich: Institut für Geodäsie und Photogrammetrie – Eidgenössische Technische Hochschule Zürich.

Ellenberg, A., Branco, L., Krick, A., and Bartoli, I. (2014). "Use of Unmanned Aerial Vehicle for Quantitative Infrastructure Evaluation", *Journal of Infrastructure Systems*, 21(3), 04014054.

Eschmann, C., Kuo, C. M., Kuo, C. H., and Boller, C. (2012), "Unmanned Aircraft Systems for Remote Building Inspection and Monitoring", paper presented at the Proceedings of the 6th European Workshop on Structural Health Monitoring, 4–6 September, Dresden, Germany.

European Aviation Safety Agency. (2019). "Draft AMC and GM to Regulation./. [IR] Laying Down Rules and Procedures for the Operation of Unmanned Aircraft and to Its Annex (Part-UAS)", Available from: www.easa.europa.eu/easa-and-you/civil-drones-rpas.

FAA. (2017). *Aeronautical Information Manual: Official Guide to Basic Flight Information and ATC Procedures*. U.S. Department of Transportation, Available from: www.faa.gov/air_traffic/publications/media/AIM_Basic_dtd_10-12-17.pdf.

Finn, R. L. and Wright, D. (2012). "Unmanned Aircraft Systems: Surveillance, Ethics and Privacy in Civil Applications", *Computer Law & Security Review*, 28(2), 184–194.

Freimuth, H., Müller, J., and König, M. (2017), "Simulating and Executing UAV-Assisted Inspections on Construction Sites", paper presented at the 34th International Symposium on Automation and Robotics in Construction (ISARC 2017).

Gallagher, K. (2017). "3 Industries Being Transformed by Drone Mapping", Available from: http://droneblog.com/2017/04/06/3-industries-being-transformed-by-drone-mapping/ (accessed 04/06/2017).

Gheisari, M. and Esmaeili, B. (2016). "Unmanned Aerial Systems (UAS) for Construction Safety Applications", *Construction Research Congress*, 2016, 2642–2650, https://doi.org/10.1061/9780784479827.263.

Gheisari, M. and Esmaeili, B. (2019). "Applications and Requirements of Unmanned Aerial Systems (UASs) for Construction Safety", *Safety Science*, 118, 230–240.

Gheisari, M., Irizarry, J., and Walker, B. N. (2014). "UAS4SAFETY: The Potential of Unmanned Aerial Systems for Construction Safety Applications", in *Proceedings of Construction Research Congress*, pp. 1801–1810.

Gheisari, M., Karan, E. P., Christmann, H. C., Irizarry, J., and Johnson, E. N. (2015), "Investigating Unmanned Aerial System (UAS) Application Requirements within a Department of Transportation", paper presented at the Transportation Research Board 94th Annual Meeting, 11–15 January, Washington DC, United States.

Golparvar-Fard, M., Peña-Mora, F., Arboleda, C. A., and Lee, S. (2009a). "Visualization of Construction Progress Monitoring with 4D Simulation Model Overlaid on Time-Lapsed Photographs", *Journal of Computing in Civil Engineering*, 23(6), 391–404.

Golparvar-Fard, M., Peña-Mora, F., and Savarese, S. (2009b). "D4AR-A 4-Dimensional Augmented Reality Model for Automating Construction Progress Monitoring Data Collection, Processing and Communication", *Journal of Information Technology in Construction (ITCON)*, 14, 129–153.

Goodman, J. (2017). "A New Use for Drones: On-Site Safety", Available from: www.builderonline.com/building/building-science/a-new-use-for-drones-on-site-safety_o (accessed 02/22/2017).

Goodrich, M. A. and Schultz, A. C. (2007). "Human-Robot Interaction: A Survey", *Foundations and Trends in Human-Computer Interaction*, 1(3), 203–275.

Groetelaars, N. J. and Amorim, A. L. D. (2012). "Dense Stereo Matching (DSM): conceitos, processos e ferramentas para criação de nuvens de pontos por fotografias", in SIGRADI 2012 – Congreso de La Sociedad Iberoamericana de Gráfica Digital, 16, 2012, Fortaleza. pp. 361–365.

Teal Group (2016). World Civil Unmanned Aerial Systems Market Profile & Forecast 2016 Edition. Available from the Teal Group Corporation at: www.tealgroup.com (accessed 10/16/2016).

Harris, L. (2015). "How Drones, VR and BIM Are Improving Construction Jobsite Safety", Available from: www.gray.com/news/blog/2015/05/04/how-drones-vr-and-bim-are-improving-construction-jobsite-safety (accessed 05/04/2015).

Howard, J., Murashov, V., and Branche, C. M. (2017). "Unmanned Aerial Vehicles in Construction and Worker Safety", *American Journal of Industrial Medicine*, 61(1), 3–10, https://doi.org/10.1002/ajim.22782.

Hubbard, B., Wang, H., Leasure, M., Ropp, T., Lofton, T., and Hubbard, S. (2015). "Feasibility Study of UAV Use for RFID Material Tracking on Construction Sites", in *51st ASC Annual International Conference Proceedings*.

Irizarry, J. and Costa, D. B. (2016). "Exploratory Study of Potential Applications of Unmanned Aerial Systems for Construction Management Tasks", *Journal of Management in Engineering*, 32(3), 05016001–05016001.

Irizarry, J., Gheisari, M., and Walker, B. N. (2012). "Usability Assessment of Drone Technology as Safety Inspection Tools", *Journal of Information Technology in Construction (ITCON)*, 17(12), 194–212.

Jenkins, D. and Vasigh, B. (2013). "The Economic Impact of Unmanned Aircraft Systems Integration in the United States", pp. 1–40, www.auvsi.org/our-impact/economic-report.

Karan, E. P., Christmann, C., Gheisari, M., Irizarry, J., and Johnson, E. N. (2014). "A Comprehensive Matrix of Unmanned Aerial Systems Requirements for Potential Applications within a Department of

Transportation", in *Construction Research Congress in Atlanta, Georgia, 2014*, American Society of Civil Engineers, pp. 964–973.

Kerle, N., Fernandez Galarreta, J., and Gerke, M. (2014), "Urban Structural Damage Assessment with Oblique UAV Imagery, Object-Based Image Analysis and Semantic Reasoning", paper presented at Proceedings of the 35th Asian Conference on Remote Sensing, 2014.

Kim, S., Irizarry, J., Costa, D. B., and Mendes, A. T. (2016), "Lessons Learned from Unmanned Aerial System-Based 3D Mapping Experiments", in 52nd ASC Annual International Conference Proceedings.

Knight, W. (2015). "New Boss on Construction Site Is a Drone", Available from: www.technologyreview. com/s/540836/new-boss-on-construction-sites-is-a-drone/ (08/26/2015).

Lin, J. J., Han, K. K., and Golparvar-Fard, M. (2015), "A Framework for Model-Driven Acquisition and Analytics of Visual Data Using UAVs for Automated Construction Progress Monitoring", in *Computing in Civil Engineering in Austin, TX, 2015*, American Society of Civil Engineers, pp. 156–164, https://doi.org/10.1061/9780784479247.020.

Liu, P., Chen, A. Y., Huang, Y., Han, J., Lai, J., Kang, S., Wu, T., Wen, M., and Tsai, M. (2014). "A Review of Rotorcraft Unmanned Aerial Vehicle (UAV) Developments and Applications in Civil Engineering", *Smart Structures and Systems*, 13(6), 1065–1094.

Martin, J., Edwards, H. H., Burgess, M. A., Percival, H. F., Fagan, D. E., Gardner, B., Ortegaortiz, J. G., Ifju, P., Evers, B. S., and Rambo, T. J. (2012). "Estimating Distribution of Hidden Objects with Drones: From Tennis Balls to Manatees", *PLoS One*, 7(6), e388821–e388828.

Merino, L., Caballero, F., Martínezdedios, J. R., Maza, I., and Ollero, A. (2012). "An Unmanned Aircraft System for Automatic Forest Fire Monitoring and Measurement", *Journal of Intelligent and Robotic Systems: Theory and Applications*, 65(1–4), 533–548.

Merz, T. and Kendoul, F. (2011), "Beyond Visual Range Obstacle Avoidance and Infrastructure Inspection by an Autonomous Helicopter", in *IEEE/RSJ International Conference on Intelligent Robots and Systems in San Francisco, CA, 2011*, IEEE, pp. 4953–4960, https://doi.org/10.1109/IROS.2011.6094584.

Metni, N. and Hamel, T. (2007). "A UAV for Bridge Inspection: Visual Servoing Control Law with Orientation Limits", *Automation in Construction*, 17(1), 3–10.

Michael, N., Shen, S., Mohta, K., Mulgaonkar, Y., Kumar, V., Nagatani, K., Okada, Y., Kiribayashi, S., Otake, K., Yoshida, K., and Ohno, K. (2012). "Collaborative Mapping of an Earthquake-Damaged Building Via Ground and Aerial Robots", *Journal of Field Robotics*, 29(5), 832–841, https://doi.org/10.1002/rob.21436.

Murphy, R. R., Steimle, E., Griffin, C., Cullins, C., Hall, M., and Pratt, K. (2008). "Cooperative Use of Unmanned Sea Surface and Micro Aerial Vehicles at Hurricane Wilma", *Journal of Field Robotics*, 25(3), 164–180, https://doi.org/10.1002/rob.20235.

Newcome, L. R. (2004). *Unmanned Aviation: A Brief History of Unmanned Aerial Vehicles*. Reston, VA: AIAA.

Nicas, J. (2015). "Drones' Next Job: Construction Work", Available from: www.wsj.com/articles/drones-next-job-construction-work-1421769564 (accessed 01/20/2015).

Opfer, N. D. and Shields, D. R. (2014). "Unmanned Aerial Vehicle Applications and Issues for Construction", in *Proceedings of 121st ASEE Annual Conference & Exposition*, Indianapolis, USA.

Pamintuan-Lamorena, M. (2014). "Drones Used to Measure Radiation in Fukushima Nuclear Plant", Available from: http://japandailypress.com/drones-used-to-measure-radiation-in-fukushima-nuclear-plant-2743074/ (accessed 01/27/2014).

Pereira, A. C. and Romero, F. (2017). "A Review of the Meanings and the Implications of the Industry 4.0 Concept", *Procedia Manufacturing*, 13, 1206–1214.

Pereira, F. C. and Pereira, C. E. (2015). "Embedded Image Processing Systems for Automatic Recognition of Cracks Using UAVs", *IFAC-PapersOnLine*, 48(10), 16–21, https://doi.org/10.1016/j.ifacol.2015.08.101.

Perttula, P., Korhonen, P., Lehtela, J., Rasa, P., Kitinoja, J., Makimattila, S., and Leskinen, T. (2006). "Improving the Safety and Efficiency of Materials Transfer at a Construction Site by Using an Elevator", *Journal of Construction Engineering and Management*, 132(8), 836–843.

Pinto, A., Nunes, I. L., and Ribeiro, R. A. (2011). "Occupational Risk Assessment in Construction Industry – Overview and Reflection", *Safety Science*, 49(5), 616–624.

Pratt, K. S., Murphy, R. R., Burke, J. L., Craighead, J., Griffin, C., and Stover, S. (2008), "Use of Tethered Small Unmanned Aerial System at Berkman Plaza II Collapse", in *2008 IEEE International Workshop on Safety, Security and Rescue Robotics in Sendai, Japan, 2008*, IEEE, pp. 134–139, https://doi.org/10.1109/SSRR.2008.4745890.

Puri, A. (2005). *A Survey of Unmanned Aerial Vehicles (UAV) for Traffic Surveillance*, University of South Florida, Tampa, FL.

Rathinam, S., Kim, Z. W., and Sengupta, R. (2008). "Vision-Based Monitoring of Locally Linear Structures Using an Unmanned Aerial Vehicle", *Journal of Infrastructure Systems*, 14(1), 52–63.

Remondino, F. (2011). "Heritage Recording and 3D Modeling with Photogrammetry and 3D Scanning", *Remote Sensing*, 3, 1104–1138.

Rios-Morentin, D. (2011). "UAS Based Surveillance for Border Control Against Irregular Migration and Trafficking: Spanish Scenario", Proceedings of AUVSI Unmanned Systems North America Conference, 1417–1428, Washington, USA.

Roca, D., Lagüela, S., Díaz-Vilariño, L., Armesto, J., and Arias, P. (2013). "Low-Cost Aerial Unit for Outdoor Inspection of Building Façades", *Automation in Construction*, 36 December, 128–135.

Sankarasrinivasana, S., Balasubramaniana, E., Karthika, K., Chandrasekarb, U., and Gupta, R. (2015). "Health Monitoring of Civil Structures with Integrated UAV and Image Processing System", *Procedia Computer Science*, 5, 508–515.

Siebert, S. and Teizer, J. (2014). "Mobile 3D Mapping for Surveying Earthwork Projects Using an Unmanned Aerial Vehicle (UAV) System", *Automation in Construction*, 41, 1–14, https://doi.org/10.1016/j.autcon.2014.01.004.

Stamp. (2013). "The Drones of the Future May Build Skyscrapers", Available from: www.smithsonianmag.com/arts-culture/the-drones-of-the-future-may-build-skyscrapers-18390584/ (accessed 02/15/2013).

Teizer, J. (2015). "Status Quo and Open Challenges in Vision-Based Sensing and Tracking of Temporary Resources on Infrastructure Construction Sites", *Advanced Engineering Informatics*, 29, 225–238.

Tomita, H., Takabatake, T., Sakamoto, S., Arisumi, H., Kato, S., and Ohgusu, Y. (2017). "Development of UAV Indoor Flight Technology for Building Equipment Works", in *Proceedings of the International Symposium on Automation and Robotics in Construction, 2017*, Vilnius Gediminas Technical University, Department of Construction Economics & Property, Vilnius, Vol. 34, pp. 1–6.

Turner, D., Lucieer, A., and Wallace, L. (2014). "Direct Geo-Referencing of Ultrahigh-Resolution UAV Imagery", *IEEE Transactions on Geoscience and Remote Sensing*, 52(5), 2738–2745.

U.S. Department of Transportation. (2016). Advisory Circular – Small Unmanned Aircraft Systems AC 107-2. Available from the Federal Register at: https://federalregister.gov/a/2016-15079 (accessed 08/16/2016).

Wang, W.-C., Weng, S.-W., Wang, S.-H., and Chen, C.-Y. (2014). "Integrating Building Information Models with Construction Process Simulations for Project Scheduling Support", *Automation in Construction*, 37, 68–80.

Wefelscheid, C., Hänsch, R., and Hellwich, O. (2011). "Three-Dimensional Building Reconstruction Using Images Obtained by Unmanned Aerial Vehicles", *ISPRS Zurich 2011 Workshop in Zurich, Switzerland*, 2011, Copernicus GmbH, Gottingen, XXXVIII-1/C22, 183–188, https://doi.org/10.5194/isprsarchives-XXXVIII-1-C22-183-2011.

Wen, M. C., and Kang, S. C. (2014), "Augmented Reality and Unmanned Aerial Vehicle Assist in Construction Management", in *2014 International Conference on Computing in Civil and Building Engineering in Orlando, Florida, 2014*, American Society of Civil Engineers, https://doi.org/10.1061/9780784413616.195.

Willmann, J., Augugliaro, F., Cadalbert, T., D'Andrea, R., Gramazio, F., and Kohler, M. (2012). "Aerial Robotic Construction Towards a New Field of Architectural Research", *International Journal of Architectural Computing*, 10(3), 439–459.

Yamamoto, T., Kusumoto, H., and Banjo, K. (2014), "Data Collection System for a Rapid Recovery Work: Using Digital Photogrammetry and a Small Unmanned Aerial Vehicle (UAV)", in *International Conference on Computing in Civil and Building Engineering in Orlando, FL, 2014*, American Society of Civil Engineers, pp. 875–882, https://doi.org/10.1061/9780784413616.109.

Zhou, S. and Gheisari, M. (2018). "Unmanned Aerial System Applications in Construction: A Systematic Review", *Construction Innovation*, 18(4), 453–468.

Zhou, Z., Irizarry, J., and Lu, Y. (2018). "A Multidimensional Framework for Unmanned Aerial System Applications in Construction Project Management", *Journal of Management in Engineering*, 34(3), 04018004.

Zhou, Z. and Mi, C. (2017). "Social Responsibility Research within the Context of Megaproject Management: Trends, Gaps and Opportunities", *International Journal of Project Management*, https://doi.org/10.1016/j.ijproman.

15

FUTURE OF ROBOTICS AND AUTOMATION IN CONSTRUCTION

Borja Garcia de Soto and Miroslaw J. Skibniewski

15.1 Aims

- History and evolution of robotics and automation in the construction industry.
- Definition of construction automation.
- The possibilities that Construction 4.0 brings to the integration of robotics and automation to the construction industry.
- Challenges and opportunities fueled by the adoption of Construction 4.0.

15.2 Introduction

The idea of using robots and automating construction sites is not new. The first research and publications on construction robotics date back to the 1970s in the former Soviet Union (Araksyan & Volkov, 1985; Frenkel, 1987, 1988; Vilman, 1989). In the 1980s, the interest in this field extended to other geographical locations. In the US, Carnegie Mellon University started independent research in construction robotics. Different centers were promoted by the Robotics and Field Sensing Committee of the ASCE. In Europe, different institutions were also working in this field. France (Centre Scientifique et Technique du Batiment), Germany (the Fraunhofer Institute for Production Automation, Technical University of Karlsruhe, and the Technical University of Munich), the United Kingdom (Lancaster University, City University in London, and the former Bristol Polytechnic), and Spain (Universidad Carlos III de Madrid) (Budny et al., 2009). In Japan, construction companies were leading the research and application of robotic systems and automation. The *Big Five*[1], as they were referred to, invested back in the 1980s about 1% of their annual gross revenue (about $15 billion) for research and development, although not all of it was to fund projects in robotics and automation (Cousineau & Miura, 1998). Examples of early research in automation and robotics in construction in Japan includes Hasegawa (1984), Shimomura and Sonoda (1984), Yoshida et al. (1984) or Kano and Tamura (1984), which mostly focused on single task robots. By the mid-1980s, robotic systems were developed and introduced for inspection tasks on a radioactively contaminated building site. In 1984, the first International Symposium on Automation and Robotics in Construction (ISARC) was hosted by Carnegie Mellon University and had seven papers presented. Since then, there have been 36 editions of ISARCs, with an increasing trend in the number of

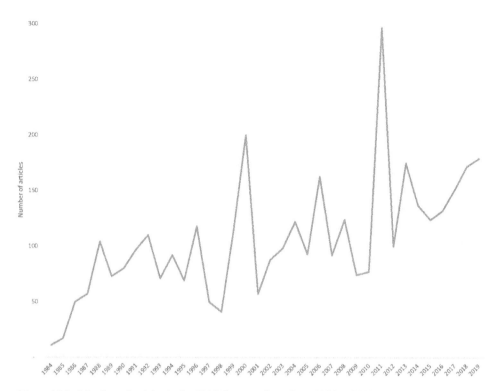

Figure 15.1 Number of articles in the ISARC proceedings from 1984 to 2019

publications and a peak of 297 in the 28th ISARC in Seoul (Figure 15.1). As of 2019, more than 3,800 articles with the latest developments related to the topic of automation and robotics in construction have been included in the ISARC proceedings. Interested readers are referred to the IAARC publication site[2] where all ISARC proceedings are available. Another major source is *Automation in Construction*, an international research journal published by Elsevier since 1992.

Warszawski (1984a) published one of the first critiques on the use of robots in the building sector and proposed different robot configurations to address different construction tasks. Skibniewski and Hendrickson (1985) showed the economic feasibility of a robotic system to perform cleaning and maintenance tasks, using the clearing of concrete formwork as an example. In 1986 Skibniewski et al. provided a review of the existing and foreseeable robotic systems in the area of construction finishing works, and explored their potential use and impact to the cost of design elements of concoction projects. Skibniewski (1988a) proposed an expert system for decision support in regard to implementing advanced robotic technology on the construction site. By 1992, the first full-scale application of construction automation was materialized by contractors in Japan through the development of the so-called sky factories (Bock & Linner, 2016a, Cousineau & Miura, 1998). Haas et al. (1995) summarized the progress of robotics in civil engineering from the early 1980s to the mid-1990s. Their study highlighted that despite the significant progress made, the lack of investment from industry and government was one of the main reasons preventing their development. An overview of the history and evolution of automation and robotics in construction and key milestones is shown in Figure 15.2. The curve shows a qualitative representation of the interest and effort made in this field.

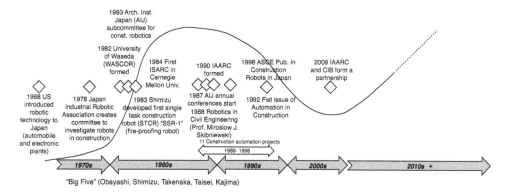

Figure 15.2   History of robotics in construction (Cousineau & Miura, 1998, IAARC, n.d.)

Since the beginning, the primary goals and motivations of using automation and robotics have been (1) to have an economic gain (i.e. improve productivity, reduce manpower, reduce construction time, and reduce cost), (2) to improve the quality of the finished product, (3) to improve working conditions (i.e. relieve workers from physically demanding work, enhance the safety and work environment of construction sites, and close the gap in the shortage of skilled labor), and (4) to take advantage of new technologies by incorporating them in the different phases of construction projects (Budny et al., 2009). Although some of the goals have been met (e.g. quality), others did not achieve the expected outcomes.

The primary limitations encountered in the early stages included the fact that economic gains were not as apparent as those observed in other industrial applications. For the most part, construction robots failed to increase productivity, and reduce construction durations, cost, or human resources. In addition, the construction technology and legal environment in the 1980s and 1990s were not favorable and one of the reasons why a robotic technology revolution did not happen then.

Fast-forward almost four decades, and when you look around on construction sites, you will see that still, the use of robots and construction automation is very limited or non-existent. The industry still struggles to pass the prototype and research stages, and their development and applications are still considered experimental and mostly conducted by universities and research centers. The main reasons are mostly the same as those that prevented their implementation in the 1980s and 1990s. Currently, the implementation of robots in construction sites is still limited. Nonetheless, their use will undoubtedly increase as more cost-effective applications are found (Garcia de Soto et al., 2018a).

15.2.1 Construction 4.0: a new push towards automation and robotic systems in construction

Significant technological advancements are pushing the architecture, engineering, construction, and facilities management (AEC/FM) industry towards digitalization and automation (known as Construction 4.0). Among enabling technologies available today that did not exist in the 1980s or 1990s are mobile and cloud computing/telecommunications, big data and deep learning, wireless sensor networks, and BIM. Thoughtful implementation of these, together with classic and modern robotic technology concepts (e.g. collaborative, swarm), can make a significant difference in the success of construction robot applications today as compared with what was available four decades ago. This makes the future of construction automation and robotics promising (Garcia de Soto, 2019). According to the International Federation of Robotics (IFR, 2018a), the global operational stock of industrial robots will continue to grow at around 16% on average per year until 2021, reaching to almost four million robots by that year. In the sector of service robots (professional, personal/domestic use), the rate of increase is also significant with projected sales raising from 109,500 units in 2017 to 736,600 between 2019 and 2021 (IFR, 2018b). Therefore, it is apparent that machines are here to stay, and that machine and human interactions will continue to grow.

It is clear that automation and robotics in construction can increase production rates in the building industry not only because they lead to significant time saving for complex designs, but also because they exhibit the ability to transfer design data directly to 1:1 assembly operations and automated construction (Keating & Oxman, 2013). However, large-scale applications are still limited and face challenges of changing conventional construction processes and roles of project participants. Initial attempts have been made in real practice to evaluate the benefits for the construction sector. For instance, Gramazio Kohler Research at ETH Zurich has accomplished different building demonstrators constructed with robotic technologies. The brick facade of the Gantenbein Vineyard showed the possibilities of computational design and robotic construction for the prefabrication of complex multi-functional brick structures. As the robot could be driven directly by the design data, without having to produce additional implementation drawings, the designers were able to work on the design of the facade until the moment of starting production (Gramazio & Kohler, 2008). A more recent project "The Sequential Roof" successfully verified the potential of additive digital fabrication (dfab) processes for the prefabrication of complex timber structures at full building scale. This robotically assembled 2,300 square meter roof is formed by 120 timber trusses, each one produced in 12 hours. The development of robust computational design and automated construction framework allowed a tenfold reduction in construction time (Willmann et al., 2016). Contributions have also been made for developing concrete structures, especially for non-standard building elements. For example, the concrete printing process developed at Loughborough University consisted of the additive fabrication of full-scale building elements such as panels and walls with the use of a gantry robot. According to Lim et al. (2012), this process enables design freedom, precision of manufacture with functional integration, and elimination of labor-intensive molding. There have been successful full-scale applications (Labonnote et al., 2016), the most recent one related to the "3D Printed Habitat Challenge" a competition that commenced in 2015 organized by NASA (n.d.).

15.3 Classification

When we think of robots, we often imagine them as human-like images. In reality, however, construction robots look more like machines and are less capable than people (Cousineau & Miura, 1998). Construction robots vary in sophistication. They range from simple tools to fully automated devices. The terms for construction automation and robotics in construction have many general definitions, and different people see them at different levels of generality (Chen et al., 2018). According to the International Association for Automation and Robotics in Construction, Professor Sangrey said at the first ISARC, held in Pittsburgh in 1984 that "robotics in construction represents one of a group of related applications for robotics in unstructured environments" (IAARC, n.d.). Skibniewski and Russell (1991) used the term construction automation to principally mean the execution of construction tasks using robots.

Their work showed that the automation of construction tasks requires substantial adjustments to the construction schedule and shifting of project recourses. Skibniewski (1992) defined construction automation as the "engineering or performance of any construction process, on-site or off-site, by means of teleoperated, numerically controlled, semiautonomous, or autonomous equipment." He also referred to construction robotics as the "advanced construction equipment exhibiting any level of capability related to teleoperation, sensory data collection and processing, numerically controlled, or autonomous task performance." More recent definitions include Jung et al. (2013) who referred to "construction automation" as a machine-centered construction factory technology for applying robotic systems on the construction field. A more limited definition by Vähä et al. (2013) describes "construction automation" as the automatic assembly method enabled by computer numerical control and real-time sensing technologies. Bock (2015) proposed a general definition of "construction automation" as a new set of technologies and processes that will change the whole course and idea of construction in a fundamental way. Since the concept of Industry 4.0 has been introduced as a popular term for digitalization and automation of the manufacturing environment, the definition of construction automation has been extended to include information modeling and digitalization (Oesterreich & Teuteberg, 2016). Some researchers argue that "construction automation" is the integration of computer-aided design and robot-based on-site technologies for simplification of overall activities (Keating & Oxman, 2013). A few of them, such as Willmann et al. (2016), now use the term "digital fabrication" as a synonym for "construction automation", particularly when referring to customized building construction (Chen et al., 2018).

On-site construction automation aims to bring fabrication processes on construction sites. Sousa et al. (2016) classified on-site technologies in three main categories: large-scale robotic structures, mobile robotic arms, and flying robotic vehicles. A well-known example from the first category is Contour Crafting, a robotic structure for 3D printing large-scale construction, developed at the University of Southern California (Khoshnevis, 2004). An example of a mobile robot for on-site construction is the semi-automated mason (SAM) developed by construction Robotics (Sklar, 2015), or the "In situ Fabricator" (IF), developed at ETH Zurich (Giftthaler et al., 2017). Finally, the use of flying robots in construction is a novel technique developed to avoid mobility constraints and the need for cranes on construction sites. Imperial College London developed an application of these technologies for polyurethane foam deposition (Hunt et al., 2014).

On the other hand, off-site digital fabrication aims to custom design and prefabricate large-scale complex architectural elements off-site. Among existing additive dfab technologies, the most common for prefabrication include gantry robots, fixed robotic arms, and 3D printers. An example of additive prefabrication with a fixed robotic arm is the project DEMOCRITE from XtreeE and ENSA Paris-Malaquais. The project aims to construct complex concrete structural elements with increased performance and material optimization (Gosselin et al., 2016). Finally, the use of 3D printers is currently investigated for prefabrication of architectural elements. The project D-Shape, developed by Enrico Dini uses this technology for 3D printing sand structures through a binder-jetting process (Cesaretti et al., 2014). Construction robots have characteristics different from both industrial and civil engineering robots. In addition, robotic technologies can be classified as single-task robots (STRs) and automated construction systems. For more information about the classification of robots, the reader is referred to Chapter 16 (Robots in Indoor and Outdoor Environments) of this Handbook.

15.3.1 Single-task robots vs. construction automation

Single-task robots are designed to imitate skilled labor. They perform a specific job, such as concrete finishing or welding. The main purpose of STR is to alleviate skilled labor for those jobs considered "3k" or "kitsui" (difficult), "kitanai" (dirty), and "kiken" (dangerous) (Cousineau & Miura, 1998). Many single-task robots have been introduced in construction, although only a few are used in construction projects. Examples include robots for finishing concrete work (Skibniewski & Kunigahalli, 2008), spray painting, rebar placement, tile inspection, or material handling (Bock & Linner, 2016b). IAARC (1998) provided a compilation of robots and automated machines in construction, covering the following construction applications: demolition, surveying, excavation and earthmoving, paving, tunneling, concrete transportation and distribution, concrete-slab screeding and finishing, cranes and autonomous trucks, welding and positioning of structural steel members, fireresisting and paint spraying, inspection and maintenance, and integrated building construction. Budny et al. (2009) provide a comprehensive list, and detailed description, of machinery found in construction sites. They define such machines as devices used to increase or replace human physical force in carrying out construction tasks to create mechanized sites to conduct manufacturing processes in construction. Examples of such machines include cranes, site lifts, or equipment for finishing work. Some examples of recent single-task robots are shown in the next section.

Construction automation uses principles of industrial automation and typically have the following components: on-site factory protected from the environment, automated jack system, automated material handling and tracking system, and a centralized information control system (Bock & Linner, 2015, Cousineau & Miura, 1998, Skibniewski & Kunigahalli, 2008). Japanese construction companies pioneered the entire robotic high-rise building construction system, e.g. Obayashi Corporation "Big Canopy" system, Shimizu Corporation "Smart" system. A summary of early examples of sky factories is shown in Table 15.1. Readers interested in automated systems are directed to Bock and Linner (2016a). For a high-rise building, the final product is a building, and the repeated unit is one story. The factory is built on site and jacked up as each floor is completed to begin work on the next floor. To get fruitful results, a total robotized construction system, including the design and construction method, should be developed (Ueno et al., 1986). Using just-in-time delivery systems, bar-coded components, and computerized information management systems, construction automation is improving with repeated use.

Table 15.1 Buildings constructed by Japanese contractors using construction automation (adapted from Cousineau & Miura, 1998)

Company	System	Building type	Structure type	No. Stories	Year
Takenaka	Push-Up	Office	Steel	N/A	1989–1991
Shimizu	SMART System	Office	Steel	20	1991–1994
Obayashi	ABCSystem	Residential	Steel	10	1991–1994
Taisei	T-Up	Office	Steel	34	1992–1994
Maeda	MCCS	Office	Steel	10	1992–1994
Takenaka	Push-Up	Office	Steel	14	1993-1995
Akatsuki 21	Fujita	Office	Steel	16	1994–1996
Shimizu	SMART System	Office	Steel	30	1994–1997
Kajima	AMURAD	Residential	Precast conc.	9	1995–1996
Obayashi	Big Canopy	Residential	Precast conc.	26	1995–1997
Maeda	MCCS	Office	Steel	8	1995–1998

15.4 Current status/examples

More recently, novel robotic construction technologies and processes have been developed, and their potential contribution to improving the productivity of the building industry should be evaluated. Some current examples of STRs include the Semi-Automated Mason (SAM100), the TyBot rebar-tying robot, the In situ Fabricator, or the HRP-5P humanoid bot. A recent example of construction automation includes the East Village No.8 project by the Mace Group, using the concept of sky factory originally used by Japanese construction firms in the 1990s.

New robotic systems are becoming technically and economically possible in the construction industry, and it is expected that they will gradually be used in the industry as more cost-effective solutions are found. In addition, the current legal environment, building codes, and regulations are now more inclined to allow flexibility to incorporate the changes in the industry, giving construction robots and automation a higher chance to succeed. The examples shown below are meant to provide an overview of current systems and applications and are not meant to be an exhaustive list.

15.4.1 Semi-Automated Mason (SAM)

The Semi-Automated Mason (SAM) is a brick laying robot designed and engineered by Construction Robotics. It is the first commercially available bricklaying robot for on-site masonry construction. It is designed to work collaboratively with the mason, which helps with the reduction of health and safety impact on the workforce (Figure 15.3). According to Construction Robotics (n.d.a), it increases masons' productivity by three to five times while reducing lifting by 80%. Since it was first used in construction sites in 2015, it has been used in 15 projects (Table 15.2).

Figure 15.3 General view of SAM and mason worker in the Delbert Day project, MO, USA in 2016 (courtesy of Construction Robotics, 2016)

Table 15.2 Projects in which SAM has been used since its first application in 2015 (source: Construction Robotics, SAM's portfolio of work, n.d.b)

Project name	Location	Owner	General Contractor	Mason Contractor	Architect	SAM On Site	SAM Bricks	Brick Type
UMHHC BRIGHTON HEALTH CENTER SOUTH	Brighton, MI	The University of Michigan Hospitals and Health Centers	The Christman Company	Leidal & Hart Mason Contractors	HKS Architects	Jul-17	17,000+	Utility
Erlanger Medical Office Building	Chattanooga, TN	Johnson Development	Freese Johnson	Jenkins Masonry	Wakefield Beasley	May-18	25,000+	Utility
POFF Federal Building	Roanoke, VA	General Services Administration	HOAR Construction	United Building Envelope Restoration	MTFA Architects	Jul-18	250,000+	Modular
Shenandoah University – Athletics & Events Center	Winchester, VA	Shenandoah University	Howard Shockey & Sons Inc.	J. D. Long Masonry	ESa Architects	Jun-17	80,000+	Modular
Ford Driving Dynamics Lab	Wayne, MI	Ford	N/A	Davenport Masonry	Albert Kahn Associates, Inc.	Apr-17	15,000+	Modular
Cavalier Greene	Corunna, MI	The Woda Group	Woda Construction, Inc.	Davenport Masonry	Hooker Dejong, Inc.	Feb-17	N/A	Modular
Purdue University, Flex Lab Facility	West Lafayette, IN	Purdue University	Pepper Construction	Wilhelm Construction	Jacobs/Ennead	2017	60,391	Modular
Towneplace Suites	Cranberry Springs, PA	Cranberry Springs Development Group I, L.P.	N/A	Arch Masonry	N/A	Aug-16	9,043	Utility
Aldi	Pulaski, NY	Aldi Inc.	JCS Construction Resources Inc	C&D LaFace Construction	N/A	Jun-16	33,340	Belden – Modular
Delbert Day Middle School	Rolla, KS	Phelps County Regional Medical Center	McCarthy Building Companies, Inc.	Heitkamp Masonry	N/A	Apr-16	24,950	Modular
Middle School	Goodlettsville, TN	Metro Nashville Public Schools	Hardaway Construction	Wasco Inc.	N/A	Feb-16	13,803	Utility
Columbus McKinnon	Getzville, NY	Columbus McKinnon Corporation	Unliand	Brawdy Construction	N/A	Nov-15	15,000	Utility
Potomac Valley Brick Showroom	Frederick, MD	Potomac Valley Brick	Merritt Construction	R. W. Sheckles	N/A	Sep-15	3,846	Belden Utility
The Lab School	Washington, DC	The Lab School of Washington, DC	Clark Construction	Manning Construction	N/A	Aug-15	10,695	Norman & Closure
Laramie High School	Laramie, WY	Albany County School District #1	Haselden Construction	Soderberg Masonry	N/A	Jun-15	7,000	Utility

15.4.2 *TyBot rebar-tying robot*

TyBot is an autonomous robotic arm rigged to a gantry crane developed by Advanced Construction Robotics (ACR). It is a single task robot that can tie together steel reinforcement bars (Figure 15.4) and provides a good alternative for a tedious, laborious, unpopular, and accident-prone activity. According to ACR (n.d.), TyBot can replace a rebar-tying crew. Only one technician is required to monitor performance, reload the tie wire spool, and ensure that the robot operates safely.

15.4.3 *In situ Fabricator*

The In situ Fabricator (IF) is a semiautonomous, mobile robot designed explicitly for additive construction on-site (Figure 15.5). The IF is a research project from Gramazio Kohler Research, ETH Zurich and the National Centre of Competence in Research "Digital Fabrication". The height of the IF is the same as a standard wall and has a total weight of 1.4 tons. The IF robot is equipped with tracks driven by hydraulic motors, which can achieve a speed of 5 km/h. It is physically capable of moving on a non-flat terrain with obstacles found on a typical construction site. Moreover, it can be equipped with different tools or end effectors to perform a wide range of building tasks. The IF is equipped with a camera-based sensing system for global localization of the robot in the construction site and local detection of the element being built. The system can process architectural design decisions using Python code and then execute task loops over the whole building process. The camera sensing allows to check between true measurements of the structure during build-up and provide less than 5 mm positioning accuracy at the end effector based on the architectural design data (Giftthaler et al., 2017).

Figure 15.4   Single task robot to tie rebar in a bridge project. (courtesy of TyBot LLC)

Figure 15.5   In situ Fabricator (courtesy of Gramazio Kohler Research, ETH Zurich)

15.4.4 HRP-5P humanoid bot

Researchers at the National Institute of Advanced Industrial Science and Technology (AIST) in Japan, have developed a humanoid robot prototype, HRP-5P, intended to autonomously perform heavy labor or work in hazardous environments (Kaneko et al., 2019). The use of HRP-5P, as a development platform in collaboration between industry and academia. Although some robotics experts disagree with the practicality of this type of humanoid robots and indicate that they are not realistic at least for next 30 years, examples such as HRP-5P promote R&D toward practical application of humanoid robots in construction sites. Key elements of the prototype include autonomous behavior with the ability to map the surrounding environment and perform object recognition (AIST, 2018). Specifically, this work involves the following operations (Figure 15.6):

1. Generate a 3D map of the surrounding environment, detect objects, and approach the workbench.
2. Lean against the workbench, slide one of the stacked gypsum boards to separate it, and then lift it.
3. While recognizing the surrounding environment, carry the gypsum board to the wall.
4. Lower the gypsum board and stand it against the wall.
5. Using high-precision AR markers, recognize and pick up a tool.
6. Holding a furring strip to keep HRP-5P itself steady, screw the gypsum board into the wall.

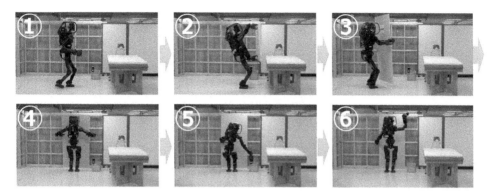

Figure 15.6 Autonomous installation of gypsum board by HRP-5P (courtesy of AIST; AIST, 2018)

15.4.5 Sky factories – Mace Group: East Village No. 8

The concept of sky factories to promote construction automation can be traced to the first applications of construction automation in Japan when in the 1990s they used "sky factories" for the construction of high-rise buildings (Cousineau & Miura, 1998; Bock & Linner, 2016a). A recent application of the sky factory concept was used by the Mace Group, in the UK, for the construction of the East Village No. 8 project that started in 2016. The six-story sky factory (Figure 15.7a) created an indoor construction site – reduced noise and safety risks, and prevented environmental delays (Mace Group, n.d.). Using prefabrication and the factory-style construction method contributed to a faster (each floor was constructed in one week) and safer construction delivery. Key benefits of this project included (1) Off-site innovation: 98% of this project's entire infrastructure was built off-site. This reduced the number of site vehicles on local roads by 30% and reduced site waste by 75%; (2) Reduced risk: The factory space overhung the building's floor plate, eliminating any need to work at height. The waterproof factory meant construction could continue even in adverse weather conditions (Figure 15.7b); (3) Controlled sequencing: Subcontractors had full control of a floor of the building for a week, which helped eliminate defects and increased productivity. At its peak, 32 trade handovers took place in one week; and (4) Continuous improvement and lessons learned: The final levels of the two towers took just 39 hours to complete. At every stage of the project, better ways of working with the sky factory were developed, continually evolving, and sharing lessons learned across the project team.

15.5 Main challenges and future directions

Despite current technological advancements, construction robots still face several challenges. For example, they must be able to handle different elements which vary in weight and size, making balance control nontrivial. They should be able to navigate through generally unstructured environments autonomously. Although significant achievements have been made in the area of machine vision and sensors (ultrasound, laser, or gyrocompass sensors) so that robots can correct their course and adapt to changing environments, there are still accuracy and safety issues that need to be addressed in real construction sites. In addition, they have to be able to cope with the environment. Construction robots working in construction sites will be exposed

Figure 15.7 Exterior view of sky factory (a) and general view of the interior workspace (b) (courtesy of Mace Group)

to dust, rain, wind, and extreme temperatures; none of these conditions are favorable for robotic systems. One of the main challenges that needs to be addressed has to do with the variability of tasks that construction robots should be able to perform, bringing an additional layer of complexity when adjusting the robots to account for that variety, robots have a sequence or digital control rather than the playback control found in most industrial robots. Robots with manipulators often have simple coordinate systems to direct three-dimensional motion.

The standardization of components and processes will be fundamental to achieve the level of automation of the automotive industry (Fujinami et al., 1995). However, such standardization is hard to find in the AEC industry. Although a "standard" building is plausible, it is architecturally undesirable. But perhaps the most pressing challenges have to do with the societal implications, in particular as they relate to changes in the workforce.

In many regions, building codes and officials make it very difficult to move forward the use of 3D-printed structures. To take full advantage of automation's potential, municipal regulations and building codes must keep pace (Garcia de Soto, 2019). Design changes should be permitted by building codes to accommodate the use of robots for construction tasks (Skibniewski, 1992). In Dubai, for example, new regulations require that by 2025, every building must be constructed with 25% of its material from 3D-printed sources (Dubai Future Foundation, n.d.).

In addition, the nature of Construction 4.0 makes the consideration of cybersecurity of paramount importance. These two elements are discussed below.

15.5.1 *Implications for the workforce*

There is no doubt that Construction 4.0 will attract a new tech-savvy generation of workers to the construction sector. It is expected that unpleasant aspects of construction work (e.g. working in dangerous, dirty, and difficult conditions) will be automated, leading to an improvement in job satisfaction for workers. Since it is anticipated that the use of robotic systems and on-site automation will start with unsafe and unappealing tasks for workers, there should be a general acceptance from policymaking institutions and labor organizations. Also, since perceptions of the work being physically too demanding will no longer be valid, there is also an opportunity to increase the share of women working in the construction industry.

The organizations will also suffer modifications. There will be a movement from current fragmented projects to project-based integrations (enabled through digitalization), and eventually to a platform-based integration (based on personalization) as a way to cope with the new roles and increased levels of collaboration, coupled with the amplified involvement of the owners (enabled through the platform). Although the transition (or short term) will be characterized by the adoption of conventional structures trying to incorporate critical elements from Construction 4.0, the long term view suggests a clear departure from fragmented organizational structures towards platform-based structures to support full integration between planning and construction.

The fact that the construction industry is getting ready for the Fourth Industrial Revolution, with many opportunities to innovate, is stimulating and can become attractive to new generations. Further research is needed to evaluate the impacts of Construction 4.0 to the functional division, supply chain, organizational structures, and business models (with a particular emphasis on cybersecurity), as well as the project deliveries and contract strategies of the AEC/FM industry, and to assess additional social impacts, such as changes in education and training schemes (Garcia de Soto et al., 2019).

15.5.1.1 *Human-machine interaction*

It is apparent that through Construction 4.0, more robots will be introduced in the work environment. It is expected that robotic technologies and conventional construction will coexist (Garcia de Soto et al., 2018b), this means that the human-machine relationship will continue to evolve, making the human interaction with those systems a very important component. A significant amount of research has been made to get a better understating of how machines and artificial intelligent systems should act. Trafton et al. (2006) indicated that the way in which an artificial system acts has a profound effect on how people act toward the system. Similarly, the way an artificial system thinks, has a profound effect on how people interact with the system. We need new ways of thinking about how decisions are made and how strategic interactions take place to avoid problems that occur when machines and humans work at cross purposes and do not collaborate or coordinate. An apparent reason why the collaboration could fail or be sabotaged has to do with the generalized perception of superiority of robots and their impact on the future of jobs. Frey and Osborne (2017) estimated that around 47% of total US employment has a high risk of computerization by the 2030s, while the estimations by Arntz et al. (2016) were quite a bit lower, only 10%. The findings in Berriman (2017) are somewhere in between, estimating that 35% of US jobs are in danger of being lost to the robots. Most studies have minimized the potential effects of automation on job creation, and have tended to ignore other relevant trends, including globalization, population aging, urbanization, and the rise of the green economy (Bakhshi et al., 2017). Although some studies and projections are pessimistic about the impacts on labor (Frey & Osborne, 2017), others give a more optimistic view (Arntz et al., 2016, OECD, 2016). In any case, it is clear to understand why humans have exhibited a variety of attitudes towards robots and AI; in times, fearful and skeptical, in others, welcoming and cooperative.

15.5.2 Cybersecurity awareness

One of the key elements of the benefits of Construction 4.0 is the digitalization of data. The amount of data available during the different phases of a project can be significant, and the digitalization of the industry must be prepared to handle this data in a secure way. According to the International Telecommunication Union (ITU), cybersecurity is defined as the

> collection of tools, policies, security concepts, security safeguards, guidelines, risk management approaches, actions, training, best practices, assurance and technologies that can be used to protect the cyber environment and organization and user's assets. Organization and user's assets include connected computing devices, personnel, infrastructure, applications, services, telecommunications systems, and the totality of transmitted and/or stored information in the cyber environment.
>
> *(ITU, n.d.)*

Many of these elements are currently applicable to the construction sector and should be considered. In addition, current normative and regulations must also be observed. This means that the construction sector has to comply with the protection of the data it stores. For example, the introduction of the General Data Protection Regulation (GDPR) in Europe in May 2018 (Regulation (EU), 2016), requires improved cybersecurity for the operators of essential services, which includes construction projects using digitally built environments, including digital infrastructure (i.e. smart cities) and intelligent buildings. In the UK, the Publicly Available Specification (PAS) 1192 5 (BSI, 2015) provides a framework to ensure that information is shared in a security-minded way and to enable the reliability and security of digitally built assets, keeping in mind that the data stored about built assets could be used by those with malicious intent. The Code of Practice for Cyber Security in the Built Environment (IET, 2014), addressed how cybersecurity should be considered throughout a building's life cycle with a focus on building-related systems and all connections to the broader cyber environment. Stakeholders involved in the AEC/FM industry, and particularly construction contractors, should implement strategies and educate employees to secure the data related to their projects (Mantha & Garcia de Soto, 2019). For additional information, the reader is referred to Chapter 22 (Cyber threats and actors confronting the Construction 4. 0) of this Handbook.

15.6 Conclusion

Although it will take more time to assess how digitalization and automation might affect the construction supply chains, business models, employment, cybersecurity, and AEC project delivery, there is a growing consensus that new processes and technologies are critical to improving construction productivity.

Particular attention should be made to the transition phase, in which human-robot interaction will play an essential part. In addition, close consideration should be given to conflicts that will arise regarding the conventional and new delivery systems, organizational structures, and social implications, as these will have a profound influence the future of construction automation and robotics.

15.7 Summary

- Construction 4.0 can make a significant difference in the successful integration of construction robot applications.
- New robotic systems are becoming technically and economically possible and it is expected that they will gradually be used in the industry as more cost-effective solutions are found.
- The struggles faced by the construction industry to pass the prototype and research stages can be overcome by a thoughtful implementation of classic and modern robotic technology concepts.
- In general, the current legal environment, building codes, and regulations allow flexibility to incorporate the changes needed to give construction robots and automation a higher chance to succeed.

Notes

1 Takenaka, Kajima, Taisei, Obayashi, and Shimizu.
2 www.iaarc.org/publications/search.php.

References

ACR. (n.d.). TyBot: Frequently asked questions. *Advanced Construction Robotics*. Available at https://tybotllc.com/faq. Accessed on April 28, 2019.

AIST. (2018). *Development of a Humanoid Robot Prototype, HRP-5P, Capable of Heavy Labor. National Institute of Advanced Industrial Science and Technology*. Available at: https://aist.go.jp/aist_e/list/latest_research/2018/20181116/en20181116.html. Accessed on January 29, 2019.

Araksyan, V. & Volkov, V. (1985). *Mechanization and Automation of Heavy and Labor-Intensive Works*. Moscow: Znanye. 64, in Russian.

Arntz, M., Gregory, T., & Zierahn, U. (2016), The risk of automation for jobs in OECD countries: A comparative analysis. OECD Social, Employment and Migration Working Papers, No. 189, OECD Publishing, Paris. DOI; 10.1787/5jlz9h56dvq7-en. Accessed on April 24, 2018.

Bakhshi, H., Downing, J., Osborne, M., & Schneider, P. (2017). *The Future of Skills: Employment in 2030*. London: Pearson and Nesta.

Berriman, R. (2017) Will robots steal our jobs? The potential impact of automation on the UK and other major economies. *PwC. Part of the UK Economic Outlook* March, 2017. Available at: http://pwc.co.uk/economic-services/ukeo/pwc-uk-economic-outlook-full-report-march-2017-v2.pdf. Accessed on October 15, 2018.

Bock, T. (2015). The future of construction automation: Technological disruption and the upcoming ubiquity of robotics. *Automation in Construction*, 59, 113–121. doi: 10.1016/j.autcon.2015.07.022.

Bock, T. & Linner, T. (2015). *Robotic Industrialization: Automation and Robotic Technologies for Customized Component, Module, and Building Prefabrication*. Cambridge: Cambridge University Press. doi: 10.1017/CBO9781139924153.

Bock, T. & Linner, T. (2016a). *Site Automation: Automated/Robotic On-Site Factories*. Cambridge: Cambridge University Press. doi: 10.1017/CBO9781139872027.

Bock, T. & Linner, T. (2016b). *Construction Robots: Elementary Technologies and Single-Task Construction Robots*. Cambridge: Cambridge University Press. doi: 10.1017/CBO9781139872041.

BSI, PAS 1192–5. (2015). *Specification for Security Minded Building Information Modelling, Digital Built Environments and Smart Asset Management*. London: British Standards Institution. Available at: https://shop.bsigroup.com/ProductDetail/?pid=000000000030314119. Accessed on April 9, 2019.

Budny, E., Chłosta, M., Meyer, H. J., & Skibniewski, M. J. (2009) Construction machinery. In: K. H. Grote and E. Antonsson (eds) *Springer Handbook of Mechanical Engineering*. Berlin, Heidelberg: Springer Handbooks. Springer. doi: 10.1007/978-3-540-30738-9_14.

Cesaretti, G., Dini, E., De Kestelier, X., Colla, V., & Pambaguian, L. (2014). Building components for an outpost on the Lunar soil by means of a novel 3D printing technology. *Acta Astronautica*, 93, 430–450. doi: 10.1016/j.actaastro.2013.07.034.

Chen, Q., Garcia de Soto, B., & Adey, B. T. (2018). Construction automation: Research areas, industry concerns and suggestions for advancement. *Automation in Construction*, 94, 22–38. doi: https://doi.org/10.1016/j.autcon.2018.05.028.

Construction Robotics. (2016). Portfolio: Delbert Day – Rolla, KS (image number three). Available at: www.construction-robotics.com/portfolio-items/delbertdayrolla/. Accessed on April 12, 2019.

Construction Robotics. (n.d.a). SAM100. *Semi-Automated Mason*. Available at https://construction-robotics.com/sam100. Accessed on April 2, 2019.

Construction Robotics. (n.d.b). SAM's portfolio of work. Available at: www.construction-robotics.com/recent-works/. Accessed on April 12, 2019.

Cousineau, L. & Miura, N. (1998). *Construction Robots: The Search for New Building Technology in Japan*. Reston, VA: ASCE Publications. ISBN 0-7844-0317-1 p. 130.

Dubai Future Foundation. (n.d.). Dubai 3D printing strategy: Our goals. Available at: www.dubaifuture.gov.ae/our-initiatives/dubai-3d-printing-strategy. Accessed on April 11, 2019.

Frenkel, G. Y. (1987). *Robotization of Work Processes in Construction*. Moscow: Stroyizdat. 174, in Russian.

Frenkel, G. Y. (1988). *Application of Robotics and Manipulators in the Construction Industry: Construction and Progress in Science and Technology*. Moscow: Znanye. 64, in Russian.

Frey, C. B. & Osborne, M. A. (2017). The future of employment: How susceptible are jobs to computerisation? *Technological Forecasting and Social Change*, 114, 254–280.

Fujinami, Y., Mitsuoka, H., Suzuki, A., & Kimura, T. (1995). Construction robotics and automation research at Taisei. *Computer Aided Civil and Infrastructure Engineering*, 10 (6), 401–413. doi: 10.1111/j.1467-8667.1995.tb00300.x.

Garcia de Soto, B. (2019). The construction industry needs a robot revolution. *WIRED*. Available at: www.wired.com/story/the-construction-industry-needs-a-robot-revolution. Accessed on April 6, 2019.

Garcia de Soto, B., Agustí-Juan, I., Hunhevicz, J., Joss, S., Graser, K., Habert, G., & Adey, B. T. (2018a). Productivity of digital fabrication in construction: Cost and time analysis of a robotically built wall. *Automation in Construction*, 92, 297–311. doi: 10.1016/j.autcon.2018.04.004.

Garcia de Soto, B., Agustí Juan, I., Joss, S., & Hunhevicz, J. (2019). Implications of Construction 4.0 to the workforce and organizational structures. *International Journal of Construction Management*. doi: 10.1080/15623599.2019.1616414.

Garcia de Soto, B., Agustí Juan, I., Joss, S., Hunhevicz, J., Habert, G., & Adey, B. T. (2018b). Rethinking the roles in the AEC industry to accommodate digital fabrication. Sixth Creative Construction Conference 2018, CCC 2018., June 30–July 3, 2018, Ljubljana, Slovenia doi: 10.3311/CCC2018-012.

Giftthaler, M., Sandy, T., Dörfler, K., Brooks, I., Buckingham, M., Rey, G., Kohler, M., Gramazio, F., & Buchli, J. (2017). Mobile robotic fabrication at 1:1 scale: The In situ fabricator. *Construction Robotics*, 1–12. doi: 10.1007/s41693-017-0003-5.

Gosselin, C., Duballet, R., Roux, P., Gaudillière, N., Dirrenberger, J., & Morel, P. (2016). Large-scale 3D printing of ultra-high performance concrete – a new processing route for architects and builders. *Materials & Design*, 100, 102–109. doi: 10.1016/j.matdes.2016.03.097.

Gramazio, F. & Kohler, M. (2008). *Digital Materiality in Architecture*. Baden, Switzerland: Lars Müller Publishers. Available at: https://lars-mueller-publishers.com/digital-materiality-architecture.

Haas, C., Skibniewski, M., & Budny, E. (1995). Robotics in civil engineering. *Computer-Aided Civil and Infrastructure Engineering*, 10, 371–381. doi: 10.1111/j.1467-8667.1995.tb00298.x.

Hasegawa, Y. (1984) Robotization of construction work. Proceedings of the 1st ISARC, Pittsburgh, USA. doi: 10.22260/ISARC1984/0009

Hunt, G., Mitzalis, F., Alhinai, T., Hooper, P. A., & Kovac, M. (2014). 3D printing with flying robots. IEEE International Conference on Robotics and Automation (ICRA), Hong Kong, pp. 4493–4499. doi: 10.1109/ICRA.2014.6907515.

IAARC. (1998). Robots and automated machines in construction. *International Association for Automation and Robotics in Construction*, Available at: https://iaarc.org/a_publications_IAARC_catalogue.pdf. Accessed on March 15, 2019.

IAARC. (n.d.). The history of IAARC. International Association for Automation and Robotics in Construction. Available at: https://iaarc.org/pe_about.htm. Accessed on March 12, 2019.

IET. (2014). Standards, code of practice for cyber security in the built environment, Institution of Engineering and Technology. Available at: https://electrical-shop.theiet.org/books/standards/cyber-cop.cfm. Accessed on February 12, 2019.

International Federation of Robotics (IFR). (2018a). Word Robotics 2018. Available at: https://ifr.org/ Accessed on April 24, 2019.

International Federation of Robotics (IFR). (2018b). IFR press conference. October 18 2018, Tokyo, Japan. Available at: https://ifr.org/downloads/press2018/WR_Presentation_Industry_and_Service_Robots_rev_5_12_18.pdf. Accessed on April 24, 2019.

ITU, International Telecommunication Union. (n.d.). Definition of cybersecurity, Referring to ITU-T X.1205. *Overview of cybersecurity*. Available at: https://itu.int/en/ITU-T/studygroups/com17/Pages/cybersecurity.aspx. Accessed on February 12, 2019.

Jung, K., Chu, B., & Hong, D. (2013). Robot-based construction automation: An application to steel beam assembly (Part II). *Automation in Construction*, 32, 62–79. doi: 10.1016/j.autcon.2012.12.011.

Kaneko, K., Kaminaga, H., Sakaguchi, T., Kajita, S., Morisawa, M., Kumagai, I., & Kanehiro, F. (2019). Humanoid robot HRP-5P: An electrically actuated humanoid robot with high-power and wide-range joints. *IEEE Robotics and Automation Letters*, 4 (2), 1431–1438. doi: 10.1109/LRA.2019.2896465.

Kano, N. & Tamura, Y. (1984). A new management tool for robotized construction projects – Application of computer graphics in construction planning and scheduling. Proceedings of the 1st ISARC, Pittsburgh, USA. doi: 10.22260/ISARC1984/0002.

Keating, S. & Oxman, N. (2013). Compound fabrication: A multi-functional robotic platform for digital design and fabrication. *Robotics and Computer-Integrated Manufacturing*, 29 (6), 439–448. doi: 10.1016/j.rcim.2013.05.001.

Khoshnevis, B. (2004). Automated construction by contour crafting – related robotics and information technologies. *Automation in Construction*, 13 (1), 5–19. doi: 10.1016/j.autcon.2003.08.012.

Labonnote, N., Rønnquist, A., Manum, B., & Rüther, P. (2016). Additive construction: State-of-the-art, challenges and opportunities. *Automation in Construction*, 72, 347–366. doi: 10.1016/j.autcon.2016.08.026.

Lim, S., Buswell, R. A., Le, T. T., Austin, S. A., Gibb, A. G., & Thorpe, T. (2012). Developments in construction-scale additive manufacturing processes. *Automation in Construction*, 21, 262–268. doi: 10.1016/j.autcon.2011.06.010.

Mace Group. (n.d.). One big leap for construction. Available at https://macegroup.com/projects/east-village-no8. Accessed on February 12, 2019.

Mantha, B. & Garcia de Soto, B. (2019). Cybersecurity risk identification and vulnerability assessment in the construction industry. In Proceedings of the Eighth Creative Construction Conference (CCC 2019), 29 June–2 July, 2019, Budapest, Hungary.

NASA. (n.d.). NASA's centennial challenges: 3D-printed habitat challenge. Available at: https://nasa.gov/directorates/spacetech/centennial_challenges/3DPHab/index.html. Accessed on April 12, 2019.

OECD. (2016). Automation and independent work in a digital economy. *Policy Brief on The Future of Work*. OECD Publishing, Paris. Available at https://oecd.org/employment/Policy%20brief%20-%20Automation%20and%20Independent%20Work%20in%20a%20Digital%20Economy.pdf. Accessed on April 29, 2017.

Oesterreich, T. D. & Teuteberg, F. (2016). Understanding the implications of digitisation and automation in the context of Industry 4.0: A triangulation approach and elements of a research agenda for the construction industry. *Computers in Industry*, 83, 121–139. doi: 10.1016/j.compind.2016.09.006.

Regulation (EU). (2016). 679. General data protection regulation (protection of natural persons with regard to the processing of personal data and on the free movement of such data, and repealing directive 95/46/EC). *European Parliament*. Available at: https://eur-lex.europa.eu/eli/reg/2016/679/oj. Accessed on April 12, 2019.

Shimomura, Y. & Sonoda, T. (1984). Tunnelling by robots – Shield driving automatic control system. Proceedings of the 1st ISARC, Pittsburgh, USA. doi: 10.22260/ISARC1984/0007.

Skibniewski, M. J. (1988a). Framework for decision-making on implementing robotics in construction. *Journal of Computing in Civil Engineering*, 2 (2), 188–201. doi: 10.1061/(ASCE)0887-3801(1988)2:2(188).

Skibniewski, M. J. (1988b). *Robotics in Civil Engineering*. Southampton, Boston, New York: Computational Mechanics Publications and Van Nostrand Reinhold.

Skibniewski, M. J. (1992). Current status of construction automation and robotics in the United States of America. 9th International Symposium on Automation and Robotics in Construction (pp. 17–24). doi: 10.22260/ISARC1992/0003.

Skibniewski, M. J., Derrington, P., & Hendrickson, C. (1986). Cost and design impact of robotic construction finishing. Proceedings of the 3rd ISARC, Marseille, France. doi: https://doi.org/10.22260/ISARC1986/0038.

Skibniewski, M. J. & Hendrickson, C. T. (1985). Evaluation method for robotics implementation: Application to concrete form cleaning. Proceedings of the 2nd ISARC, Pittsburgh, USA. doi: 10.22260/ISARC1985/0014.

Skibniewski, M. J. & Kunigahalli, R. (2008). *Automation in Concrete Construction. Concrete Construction Engineering Handbook*. Chapter 18 CRC Press Boca Raton: Taylor & Francis Group. doi: 10.1201/9781420007657.

Skibniewski, M. J. & Russell, J. S. (1991). Construction robot fleet management system prototype. *Journal of Computing in Civil Engineering*, 5 (4), 444–463. doi: 10.1061/(asce)0887-3801(1991)5:4(444).

Sklar, J. (2015). Robots lay three times as many bricks as construction workers. *MIT Technology Review*. Available at: https://technologyreview.com/s/540916/robots-lay-three-times-as-many-bricks-as-construction-workers/. Accessed: April 24, 2017.

Sousa, J. P., Palop, C. G., Moreira, E., Pinto, A. M., Lima, J., Costa, P. and Moreira, A. P. (2016). The SPIDERobot: A cable-robot system for on-site construction in architecture. *Robotic Fabrication in Architecture, Art and Design* (pp. 230–239). Springer International Publishing. doi: 10.1007/978-3-319-26378-6_17.

Trafton, J. G., Schultz, A. C., Cassimatis, N. L., Hiatt, L. M., Perzanowski, D., Brock, P. D., Bugajska, M. D., & Adams, W. (2006) Communicating and collaborating with robotic agents. In: *Cognition and Multi-agent Interaction: From Cognitive Modeling to Social Simulation*. Washington, DC: Naval Research Laboratory, Navy Center for Applied Research in Artificial Intelligence (NCARAI). Available at https://apps.dtic.mil/dtic/tr/fulltext/u2/a480639.pdf.

Ueno, T., Maeda, J., Yoshida, T., & Suzuki, S. (1986) Construction robots for site automation. In: *CAD and Robotics in Architecture and Construction* (pp. 259–268). Boston, MA: Springer. doi: https://doi.org/10.1007/978-1-4684-7404-6_26.

Vähä, P., Heikkilä, T., Kilpeläinen, P., Järviluoma, M., & Gambao, E. (2013). Extending automation of building construction – Survey on potential sensor technologies and robotic applications. *Automation in Construction*, 36, 168–178. doi: 10.1016/j.autcon.2013.08.002.

Vilman, Y. A. (1989). *Fundamentals of Robotization in Construction*. Moscow: Vysshaya Shkola. 271, in Russian.

Warszawski, A. (1984a). Application of robotics to building construction. In Proceedings of the 1st International Symposium on Automation and Robotics in Construction (ISARC), Pittsburgh, USA. doi:10.22260/ISARC1984/0003.

Willmann, J., Knauss, M., Bonwetsch, T., Apolinarska, A. A., Gramazio, F., & Kohler, M. (2016). Robotic timber construction – Expanding additive fabrication to new dimensions. *Automation in Construction*, 61, 16–23. doi: 10.1016/j.autcon.2015.09.011.

Yoshida, T., Ueno, T., Nonaka, M., & Yamazaki, S. (1984). Development of spray robot for fireproof cover work. Proceedings of the 1st ISARC, Pittsburgh, USA. doi: 10.22260/ISARC1984/0006.

16

ROBOTS IN INDOOR AND OUTDOOR ENVIRONMENTS

Bharadwaj R. K. Mantha, Borja Garcia de Soto, Carol C. Menassa, and Vineet R. Kamat

16.1 Aims

- Motivate the need and role of robotics in the successful implementation of Construction 4.0.
- Develop a taxonomy for construction robots.
- Describe the key technical requirements of robotic systems.
- Discuss in detail a case study application regarding a real physical robotic deployment.

16.2 Introduction

The construction industry is an essential driver for the economy of a country and accounts for a considerable amount of its Gross Domestic Product (GDP) (WEF, 2016; UAEMOE, 2017). According to the Global Construction 2030 report (GCP&OE, 2016), the volume of construction output will grow by 85% to $15.5 trillion worldwide by 2030, with China, India and the US leading the way. Given the contributions of this industry to the economy of a country, even small advances and improvements have the potential of industry and worldwide impact, and this is exactly what is expected to happen as the industry trends towards the Fourth Industrial Revolution.

Contrary to the general trend of being resistance to change and under-digitized, construction is progressively making strides towards embracing new technologies such as robotics, big data analytics, blockchain, machine learning, artificial intelligence, 3D printing, the Internet of Things (IoT), virtual reality, and augmented reality to name a few. This is referred to as the Construction 4.0 which is the construction industry's surrogate of Industry 4.0. The aim thus is to have connected systems at every stage in the life cycle of a construction project starting from the bidding phase to the end of life including the operation and maintenance. Experts believe that successful implementation of the same will have the ability to positively impact the whole life cycle of a construction project including initiation, maintenance, and end of life phases. Particularly, researchers and construction professionals suggest that robotics forms an integral part of the Construction 4.0 framework (Oesterreich and Teuteberg, 2016).

A robot can be defined as any actuated system capable of performing tasks or actions for people with a certain degree of autonomy (i.e. programmable in more than two axes) (IFR,

2016). Robots have become increasingly pervasive in our day-to-day lives. Researchers and practitioners suggest that these robotic systems will quickly become omnipresent (Pinto, 2016). This is mainly attributed to the numerous advantages they offer such as a) Productivity improvement: can perform tasks significantly faster without getting tired (e.g. laying bricks) (Sklar, 2015), b) Safety improvement: can work in harsh and unsafe environments where humans are unwilling or unable to work (e.g. gas pipe inspection) (Liu et al., 2011), c) Cost-effectiveness: in the long run, robots are more economical than human counterparts (e.g. performance evaluation of existing old buildings) (Mantha et al., 2018), d) Quality improvement: robots have the capacity of being more precise and accurate than humans (e.g. building structural monitoring and inspection) (Liu et al., 2011). However, there are some challenges associated with the deployment of these robotic systems such as privacy, reliability, and acceptance. Despite these, the demand for these robotic systems is exponentially increasing.

For example, there is a tenfold raise in the venture capital investment in robotics in a span of five years and this still continues to increase (The Robot Report, 2017, 2019). More importantly, the world robotics executive summary released in 2018 estimates around 37 billion USD in sales for the professional service robot installations alone between 2019 and 2021 (IFR, 2018). In addition, a study published by McKinsey shows that the price of these robotics systems continues to drop (almost halved), whereas, the labor costs have consistently increased (Tilley, 2017).

Several different types of robotic systems were studied, explored, suggested, and developed for various construction-related applications. Examples include a brick laying robot (FBR, 2019), a steel beam assembly robot (Jung et al., 2013), a progress monitoring drone (Skycatch, 2019), an as-built modeling robot (Feng et al., 2015), a building performance monitoring robot (Mantha et al., 2018), and many more. Though there exist quite a few number of such prototypes, robots are still not widely represented in the construction industry. This is intuitive given the customized product range, highly unstructured and harsh environments unlike other sectors (e.g. automotive). Thus, there is a strong need to investigate the potential of these systems, especially for the change-resistant construction industry. Several fundamental challenges (e.g. technical, social, and economical) are still yet to be addressed before fully taking advantage of the full potential of these systems. Particularly, investigation of cybersecurity implications and related challenges has received very less attention in the literature (Mantha and Garcia de Soto, 2019a). The overarching goal of this chapter is to briefly discuss history of robotics, underlying concepts, and corresponding construction-domain specific applications. Section 16.3 shows a taxonomy for construction robots based on the Construction 4.0 framework. Section 16.4 elaborates on some of the key fundamental capabilities for robotic systems, and Section 16.5 details a built environment application for autonomous mobile robots.

16.3 Classification of robots

As per the brief definition of the robot discussed previously, it can be understood that the history of robotics is directly or indirectly intertwined with the history of science and technology such as mechanical, electrical, electronic, and computing. Though the idea of robot can be linked back to medieval times, the word was coined only in 1921 by Karel Capek (RobotShop, 2008). After that, the word "robotics" was coined that relates to the technology of robots (RobotShop, 2008). In just a few years, industrial robots gained momentum and significantly stayed at the forefront of robotic technology and innovation. However, due to several reasons such as reluctance to adopt new technologies, the research, and development of construction robotics boomed only in the 1980s in Japan (Kangari and Yoshida, 1989). Experts suggest that with the advancements in computing capabilities and cheaper sensor technologies, research and development activities are currently booming. Numerous prototypes are being developed, and a phenomenal amount of interest and funding is being made due to the progressive acceptance in the construction industry.

Given the complexity of robot and robotic technologies, it is not possible to either define or classify robotic systems in a unique way. To address this, researchers and practitioners resorted to several different variations of descriptions, methods, and techniques. That is, robots can be classified based on several factors such as power source, shape, size, type of application, degrees of freedom, mechanical structure, workspace geometry, motion characteristics, and cognitive ability. For example, a classification based on mechanical structure includes fixed robots (e.g. manipulators) and mobile robots (e.g. iRobot Roomba Vacuum Cleaner) (Reeler, 2017). An example sub classification of mobile robots such as wheeled, tracked, and legged robots is shown in Figure 16.1.

Another classification can be based on the levels of autonomy (i.e. cognitive ability) or human involvement such as tele-operated (e.g. operated or fully controlled by humans), semi-autonomous (partial human involvement), and fully autonomous robots (no human involvement) (Choi et al., 2014). In the current context, robots are classified based on work environment such as indoor and outdoor robots. As the name suggests, indoor robots are those that work in the built environment, and outdoor robots work otherwise.

Furthermore, each of these is categorized based on two of the most significant phases[1] of construction life cycle namely construction and demolition phase (C&D), and operation and maintenance (O&M) phase. Though other life cycle phases such as concept and design consist of automation technologies, specifically the concept of robotics cannot be directly attributed to them. Thus, the classification maps outdoor and indoor robots to C&D and O&M phases respectively. Figure 16.2 shows a further classification of these based on the type of tasks performed by each of these robots.

16.4 Key fundamental capabilities

Any construction robot needs to possess a few fundamental abilities to perform tasks which are agnostic of the type of work environment or task requirement. Four of the key capabilities include, a) Task allocation – ability to optimally divide the desired set of tasks among the available robots (in case of multiple robots); b) Route Planning – ability to plan a route to intended destination location(s); c) Localization – ability to identify the current location in a global or local frame of reference; and d) Navigation – ability to direct itself to the intended destination location(s). Depending on the objective and type of application, each of the robots might have

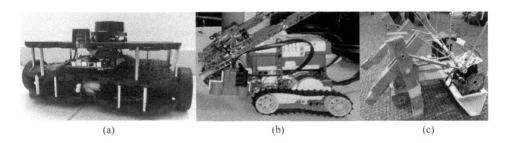

(a)   (b)   (c)

Figure 16.1   Different subtypes of robots namely a) wheeled, b) tracked, and c) legged depending on the mechanical structure of the robot

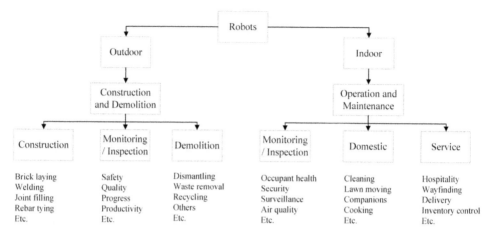

Figure 16.2 Taxonomy of robots based on work environment (i.e. indoor and outdoor), task requirements (e.g. domestic tasks), and construction life cycle phase.

to fulfill some or all of these functionalities. For example, in most cases, a fixed industrial style manipulator robot need not navigate in the given environment. On the contrary, successful task accomplishment of a mobile robot significantly depends on navigation and maneuvering capabilities. These four fundamental capabilities are briefly described in the sub sections below in the context of indoor and outdoor construction robots. Specifically, few of the respective methods and algorithms along with their advantages and limitations are discussed.

16.4.1 Task allocation and route planning

Task allocation means the optimal division of a given set of tasks among the available robots or determination of an optimal number of robots for the desired set of spatiotemporal tasks. That is tasks that need to be performed in different locations in a time-constrained way. Optimized route planning (in case of single and multiple robots) and task allocation (in case of multiple robots) to achieve dynamic goals are not only critical but also challenging especially for multi-level indoor structures because of the complex geometry. A tour can be simply described as a route to visit all the desired locations originating and ending at the same location whereas a path is defined as a route to cover one or multiple destination locations without necessarily returning back to the start location (Bellman, 1962). It becomes computationally intensive to determine the most optimal tour as the geometric complexity of the built environment increases (LaValle and Kuffner, 2001). The problem of finding an optimal tour is well studied. For example, the Travelling Salesman Problem (TSP) would be applicable when a tour is planned for a single traveler (in this study a single robot). Similarly, when the tour is planned for multiple robots, the multiple Travelling Salesman Problem (mTSP) could be used. Both are a variant of the Vehicle Routing Problem (VRP) (Rao and Biswas, 2008; Ponraj and Amalanathan, 2014). A more generalized version of the above mentioned single depot mTSP is called Multi-depot mTSP (MmTSP). As the name suggests, in the case of MmTSP, there exist multiple depots and salesmen (or robots) at each depot. Depending on where the robots return at the end of their tour, MmTSP can be further divided into two types namely fixed destination MmTSP and nonfixed destination MmTSP (Kara and Bektas, 2006). In the former, all the robots start and end their route at their respective depots whereas in the latter, they can start and end their tours at different depots.

An example problem statement for such task oriented robots can be as follows (Mantha and Garcia de Soto, 2019c).

Consider an indoor building network consisting a total of "D" depots (e.g. charging stations or service locations), "x" tasks (distributed across different locations), and "m" robots. The goal is to take advantage of the available resources (i.e. robots) and determine an optimal path to visit the task requirement locations. That is identify a total of "m" tours (one for each robot) to visit all the task requirement locations (i.e. a total of x locations as mentioned above) so that the total distance traveled by all these robots is optimized (i.e. minimized in this context). Since the problem statement considers a fixed destination MmTSP, each of these robots start and end at their respective depots. Representing this mathematically, the term that needs to be optimized is the product of the edge distance and the number of times the respective edge has been visited any of the robots as shown in Equation 16.1 (adapted from Mantha and Garcia de Soto, 2019c). The edges here refer to the links (e.g. corridors, stairs, and elevators) formed based on the graphical network generated from the indoor building environment (e.g. floor plan). Further details regarding the graph network generation can be found in Mantha and Garcia de Soto (2019c).

$$\sum_{a=0}^{n}\sum_{b=0}^{n}\sum_{c \in D} \alpha_{abc} d\{N_a, N_b\}$$

(16.1)

Where $\alpha_{abc} \geq 0$ is the total number of times the edge between nodes N_{ai} and N_b is visited by all the robots. c is one of the D depots. is the respective distance between the nodes N_a and N_b. The total number of vertices or commonly referred to as nodes in the indoor graph network are represented by n. There can be other variants for this problem statement by varying factors such as number of base and destination depots.

The determination of the most optimal solution is not only tedious but also costly due to a polynomial time computational complexity. For example, the algorithm takes k steps (polynomial function of n) to determine the most optimal path for a network consisting of n nodes (Gorbenko et al., 2011). As the number of nodes increase, the computational complexity increases exponentially. An example of a graphical network representation is shown in Figure 16.3.

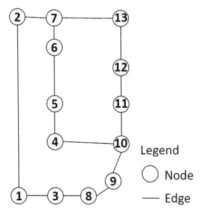

Figure 16.3 Graphical node network representation consisting of 13 nodes and 14 edges

In an aim to address the above mentioned issue, researchers attempted to determine the near optimal solution which is as close as possible to the most optimal solution. For instance, several alternative methods such as a heuristic approach were suggested for Unmanned Aerial Vehicles (UAVs) route planning particularly focusing on the outdoor environments. These methods cannot be directly adopted for indoor environments because the networks formed based on the indoor environments are incomplete due to access restrictions (e.g. presence of walls which restrict the direct access between two locations) unlike outdoor environments (Ryan et al., 1998; Avellar et al., 2015; Yakici, 2016; Coelho et al., 2017).

In another study conducted by Yu et al. (2002), an approach based on Genetic Algorithm (GA) was proposed to address the autonomous mobile robot route planning. Similarly, Gorbenko et al. (2011) suggested a route planning approach for indoor service robots and showed that it is an Nondeterministic Polynomial (NP) complete optimization problem. Both these approaches assume that each robot is accountable for a distinctive Hamiltonian path. A network is said to consist of a Hamiltonian path if each and every node can be visited exactly once without revisiting a node again. However, this might not be possible for networks formed based on indoor environments where situations like obstructions or end of hallways might force the robot to retrace some of the path already traversed and revisit some nodes along the way. Mantha et al. (2017) addressed some of these limitations and proposed an optimization approach to solve single-depot mTSP. That is, all the robots start and end their respective tours at one single identified depot. However, Mantha and Garcia de Soto (2019c) extended this further and proposed an iterative optimization algorithm to solve the fixed distance MmTSP. That is, all the robots start and end their tours at distinct identified or determined depots.

16.4.2 *Localization and navigation*

In the current context, localization is defined as the robot's ability to identify its current location in a given global frame of reference (Levitt and Lawton, 1990). For example, a robot being able to recognize its current location to be in room 201, or knowing its location and orientation in the global coordinate reference system. The robot's navigation can be briefly defined as the robot's ability to plan a course of action to reach the destination location while accurately localizing itself in its frame of reference at strategic locations (Levitt and Lawton, 1990).

Based on the present context, robot localization can be briefly described as the ability of any given robot to detect (or recognize) its existing location in the global reference frame (Levitt and Lawton, 1990). That is, the ability of a robot to identify the real-time (or strategic time based) location including position (e.g. in front of room 207) and orientation (e.g. headed north). Furthermore, navigation of the robot can be described briefly as the capacity of the robot to formulate an action plan to visit the target destinations of interest while being able to successfully localize at the strategic locations (Levitt and Lawton, 1990).

Several indoor and outdoor localization and navigation techniques have been explored previously. Literature suggests that every method has advantages and limitations. Some of the previous approaches explored for robot localization include Global Positioning System (GPS), Wireless Local Area Network (WLAN), Radio Frequency Identification (RFID), Ultra-Wide Band (UWB), Inertial Measurement Unit (IMU), Bluetooth, cameras, and lasers. Though GPS works better outdoors, it suffers from accuracy and Line Of Sight (LOS) problems indoors (Khoury and Kamat, 2009). Alternatively, Wi-Fi, Bluetooth, and RFID based localization were explored for indoor localization. Wi-Fi is an economical solution because most of the existing infrastructure consists of wireless nodes required for localization. However, it suffers from a significant error in localization accuracy (Torres-Solis et al., 2010; Montañés et al., 2013). Bluetooth and RFID based localization tends to be expensive, time-consuming and also have space constraints because of the requirement of wireless infrastructure deployment indoors (Raghavan et al., 2010). Similarly, UWB based systems require a large number of receivers making it inconvenient and infeasible (due to space constraints) indoors (Montañés et al., 2013).

Laser scanner and natural marker (camera) based techniques eliminate the need to instrument the physical space but they are highly expensive, sensitive to obstructions and lighting conditions and require high computational capabilities (Burgard et al., 1998; Bar-Shalom et al., 2001; Thrun et al., 2005; Habib, 2007; Feng and Kamat, 2012). To summarize, common disadvantages affecting a majority of the reviewed methods include low accuracy, significant upfront investments, high computational requirements, and complex instrumentation of the indoor environment.

One of the vision-based methods using fiducial markers, however, is particularly immune to the disadvantages mentioned above afflicting other methods. Fiducial markers (Olson, 2011), as shown in Figure 16.4, offer high accuracy in determining and estimating their relative 3D pose in an environment, require relatively less computing capabilities, are cost-effective and are easy to install (Iwasaki and Fujinami, 2012).

In addition, fiducial markers can store virtual information regarding a multitude of things such as information regarding physical location (floor and room level information), emergency evacuation directions, indoor navigational information, and inspection related data regarding building systems helpful for facility managers (Feng and Kamat, 2012). Feng and Kamat (2012) have demonstrated how markers having virtual information and navigational directions can help humans navigate indoors. To take this further, Mantha et al. (2018) utilized the virtual location information (for localization), navigational direction (for navigation), and 3D pose estimates (for drift correction) to achieve autonomous behavior of the mobile indoor robot. Furthermore, these same markers were used for achieving autonomous navigation of excavators with the help of multiple camera marker networks (Feng et al., 2015). Generalizing the above mentioned approaches, Mantha and Garcia de Soto (2019b) developed a framework in the form of a process flowchart to systematically design a reliable fiducial marker network for achieving autonomous robot navigation. That is, to design the desired robot, determine required sensors, and create the optimal marker network map. A brief overview of this process is shown in Figure 16.5, and described in the sub section below based on the excerpts from Mantha and Garcia de Soto (2019b).

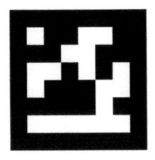 

Figure 16.4   Two of the AprilTags from 36h11 series

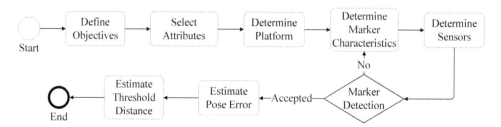

Figure 16.5   Overview of the marker network map design process for autonomous indoor robotic navigation (adapted from Mantha and Garcia de Soto, 2019b)

16.4.2.1 *Marker network map design process*

Define objectives – It is assumed that the facility manager is responsible for defining the objectives pertaining to the application context and the corresponding inputs. The objectives can be derived from the targeted action or intended tasks that need to be performed. For example, consider an indoor environmental monitoring robot. The intended action is to gather data regarding ambient parameters such as temperature, relative humidity, and lighting, and compare it with the standard parameter range or occupant preferences. Other relevant inputs can be the locations in the building and the frequency at which the data needs to be collected.

Select attributes – Attributes primarily represent the physical properties and characteristics of the built environment such as facility type, floor plans, equipment, surface geometry, flooring type, thermal zones, lighting, acoustics, and other services. In the ambient parameter monitoring example discussed, some of the relevant attributes can be indoor temperature, relative humidity, light intensity, sound levels, and indoor air quality.

Determine platform – As the name suggests, the goal of this step is to determine the ideal robotic platform based on the defined objectives. As discussed previously, the classification of robots can be dependent on several things such as the type of work, mechanical structure, or morphology (ISO, 2012). The goal is to determine the mechanical structure of the robot whether it is a mobile robot with wheels, legs, wings, tracks, or any other existing automated platform. For an occupant comfort monitoring and data collection application example, a mobile robot with wheels and a display platform for interaction might be ideal. Though a legged locomotion-based robot might also serve the purpose, a mobile robot with wheels might outpace the legged robots and can be more efficient (faster) in collecting data.

Determine marker characteristics – Marker characteristics refer to the type, size, library size, and placement of the markers. Several different types of markers were developed and studied by researchers such as planar markers, 2D bar codes, ARToolKit, BinARyID, AprilTags, ArUco, and ChiliTags (Knyaz and Sibiryakov, 1998; Olson, 2011; Bonnard et al., 2013; Romero-Ramirez et al., 2018). Some of these markers are shown in Figure 16.6. The marker library (or dictionary) size is the number of unique markers available in chosen given marker type. For example, AprilTags have more than 4,000 unique codes, whereas ARTags have about 2,000 (Olson, 2011). Different marker placement techniques such as wall mounted, ceiling mounted, and floor mounted have been explored for built environment settings (Röhrig et al., 2012; Shneier and Bostelman, 2015; Mantha et al., 2018). However, it has to be noted that each of these methods has its own advantages and disadvantages depending on the built environment attributes such as ceiling heights, occlusions, and occupant actions such as occupant movement.

ARToolKit   2D Bar Code   AprilTag

Figure 16.6   Example representation of a few different markers (adapted from Mantha and Garcia de Soto, 2019b)

Determine sensors – The type of sensors required by any robot can be broadly divided into two categories. The first set of sensors which are responsible for the purpose of achieving the key technical capabilities such as localization, navigation, pose estimation, and drift correction. The second set of sensors required on the robot are directly related to the type of the application. Generally, marker-based pose estimation systems (vision based) offer several advantages compared to other approaches as that they do not require expensive cameras. The type of cameras available on mobile phones these days can be used. Typically, most of the robotic systems come with a built-in camera. These cameras are of sufficient quality to be used for the purpose of marker-based localization and pose estimation (Romero-Ramirez et al., 2018).

Marker detection – Though the detection algorithm varies from one type of marker to another, a general algorithmic process is as follows. First, images are captured at a very high rate and are analyzed for the presence of a marker. This process is called segmenting. Second, the computer decodes the information from the markers in the form of 1s and 0s and determines the unique identification of the marker by cross-referencing (matching) with the library of markers.

Estimate pose error – One of the most important factors that have a direct impact on whether the robot will successfully navigate to desired locations or not, is the relative pose error accumulation of the robot. This is because there is always a difference between the robot's actual and ideal pose. It is important to estimate this at regular intervals and rectify it accordingly. Relative pose between camera and marker can be determined by a total of six components, three of which correspond to translation (e.g. x, y, and z) and three of which correspond to the rotation (e.g. roll, pitch, and yaw angles). However, depending on the type of application, only some of these six components might be required. For example, in case of ambient data collection, only the lateral distance (distance between the camera and the marker) is extracted and hence the pose error is estimated only using the variance of this parameter with respect to the ground truth values.

Estimate threshold distance – At this stage, the robot's camera can detect markers (landmarks), localize itself in the built environment, and navigate based on the relative pose. Since there will be errors accumulated along the navigational path, it is necessary to place the markers at strategic locations to rectify the errors and ensure the robot reaches its next intended destination without drifting too much (e.g. colliding with the wall). So, in addition to placing markers at the locations of interest, additional markers need to be placed along the way. The objective of this specific task is to determine the threshold distance (d_{th}) between any two successive markers. That is, determine the maximum distance between two consecutive markers along the navigational path of the robot.

16.5 Case study application

The objective of this section is to discuss a case study performed by Mantha et al. (2018) where actionable building information (i.e. ambient data such as temperature, humidity, and lighting which represent the building state) is gathered using a mobile agent (i.e. robot) based data collection. To achieve this, a mobile robot should be able to navigate to various strategic locations in a building space within a stipulated time frame to get a valid snapshot of the building's state. Data is collected and stored with the help of onboard sensors and a computer on the mobile robot. Since there is only one robot and the user defines the path, task allocation and path planning do not need to be performed. Therefore, localization and navigation are the only relevant key technical capabilities discussed in section 16.4.

The following sections provide a brief overview of the proposed technical approach relating to the design of the robot, localization/navigation algorithms, and data collection. The case study presented is a compilation of excerpts from one of the authors' previous studies (Mantha et al., 2018). The following is a condensed version for easier understanding of the reader.

16.5.1 Ambient data collection in buildings using autonomous mobile indoor robots

Recent studies have focused on autonomous indoor robots for achieving various tasks such as assisting elderly people (e.g. the health care industry) (Linner et al., 2015), butler robots (e.g. the hospitality industry) (Boltr, 2014), and helping visitors to navigate indoor environments (e.g. in large commercial buildings) (Biswas and Veloso, 2010). Though autonomy is common in all the robots mentioned above, applications, sensors used, actuation mechanisms, and the respective algorithms differ amongst each other. For example, Linner et al. (2015) use Simultaneous Localization and Mapping (SLAM) for navigation whereas Biswas and Veloso (2010) utilize the Wi-Fi infrastructure for localization and navigation. This case study proposes landmark based localization and navigation which uses fiducial markers (AprilTags) developed by Olson (2011) as landmarks for autonomous mobile robotic data collection process. The aim is to achieve autonomous navigation capabilities with sparse localization instead of continuous localization as opposed to the previous studies. Few of the factors that need consideration and additional investigation are the type of data to be collected (accordingly the type of sensors to be placed on the mobile robot), the frequency of data collection (how frequently the data needs to be collected at every location), the waiting time at each location where data is collected, the number of mobile robots required to monitor the entire building (depending on the size of the buildings), and the optimization of the travel time and path.

16.5.1.1 Design of the mobile robot

The robotic platform, equipped with the iCreate base (widely known as TurtleBot) was chosen as the mobile data collection platform and sensors such as Cozir® CM 0199 (for temperature,

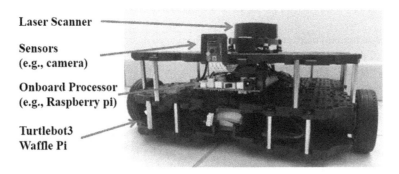

Laser Scanner

Sensors
(e.g., camera)

Onboard Processor
(e.g., Raspberry pi)

Turtlebot3
Waffle Pi

Figure 16.7   Components of the mobile robot used for ambient data collection in buildings

humidity, and CO2 levels), HOBO U12 (for light and occupancy levels), Lutron (for natural light levels), and NinjaBlocks (for airspeed) were used for the data collection. Figure 16.7 shows the mobile robot with some of the following components, 1) TurtleBot – for navigating the indoor environment; 2) Onboard computing – to communicate with the TurtleBot; 3) Camera – for the TurtleBot to detect fiducial markers, localize, and estimate its relative pose in the indoor environment; 4) Remote laptop – to execute the corresponding autonomous navigation programs wirelessly; and 5) Sensors – for monitoring and data collection of various ambient parameters.

In order for the mobile robot to autonomously navigate indoors and ambient data, it needs to: 1) localize in the indoor environment, 2) navigate to the intended data collection locations, 3) collect the respective data (such as temperature, humidity, and light intensity), and 4) geotag (record the collected data with the physical location) for further analysis. Though the steps are sequential, it is an iterative loop which signifies that the data collection task needs to be performed for all the required locations in the building. That is, after the data is geotagged at a particular location, the robot continues its journey to the next data collection location. Each of these steps and the associated algorithms is described below.

16.5.1.2 Localization

The general process of fiducial marker based localization is described below including setting up the robot (e.g. installing the sensors on the robot) and environment (e.g. installing the markers in the building).

INITIAL SETUP

First, all the required software packages such as Robotic Operation System (ROS), ROS developer kit, TurtleBot software, and network connectivity need to be installed on the mobile robot's netbook and remote laptop or wireless controller. Second, a bidirectional communication has to be established between netbook and the remote laptop and then the netbook is placed on the iCreate base. Finally, the remote laptop is connected to the mobile robot's netbook with the help of Secure Socket Shell (SSH) connection in the terminal. More information regarding detailed installation, software packages, robot hardware, and sensor driver instructions can be found in Quigley et al. (2009).

MARKER PLACEMENT

For this study, unique fiducial markers are required to be placed at strategic locations along the navigational path of the robot (e.g. corridors, entrances to rooms, etc.) as shown in Figure 16.8 to form a Marker Network Map (MNM). These strategic locations include examples such as the end of the corridors, the intersection of hallways, elements that allow the transfer of users from one floor to another (e.g. staircase and elevator), and entrances to rooms. The MNM can then be used to generate a graphical node network G = {N, E} where N includes nodes representing locations of markers and E represents edge links connecting these nodes (e.g. corridors and stairs). Each edge can have edge weight attributes that can represent quantities such as distance or time. The network formed henceforth can be used to determine the optimal paths in the building or reroute in case of maintenance or emergency situations.

MARKER DETECTION

As discussed in section 16.4.2.1 markers are detected with the help of cameras and underlying vision based algorithms. Then, virtual information stored in the marker along with relative pose information is extracted and interpreted (Olson, 2011; Feng and Kamat, 2012).

OTHER CONSIDERATIONS

Three of the important and interrelated factors which need attention are the density of the makers (that needs to be deployed in the environment), the distance between the wall and the robot's ideal path, and the drift accumulation. For example, if the robot's ideal path is closer

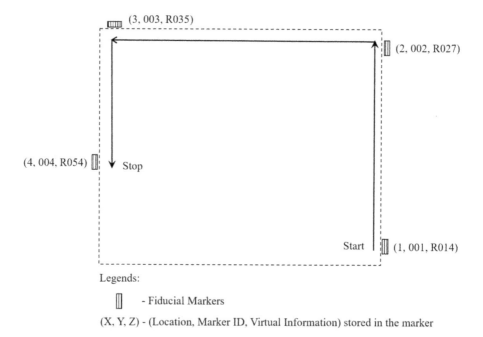

Figure 16.8   Marker Network Map (MNM) with virtual information regarding the location stored in each of the markers (adapted from Mantha et al., 2018)

to the wall, the allowable drift accumulation will be significantly less (compared to the robot's path being on the centerline of the corridor) to avoid the robot colliding with the wall. This requires a more frequent drift correction. This will result in the need for a higher density of markers (i.e. additional markers) to help in estimating the drift. The number of markers that needs to be instrumented in the building increases with the density of the marker network. To optimize the density of the marker network, the allowable drift should be maximized. For indoor building conditions, allowable drift can be maximized if the ideal path of a robot follows the centerline of the corridor (discussed in more detail in the drift correction sub sections in navigation section 16.5.1.3 of this case study).

16.5.1.3 Navigation

The overview of the navigational algorithm logic is represented as a flowchart in Figure 16.9. First, the onboard traditional RGB camera continuously captures images that might potentially contain a known fiducial marker. The images are processed by the marker recognition module (an algorithm which detects the presence of the marker as previously discussed in the technical approach section of the case study). If a known marker is detected by the marker recognition module, the ID and relative pose of the robot with respect to the fiducial marker (in the camera's reference frame) is outputted by the module. Each ID is associated with a physical location in the indoor environment as shown in Figure 16.8. The current location of the robot based on the information mentioned above is estimated and pose correction is calculated based on the drift correction algorithms discussed below. The current approach of navigation can be termed as treasure-hunt-based navigation because the robot traverses the path from one marker to another marker with the help of clues provided at each marker. This means that the robot will follow the previously known navigational instruction (or clue provided by the last seen marker) until it finds a new marker. To avoid the cases of obstruction (robot not being able to see one or two markers in the navigation path), it can be easily programmed to store information regarding next couple of markers and still traverse the navigational path accordingly. Previous studies have not utilized 3D relative pose information and virtual directional information from fiducial markers for indoor robot navigation, thus the presented method contributes to advancement to the field.

DRIFT CORRECTION

One of the most important factors that have a direct impact on whether the robot will traverse the entire path is drift accumulation of the robot. Drift is defined as the distance between the actual location and the ideal location of the robot at any given point of time. Since the drift is corrected at each marker location along the navigation path, the maximum drift in the entire path is guided by the marker to marker distance. Marker to marker distance is the distance between any two consecutively placed markers in the fiducial marker network. If the robot's ideal path is along the centerline in a building's corridor, then the maximum allowable drift in any direction (+ve or − ve) has to be less than half the width of the corridor as shown in Figure 16.10. This is to also account for the size of the robot itself.

For example, the robot might collide with the wall before reaching the next marker if it drifts an amount of w/2 on either side of the robot's ideal path. Thus, a confined path which is safe for the robot navigation (i.e. it would not collide with the walls or lose track of the markers during its path) has to be determined based on the width of the corridor mainly because of the possible errors in estimation of robot's current pose, accumulated drift, and turning angle.

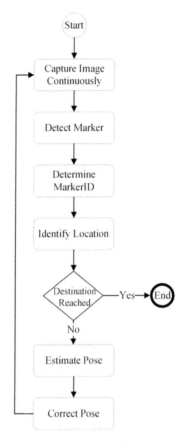

Figure 16.9 Autonomous indoor treasure hunt based navigation algorithm with the help of a network of fiducial markers (adapted from Mantha et al., 2018)

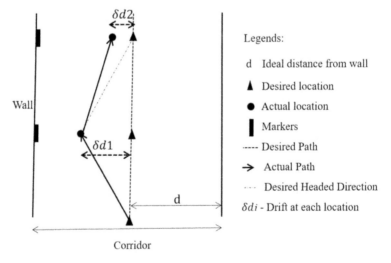

Figure 16.10 Illustration of drift accumulation and marker to marker distance for an indoor mobile robot (adapted from Mantha et al., 2018)

As discussed, the robot's drift needs to be corrected at every marker location to reach the targeted destination location successfully. The current drift correction algorithm proposed in this case study is based on the relative pose information estimated with the help of the onboard camera, computer, the known marker's pose, and the known location in the fiducial markers network as per the algorithm discussed in the localization section. The algorithm returns 3D relative pose information (as shown in Equation 16.2) in the camera reference frame with respect to the marker. The entire matrix is referred to as a homogenous transformation matrix in which R (3*3) denotes rotation matrix, and T (3*1) denotes translation matrix. From that, the lateral distance information (distance between the camera and marker), d as shown in Figure 16.10, is extracted and the drift at each location ($\delta d\ i$) is estimated. After that, the pose correction (turning angle (αi) in this case) is calculated with the help of few trigonometric expressions and equations. Further details regarding the equations can be found in Mantha et al. (2018). Since the drift at each location might be different, the turning angle at each location also differs. Although Feng and Kamat (2012) used the relative pose estimate to determine navigational direction, estimating and correcting drift has not been addressed in prior related work.

$$H = \left[R_{3*3}\ T_{3*1} \right]_{3*4}$$

$\qquad\qquad\qquad\qquad\qquad\qquad\qquad\qquad\qquad\qquad\qquad\qquad$ (16.2)

Where:

H is the part of Homogeneous transform matrix returned by the marker detection algorithm. R_{3*3} is the rotation matrix which consists of nine elements and T_{3*1} consists of three elements encompassing the three-dimensional relative rotational and translational information.

16.5.1.4 Data collection

The immediate next step once the robot is cognizant of its surroundings and capable of navigating indoors is collecting data. Data collection is the process of gathering required information regarding various parameters of interest in a timely or systematic manner for testing a hypothesis or analyzing a research problem (NIU, 2016). The traditional way of collecting data is with the help of multiple sets of the same sensors in the areas/rooms/locations of interest. Whereas, this case study uses a single set of onboard sensors and a mobile platform to gather data at all the areas/rooms/locations of interest. In the context of this case study, ambient parameters such as temperature, humidity, indoor air quality, and light intensity are collected in buildings. The timely manner of data collection can be referred to as the frequency of data collection which is a function of the robot speed and the required amount of data from a particular location.

16.5.1.5 Geotagging

Geotagging is the process of associating the information gathered (data) with the corresponding location such as latitude, longitude, specific name, or a specific location. In this study, geotagging is associating the ambient data collected along with its location information such as room number and floor number. To achieve this, a programmable interface is developed which bridges the communicated physical location information (as given by the fiducial markers) with the sensor data (obtained from the data collection). A Python program is written which subscribes the published ROS data regarding the location of the robot (given by the fiducial marker as discussed in the localization section), concatenates it with the retrieved sensor data along with the time stamp, and exports the data to an Excel file locally stored in the onboard netbook. This data can be stored in a local server or it can be updated to an online big data set interface for further real-time processing/analysis or can be stored as a data repository.

16.6 Conclusion

After a long history of under-digitization and resistance to change, construction is progressively moving towards adopting new technologies such as automation and robotics. Robotics offers significant potential to address some of the significant challenges faced in construction relating to productivity, safety, quality, cost, and time. Based on the characteristics of the work environment, construction robots can be classified into indoor and outdoor robots which directly correspond to the O&M and C&D phases of construction life cycle respectively. Any construction robot at least needs to have four fundamental capabilities such as task allocation, route planning, localization, and navigation for any given application. This is demonstrated with the case study example application namely autonomous mobile robotic data collection. Several other concepts and prototypes have been explored, investigated, and studied by researchers and practitioners. However, there lacks a widespread deployment of these systems especially for construction applications. This is possibly due to several reasons such as educational (e.g. lack of awareness), economical (e.g. lack of investment), social (e.g. fear of losing jobs), technological (e.g. lack of trust), and many more. Thus, professionals from industry and academia need to collaborate closely to further investigate the reasons and bridge this gap to make construction robots a reality. While addressing these gaps, it is also of paramount importance to address cybersecurity-related challenges related to the adoption of Construction 4.0. That is, to identify cyber vulnerabilities, investigate risks, develop threat models, and improve the safety and security of our construction sites and built infrastructure systems.

16.7 Summary

- The role of robots in the successful implementation of Construction 4.0 is discussed, motivated, and established.
- A taxonomy for construction robots is developed based on the work environment (i.e. indoor and outdoor) and application type. Each of these is categorized based on the use of these systems in different phases of construction.
- Four fundamental capabilities of robotic systems namely task allocation, route planning, localization, and navigation are described in detail.
- A real case study implementation of an autonomous mobile robotic to collect data inside buildings is discussed in detail.

Notes

1   These are the most significant phases in terms of time, cost, and resources spent compared to the other phases in the life cycle of a construction project (UNEP, 2007).

References

Avellar, G., Pereira, G., Pimenta, L., and Iscold, P. (2015). "Multi-UAV routing for area coverage and remote sensing with minimum time." *Sensors*, *15*(11), 27783–27803. doi:10.3390/s151127783.

Bar-Shalom, Y., Li, X. R., and Kirubarajan, T. (2001). "Estimation with applications to tracking and navigation: Theory algorithms and software." John Wiley and Sons, New York, doi.10.1002/0471221279.

Bellman, R. (1962). "Dynamic programming treatment of the travelling salesman problem." *Journal of the ACM, 9*(1), 61–63. doi:10.1145/321105.321111.

Biswas, J. and Veloso, M. (2010). "Wifi localization and navigation for autonomous indoor mobile robots." *In Robotics and Automation (ICRA), 2010 IEEE International Conference on* (pp. 4379–4384). IEEE. doi:10.1109/ROBOT.2010.5509842

Boltr. (2014). "SaviOne robot butler starts room service deliveries this week." www.savioke.com/ Accessed: 04/12/2015.

Bonnard, Q., Lemaignan, S., Zufferey, G., Mazzei, A., Cuendet, S., Li, N., Özgür, A., and Dillenbourg, P. (2013). "Chilitags 2: Robust fiducial markers for augmented reality and robotics" http://chili.epfl.ch/software.

Burgard, W., Cremers, A. B., Fox, D., Hähnel, D., Lakemeyer, G., Schulz, D., Steiner, W., and Thrun, S. (1998). "The interactive museum tour-guide robot." *AAAI*, Madison, Winsconsin, USA, 11–18.

Choi, J. J., Kim, Y., and Kwak, S. S. (2014). "The autonomy levels and the human intervention levels of robots: The impact of robot types in human-robot interaction." *IEEE International Symposium on Robot and Human Interactive Communication* (pp. 1069–1074). doi:10.1109/ROMAN.2014.6926394.

Coelho, B., Coelho, V., Coelho, I., Ochi, L., Zuidema, D., Lima, M., and Da Costa, A. (2017). "A multi-objective green UAV routing problem." *Computers & Operations Research, 88*, 306–315. doi:10.1016/j.cor.2017.04.011.

FBR – Fast Brick Robotics. (2019). "Robot construction is here." www.fbr.com.au/view/hadrian-x Accessed: 18/02/2019.

Feng, C. and Kamat, V. R. (2012). "Augmented reality markers as spatial indices for indoor mobile AECFM applications." *In Proceedings of 12th International Conference on Construction Applications of Virtual Reality (CONVR 2012)* (pp. 224–235).

Feng, C., Xiao, Y., Willette, A., McGee, W., and Kamat, V. R. (2015). "Vision guided autonomous robotic assembly and as-built scanning on unstructured construction sites." *Automation in Construction, 59*, 128–138. doi:10.1016/j.autcon.2015.06.002.

GCP & OE – Global Construction Perspectives and Oxford Economy. (2016). "Global construction market to grow $8 trillion by 2030: Driven by China, US and India." www.ice.org.uk/ICEDevelopmentWebPortal/media/Documents/News/ICE%20News/Global-Construction-press-release.pdf Accessed: 18/02/2019.

Gorbenko, A., Mornev, M., and Popov, V. (2011). "Planning a typical working day for indoor service robots." *IAENG International Journal of Computer Science, 38*(3), 176–182.

Habib, M. K. (2007). "Robot mapping and navigation by fusing sensory information." *INTECH Open Access Publisher*.

IFR. (2016). "Introduction in to service robots." https://ifr.org/img/office/Service_Robots_2016_Chapter_1_2.pdf Accessed: 7/3/2019.

IFR. (2018). "Executive survey world robotics 2018 service robots." https://ifr.org/downloads/press2018/Executive_Summary_WR_Service_Robots_2018.pdf Accessed: 10/03/2019.

ISO. (2012). Robots and robotic devices — Vocabulary iso.org Accessed: 29/01/2019.

Iwasaki, M. and Fujinami, K. (2012). "Recognition of pointing and calling for industrial safety management." *In Proceedings of the 2012 First ICT International Senior Project Conference and IEEE Thailand Senior Project Contest* (pp. 50–53).

Jung, K., Chu, B., and Hong, D. (2013). "Robot-based construction automation: An application to steel beam assembly (Part II)." *Automation in Construction, 32*, 2013 62–79. doi:10.1016/j.autcon.2012.12.011.

Kangari, R. and Yoshida, T. (1989). "Prototype robotics in construction industry." *Journal of Construction Engineering and Management, 115–2*. doi:10.1061/(ASCE)0733-9364(1989)115:2(284).

Kara, I. and Bektas, T. (2006). "Integer linear programming formulations of multiple salesman problems and its variations." *European Journal of Operational Research, 174*(3), 1449–1458. doi:10.1016/j.ejor.2005.03.008.

Khoury, H. M. and Kamat, V. R. (2009). "Evaluation of position tracking technologies for user localization in indoor construction environments." *Automation in Construction, 18*(4), 444–457. doi:10.1016/j.autcon.2008.10.011.

Knyaz, V. A. and Sibiryakov, R. V. (1998). "The development of new coded targets for automated point identification and non-contact surface measurements, 3D surface measurements." *IAPRS*, *23*(5), 80–85.

LaValle, S. M. and Kuffner, J. J. (2001). "Randomized kinodynamic planning." *The International Journal of Robotics Research*, 20(5), 378–400. doi:10.1177/02783640122067453.

Levitt, T. S. and Lawton, D. T. (1990). "Qualitative navigation for mobile robots." *Artificial Intelligence*, *44*(3), 305–360. doi:10.1016/0004-3702(90)90027-W.

Linner, T., Güttler, J., Bock, T., and Georgoulas, C. (2015). "Assistive robotic micro-rooms for independent living." *Automation in Construction*, *51*(2015), 8–22. doi:10.1016/j.autcon.2014.12.013.

Liu, B., Luk, F., and Tong Y. C. (2011). "Application of service robots for building NDT inspection tasks." *Industrial Robot: An International Journal*, 38(Issue 1), 58–65. doi:10.1108/01439911111097850.

Mantha, B. and Garcia de Soto, B. (2019a). "Cyber security challenges and vulnerability assessment in the construction industry." In *Proceedings of the 2019 Creative Construction Conference*. 29 June–2 July 2019, Budapest, Hungary.

Mantha, B., Menassa, C., and Kamat, V. (2018). "Robotic data collection and simulation for evaluation of building retrofit performance." *Automation in Construction*, *92*, 88–92. doi:10.1016/j.autcon.2018.03.026.

Mantha, B. R. and Garcia de Soto, B. (2019b). "Designing a reliable fiducial marker network for autonomous indoor robot navigation." In *Proceedings of the 36th International Symposium on Automation and Robotics in Construction (ISARC)*. May 21–24, 2019, Banff, AB, Canada. doi:10.22260/ISARC2019/0011.

Mantha, B. R. and Garcia de Soto, B. (2019c). "Task allocation and route planning for robotic service networks with multiple depots in indoor environments." In *ASCE International Conference on Computing in Civil Engineering (i3CE)*. 17–19 June 2019, Georgia Institute of Technology, Atlanta, Georgia, USA (2019). doi: 10.1061/9780784482438.030.

Mantha, B. R., Menassa, C., and Kamat, V. (2017). "Task allocation and route planning for robotic service networks in indoor building environments." *Journal of Computing in Civil Engineering*, *31*(5), 04017038. doi:10.1061/(ASCE)CPP.1943-5487.0000687.

Montañés, J., Rodríguez, A., and Prieto, I. S. (2013). "Smart indoor positioning/location and navigation: A lightweight approach." *IJIMAI*, *2*(2), 43–50.

NIU (2016). "Responsible conduct of research" https://ori.hhs.gov/education/products/n_illinois_u/dfront.html Accessed 25/04/2016.

Oesterreich, T. D. and Teuteberg, F. (2016). "Understanding the implications of digitisation and automation in the context of Industry 4.0: A triangulation approach and elements of a research agenda for the construction industry." *Computers in Industry*, *83*, 121–139. doi:10.1016/j.compind.2016.09.006.

Olson, E. (2011). "AprilTag: A robust and flexible visual fiducial system." *Proceedings of IEEE International Conference on Robotics and Automation*, pp. 3400–3407. 10.1109/ICRA.2011.5979561.

Pinto, J. (2016) "Intelligent robots will be everywhere" automation.com Accessed: 09/01/2019.

Ponraj, R. and Amalanathan, G. (2014). "Optimizing multiple travelling salesman problem considering the road capacity." *Journal of Computer Science*, *10*(4), 680. doi:10.3844/jcssp.2014.680.688

Quigley, M., Conley, K., Gerkey, B., Faust, J., Foote, T., Leibs, J., and Ng, A. Y. (2009). "ROS: An open-source Robot Operating System." *ICRA Workshop on Open Source Software*, *3*(3.2), 5.

Raghavan, A. N., Ananthapadmanaban, H., Sivamurugan, M. S., and Ravindran, B. (2010). "Accurate mobile robot localization in indoor environments using Bluetooth." In *Robotics and Automation (ICRA), 2010 IEEE International Conference on* (pp. 4391–4396). IEEE. doi:10.1109/ROBOT.2010.5509232.

Rao, J. and Biswas, S. (2008). "Joint routing and navigation protocols for data harvesting in sensor networks." *IEEE Conference on Mobile Ad Hoc and Sensor Systems* (pp. 143–152). doi:10.1109/MAHSS.2008.4660041.

Reeler. (2017). "Analysis of existing robots classifications." http://reeler.eu/fileadmin/user_upload/REELER/Annex_3_Analysis_of_existing_robots_classifications.pdf Accessed: 09/01/2019.

RobotShop. (2008). "History of robotics: Timeline" www.robotshop.com/media/files/pdf/timeline.pdf Accessed: 11/03/2019.

Röhrig, C., Kirsch, C., Lategahn, J., Müller, M., and Telle, L. (2012). "Localization of autonomous mobile robots in a cellular transport system." *Eng Lett*, *20*, 2.

Romero-Ramirez, F., Muñoz-Salinas, R., and Medina-Carnicer, R. (2018). "Speeded up detection of squared fiducial markers." *IVC*, 38–47 (46). doi:10.1016/j.imavis.2018.05.004.

Ryan, J. L., Bailey, T. G., Moore, J. T., and Carlton, W. B. (1998), "Reactive tabu search in unmanned aerial reconnaissance simulations." *Winter Simulation Conference*, 873–880. doi: https://doi.org/10.1109/WSC.1998.745084.

Shneier, M. and Bostelman, R. (2015). "Literature review of mobile robots for manufacturing, *NIST*." nist.gov Accessed: 23/01/2019.

Sklar, J. (2015). "Robots lay three times as many bricks as construction workers." www.technologyreview.com/s/540916/robots-lay-three-times-as-many-bricks-as-construction-workers/ Accessed: 09/01/2019.

Skycatch. (2019). "Explore1 high accuracy drone." www.skycatch.com/solution/hpp/explore1/ Accessed: 18/02/2019.

The Robot Report. (2017). "2016 best year ever for funding robotics startup companies." www.therobotreport.com/2016-best-year-ever-for-funding-robotics-startup-companies/ Accessed: 09/01/2019.

The Robot Report. (2019) "Robotic investments February 2019 – Demaitre, E." www.therobotreport.com/february-2019-robotics-investments/ Accessed: 07/03/2019.

Thrun, S., Burgard, W., and Fox, D. (2005). *Probabilistic Robotics*. MIT Press, Cambridge, MA.

Tilley. (2017) "Automation, robotics and the factory of the future" www.mckinsey.com/business-functions/operations/our-insights/automation-robotics-and-the-factory-of-the-future Accessed: 09/01/2019.

Torres-Solis, J., Falk, T. H., and Chau, T. (2010). "A review of indoor localization technologies: Towards navigational assistance for topographical disorientation." *Ambient Intelligence*, 51–84. 2010. doi:10.5772/8678.

UAEMOE – United Arab Emirates Ministry of Economy. (2017) "Annual economic report 2017" www.economy.gov.ae/EconomicalReportsEn/MOE%20Annual%20Report%202017_English.pdf Accessed: 18/02/2019.

United Nations Environment Programme (UNEP). (2007). "Buildings can play a key role in combating climate change." www.sciencedaily.com/releases/2007/04/070407150947.htm Accessed: 18/02/2019.

WEF – World Economic Forum. (2016). "Shaping the future of construction a breakthrough in mindset and technology." www3.weforum.org/docs/WEF_Shaping_the_Future_of_Construction_full_report__.pdf Accessed: 18/02/2019.

Yakıcı, E. (2016). "Solving location and routing problem for UAVs." *Computers and Industrial Engineering*, *102*, 294–301. doi:10.1016/j.cie.2016.10.029.

Yu, Z., Jinhai, L., Guochang, G., Rubo, Z., and Haiyan, Y. (2002), "An implementation of evolutionary computation for path planning of cooperative mobile robots." *4th World Congress on Intelligent Control and Automation*, IEEE, Vol. 3, (pp. 1798–1802). doi: https://doi.org/10.1109/WCICA.2002.1021392.

17

DOMAIN-KNOWLEDGE ENRICHED BIM IN CONSTRUCTION 4.0

Design-for-safety and crane safety cases

Md. Aslam Hossain, Justin K. W. Yeoh, Ernest L. S. Abbott, and David K. H. Chua

17.1 Aims

- Discuss the importance of BIM to Construction 4.0.
- Identify the benefits of domain knowledge to BIM and Construction 4.0.
- Identify two forms of domain knowledge.
- Demonstrate the application of domain knowledge enriched BIM for Design-for-Safety.
- Demonstrate the application of domain knowledge enriched BIM in automating tower crane safety compliance.

17.2 Introduction

17.2.1 Construction 4.0

Construction 4.0 can be thought of as describing the trends of digitalization and automation in the Architectural, Construction, Engineering, and Facilities Management (AEC/FM) industry. Adopting Construction 4.0 technologies is expected to bring about significant benefits to the entire industry. Some of these benefits include enhanced productivity, cost efficiency, and better quality products. These can be realized through tighter integration with Internet-of-Things (IoT) and even Internet-of-People (IoP), achieving enhanced collaborations between project stakeholders (PricewaterhouseCoopers, 2016). Studies have suggested that embracing Construction 4.0 will spur the AEC/FM industry to transform from the current manual labor intensive, inefficient and potentially hazardous industry, into a technology-focused, data-driven industry (Schober, Hoff and Sold, 2015).

Despite the aforementioned benefits, Construction 4.0 has significant challenges and obstacles to overcome:

- *Nature of AEC/FM industry.* The nature of the AEC/FM industry is fragmented and relationships between stakeholders are often adversarial. This prevents digital integration from occurring as stakeholders withhold information from one another. Existing workflows must also be adapted or even completely overhauled to accommodate digital integration.
- *Lack of data interoperability.* The aforementioned fragmentation between stakeholders is exacerbated by the use of different software platforms, preventing data interoperability. Today, open standards like Industry Foundation Classes (IFC) and Construction Operations Building Information Exchange (COBie) are available to mitigate some of these data interoperability issues. However, more needs to be done.
- *Rapid change leads to increased risk.* The rapid evolution of Construction 4.0 tools means that the AEC/FM industry is often not ready to exploit them to the fullest potential. Companies who are first movers may open themselves up to the risk of technological change or suffer inefficiencies due to unfamiliarity with the tools and associated workflows. This may put them at a significant disadvantage in the short term, and a proper change management plan must be in place to mitigate this.
- *Change of skillsets and mindsets.* For the adoption of Construction 4.0 to occur, the traditional skillsets of AEC/FM practitioners need to be augmented. To survive in the Construction 4.0 future, AEC/FM practitioners must be fluent in the use of digital tools to achieve operational efficiency. More importantly, the culture and mindset of the AEC/FM practitioners must be changed so that they are more receptive to the rapid changes required by Construction 4.0.

The above challenges are further exacerbated by the broad definitions applied to the term Construction 4.0. For example, Singapore's adaptation of Construction 4.0, termed "Integrated Digital Delivery", places its emphasis on digitalization as a key driver for productivity enhancement through design, fabrication, construction and asset management (Building Construction Authority (BCA) Singapore, 2018). A European interpretation of Construction 4.0 identifies the linkage between IT infrastructure, data, and robotics as being important to mimic industrial manufacturing processes (European Construction Industry Federation, 2017). These broad definitions provide a guideline for industry practitioners but lack specifics on use cases for Construction 4.0. One objective of this chapter is to demonstrate specific applications for Construction 4.0 in the domain of construction safety.

17.2.2 *The role of BIM in Construction 4.0*

To characterize the features of Industry 4.0 (and by extension, Construction 4.0) systems, Hermann, Pentek and Otto (2016) set out the following design principles:

- *Interconnection* between buildings, machines, devices, sensors, and people.
- *Information transparency* as a key attribute of context-aware information so that decision makers can make appropriate decisions.
- *Technical assistance* via the aggregation and visualization of comprehensive information to facilitate enhanced decision making by decision makers.

Similarly, Oesterreich and Teuteberg (2016) also identified the following as key features a successfully implemented Construction 4.0 system should achieve:

- *Horizontal integration* across the construction value chain. This implies an integration of data flows between stakeholders across the construction value chain.
- *End-to-end digital integration* of engineering to achieve mass customization, i.e. maximizing economies of scale in production whilst delivering customized products.
- *Vertical integration* and networked systems within the company. This implies a single source of virtual information and data for the construction project must be available.

Studies claim the above can be achieved through the deployment of several technologies, including Internet-of-Things (IoT), Cloud Computing, and Cyber-Physical Systems (Lu, 2017). Apart from these technologies, Building Information Modeling (BIM) is emerging as one key technology for digitalization of the construction industry (Dallasega, Rauch and Linder, 2018).

Today, BIM is already an important tool for the AEC/FM industry, and in many countries, it is already compulsory to use BIM, especially to deliver public infrastructure projects. Serving as a common source of built environment data, it is an information repository about the project, accessible to all project stakeholders. As such, it is able to connect the virtual built environment with the decision makers and stakeholders. With a structured BIM process and sufficient level of development, it is also able to deliver information transparency, such as allowing decision makers to view material quantities with respect to the construction schedule.

As BIM tools advance in features, and AEC/FM practitioners become savvier and more aware of the potential of BIM, shifts in the decision-making process can be expected. For example, cost and quality decisions can now be made further upstream by clients as they are now able to better visualize the information from BIM. Also, a shift towards offsite construction will fundamentally alter the existing construction supply chain. With the increased use of BIM, it is inevitable that Construction 4.0 will have a great impact on the AEC/FM industry.

17.2.3 *The importance of domain knowledge to achieving Construction 4.0*

To facilitate the following discussion, a framework reimagined by Rowley (2007) called the DIKW Pyramid (see Figure 17.1) is introduced. In this framework, data refers to basic descriptions of objects and their properties. These data objects lack context and are merely containers of data. Computer-aided drawings are examples of data objects. Information is a construct that is generated by processing data and making sense of them. Often this entails embedding basic semantic information about the object so that classification, selection, or sorting is possible. BIM is a technology that resides in this layer.

Knowledge and Wisdom are the two higher levels of the framework. Knowledge refers to the act of making sense of information, and deriving further know-how or actionable insights. From the perspective of the built environment, such actionable insights must be accompanied by the realization of the exact situations, contexts, or specific project conditions in which to apply these insights. Hence, wisdom refers to the understanding and appreciation of applying knowledge in different contexts. From a Construction 4.0 perspective, it is the authors' opinion that achieving wisdom is perhaps still an ideal and often unattainable.

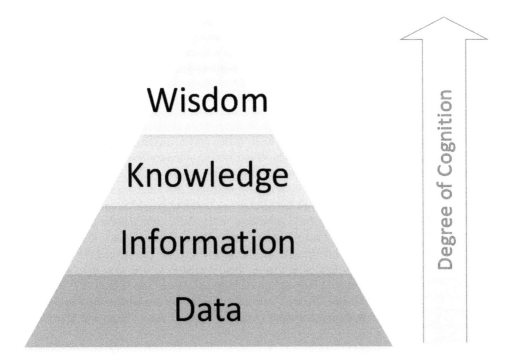

Figure 17.1   The DIKW pyramid

Domain knowledge enriched BIM seeks to advance this limitation of current BIM technology. The current BIM technology lacks the cognition necessary to automate the decision-making process. It is only through domain knowledge enriched BIM that contextual decision making can be supported.

Knowledge may be classified as explicit or tacit. Explicit knowledge deals with knowledge that can be recorded or communicated through mediums. It is easily and quickly transmitted from one to another. Tacit knowledge is the opposite of explicit knowledge in that it is difficult to transmit. Experience is an example of tacit knowledge. Hence, two different approaches may need to be adopted if knowledge is to be encoded into BIM.

From the above discussion, it is evident that while guidelines to adopt Construction 4.0 exist, specific use cases have not been clearly articulated. Moreover, despite the increased adoption of BIM, an important component in the form of domain knowledge is often not available. Without domain knowledge, contextual decision making cannot be effectively carried out. Hence, the objective of this chapter is to demonstrate specific applications of domain knowledge enriched BIM in the context of Construction 4.0.

Two specific applications have been chosen to demonstrate how tacit and explicit knowledge can be dealt with to enhance BIM. The first application looks at how tacit safety knowledge can be incorporated into BIM to enable Design-for-Safety which is adapted from Hossain et al. (2018). The outcome is a new framework to automatically carry out BIM-integrated safety risk reviews from multiple contextual building designs. The second application explores a framework to show how regulatory compliance for crane safety planning can be achieved using Crane Information Models that are explicit representations of safety knowledge (adapted from Yeoh, Wong and Peng, 2016).

17.3 Tacit knowledge application 1: Design-for-Safety knowledge enrich BIM for risk reviews

17.3.1 Background and problem statement

Statistics have shown that construction fatalities represent 36% of all work-related fatalities in Singapore, 27% and 18% in the UK and the USA, respectively (B.o.L. Statistics, 2013; MOM, 2016; HSE, 2018). Safety is a major concern in the construction industry. However, there is a growing understanding that many hazards that may arise during construction, operation, maintenance, and repair works, can be eliminated or at least alleviated through careful consideration during design phase, i.e. before construction begins. This effort is commonly known as Prevention-through-Design (PtD) or Design-for-Safety (DfS). However, difficulties in implementing PtD or DfS are lack of formal training in safety standards and best practices (Zarges and Giles, 2008; Tymvios and Gambatese, 2016; Toh, Goh and Guo, 2017). This means the success of identifying a hazard during design depends heavily upon the designer's ability, experience, and safety knowledge. This cannot always be guaranteed. Therefore, a knowledge-based design for safety library would help designers to take necessary action in addressing construction hazards in their design (Gambatese, Hinze and Haas, 1997; Nguyen et al., 2014). However, not many effective software solutions are available to help designers to make construction safety design decisions (Martínez-Aires, López-Alonso and Martínez-Rojas, 2018).

A comprehensive library of DfS knowledge along with safety rules is necessary to assist designers to account for construction-related hazards during their design. Nguyen et al. (2014) described a knowledge formalization framework to represent DfS knowledge. Moreover, DfS knowledge is to be integrated with BIM platform since BIM has been found to be the kernel of information management shared by all stakeholders involved in a construction project's life cycle starting from design, construction, operation, maintenance and finally demolition.

This part of the chapter presents a structured rule-based DfS knowledge library and is then integrated with a BIM platform providing a BIM-integrated risk review system so that designers can check the risk of their design elements against the safety rules and address them accordingly. However, not all risks can be eliminated with design, and hence, the residual risk must be addressed during the actual construction, operation and maintenance. Accordingly, a Risk Register system is built to capture all identified risks along with their mitigation status and relate this to BIM to enable progressive monitoring of the risk from design to construction, operation, and maintenance. The proposed BIM-integrated risk review system (Hossain et al., 2018) will assist the designer in identifying potential hazards along with design suggestions to mitigate or at least lower the risk level, and the Risk Register system will provide a guideline on necessary actions to be taken during the actual construction, operation and maintenance works.

17.3.2 Knowledge representation structure and mechanism

To achieve the BIM-integrated risk review system, the DfS knowledge representation structure and mechanism are discussed first. Often such DfS knowledge are tacit, and not properly documented nor expressed. As such, the challenge is representing the knowledge to handle multiple contextual scenarios. To address this, the following rule based framework is proposed.

17.3.2.1 Rule-based DfS knowledge library

Designers often face difficulties in identifying the hazards/risks that may arise at different work phases by their design elements. Particularly, designers are unaware of the conditions (work activity) and constraints (physical parameter) that pose a risk by a design element. For example, a simple precast beam can pose a risk of "Transportation and breakage issue" and "Falling object" if the length is too long or lifting points are not provided at the appropriate location (one-fifth of the beam length measured from the edge). Often, designers miss the requirement of the beam length and provision of lifting point for safe lifting, transportation, and installation of the beam. To help designers with DfS knowledge, the following rules can be set, as in Figure 17.2.

A checking system is necessary by setting safety rules in a DfS knowledge library. Nevertheless, the setting of safety rules requires expert knowledge since different design elements may pose different risks under different conditions and constraints. Different "required design" features may need to be applied to mitigate the risk based on the condition and constraint. A structured framework is necessary to systematically capture DfS knowledge. A six-level hierarchical taxonomy is developed to better capture the safety knowledge, as shown in Figure 17.3.

The first level refers to the type of design discipline such as Architectural, Structural, M&E, Geotechnical, and Temporary Structure and so on. Level two represents the design elements of various design topics. Third level states condition or work activity applies to the design element at different work phases (construction, operation or maintenance). For example, a slab opening can pose risk during construction until a railing is provided around the opening. However, there will be no safety issue for the opening during the service life of the building once railing is installed.

Level four describes the physical parameter that sets the constraint and can make a design element hazardous. The constraint can be a physical constraint (size, geometry or weight), material constraint (a specific type of material) or property constraint (describe an additional attribute of the design element such as IsExternal, IsLoadBearing, HasOpening, HasLiftingPoint and so on). The fifth level represents the risk issues associated with the condition and constraint of the design element. The final level captures the required design feature which is the design solution(s) to mitigate the risk issue. If required design feature is not satisfactory, some changes need to be made to the physical parameter of the concern design element or other nearby object(s) to address the safety issue.

Following the six-level taxonomy hierarchical model, a structure of DfS rules can be developed as shown in Figure 17.4. The rule structure has two major parts: Check for Hazard; and Mitigation Measures. The first part sets the checking criteria against any potential design risk and the second part describes the DfS required design feature to mitigate the risk.

The general structure of the DfS required design feature can be elaborated as shown in Figure 17.5. Here the design element can be the same element for which a rule is set or can be a different design element to address the risk. For example, an opening on a slab will pose

---

- *Constraint 1: Length > x;* where *x* is the specified length of the beam that depends on how it will be carried and installed.
- *Constraint 2: HasLiftingPoint IS False*

Figure 17.2 Example of rules for DfS

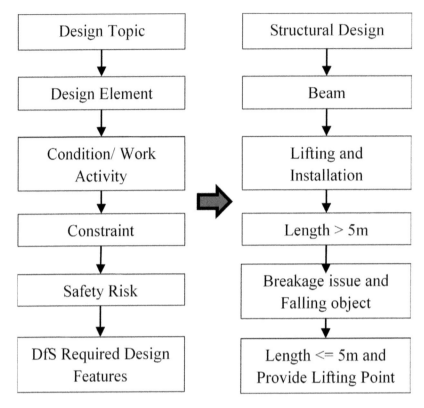

Figure 17.3   Six-level DfS taxonomy hierarchy

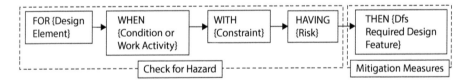

Figure 17.4   Structure of DfS rule

risk of falling through the opening. One of the design solutions to address this risk could be designing a railing around the opening with appropriate height (e.g. 1 m). Hence, design element railing in required design feature is different from the concern design element slab for which the rule is set for. Parameter constraints are similar to the constraints described for the first part of the rule. For example, physical constraint for the railing should be "*Railing Height ≥ 1 m*". Mitigation narrative can be selected from a list of design suggestions that applies to mitigate a specific risk. Whereas, functional constraint is defined as the functionality that must be performed by the design element. An example of functional constraint can be the load carrying capacity of the railing. If the railing is to carry a specific load, its design should be done accordingly such as railing material, size and joint with the slab. Finally, additional comment is a free format text that allows the user a high degree of flexibility. A design element may have all or some of the components as its DfS required design feature.

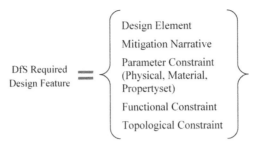

Figure 17.5 Structure of DfS required design feature

Referring to Figure 17.4, when a design element meets the first part of the DfS rule meaning that a risk has been identified then the rule checks whether the required design feature is satisfactory in mitigating the risk. It should be noted that even if a risk is identified for a design element based on the constraint(s), the design element may not need additional design consideration if other constraints as defined by the DfS required design feature are present for that particular risk (an explanation has been provided later in the following case study). If the design element fails the criteria of required design feature or if there is no mitigation measure present then the risk is to be handled manually. Whenever a design element meets the checking criteria of the first part of a rule it is to be recorded in a Risk Register irrespective of satisfying the criteria for required design feature or not. This is because if the mitigation measure is met, then the Risk Register will be a guideline to follow up the risk during the execution of actual work. On the other hand, if no mitigation measure is present, then it will draw the designer's attention to address the safety issue or the residual risk is to be handled manually by the respective stakeholder.

17.3.2.2 Classification of DfS rules

In order to manage the DfS rule, three types of rules were proposed by Hossain et al. (2018). These were Atomic Rule, Meta-Element Rule, and Meta Rule. Atomic Rules are very low-level rules, and every constraint associated with a design element makes one Atomic Rule. These rules form the basis for checking a building model against any design risk. Structure of the Atomic Rule is the one shown in Figure 17.4. Among the five clauses (FOR, WHEN, WITH, HAVING and THEN) of the rule structure, WITH clause sets the constraint for an Atomic Rule which is used to check if there is any risk imposed by the design element. Constraints are associated with Condition or Work Activity under WHEN clause. When a Constraint is met for a design element (FOR clause) then a Risk is identified (HAVING clause). Required Design Feature is set under THEN clause to mitigate the identified risk.

The Atomic rule is useful in defining a specific rule, or rules, for a design element and checking against the risk. However, there could be several hundred Atomic Rules for a building facility and the constraint value for an Atomic Rule can be different based on the site condition, equipment capacity, working environment and so on. Moreover, there are some conditions and constraints that are common to more than one design element. Hence, defining Atomic Rules one by one and managing those rules can be tedious. In order to better manage the rules, concept of Meta Element Rule and Meta Rule are introduced. Meta Element Rule is a collection of few Atomic Rules consisting of some conditions, constraints, and risks associated with a design element. Finally, Meta Rule is a high-level rule, and it is also a collection of few Atomic Rules; however, it applies for several design elements with a common condition, constraint, and risk. Such categorizations give the flexibility to manage rules and checking safety issues for a specific design element or identifying the design elements that might have the same risk under common working condition/constraint.

17.3.3 Integrating DfS knowledge in BIM

17.3.3.1 Building the DfS knowledge library

To build the DfS rule-based knowledge library, a set of primary knowledge-based libraries is necessary following the six-level DfS taxonomy hierarchy of Figure 17.3. The first library is a list of design topics or design disciplines that includes Architectural, Structural, M&E, Geotechnical, Site and Environmental, and Temporary Structure and Machine. The second library is the list of design elements for each design topic. Since the DfS knowledge library is to be used by the designers for their design elements, the list in the library is to be rigorous enough to be picked by the designers. IFC (an ISO standard open platform, ISO 16739:2013) elements have been stored in the library since these are worldwide recognized unified elements.

The third library comprises a list of conditions or work activities (referred as Meta Condition) associated with different work phases of a building's life cycle. The examples of condition narratives are *working height, mechanical lifting, overhead work* etc. The Meta Condition library has been built from expert knowledge and existing literature. Nevertheless, new conditions can be added in the library by the respective users.

The fourth library contains three tables for physical constrain, material constraint and property set constraint, respectively. Physical constraint can be set from a list of standard IFC "Quantity Use Definition" for design element. For Example, standard IFC physical quantities for beam are NominalLength, CrossSectionArea, OuterSurfaceArea, TotalSurfaceArea, GrossVolume, NetVolume, GrossWeight and NetWeight. Detail description for each quantity can be found in IFC2x3 and IFC4 definition. However, other physical parameters that are not in the "Quantity Use Definition" can easily be incorporated in defining the DfS rule. For instance, height of beam (WorkingHeight) in a building is not a standard IFC quantity but this is necessary to define DfS rule for working at height. Hence, WorkingHeight is incorporated in the physical constraint library. The material library is the list of materials used in building construction such as concrete, steel, glass and so on and maintained by the user. Whereas, propertyset constraints are also set from a list of standard IFC propertyset as well any user defined property which can be incorporated in a similar way to physical constraints.

The fifth library is the list of risk narrative that may be encountered during the life cycle of building projects. Figure 17.6 shows a screen shot of defining risk narrative in the risk library. Finally, the sixth library is the mitigation narratives that consists both design suggestions and construction suggestions. Design suggestions are part of Required Design Feature; whereas, as mentioned earlier, the construction suggestions are recorded in a Risk Register system to address the residual risk in actual construction. Figure 17.7 is a sample of mitigation narratives list for design element slab. Mitigation narratives can be similarly listed for other design elements.

The aforementioned primary libraries are the basis for developing DfS rule-based knowledge library. Figure 17.8 depicts the language structure for defining Atomic Rule using the primary knowledge libraries. The language of the rule is an English-like language that is used to interrogate BIM compliant digital models of buildings. The right-hand side of Figure 17.8 is the

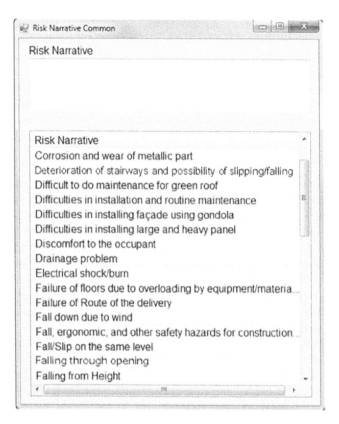

Figure 17.6 List of risk narratives

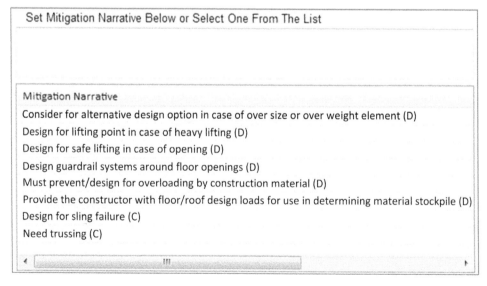

Figure 17.7 Defining mitigation narratives ("D" stands for design suggestion and "C" stands for construction suggestion)

Dictionary of DfS rule used for design elements under a design topic. Corresponding Condition, Constraint, Risk and Required Feature can be set that applies to a design element and makes an Atomic Rule. Similarly, Meta Condition is set with Constraints from the library of physical, material and property set constraints in order to define "Meta Element Rules" and "Meta Rules".

In this way, primary knowledge libraries are used to build the DfS rule-based knowledge libraries. The libraries have been developed in such a way that those can be further elicited by input from the users such as architects, engineers, consultants, constructors or others.

17.3.3.2 Integrating the DfS knowledge library with BIM

The DfS knowledge library is meant to be used by designers to check their design for risks and mitigate them early in the design phase. It is known that BIM becomes the core in managing any building facilities and safety issues should be linked with the BIM model so that it can be visualized and addressed (Wetzel and Thabet, 2015). Accordingly, a BIM-integrated intelligent risk review system is developed using the DfS rule-based knowledge library. The system comprises three key components as shown in Figure 17.9. The first is a design for safety knowledge (DfSK) library which covers DfS rules for design related hazards. The second is a reasoning engine to parse the safety rules for an intelligent design for safety review/ assessment using BIM compliant building model. The third is a BIM-integrated Risk Register that facilitates the management and control of the safety issues identified.

The intelligent DfS review system works through design for safety language (DfSL) similar to the structure shown in Figure 17.8 and is used to examine BIM compliant building models for potential design risks. The core of the checking system is a Reasoning Engine which reads BIM compliant digital model in IFC format and comprises a set of safety checking algorithms. The reasoning engine compares the physical parameters of the facility in BIM (either readily available in BIM model or some inferences are done for connectivity, distance, height and so on) against the DfS rules. The resultant output is the list of identified design risks/hazards which are visualized and tagged in the BIM compliant 3D model and recorded in a Risk Register.

After identifying risks along with mitigation measure (if available) in the design phase using the intelligent DfS review system, a Risk Register system is necessary to follow up for residual risk during construction, operation and maintenance. It is understandable that not every risk can be addressed/mitigated in the design phase and need to be handled during the actual execution of the work. Nevertheless, identifying risks early in the design phase would help the constructors prepare for those risks well before commencing their work. This will help in avoiding any delay that may arise to arrange necessary preparation for the risk. Hence, a Risk Register system would be useful to store all the risk histories such as the possible source of risk, action taken by the designer and necessary follow-up to be made for any residual risk.

Figure 17.8 Language structure of defining Atomic Rules

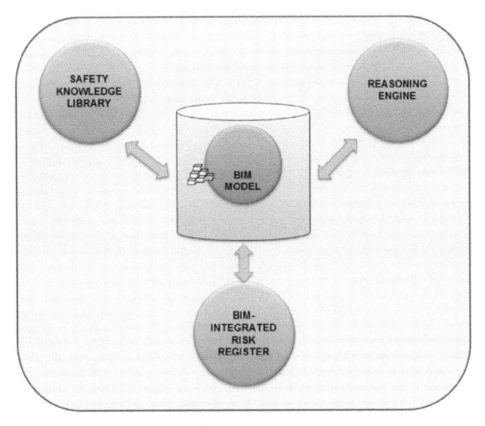

Figure 17.9 Key components for DfS review system

In the proposed DfS review system, a Risk Register has been integrated in the BIM plat-form so that when a designer works on his/her design element with the DfS rule-based knowl-edge library, all the identified risks, action taken to mitigate the risks and any other suggestions to follow up for residual risk are linked with that design element and are stored in a database for future reference.

17.3.4 Case study: DfS knowledge library for safety risk reviews

After developing the DfS rule-based knowledge library and integrating with the BIM platform, it has been tested with a BIM model to illustrate the effectiveness of the developed system. The BIM model is a five-story building (see Figure 17.10) for which parameters (physical, material and property set constraints) for some of the design elements have been purposely set in order to check against some predefined DfS rules. For example, some precast (material constraint) beams and columns are added in the BIM model with length >12 m (physical constraint), some windows are <0.75 m in width and so on. Moreover, some of the precast beams and columns have property set constraint "HasLifting Point" in order to lift the precast element by crane, and most of the windows have the property set constraint "IsSwinging" in order to enable easy cleaning and maintenance except few.

Figure 17.10 Illustrative BIM

Figure 17.11 (a) and (b) show two examples of some basic DfS rule-based checking for "Structural" design elements beam and column. When the checking is done against DfS rules, the model becomes semi-transparent highlighting those design elements that meet the DfS rule; i.e. some risk has been identified. In the first checking (Figure 17.11a), the constraint for beam and column has been set as length >12 m. Consequently, the Figure highlights the precast beams and columns those are >12 m length. Here, the elements are too big to lift and the risk is "Falling Object" if proper lifting point is not provided. Hence, the required design feature in order to mitigate the risk of "Falling Object" would be "HasLiftingPoint IS True". For the next checking (Figure 17.11b), the DfS rule is set as follows:

FOR{Beam}WHEN{Lifting} WITH{Length > 12} HAVING{Falling Object} THEN{Has-LiftingPoint IS True}.

In the BIM model, lifting points have been provided for some beams and columns and this has been done with property set constraint. Subsequently, only those beams and columns are highlighted that match the rule. In this case, if a beam/column is >12 m length but it has lifting point (which is the required design feature for the risk of over length) will not be highlighted.

After identifying and addressing the design deficient risk with the proposed risk review system, this information is to be stored in a Risk Register system for future references. Moreover, all the risks identified may not be mitigated with design. Those risks are to be addressed during actual construction. Risk Register system stores the residual risk as well and can be passed to the constructor for their action. For instance, Design for Safety Guidelines by WSH Council, Singapore (WSHCouncil, 2011) suggests maintaining a Risk Register to record the risk that cannot be mitigated through design changes. Such early identification would help a constructor to be prepared for risk in advance.

Figure 17.12 depicts a Risk Register for the design element "Window," which is under design discipline "Architectural". For illustration purposes, width of some windows for the case example BIM model has been set < 0.75 m for which risk is "Fire escape". Whereas, propertyset constraint "IsSwinging IS False" is set for windows at the backside of the building

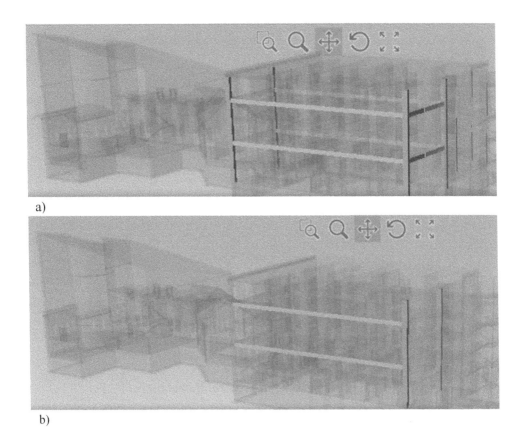

a)

b)

Figure 17.11   Rule-based checking for design element beam and column

for which risk could be "Falling from height" during operation and maintenance. Consequently, windows are highlighted in BIM model that fall within DfS rules and corresponding risk with possible mitigation measures are stored in Risk Register. Now, the architect could reconsider the window size with width >0.75 m to mitigate the risk of "Fire escape" or he/she may ignore the risk if an alternative fire escape is designed. Accordingly, he/she could put a comment in the Risk Register mentioning that an alternative fire escape has been designed and the risk can be considered as mitigated/addressed. The design element for which risk has been addressed can be marked in different (green) color in the Risk Register. Similarly, for windows with no swinging, the architect could select windows with swinging so that cleaning can be done from inside building eliminating the risk of "Falling from height". Alternatively, if the architect does not want to change his/her window selection for some reason, the risk could be mitigated by installing equipment for a gondola so that cleaning can be done from outside of building. This action is to be noted in the Risk Register with the responsible person so that the residual risk can be addressed during actual construction. In this case, the window will not be turned to green since the risk has not been addressed yet. In this way, the Risk Register facilitates recording and tracking of responsible person, verifying person and additional comments for follow up. Other safety issues can be similarly checked and managed using the proposed BIM-integrated DfS knowledge library and Risk Register system.

Figure 17.12   Risk Register for the case example

17.4 Explicit knowledge application 2: BIM-based tower crane safety compliance system

17.4.1 Background and problem statement

Fixed cranes (comprising Tower Cranes and Luffing Jib Cranes) are commonly deployed in high-rise projects. Significant construction challenges and compliance issues arising from the use of these cranes may be present in a densely built-up urban environment.

Cranes today, represent the single biggest equipment investment deployed on a construction site (Al-Hussein et al., 2006), and are also one of the main sources of hazards on site (Shapira and Lyachin, 2009). Statistics from the Workplace Safety and Health Institute (Singapore) (Workplace Safety and Health Institute Singapore, 2015) show that crane related incidents accounted for almost 20% of all fatalities in the construction industry from 2011 to 2014.

Several factors for the dismal safety record of static crane operations have been put forth (Workplace Safety and Health Council (Singapore), 2009):

- *Manpower*. Issues with manpower can be attributed to poor training leading to inadequate knowledge or information, willfully violating rules and regulations, and human error.
- *Machine*. Machine issues contribute to crane failure due to poorly maintained mechanical and electrical components and poor choice of machines for the job.
- *Medium*. These are related to the specific project context, including challenging environmental and site constraints.
- *Management*. Management oversight can arise from a lack of supervision, poor implementation of safe work practices, and a failure to plan.

Many of the problems identified above can be addressed during the crane planning stage with relevant information and data. The current challenge is that data to address the above issues reside in separate external repositories and/or databases, and checking the data manually requires significant time and effort. A possible application of Construction 4.0 to address some of the factors highlighted above during the crane planning stage is proposed. In this proposed system, crane information and safety compliances are explicitly modeled to create digital twins of tower cranes that is referred to as Crane Information Models. These models are then placed within BIM, creating BIMs with enhanced crane safety information. This will facilitate the crane layout planning and compliance checking processes.

The proposed system architecture of the tower crane safety compliance system is shown in Figure 17.13. The overall system architecture comprising three inputs: Machine (Tower Crane Information), Manpower (Operator records), and Medium (Contextual Project Information in the form of BIM). A querying and reasoning mechanism is used to check the compliance of the tower cranes to the existing safety regulations, and also generate a plan depicting the feasible crane configuration and layout.

17.4.2 Knowledge representation structure and mechanism

17.4.2.1 Describing and modelling spaces in Crane Information Models

To represent the required regulatory knowledge, the following taxonomy in Figure 17.14 is adopted from Yeoh, Wong and Peng (2016) to describe tower crane safety regulations. Song and Chua (2006) introduced a formal framework to represent functional construction requirements, and this framework was extended by Yeoh and Chua (2014) to incorporate non-functional

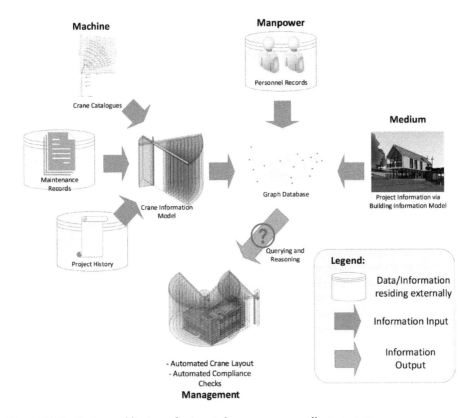

Figure 17.13 System architecture of automated tower crane compliance system

requirements as well. These construction requirements may be described as capabilities and conditions for which the construction process must conform to. The crane safety compliance requirements considered in this chapter are a subset of these construction requirements.

The knowledge in the regulations can be made explicit by modeling the spaces in BIM. This is because the regulations are typically related to how spaces interact. For example, if a part of the building coincides with the crane clearance, this is a violation. To achieve this, the characterization of space utilization in Chua, Yeoh and Song (2010) was adopted, and three main classes of space are identified: Hazard space, Process space and Product space. The spaces in crane safety compliance requirements originate from two sources: Building Site and Crane.

The crane and building structure are examples of a product space, and describe the physical elements of the crane and building. The crane working envelope is a type of process space. It is the union of the different possible jib positions of the crane, and its corresponding zone of operation. Building process spaces include the path and activity spaces. The interdiction spaces are spaces where no space utilization is allowed. Two types of interdiction spaces are possible: Clearances and Hazards. Clearance spaces are meant to denote protective zones against physical elements. Clearance spaces include buffer areas around physical elements. Hazard zones (spaces) are used to demarcate areas of hazardous operations. Examples of hazard zone include the zone under the counterweight of the counter-jib.

Using the above taxonomy, Crane Information Models (CIM) are introduced. These CIMs are enhanced crane models that are derived from the manufacturer's crane catalogues. In addition to that, the spaces are explicitly mapped to the crane models as shown in Figure 17.15.

17.4.2.2 Describing the relationships between spaces

Using the Resource Description Framework (RDF), the relationships in Figure 17.16 can be stored as triples of the form *<subject, predicate, object>*. Subjects and Objects may refer to the spaces or to concepts (e.g. Fixed Crane), while the predicate refers to the relationship between subject and object. For example, a possible triple is *<Building Structure, intersects, Crane Hazard Zone 2>*.

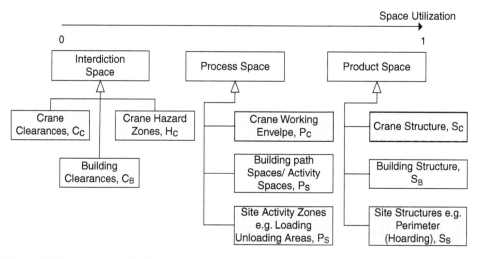

Figure 17.14 A taxonomy for fixed crane spaces

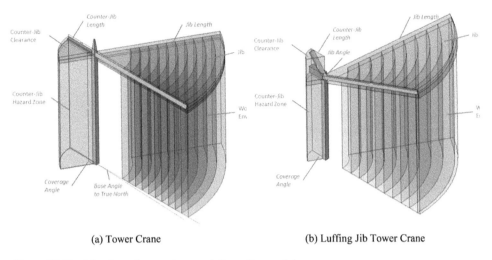

(a) Tower Crane   (b) Luffing Jib Tower Crane

Figure 17.15   Mapping of spaces to crane information models

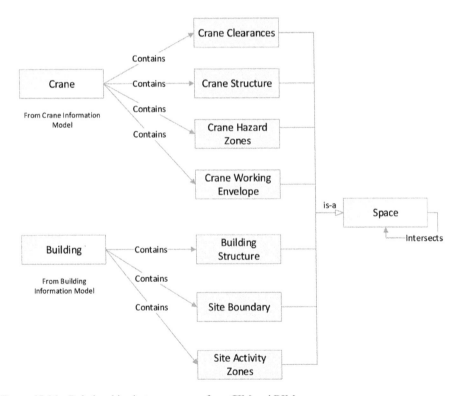

Figure 17.16   Relationships between spaces from CIM and BIM

The "intersects" relationship can be determined using a spatial interference check – a feature commonly found in BIM. This is carried out after the Crane Information Models have been collocated within BIM, and a pair-wise spatial interference check conducted between various spatial elements. The outcome of a successful spatial interference check is an assertion of the triple *<Space1, intersects, Space 2>* into a graph database.

17.4.3 Using Crane Information Models for tower crane safety compliance in BIM

Various safety and operational rules need to be checked to ensure the crane safely complies with regulations. Some common crane layout rules include the following:

- Tower Crane Workspace must be within Site Boundary.
- Hazard Zone of Crane counter-jib must not interfere with Workspace of another Crane.
- When Workspaces of two tower cranes interfere with each other, there must be a minimum 3 m height separation between jibs.
- There must be a height separation of at least 3 m from the building roof to the lowest point of the hook on the crane.

Figure 17.17 shows the workflow to check the safety compliances. This starts with collocating the CIM with the BIM. The elements from both BIM and CIM are extracted to populate the graph database. The interference check described above is used to determine the relationships between the elements. This is then entered into the graph database. Having identified and represented the basic knowledge, it is now possible to model safety non-compliances as SPARQL queries to the graph database. An example of the query is shown in the figure.

17.4.4 Case study: tower crane safety compliance checking

An academic building was used as a case study to validate the tower crane safety compliance analysis as well as to demonstrate the applicability of CIM. The building comprised

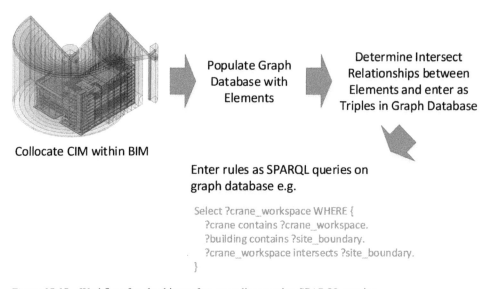

Figure 17.17 Workflow for checking safety compliance using SPARQL queries

two adjoining blocks. The first block is a ten-story structure comprising primarily precast elements, of which there are about 210 pieces per story. The second is a six-story carpark, with about 70 precast elements per story. The largest element, a post tensioned deep beam situated at the auditorium, weighed about 8 tons. The site was also subject to a height restriction of 80 m. The proposed lifting plan is shown in Figure 17.18. Two tower cranes and a luffing jib crane with specific characteristics as shown in the figure were analyzed.

Figure 17.19 shows the plan view of the project site. From this plan view, the site boundary has not been exceeded by the proposed workspace envelopes or the various clearances required by the crane. The crane positions are adequate, with no clashes occurring between the counter-jib clearances. Also, the coverage requirement is met, with all elements within the workspace envelopes.

The 8 ton deep beam is situated at the intersection of the workspace envelopes of TC1 and TC2. An inspection of the load capacities of the discretized workspace envelopes reveals TC2 does not have adequate capacity to hoist. However, TC1 is found to be adequate.

Inspecting the elevation view of the model (Figure 17.20), TC1 and TC2 fulfills the requirement of minimum jib height separation. Also, TC1 meets the minimum building height separation from the ten-story block. Similarly, TC3 meets the building height separation requirement from the six-story carpark. The total height of the cranes was also found to be adequate within the permissible site limit.

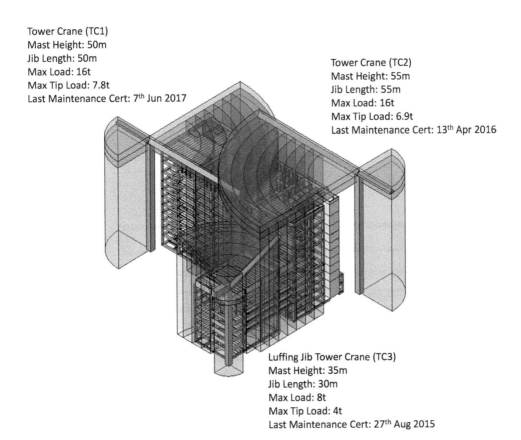

Tower Crane (TC1)
Mast Height: 50m
Jib Length: 50m
Max Load: 16t
Max Tip Load: 7.8t
Last Maintenance Cert: 7th Jun 2017

Tower Crane (TC2)
Mast Height: 55m
Jib Length: 55m
Max Load: 16t
Max Tip Load: 6.9t
Last Maintenance Cert: 13th Apr 2016

Luffing Jib Tower Crane (TC3)
Mast Height: 35m
Jib Length: 30m
Max Load: 8t
Max Tip Load: 4t
Last Maintenance Cert: 27th Aug 2015

Figure 17.18 Proposed layout with Crane Information Models

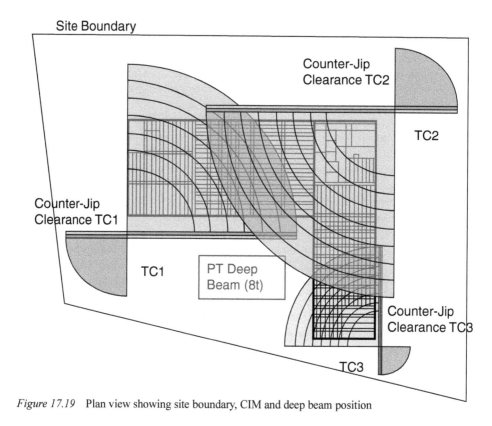

Figure 17.19   Plan view showing site boundary, CIM and deep beam position

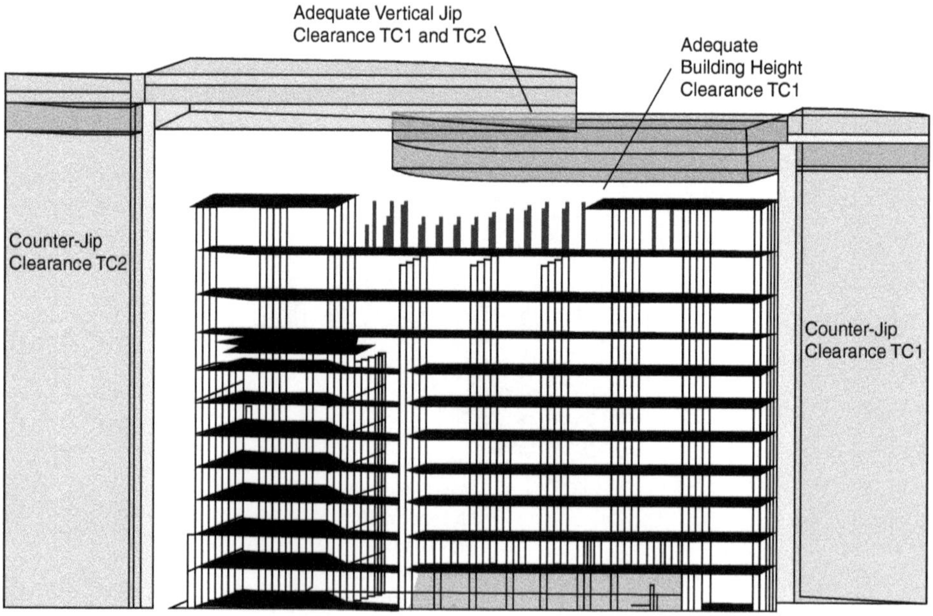

Site Boundary

Counter-Jip
Clearance TC2

TC2

Counter-Jip
Clearance TC1

TC1

PT Deep
Beam (8t)

Counter-Jip
Clearance TC3

TC3

Adequate Vertical Jip
Clearance TC1 and TC2

Adequate
Building Height
Clearance TC1

Counter-Jip
Clearance TC2

Counter-Jip
Clearance TC1

Figure 17.20   Elevation view showing CIM and building heights with clearances

17.5 Conclusion

Construction 4.0 is going to cause a major paradigm shift in the AEC/FM industry. It is expected that new tools, processes and skillsets will be required by practitioners. This chapter has established the importance of BIM as a key enabler of Construction 4.0, and how it will serve a major role in becoming a vertical and horizontal integrator across the entire construction value chain. Despite its vital importance, the following gaps are identified:

- Construction 4.0 has various guidelines but specific use cases and applications have not been articulated.
- BIM is inadequate as a tool for actionable insight and contextual decision making; domain knowledge enriched BIM is required.
- Two approaches of enhancing BIM knowledge are possible, depending on the type of knowledge: tacit and explicit.

This chapter demonstrates both approaches through application examples. The first demonstrates how tacit knowledge, like Design-for-Safety knowledge, can be represented. The structures of DfS rules have been presented that allow the various contexts of the knowledge to be captured. This chapter further shows how these rules can be integrated with a BIM based environment to provide guidance to designers.

The second approach demonstrates how knowledge can be made explicit, and used for tower crane safety compliances. Crane Information Models are introduced as a domain knowledge enriched representation of tower cranes in BIM, and further show how these can be represented and reasoned with to determine if tower crane regulatory and safety compliances have been met.

In future, as Construction 4.0 becomes more pervasive, the deluge of building and project information will require more decentralized decision making either by human decision makers, or autonomously. A hybrid of both domain knowledge enriched BIM approaches could be possible, depending on the complexity of the problem.

17.6 Summary

- BIM is a key enabler for Construction 4.0.
- Domain knowledge enriched BIM is required for contextual decision making, and is necessary for the future of Construction 4.0.
- Domain knowledge exists as both tacit and explicit knowledge, and different approaches may exist to represent them.
- Tacit domain knowledge enriched BIM for Design-for-Safety is possible by using rules to capture the contextual conditions.
- Explicit domain knowledge enriched BIM for automating tower crane safety and regulatory compliance can be achieved by explicitly representing the information in Crane Information Models.

References

Al-Hussein, M., Athar Niaz, M., Yu, H. and Kim, H. (2006). "Integrating 3D visualization and simulation for tower crane operations on construction sites", *Automation in Construction*, 15(5), pp. 554–562. doi: 10.1016/j.autcon.2005.07.007.

B.o.L. Statistics (2013) *Census of Fatal Occupational Injuries*.

Building Construction Authority (BCA) Singapore (2018). *Integrated Digital Delivery (IDD)*. Available at: www.bca.gov.sg/IntegratedDigitalDelivery/Integrated_Digital_Delivery.html (Accessed: 27 February 2019).

Chua, D. K. H., Yeoh, K. W. and Song, Y. (2010). "Quantification of spatial temporal congestion in four-dimensional computer-aided design", *Journal of Construction Engineering and Management*, 136(6), pp. 641–649. doi: 10.1061/(ASCE)CO.1943-7862.0000166.

Dallasega, P., Rauch, E. and Linder, C. (2018). "Industry 4.0 as an enabler of proximity for construction supply chains: a systematic literature review", *Computers in Industry*, 99(April), pp. 205–225. doi: 10.1016/j.compind.2018.03.039.

European Construction Industry Federation (2017). *Construction 4.0: Challenges and Opportunities*. Available at: www.fiec.eu/en/themes-72/construction-40/challenges-and-opportunities.aspx (Accessed: 1 March 2019).

Gambatese, J. A., Hinze, J. W. and Haas, C. T. (1997). "Tool to design for construction worker safety", *Journal of Architectural Engineering*, 3(1), pp. 32–41. doi: 10.1061/(ASCE)1076-0431(1997)3:1(32).

Hermann, M., Pentek, T. and Otto, B. (2016). "Design principles for industrie 4.0 scenarios", *Proceedings of the Annual Hawaii International Conference on System Sciences*. IEEE, 2016–March, pp. 3928–3937. doi: 10.1109/HICSS.2016.488.

Hossain, M. A., Abbott, E. L. S., Chua, D. K. H., Nguyen, T. Q. and Goh, Y. M. (2018). "Design-for-safety knowledge library for BIM-integrated safety risk reviews", *Automation in Construction*, 94(June), pp. 290–302. doi: 10.1016/j.autcon.2018.07.010.

HSE (2018) *Workplace fatal injuries in Great Britain 2018*.

Lu, Y. (2017). "Industry 4.0: A survey on technologies, applications and open research issues", *Journal of Industrial Information Integration*, 6, pp. 1–10. doi: 10.1016/j.jii.2017.04.005.

Martínez-Aires, M. D., López-Alonso, M. and Martínez-Rojas, M. (2018). "Building information modeling and safety management: a systematic review", *Safety Science*, 101, pp. 11–18. doi: 10.1016/j.ssci.2017.08.015.

MOM (2016) *Occupational Safety and Health Division Annual Report*.

Nguyen, T. Q., Abbott, E. L. S., Chua, D. K. H. and Goh, Y. M. (2014). "Formalizing construction safety knowledge for intelligent BIM-based review of design for safety", in Aulin, R. and Ek, Å. (eds) *Achieving Sustainable Construction Health and Safety: Proceedings of CIB W099 International Conference*. Lund, Sweden, pp. 597–607.

Oesterreich, T. D. and Teuteberg, F. (2016). "Understanding the implications of digitisation and automation in the context of Industry 4.0: A triangulation approach and elements of a research agenda for the construction industry", *Computers in Industry*, 83, pp. 121–139. doi: 10.1016/j.compind.2016.09.006.

PricewaterhouseCoopers (2016) *Industry 4.0: Building the Digital Enterprise Engineering and Construction Key Findings*.

Rowley, J. (2007). "The wisdom hierarchy: representations of the DIKW hierarchy", *Journal of Information Science*, 33(2), pp. 163–180. doi: 10.1177/0165551506070706.

Schober, K.-S., Hoff, P. and Sold, K. (2015) "Mastering the transformation journey digitization in the construction industry", *Think Act: Beyond Mainstream*. Retrieved from www.rolandberger.com/publications/publication_pdf/roland_berger_tab_radically_digital_20160121.pdf.

Shapira, A. and Lyachin, B. (2009). "Identification and analysis of factors affecting safety on construction sites with tower cranes", *Journal of Construction Engineering and Management*, 135(1), pp. 24–33. doi: 10.1061/(ASCE)0733-9364(2009)135:1(24).

Song, Y. and Chua, D. K. H. (2006). "Modeling of functional construction requirements for constructability analysis", *Journal of Construction Engineering and Management*, 132(12), pp. 1314–1326. doi: 10.1061/(ASCE)0733-9364(2006)132:12(1314).

Toh, Y. Z., Goh, Y. M. and Guo, B. H. W. (2017). "Knowledge, attitude, and practice of design for safety: multiple stakeholders in the Singapore construction industry", *Journal of Construction Engineering and Management*, 143(5), p. 04016131. doi: 10.1061/(ASCE)CO.1943-7862.0001279.

Tymvios, N. and Gambatese, J. A. (2016). "Direction for generating interest for design for construction worker safety—a Delphi study", *Journal of Construction Engineering and Management*, 142(8), p. 04016024. doi: 10.1061/(ASCE)CO.1943-7862.0001134.

Wetzel, E. M. and Thabet, W. Y. (2015). "The use of a BIM-based framework to support safe facility management processes", *Automation in Construction*, 60, pp. 12–24. doi: 10.1016/j.autcon.2015.09.004.

Workplace Safety and Health Council (Singapore) (2009) *Crane Safety Analysis and Recommendation Report*.

Workplace Safety and Health Institute Singapore (2015) *Workplace Safety and Health Report 2014*.

WSH Council (2011) *Guidelines on Design for Safety in Buildings and Structures*. Available at: www.wshc. sg/wps/PA_IFWSHCInfoStop/DownloadServlet?%0AinfoStopYear=2014&infoStopID= IS2010012500120&folder=IS2010012500120&%0Afile=DfS_Guidelines_Revised_ July2011.pdf.

Yeoh, J. K. W., Wong, J. H. and Peng, L. (2016) "Integrating crane information models in BIM for checking the compliance of lifting plan requirements", in *33rd International Symposium of Automation and Robotics in Construction (ISARC2016)*. Auburn, Alabama: International Association of Automation and Robotics in Construction (IAARC), pp. 974–982.

Yeoh, K. W. and Chua, D. K. H. (2014). "Representing requirements of construction from an IFC model", in R. Issa and I. Flood (eds) *Computing in Civil and Building Engineering 2014*. American Society of Civil Engineers, pp. 331–338. doi: doi:10.1061/9780784413616.042.

Zarges, T. and Giles, B. (2008). "Prevention through design (PtD)", *Journal of Safety Research*, 39(2), pp. 123–126. doi: 10.1016/j.jsr.2008.02.020.

18

INTERNET OF THINGS (IOT) AND INTERNET ENABLED PHYSICAL DEVICES FOR CONSTRUCTION 4.0

Yu-Cheng Lin and Weng-Fong Cheung

18.1 Aims

- To introduce IoT enabled physical technologies and relative applications in Construction 4.0.
- To provide related literature reviews for RFID, UAV, WSN, and BIM technologies in Construction 4.0.
- To survey different advanced IoT applications in Construction 4.0. (such as safety management, structural health monitoring, and smart building).
- To demonstrate a case study regarding the development of a cyber physical system of applying tunnel construction safety management in Construction 4.0.

18.2 Introduction

Industry 4.0 (I4.0), also known as the Fourth Industrial Revolution, was proposed by the German Federal Government in 2011 and incorporated into the "High-Tech Strategy 2020" project. I4.0 focuses on enhancing automation, digitalization, and intelligentization. The construction industry 4.0 (C4.0) refers to the spirit of I4.0 and integrates existing engineering technologies, processes, and requirements and then builds a more adaptive, resource-efficient, intelligent construction process. The best solutions are also proposed to meet quality, cost, and safety requirements by analyzing big data. Relevant technologies, such as Cyber-Physical Systems (CPS), Internet of Things (IoT), and Cloud Computing (CC) promote "smart manufacturing" through the real-time integration of virtual and physical states and the analysis of big data, thus enabling the enhancement or innovation of production and services. To enable CPS to have an awareness of the physical industrial status and to facilitate an object's communication ability, environmental perception and networking capabilities are essential. The Wireless Sensor Network (WSN), one of the key technologies used to support IoT and CPS, comprises a large number of sensor nodes; each node is equipped with sensors to detect physical phenomena in the real world such as temperature, light, and pressure. Building Information

Modeling (BIM) prompts an evolutional change in digitalization and informatization in the construction industry, integrating a different kind of building information into a 3D model and enhancing the management and productivity in a construction project. The integration of WSN and BIM can be used to develop CPS conforming to Construction 4.0 (C4.0) for a construction project. The use of WSN also supports big data collection for applications in big data analytics and CC.

This chapter introduces IoT enabled physical technologies and their applications in Construction 4.0. Section 18.3 introduces the relevant development and reviews of IoT, I4.0, and C4.0. Section 18.4 introduces the IoT and related technologies concerning construction, which include radio-frequency identification (RFID), WSN, and BIM, and describes how they achieve the functions of CPS. Section 18.5 surveys IoT applications in various domains of the construction industry. Section 18.6 provides a case study of infrastructure safety management and details the process for developing the CPS system, and the final section presents the conclusion and discussion.

18.3 Background

Industrial 4.0 (I4.0) is described as the Fourth Industrial Revolution; the concept of I4.0 was first introduced at the Hannover Fair in Germany in 2011. Then, the German Academy of Science and Engineering founded a working group to research related knowledge, and the relevant results were referred to as principles of governance. Other countries formulated related concept policies; for example, the United States (US) developed the "Advanced Manufacturing Partnership" (AMP) and China implemented "Made in China 2025." Digitalization and intelligentization of industry can enhance the productivity and competitiveness of China.

Many scholars have defined I4.0 from different research categories. The Consortium II Fact Sheet (Consortium II 2013) defines I4.0 as "the integration of complex physical machinery and devices with networked sensors and software, used to predict, control and plan for better business and societal outcomes." Henning et al. (2013) expounded I4.0 as "a new level of value chain organization and management across the lifecycle of products." Hermann et al. (2015) defined I4.0 as "a collective term for technologies and concepts of value chain organization." During the modular structured smart factories of I4.0, CPS creates a virtual copy of the physical production system and makes decentralized decisions. The key fundamental principles of I4.0 include service orientation, intelligent production, interoperability, Cyber-Physical Production Systems (CPPS) providing cloud/big data algorithms and analysis, and communication security (Vogel-Heuser et al. 2016). I4.0 facilitates interconnection and computerization in traditional industries, which makes an automatic and flexible adaptation of the production chain and provides new types of services and business models of interaction in the value chain (Lu 2017).

The integral result indicates that the I4.0 can be summarized as a digitalized, smart, optimized, service-oriented, and interoperable production, which correlates with computerization, CPS, IoT, big data, and high technologies.

I4.0 promotes implementing the emerging concepts of CPS and IoT into the manufacturing system, which enables the design and creation of smart factories and production processes. A CPS is defined as "a mechanism that is controlled or monitored by computer-based algorithms, tightly integrated with the Internet and its users" (Monostori et al. 2016). Therefore, a CPS is a system for collaboration between computational entities that are in intensive connection with the surrounding physical world and its ongoing processes, providing and using, at the same time, data-accessing and data processing services available on the internet (Monostori et al. 2016). In CPS, physical and software components are deeply intertwined, each operating on different spatial and temporal scales, exhibiting multiple and distinct behavioral modalities, and interacting with each other in many ways that change with context (NSF 2010). Additionally, the technical design and implementation of CPS can refer to the "5C" architecture, which includes the following five levels of content and construction modes: the smart connection level, the data-to-information conversion level, the cyber level, the cognition level, and the configuration level (Lee et al. 2015).

Integration and interoperability are two key factors in I4.0 (Chen et al. 2008; Romero and Vernadat 2016). Interoperability is "the ability of two systems to understand each other and to use functionality of one another," which represents the capability of two systems to exchange data and share information and knowledge (IDABC 2005). The four levels of the architecture of I4.0 interoperability include operational (organizational), systematical (applicable), technical, and semantic interoperability. Interoperability makes I4.0 and CPS more productive and provides cost savings. Specifically, the operational interoperability illustrates the general structures of concepts, standards, languages, and relationships within CPS and I4.0. With the integration of computer and network systems, I4.0 achieves seamless cooperation across organizations and industries.

18.4 The IoT technologies

18.4.1 IoT introduction

Kevin Ashton devised the term "IoT" in 1999, with IoT adopting the RFID on supply chain monitoring under the Electronic Product Code (EPC) global architecture (Ashton 2009). In 2005, the International Telecommunication Union (ITU) published a report named "ITU Internet Reports 2005: The Internet of Things" (ITU 2005). According to the latest definition provided by the ITU (2012), "IoT is a global infrastructure for the information society enabling advanced services by interconnecting (physical and virtual) things based on, existing and evolving, interoperable information and communication technologies." The International Organization for Standardization (ISO 2018) provided the following, similar, definition: "an infrastructure of interconnected objects, people, systems and information resources together with intelligent services to allow them to handle information of the physical and the virtual world and react." According to IERC (2012), IoT is "a dynamic global network infrastructure with self-configuring capabilities based on standard and interoperable communication protocols where physical and virtual things have identities, physical attributes, and virtual personality and use intelligent interfaces, and are seamlessly integrated into the information."

Therefore, IoT can be a combination of physical and virtual states, which comprise many active physical things such as sensors, actuators, cloud services, communication and protocols, and the enterprise and user with many specific architectures, thus providing a framework and related solutions for IoT systems. In recent years, a rapid increase in sensor and communication technologies has promoted the growth of IoT, with more and more devices and sensors being deployed in the construction and industrial sectors, including transportation, safety, health, smart building, and automotive sectors. The number of sensor applications is increasing at an exponential rate, and it is estimated that there will be over 50 billion connected devices by 2020 (Gubbi et al. 2013).

18.4.2 Enabling technologies for IoT

The essence of the IoT is to facilitate the smartness and the connection of things; therefore, various kinds of technologies are used to help to improve the management performance in the virtual and physical world. For the use of bionics in the IoT, the main technologies contain perception and communication. In recent years, many advanced sensor applications have been developed to give objects a capability to perceive light, temperature, and movement sensors and RFID. These sensors give the object a sensing ability so that it can understand its physical condition, or it can recognize or be recognized, just like the vision and hearing perception of a human. Moreover, as network and communication technologies achieve a connection between things, big data can be collected automatically and analyzed, thus enabling the things to exchange information with each other, even making judgements and reacting. The main representative technologies are introduced in the following section.

18.4.2.1 RFID

RFID is the earliest technology of the IoT, which is used to provide unique identification of objects (Ashton 2009). In 1999, the MIT Auto ID center defined the original shape of the "Internet of Things" as being based on the computer internet, using RFID and wireless data technologies for communication, and constructing an IoT that covers everything in the world to enable automatic identification of items and sharing of information (Sarma et al. 2000).

RFID uses electromagnetic fields to automatically identify and track tags attached to objects. The tags contain electronically stored information (Roberts 2006). A typical RFID system comprises three components: an antenna or coil, a transceiver (with a decoder), and a transponder (RF tag) electronically programmed with unique information (Domdouzis et al. 2007). During operation, the radio signals are emitted to the tag by the transceiver through the antenna; the tag is then activated and the data on internal chip can be read (or written) so that the object can be recognized.

In general, the reading distance of RFID reaches about 100 feet depending on the power output and the radio frequency. Many tags can be read simultaneously among the radio coverage and processed by the computer system. RFID tags can be classified into two categories depending on the data storage capability: Read-Only and Read/Write Tags. Most Read-Only tags do not have a data storage function and only have a unique pre-written ID for identifying the attached object.

Based on the frequency band, RFID systems include Low Frequency (LF) systems, High Frequency (HF) systems, and Ultra High Frequency (UHF) systems (Fernández-Caramés et al. 2017), LF RFID operates at 125 kHz and reading range is about 10 cm, which is applied in industrial identification and automation; HF RFID operates at 13.56 kHz and is about 1 m reading range, and is used in ticking and payment; UHF operates at 860–960MHz and 2.45 GHz, and is about 10 m reading range, mostly applied on logistics and inventory management (Fernández-Caramés et al. 2017). RFID tags can also be classified as "Active" and "Passive." Passive tags rely on the electromagnetic field generated by the RFID reader for electrical power and to be activated. Active tags are equipped with built-in batteries, which increase the reading range (Domdouzis et al. 2007). The EPC of RFID provides a unique identification of the object, which also satisfies the recognizing requirement of big data in the IoT world. RFID has been successfully applied in many IoT applications, such as smart factories and manufacturing, supply chains and logistics, vehicle and traffic metro systems, aviation and transportation, and payment transactions, and the applications are constantly being extended and developed.

18.4.2.2 WSN

The WSN is one of the key IoT application technologies, which enables the things to have capabilities in perception and interaction. A WSN system typically comprises wireless functional sensor devices that can smartly collect and communicate environmental information and even judge and take action.

The application of WSN contains sensor and network technologies. The general application wireless network specification includes Wi-Fi, Bluetooth, and ZigBee, which each have their features and applications, such as Wi-Fi based on the 2.4 GHz frequency band and speeds reaching 11 Mbps. Bluetooth is another 2.4 GHz wireless application for personal electronic device service, while ZigBee features low speed, cost, and low power consumption. Table 18.1 compares the wireless standards for WSNs.

A typical WSN node is mainly composed of four components: a sensing unit, a processing unit, a transceiver unit, and a power unit. (1) The sensing unit includes two parts: a sensor and Analog-to-Digital Converters (ADC). The former collects sensed environmental information and converts this information into an analog signal, and the latter converts the signal into a digital signal for processing. (2) The processing unit includes the storage and processor components. The processor withdraws data from storage, interprets the packet of wireless signal, coordinates with the neighborhood nodes, and then handles the data transferring task. (3) The transceiver unit transmits data through a radio signal and sends it to the host. (4) The power supply unit provides regular power for the operation of the sensor node (Akyildiz et al. 2002).

As many IOT devices are usually set on a person or moving object and they operate individually, it is important to know how to save power and maintain sustainable operation in the WSN system design. Another important issue in the IoT system design is the transmission ability because the distance between the nodes and the configuration of network influences the data communication efficiency and the required operating power. Concerning the transmission distance and energy savings, when the sensor node is too far away from the host, the WSN needs to establish a network routing protocol by a multiple-hop relay of nodes. The transmission distance can be extended by using relays of the nodes.

In terms of network architecture, the WSN network typically comprises three roles: the coordinator, router, and end device. The coordinator acts as the host and is responsible for

Table 18.1 Comparison between Wi-Fi, Bluetooth, and ZigBee

	Wi-Fi	*Bluetooth*	*ZigBee*
Standard	IEEE 802.11b	IEEE 802.15.1	IEEE 802.15.4
Frequency	2.4G	2.4G	2.4G/868/915 MHz
Range (m)	100	10	50
Data rate	11–54 Mbps	3–10 Mbps	250 kbps
Nodes per master	32	7	65,535
Topologies	Star	Star	Star, Mesh, Tree
Battery life	Hours	1 week	>1 year
Features	Speed	Convenience	Reliability, Low cost & Low power
Application	Video, audio, pictures, files	Audio, graphics, pictures, files	Low data communication

network launching, coordinating, and data collection. The router relays the radio signal and transfers it to the next node, which extends the WSN coverage and detection distance. The end device is equipped with various functional sensors and is responsible for measuring the surrounding conditions and returning the sensed information (Farahani 2011).

As the WSN is affected by factors of circumstances, design function, and hardware, the following items should be considered in proposed WSN systems (Akyildiz et al. 2002): (1) fault tolerance of the network, (2) network scalability, (3) hardware price, (4) hardware constraints of the sensor, (5) network topology, (6) operating environment, (7) transmission media, and (8) power consumption.

Within the developing technologies of sensors, networks, and semiconductors, the application of WSN will become deeper and more diversified, thus facilitating more complete IoT systems. Additionally, when WSN combines big data and cloud computing, the applications, and the influences of IoT are almost infinite.

18.4.2.3 BIM

BIM brings the revolution of digitalization and informatization to the entire construction industry, digitalizing and parameterizing different building information and visually integrating this information into a 3D model. BIM integrates models, databases, assets, and material and spatial relation information, providing capabilities in construction simulation, progress management, cost estimation, and energy analysis. BIM is widely utilized in the Architecture, Engineering & Construction (AEC) domain (Cerovsek 2011), and it enhances not only planning, design, construction, operation, and maintenance (O&M) processes, but also the entire building project life cycle (Eastman et al. 2011). BIM is defined by the US National Institute of Building Sciences (NIBS) as "a digital representation of physical and functional characteristics of a facility" (NIBS, 2019). The ISO (2016) defines BIM as a "shared digital representation of physical and functional characteristics of any built object (including buildings, bridges, roads, etc.), which forms a reliable basis for decisions."

During the operation phase, many operating facilities, labor activities, and conditions need to be controlled and managed. The traditional monitoring method of using a human is highly laborious, and it is hard to meet the real-time monitoring requirement. In recent years, as IoT technologies have matured, many of the sensor applications enable the infrastructure to have abilities in perception and communication (Haines 2016), which promotes BIM as a potential development in O&M. In practice, IoT devices provide the status information of things, and their positional information can be linked with the BIM model so that the spatial relations and real-time statuses can be displayed simultaneously. BIM provides a framework for the IoT information to be integrated and analyzed in a way that is meaningful to the O&M of building management. Moreover, the integration of IoT information and BIM models can facilitate the CPS for the management of the construction project. The collected huge amount of data also forms a big data environment that provides sufficient information for analytics and improvement.

The integration of BIM and the IoT establishes a virtual reproduction of building projects in which IoT provides the dynamic information of people, facilities, assets, and status of the building, and BIM provides the framework for IoT information that can be systematically integrated and spatially demonstrated. The combination of IoT and BIM provides an active building model and creates a CPS application of the building project; this combination not only achieves the optimal operation and management but also enhances the application of smart buildings.

18.5 Applications of IoT in construction

In construction, as many of the workers, equipment, and materials are frequently moved in and out of the site, various situations, such as conditions of the machines, the number of workers, and the risky environmental hazards, need to be mastered and controlled. The different monitoring tasks will incur a heavy labor duty if executed by a human. However, if the IoT applications can be introduced, they can be advantageously used for obtaining various site information, to monitor the statuses of the workers, and to judge and execute emergency actions automatically in the event of an emergency. These applications will greatly increase management efficiency and effectiveness, while improving the site environment and human safety. Many of the IoT applications in construction are introduced in the following section.

18.5.1 Surveying, mapping, and security

In recent years, the unmanned aerial vehicles (UAVs) have been used and applied widely due to price declines and advances in flight control software. Photo and real-time data capture technology from the sky can be applied to many construction domains for management improvement. Common UAV or UAVS (UAV systems) designs can be classified by their flight mechanism as either fixed-wing (aircraft), rotor (helicopter), or multi-rotor aircraft, and they are usually equipped with a high-resolution camera and Global Positioning System (GPS). The measuring precision can reach a few centimeters and be controlled by cell phone. At present, many UAV applications are quite mature and have been applied in agricultural, mining, construction, ecological, and environmental domains. Several relevant applications are described as follows: (1) Progress control of construction projects: inspection and monitoring of construction is essential for assessing site conditions. The large area and precision images enable the administrator to understand the progress of the project and identify disparities between the as-built and as-planned progresses. (2) Investigation and rescues: the rapid scanning capability of UAVs can quickly perform investigation tasks after natural disasters, such as typhoons and earthquakes, instead of having humans enter the dangerous areas, and they can quickly collect disaster information and provide reference information, thus significantly improving the efficiency and safety of rescue tasks. (3) Surveying and measuring: UAV equipped with high-resolution lenses can capture highly detailed images and perform accurate distance measurements of large areas in a short amount of time. Additionally, the information can be converted into spatial surface models for topographic mapping, volume calculations, and even as a 3D digital BIM model. (4) Safety management: UAV is advantageous in monitoring the field and personnel activities in construction areas, controlling the various unsafe conditions and providing early warnings. In emergency accidents, UAVs can quickly determine the accident location and the injured person, thus enabling an immediate rescue action and improving the safety management of the construction site.

In recent years, many scholars have researched the innovative applications of UAV. Moon et al. (2019) proposed a method for generating and merging hybrid point cloud data acquired from laser scanning and UAV based imaging. Inzerillo et al. (2018) proposed a UAV based Structure from Motion (SfM) technique, which they applied to road pavement detection, assessing the potential for improving the automation and reliability of distress detection. Morgenthal et al. (2019) presented a coherent framework for automated unmanned aircraft system-based inspections of large bridges to facilitate an automated condition assessment.

18.5.2 Safety management

Owing to the reduced size and long-term operation abilities, IoT devices are increasingly being devoted to safety management. Many of the WSN devices can be set at different construction site locations, where they can be used to monitor risk factors such as fire, smog, vibration, and high noise. In an emergency event, the safety device can be activated immediately, alerting the site workers to evacuate, and can eliminate any hazards automatically, thus preventing a serious disaster. Moreover, the sensor or RFID can be embedded into personal wearing devices such as helmets, vests, or other items, to identify and locate the worker and to determine their action and vital status. When detecting an abnormal condition, the IoT devices will alert, provide feedback, and request help immediately. The IoT enables the safety administration to establish the status of the whole site and react to risks in real time.

As applications of the IoT in safety management are becoming increasingly extensive, many scholars have already conducted specific research studies. Ding et al. (2013) presented a real-time safety early warning system to prevent accidents and improve safety management in underground construction based on IoT technology. The system has been validated and verified through a real-world application at the cross passage construction site in the Yangtze Riverbed Metro Tunnel project in China. Valero et al. (2017) proposed a novel system and data processing framework to deliver intuitive and understandable motion-related information about workers using WSN.

18.5.3 Supply chain and facilities management

The practice of C4.0 delivers construction projects faster and more flexibly and provides higher quality and reduced costs, in accordance with the lean construction approach, which attempts to improve construction processes at a minimum cost and maximum value. In addition, supply chain management (SCM) is an important issue for lean construction management. The traditional SCM and logistics rely mainly on manual management, in which the efficiency is low and it is hard to track and manage assets in real-time, and thus cannot meet the requirement for SCM.

In the management of the materials and equipment in construction, the sensors and RFID tags can help to identify item status and stock quantity. Furthermore, employees can quickly collect information about the warehousing and consummation status or check the delivery schedule of the material by scanning the RFID tags. The system can also automatically calculate the stock and issue notifications for purchasing when inventory is low, which much improves the efficiency of SCM. Additionally, the sensors can monitor the operating conditions of equipment, control abnormal situations, and issue alerts for timely repair or maintenance in advance. The relevant IoT application reduces workflow delays and enhances the efficiency of site operations. Many researchers have proven the effectiveness of using IoT applications in lean construction and SCM. Hinkka and Tätilä (2013) presented an RFID tracking implementation model for technical trade and construction industries. Their approach for building a feasible model was built based on a survey of 16 manufacturing and wholesaler company interviews. Xu et al. (2018) proposed an integrated cloud-based IoT platform by exploiting the concept of cloud asset, which enabled enterprises to adopt IoT technologies economically and flexibly. Ko et al. (2016) proposed a cost-effective materials management and tracking system based on a cloud-computing service integrated with RFID for automated tracking with ubiquitous access.

18.5.4 Structural health monitoring (SHM)

The health of buildings and infrastructure is relevant to the safety of the occupants and the population; the aging of materials and structures may lead to crash events and disasters. SHM has become an important issue in building operation and maintenance. Many physical and environmental conditions of structures, such as vibration and deformation, tensile and compressive stresses, and temperature and wind speed, can be monitored continuously using IoT or WSN devices. The real-time and long-term observed conditions could be wirelessly transferred and collected for analysis. The information also helps to check the damage of structures after earthquakes or to estimate the structure health and its service life, thus increasing the building's safety and reducing maintenance costs.

Many studies have investigated the use of IoT applications for building and structure monitoring. Bae et al. (2013) reported the results of an experimental evaluation of WSN performance in the obstructed environment of a bridge structure. Park et al. (2018) presented a real-time SHM technique for a super tall building under construction (Lotte World Tower, LWT) to propose a visual modal identification method to identify mode shapes and damping ratios based on modal responses from the monitoring system. To reduce the randomness and uncertainty underlying the structural safety risk analysis in operational tunnels, Liu et al. (2018) developed a novel hybrid approach to perform a global sensitivity analysis and analyze the input-output causal relationships of the structural safety risk, thus reducing the epistemic uncertainty in tunnel structural safety management. Hasni et al. (2018) presented a novel approach to detect damage in steel frames using a hybrid network of piezoelectric strain and acceleration sensors.

18.5.5 Smart building applications

Many IoT applications have been implemented in intelligent buildings for air conditioning, electromechanical control, security, burglary alarm systems, and fire and disaster prevention systems. The systems are all connected by a network, and the users can monitor the operation of devices from anywhere in the world. In a smart building, the operating status of various systems are collected and analyzed. The IoT enables the systems to not only be controlled but to also achieve a better balance between the systems, which brings more comfort and safety for occupants and enhances the working efficiency and productivity. Additionally, considering the impact of global warming and extreme weather conditions, building energy efficiency and green functions is becoming increasingly important. In smart buildings, the equipment linked with sensors and communication devices enables the operation conditions to be monitored in real time and collected in the cloud. The various operating information, such as the use of elevators, conference rooms, and air conditioners, is then analyzed. Through the analysis, the smart building can determine the optimal operation mode, such as controlling the air conditioning temperature based on the ambient temperature and the number of occupants, which reduces the energy consumption of the equipment and achieves energy savings. In recent years, several researchers have investigated the use of IoT applications for smart buildings. Rashid et al. (2019) presented an intuitive point-and-click framework to control electrical fixtures in a smart built environment using an ultra-wide band (UWB)-based indoor positioning system. Jia et al. (2019) investigated the state-of-the-art projects and adoptions of IoT for developing smart buildings within academic and industrial contexts.

The applications of IoT are therefore widely applied in construction, architecture, and infrastructure, and for energy saving, and the utilization is not only wide but also deep. As any object, equipment, machine, and even personnel can be connected, IoT applications greatly reduce the labor needed for monitoring and thus improve management and safety. The IoT brings enormous potential and great encouragement for future applications in the construction industry, limited only by our imagination.

18.6 Case study

18.6.1 Introduction

The following case study demonstrates how to develop a CPS in the safety management of an infrastructure that monitors hazardous gas and environmental conditions using IoT devices. As many construction sites are located underground or in confined spaces, the environment usually has a high temperature and humidity, and may also contain hazardous gases, such as flammable gases that could cause a serious explosion or carbon monoxide (CO), rendering individuals unconscious. Long-term monitoring work usually requires heavy labor, and the different locations are hard to monitor simultaneously, thus resulting in an inability to effectively monitor workers and prevent accidents.

In this study, many functional WSN sensor nodes (IoT devices) were placed in an underground tunnel site, which collected the hazardous gas, temperature, and humidity information of the working environment. The collected data were transferred back to the monitoring center using a WSN self-organized network. When detecting the abnormal gas condition, the node with a control function would automatically start up a ventilator to disperse hazardous gases. The collected information was integrated into a BIM model, with the hazardous gas, temperature, and humidity conditions displayed by an active color changing from blue to red in real time to indicate the risk degree of the location being monitored. In the event of an emergency, the administrator could remotely understand the risk at the monitoring locations and make optimal decisions for the rescue task.

18.6.2 System analysis and design

To establish the CPS of the project and link with the physical conditions, two sub-systems were integrated include the WSN and the digital information model. The model and related components were established using BIM software, which visually presents the CPS and locational conditions. For data collection, a WSN system comprising various functional nodes with sensing and communication capabilities was developed to collect the site environmental information. The CPS can be established by integrating the sub-systems, and the real risk and environmental conditions were linked and demonstrated in real time.

Experts and users were interviewed to understand the existing safety management of hazardous gases on construction sites and the practical problems encountered. The findings revealed several requirements, which included enhancing the workers' safety; implementing long-term, multi-point detection and recording; enabling convenient installation and power savings; and using an expandable and intelligent function design. Since most of the current gas detection tasks on construction sites were executed by personal hand-held detectors, which was inefficient and highly risky for staff, this study proposed using the WSN and a remote CPS to improve the current situation.

A multi-tier architecture analysis method was adopted to analyze the proposed system, with different entities classified by their attributes, thus clarifying the relationships between the sub-systems and modules. The classification includes (1) the presentation layer, (2) the application layer, (3) the database layer, and (4) the sensor and communication layers. The overall CPS system, including the sensors, the digital model, and data processing, was integrated using Microsoft C# applications. Figure 18.1 shows the multi-tier architecture and data flow of the system. Table 18.2 provides descriptions of each layer.

18.6.3 System development

18.6.3.1 WSN sub-system development

The Microsoft Gadgeteer embedded system (.NET MF) was adopted as the hardware framework of the WSN node, which was a System-on-Chip (SoC) designed system proposed to be equipped with gas, temperature, humidity sensing components, and the ZigBee wireless network module. Each WSN node was designed for a specific role, and the different type of functional WSN nodes comprised the WSN system by interactive radio communication. The following four types of wireless nodes were proposed: sensor node, router, coordinator, and control node. In the network configuration, the mesh type was designed as the network

Cyber Physical System

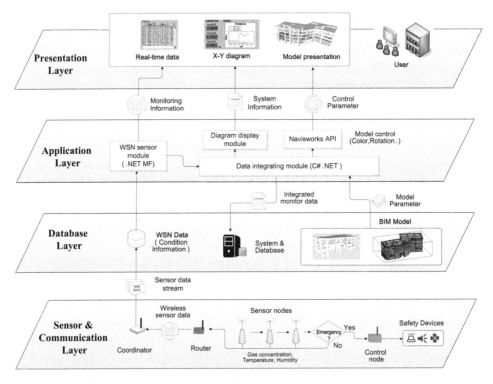

Figure 18.1 Multi-tier architecture analysis of the system

Table 18.2 The descriptions of each layer

Layer	Description	Operation
Presentation Layer	1. Real-time data collection and display 2. The Active model of CPS for safety management 3. CPS development and user interface design	1. To handle and display the real-time data returned by WSN node (digits and curve). 2. To establish the BIM model and the status indicting component for display. 3. To develop the CPS and relevant user interface by integrating BIM model and WSN information.
Sensing & Communication Layer	1. A WSN network system consisting wireless sensor nodes 2. Collection of locational hazardous gas and environmental conditions 3. The WSN smart function design	1. To develop the WSN with various functional nodes including (a) Sensor node. (b) Router. (c) Control node. (d) Coordinator. 2. To propose ZigBee-based WSN for gas, temperature and humidity detection within "mesh" type topology. 3. To design the smart functions (power saving, emergency judge and hazards removing).
Application Layer	1. Construction WSN nodes and function design 2. Establishment and control of the BIM model 3. System integration	1. To design the relevant WSN function on the single chip system (.Net MF). 2. To establish, modify the model and dynamic color control (BIM software). 3. To integrate overall system and information processing (Microsoft C#).
Data Layer	1. The data standardization and flow control 2. BIM model for spatial and conditional information integration 3. Database for sensor and system data storage	1. To design and handle the data flow including collection, transfer and integration. 2. To build a digital model to correspond with structure, assets, and monitoring location of the site. 3. To establish the database for storage of measuring records and data processing.

topology, which enables the WSN detection range to be expanded and re-links the network automatically when any node fails. Table 18.3 lists the developed WSN nodes and their functions. Figure 18.2 shows the developed sensor node.

The WSN system comprises many designed smart functions. In the monitoring task, when the hazardous gas or abnormal conditions are detected in the site, the sensor node immediately tells the control node to switch on the flash alarm and ventilation, and removes the hazardous gas automatically. The alarm gives the site workers early notice to retreat, and thus avoids a serious accident caused by gas accumulation. Meanwhile, in normal status, the sensor node records data at a lower frequency of 3 seconds/data point, which enables power saving and longer operation. Once the hazardous gas is detected, the frequency will increase to 0.3 second/data point, which keeps detailed records and provides a reference for further investigations.

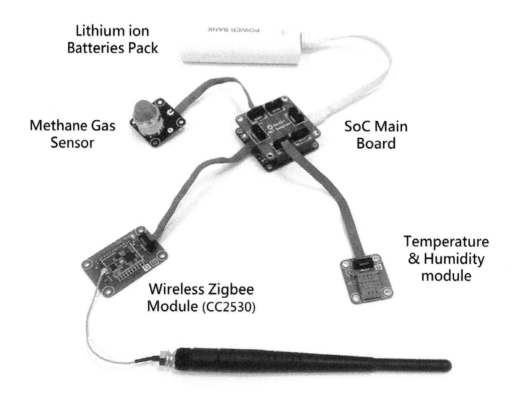

Lithium ion Batteries Pack

Methane Gas Sensor

SoC Main Board

Wireless Zigbee Module (CC2530)

Temperature & Humidity module

Figure 18.2   The developed sensor node (expanded)

Table 18.3 Developed WSN nodes and its functions

	Type of node	Function
1	Coordinator	(1) Responsible for launching the network
		(2) To coordinate the address assignment of all WSN nodes
		(3) To be a host and collect the data from all sensor nodes
2	Sensor node	(1) To monitor the hazardous gas, temperature and humidity information in each location of construction site
		(2) To ask the control node to start safety device and remove hazards when detects the abnormality
		(3) To record the detail data in emergency situation
3	Router	(1) To relay & transfer WSN transmission signal, extends communication distance and network covering range.
4	Control node	(1) Connection with safety devices (flash, alarm, and ventilator), activated in emergent situation

18.6.3.2 BIM integration

The digital model provides the visual demonstration of the physical construction site, which establishes a dimensional framework for CPS so that the measuring information can be visually displayed. The BIM model was built by Autodesk Revit and converted to a Navisworks file. To display the status of monitoring locations, spatial components were created to indicate the different conditions; meanwhile, the outside shield of the model should be properly removed to enable the inside conditional component to be displayed. The color of the component was controlled to be changed dynamically by the data returned from the WSN, including gas, temperature, and humidity conditions. Figure 18.3 shows the construction progress of the BIM model.

During the operation, the data generated by the WSN nodes was collected continuously. The data included the node's identification number, signal strength, and the detected environmental information (such as gas concentration, temperature, and humidity). The system analyzed where the data came from and identified the gas, temperature, and humidity values. The values were then referred to display the relevant color from blue to red and indicated the risk degree of the monitored location using specific model components. By this mechanism, the safety and environmental conditions of the whole site can be displayed in real time via the input data flow. Figure 18.4 shows the integrating mechanism of the BIM component and the WSN data.

18.6.4 Test and discussion

To test the performance of the proposed system, an underground construction tunnel of Mass Rapid Transit (MRT) was chosen as the experimental site. The test simulated the presence of a hazardous gas and tested the performance of the proposed CPS. For a coverage of 300 meters, the monitoring points were placed at every 100 meters, where the three nodes were set. The system was installed on a notebook, which included the CPS and digital model of the tunnel. The coordinator linked with the notebook was set at the station for receiving the detected data from the sensor node. A control node was connected with a flash alarm and ventilator, which would be activated when the WSN detected the hazardous gas among the coverage. Figure 18.5 shows the experimental layout.

During the test, the coordinator launched the network, and the sensor nodes and control node joined the network sequentially by powering on. Then, the sensor nodes were moved forward into the tunnel by staff and set at positions located 100, 200, and 300 meters from

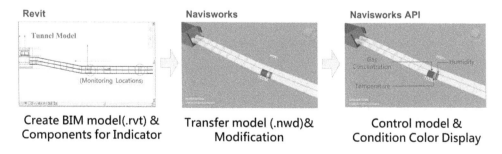

Revit	Navisworks	Navisworks API
Create BIM model(.rvt) & Components for Indicator	Transfer model (.nwd)& Modification	Control model & Condition Color Display

Figure 18.3   The BIM model constructions for the system

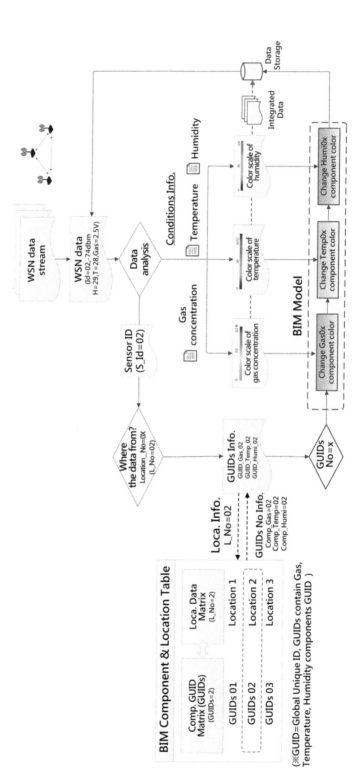

Figure 18.4 Integrating mechanism of the WSN data and BIM model component

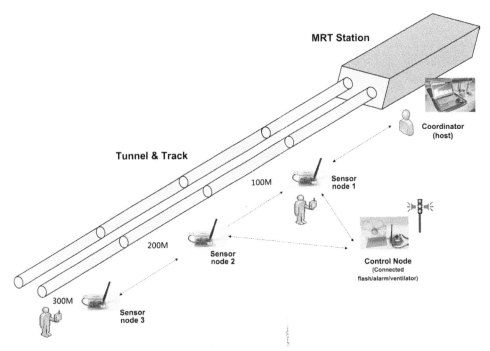

MRT Station

Tunnel & Track

Coordinator
(host)

100M

Sensor
node 1

200M

Sensor
node 2

Control Node
(Connected
flash/alarm/ventilator)

300M

Sensor
node 3

Figure 18.5   The design layout of the field test

the host. During this process, the signal strength was also recorded every 20 meters, thus observing the attenuation of the signal. When all the sensor nodes had been set in place, the staff simulated the emergency event by emitting combustible gas from a fire lighter in each monitoring position. The staff then observed the system performance to determine whether the safety devices were activated (Figure 18.6). Figure 18.7 presents the screenshot of the system operation.

The test results indicated that even in the 300 meter position, the system reacts to the gas occurrence within 1–2 seconds and displays the abnormal location on the BIM model by a red colored warning; meanwhile, the control node immediately activates the flash alarm and the ventilator to warn nearby personnel to retreat and to eliminate the hazardous gas. The attenuation test of the radio signal indicates that the transmission distance of two nodes in the tunnel is about 250 meters and 0.2 to 0.3 dBm attenuated for every meter. If a router is set between two nodes, the distance can be extended up to 450 meters.

The test result indicates that the system can be conveniently set up on a construction site and it can perform well at a detecting and monitoring task. The developed system continuously collects the environmental information and monitors the hazards of the construction site. Once the emergency event occurs, the workers can be warned in advance and retreat immediately. Additionally, the administration and rescue unit can remotely determine the risky locations and the situation of the site through CPS to make an optimal decision and take immediate rescue action. The system can smartly improve the safety management of construction sites for hazardous gas prevention, significantly decreasing the monitoring labor and enhancing the workers' safety.

Figure 18.6 The field test in the tunnel

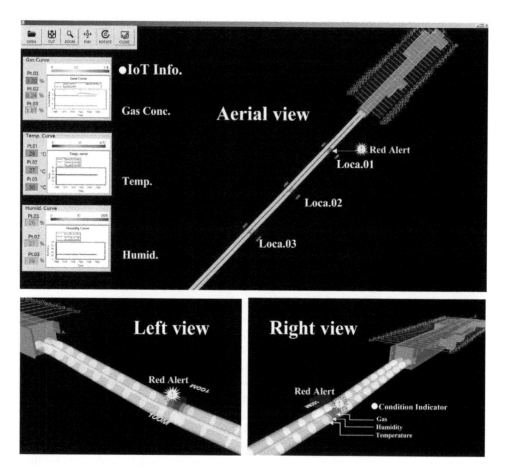

Figure 18.7 Screenshot of the operational CPS for tunnel safety management (Location 01 detected hazardous gas and demonstrates the warning in red)

18.7 Conclusion

This section describes the definitions, concepts, and framework for I4.0 and C4.0, and discusses the applications of IoT and related technologies in the construction industry. The construction industry is slower than other manufacturing industries at incorporating digitalization because of its complexity and unique characteristics. However, as the future trends of construction projects are moving toward increased size and complexity, the construction sector needs to pay more attention to IoT technologies and to enhancing its efficiency and competitive capabilities.

This article describes the use of IoT applications in the construction industry, including UAV, RFID, WSN, and BIM technologies. UAV technology improves the mapping and investigation tasks in construction projects and enables the quick survey of large areas after disasters, thus enhancing not only the efficiency but also the accuracy and safety of the task. RFID provides the remote and unique identification of things for big data applications and the IoT, which is widely applied in personnel and equipment control, logistics management, and lean construction management. WSN enables facilities and equipment to perceive and communicate, and to quickly provide the required information for big data and cloud computing analytics. Nowadays, WSN is widely applied in the fields such as traffic and localization, environment and security detection, smart buildings, and energy saving applications. BIM provides a digitalization and visualization solution for construction projects in IoT applications, providing not only static information, including assets, structures, and their spatial relations, but also powerful computing and simulation capabilities. BIM also provides a digital framework for IoT information integration and CPS establishment. The introduction of IoT technologies enables equipment and facilities to have the ability to sense, communicate, and make decisions. By integrating IoT, big data, and CPS, the digital virtual world is linked with the physical world, providing better service and predictions. IoT not only enhances the digitalization, informationization, and cyberization in the construction domain, but it also provides the required technology and solutions for smart buildings, construction, and manufacturing.

This chapter also provided an example of CPS development as applied to on-site safety management, including a description of the IoT device development, network system design, digital model establishment, and how to integrate and make them work. The developed prototype system was tested on a tunnel site, and its effectiveness was demonstrated. The results indicate that the developed system can perform the monitoring task well and is advantageous to the safety management of the site. The system can achieve the long-term, multi-locational monitoring in real time, and provides a CPS for remote management, timely rescue, early warning, and smart security, thus not only reducing the labor cost for monitoring, but also significantly enhancing the safety of workers on the site.

During the IoT introduction process, the following issues still need to be explored and solved: the required human resource and technology, the security of networks and communication, the integration and interoperability between the different systems, and the speed and the huge demands of the network. In the future, these IoT technologies will continue to impact and shape the construction industry, not only bringing new challenges and opportunities, but also promoting the industry towards facilitating the objective of Construction 4.0.

18.8 Summary

- Overview of IoT enabled physical technologies in Construction 4.0.
- Literature reviews of RFID, UAV, WSN, and BIM technologies in Construction 4.0.
- Applications of advanced IoT in Construction 4.0.
- Case study showcasing development of a cyber physical system for application in tunnel construction safety management for Construction 4.0.

References

Akyildiz, I. F., Su, W., Sankarasubramaniam, Y., and Cayirci, E. (2002). "Wireless sensor networks: A survey", *Computer Networks*, 38(4), 393–422. https://doi.org/10.1016/S1389-1286(01)00302-4.

Ashton, K. (2009). "That 'internet of things' thing", *RFID Journal*, 22(7), 97–114. Available from: www.rfidjournal.com/articles/view?4986 (accessed 2019/04/06).

Bae, S. C., Jang, W. S., and Woo, S. (2013). "Prediction of WSN placement for bridge health monitoring based on material characteristics", *Automation in Construction*, 35, 18–27. https://doi.org/10.1016/j.autcon.2013.02.002.

Cerovsek, T. (2011). "A review and outlook for a 'Building Information Model' (BIM): A multi-standpoint framework for technological development", *Advanced Engineering Informatics*, 25(2), 224–244. https://doi.org/10.1016/j.aei.2010.06.003.

Chen, D., Doumeingts, G., and Vernadat, F. (2008). "Architectures for enterprise integration and interoperability: Past, present and future", *Computers in Industry*, 59(7), 647–659. https://doi.org/10.1016/j.compind.2007.12.016.

Consortium II (2013). "Fact sheet", Available from: www.iiconsortium.org/docs/IIC_FACT_SHEET.pdf (accessed 2019/04/06).

Ding, L. Y., Zhou, C., Deng, Q. X., Luo, H. B., Ye, X. W., Ni, Y. Q., and Guo, P. (2013). "Real-time safety early warning system for cross passage construction in Yangtze Riverbed Metro Tunnel based on the internet of things", *Automation in Construction*, 36, 25–37. https://doi.org/10.1016/j.autcon.2013.08.017.

Domdouzis, K., Kumar, B., and Anumba, C. (2007). "Radio-Frequency Identification (RFID) applications: A brief introduction", *Advanced Engineering Informatics*, 21(4), 350–355. https://doi.org/10.1016/j.aei.2006.09.001.

Eastman, C., Teicholz, P., Sacks, R., and Liston, K. (2011). *BIM Handbook: A Guide to Building Information Modeling for Owners, Managers, Designers, Engineers and Contractors*, Hoboken, NJ: John Wiley & Sons.

Farahani, S. (2011). *ZigBee Wireless Networks and Transceivers*, Newton, MA: Newnes.

Fernández-Caramés, T., Fraga-Lamas, P., Suárez-Albela, M., and Castedo, L. (2017). "Reverse engineering and security evaluation of commercial tags for RFID-based IoT applications", *Sensors*, 17(1), 28. https://doi.org/10.3390/s17010028.

Gubbi, J., Buyya, R., Marusic, S., and Palaniswami, M. (2013). "Internet of Things (IoT): A vision, architectural elements, and future directions", *Future Generation Computer Systems*, 29(7), 1645–1660. https://doi.org/10.1016/j.future.2013.01.010.

Haines, B. (2016). "Does BIM have a role in the Internet of things?", Available from: https://fmsystems.com/blog/does-bim-have-a-role-in-the-internet-of-things/ (accessed 2019/04/06).

Hasni, H., Jiao, P., Alavi, A. H., Lajnef, N., and Masri, S. F. (2018). "Structural health monitoring of steel frames using a network of self-powered strain and acceleration sensors: A numerical study", *Automation in Construction*, 85, 344–357. https://doi.org/10.1016/j.autcon.2017.10.022.

Henning, K., Wolfgang, W., and Johannes, H. (2013). "Recommendations for implementing the strategic initiative INDUSTRIE 4.0", Available from: www.acatech.de/wp-content/uploads/2018/03/Final_report__Industrie_4.0_accessible.pdf (accessed 2019/04/06).

Hermann, M., Pentek, T., and Otto, B. (2015). "Design principles for Industrie 4.0 scenarios: A literature review 2015 07/ 04/2015", Available from: www.snom.mb.tu-dortmund.de/cms/de/forschung/Arbeitsberichte/Design-Principles-for-Industrie-4_0-Scenarios.pdf (accessed 2019/04/06).

Hinkka, V. and Tätilä, J. (2013). "RFID tracking implementation model for the technical trade and construction supply chains", *Automation in Construction*, 35, 405–414. https://doi.org/10.1016/j.autcon.2013.05.024.

IDABC (Interoperable Delivery of European eGovernment Services to Public Administrations, Businesses and Citizens). (2005). "European interoperability framework v 1.0", http://ec.europa.eu/idabc/en/document/3782/5584.html (accessed 2019/04/06).

IERC (European Research Cluster on the Internet of Things) (2012). "The Internet of things 2012 – New horizons", Available from: www.internet-of-things-research.eu/pdf/IERC_Cluster_Book_2012_WEB.pdf (accessed 2019/04/06).

Inzerillo, L., Di Mino, G., and Roberts, R. (2018). "Image-based 3D reconstruction using traditional and UAV datasets for analysis of road pavement distress", *Automation in Construction*, 96, 457–469. https://doi.org/10.1016/j.autcon.2018.10.010.

ISO (International Organization for Standardization) (2016). "ISO 29481-1:2016(en): Building Information Models – Information delivery manual – Part 1: Methodology and format", Available from: www.iso.org/obp/ui/#iso:std:iso:29481:-1:ed-2:v1:en (accessed 2019/04/05).

ISO (International Organization for Standardization) (2018). "ISO/IEC 30141:2018 Internet of Things (IoT) – Reference architecture", Available from: www.iso.org/standard/65695.html (accessed 2019/04/06).

ITU (International Telecommunication Union). (2005). "The Internet of Things", Available from: http://unpan1.un.org/intradoc/groups/public/documents/APCITY/UNPAN021972.pdf (accessed 2019/04/06).

ITU (International Telecommunication Union). (2012). "Series Y: Global information infrastructure, internet protocol aspects and next-generation networks, overview of the Internet of Things". Available from: www.itu.int/rec/T-REC-Y.2060/en (accessed 2019/04/06).

Jia, M., Komeily, A., Wang, Y., and Srinivasan, R. S. (2019). "Adopting Internet of things for the development of smart buildings: A review of enabling technologies and applications", *Automation in Construction*, 101, 111–126. https://doi.org/10.1016/j.autcon.2019.01.023.

Ko, H. S., Azambuja, M., and Lee, H. F. (2016). "Cloud-based materials tracking system prototype integrated with radio frequency identification tagging technology", *Automation in Construction*, 63, 144–154. https://doi.org/10.1016/j.autcon.2015.12.011.

Lee, J., Bagheri, B., and Kao, H. A. (2015). "A cyber-physical systems architecture for industry 4.0-based manufacturing systems", *Manufacturing Letters*, 3, 18–23. https://doi.org/10.1016/j.mfglet.2014.12.001.

Liu, W., Wu, X., Zhang, L., Wang, Y., and Teng, J. (2018). "Sensitivity analysis of structural health risk in operational tunnels", *Automation in Construction*, 94, 135–153. https://doi.org/10.1016/j.autcon.2018.06.008.

Lu, Y. (2017). "Industry 4.0: A survey on technologies, applications and open research issues", *Journal of Industrial Information Integration*, 6, 1–10. https://doi.org/10.1016/j.jii.2017.04.005.

Monostori, L., Kádár, B., Bauernhansl, T., Kondoh, S., Kumara, S., Reinhart, S. G., Sihn, W., and Ueda, K. (2016). "Cyber-physical systems in manufacturing", *CIRP Annals*, 65(2), 621–641. https://doi.org/10.1016/j.cirp.2016.06.005.

Moon, D., Chung, S., Kwon, S., Seo, J., and Shin, J. (2019). "Comparison and utilization of point cloud generated from photogrammetry and laser scanning: 3D world model for smart heavy equipment planning", *Automation in Construction*, 98, 322–331. https://doi.org/10.1016/j.autcon.2018.07.020.

Morgenthal, G., Hallermann, N., Kersten, J., Taraben, J., Debus, P., Helmrich, M., and Rodehorst, V. (2019). "Framework for automated UAS-based structural condition assessment of bridges", *Automation in Construction*, 97, 77–95. https://doi.org/10.1016/j.autcon.2018.10.006.

NIBS (National Institute of Building Science) (2019). "Frequently asked questions about the national BIM standard – United States", Available from: www.nationalbimstandard.org/faqs (accessed 2019/04/06).

NSF (National Science Foundation) (2010). "Cyber-Physical Systems (CPS)", Available from: www.nsf.gov/pubs/2010/nsf10515/nsf10515.htm (accessed 2019/04/06).

Park, H. S. and Oh, B. K. (2018). "Real-time structural health monitoring of a super tall building under construction based on visual modal identification strategy", *Automation in Construction*, 85, 273–289. https://doi.org/10.1016/j.autcon.2017.10.025.

Rashid, K. M., Louis, J., and Fiawoyife, K. K. (2019). "Wireless electric appliance control for smart buildings using indoor location tracking and BIM-based virtual environments", *Automation in Construction*, 101, 48–58. https://doi.org/10.1016/j.autcon.2019.01.005.

Roberts, C. M. (2006). "Radio frequency identification (RFID)", *Computers & Security*, 25(1), 18–26. https://doi.org/10.1016/j.cose.2005.12.003.

Romero, D. and Vernadat, F. (2016). "Enterprise information systems state of the art: Past, present and future trends", *Computers in Industry*, 79, 3–13. https://doi.org/10.1016/j.compind.2016.03.001.

Sarma, S., Brock, D. L., and Ashton, K. (2000) "The networked physical world" Auto-ID Center White Paper MIT-AUTOID-WH-001. Available from: http://cocoa.ethz.ch/downloads/2014/06/None_MIT-AUTOID-WH-001.pdf (accessed 2019/04/06).

Valero, E., Sivanathan, A., Bosché, F., and Abdel-Wahab, M. (2017). "Analysis of construction trade worker body motions using a wearable and wireless motion sensor network", *Automation in Construction*, 83, 48–55. http://dx.doi.org/10.1016/j.autcon.2017.08.001.

Vogel-Heuser, B. and Hess, D. (2016). "Guest editorial Industry 4.0 – Prerequisites and visions", *IEEE Transactions on Automation Science and Engineering*, 13(2), 411–413.

Xu, G., Li, M., Chen, C. H., and Wei, Y. (2018). "Cloud asset-enabled integrated IoT platform for lean prefabricated construction", *Automation in Construction*, 93, 123–134. https://doi.org/10.1016/j.autcon.2018.05.012.

19

CLOUD-BASED COLLABORATION AND PROJECT MANAGEMENT

Kalyan Vaidyanathan, Koshy Varghese, and Ganesh Devkar

19.1 Aims

- Understanding project management as construction supply chain information management.
- Identifying technology and process enablers for collaborative project management and project controls.
- Proposal for platform solutions for collaborative project management.
- Proposal of a maturity model to assist organizations transition to collaborative project management.
- Design of a questionnaire to select a Construction 4.0 compliant collaborative project management platform.

19.2 Introduction

The construction industry is one of the prime movers of any country's economy. For instance, the United Kingdom's (UK) construction industry contributes almost £90 billion to the UK economy (or 6.7%) in value added, comprises over 280,000 businesses covering some 2.93 million jobs, which is equivalent to about 10% of total UK employment (DBIS 2013). A report by Oxford Economics titled "Global Construction 2030" forecasts that the construction output will grow by 85% to USD 15.5 trillion by 2030, with three countries – China, US and India accounting for 57% of all growth (Oxford Economics 2015). Despite its importance, the construction industry is often perceived as having poor track record in terms of productivity and productivity improvements. The labor productivity in the construction industry has averaged 1% in the past two decades; a report published by McKinsey entitled "Reinventing Construction" analyses the reasons for such poor performance and suggests routes to improvements (Barbosa et al. 2017). Among the root causes identified by the report, is poor project management practices. This chapter focuses on this particular issue and attempts to provide solutions for it.

The transformation of the construction industry towards better performance has become imperative in today's demanding society. There are now increased expectations from capital project owners, in terms of quality and delivery. Added to this is the fact that construction projects are becoming larger and more complex in scope, and the timeline and cost to deliver them are becoming smaller and more stringent than ever before. The construction industry is expected to meet these expectations without compromising on quality, while ensuring environmental protection and sustainability (Flyvbjerg 2014). Such challenges place significant pressure on individual stakeholders who seek to achieve more with less investment of time and money. In addition, tighter deadlines and cost alongside fragmentation has made the construction industry dispute-prone. Often, this results in a blame game and trust deficit between stakeholders.

Specialization in and subcontracting of various aspects and tasks of construction have resulted in the interplay of multiple stakeholders in a single project; these players work together as a "virtual" organization. This set of extended stakeholders is called the "construction supply chain" (CSC) (Figure 19.1). The phrase "supply chain" in the field of manufacturing is traditionally defined as the flow of goods, services, money, and information involved in the process of converting raw materials to end products that are consumed by the end user (Simchi-levi et al. 2003). The construction supply chain, in direct analogy with this definition, may be defined as the flow of goods, information, and money for a particular project (Vaidyanathan and Howell 2007). Money flows from the owner to the contractor and tier 1, tier 2 subcontractors, who in turn pay for labor, equipment, and materials. Goods flow in the opposite direction with each subcontractor (or stakeholder) adding their value added services to ultimately deliver the project to the owner. Information flows from anyone to everyone in the supply chain.

Historically, industrial revolutions across the world heralded transformations in a variety of industries to help them meet the challenges of the times. A book by Klaus Schwab entitled *The Fourth Industrial Revolution* describes the journey of mankind from the first to the current Fourth Industrial Revolution (Schwab 2017). Before the industrial revolutions, construction was a profession that involved skilled craftsmanship. There was one master builder who was essentially architect, engineer, and construction manager. The skill was passed from one generation to another, often within families. With the advent of the industrial revolution, the need for speed in construction increased. The first and second world wars necessitated expansion of real estate and infrastructure. By this time trades were also becoming specialized and the fragmentation of the industry had begun. Vendors began to specialize in specific aspects of construction and developed expertise that helped them deliver quality solutions at competitive prices. Such fragmentation required efficient project management which began to evolve as a profession. During the third industrial revolution, technology automation tools for managing projects were created. Tools like PERT (program evaluation and review technique) and CPM (critical path method) evolved during this period. They continue to evolve as we transition to the Fourth Industrial Revolution. Now, construction project management has evolved into an independent field with companies and groups providing project management as a consulting service.

"Industry 4.0" was the term coined for the Fourth Industrial Revolution at the Hannover Fair in 2011 (Lasi et al. 2014). The construction industry adopted the concept and "Construction 4.0" has caught the attention of policy makers, academicians and practitioners. While

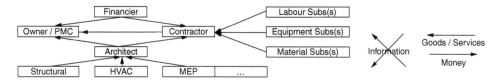

Figure 19.1 Construction supply chain

discussing the impact of the Fourth Industrial Revolution on the construction industry, a report by the World Economic Forum describes the top ten disruptive technologies in Construction 4.0 (WEF 2018). Among these disruptive technologies, cloud-based collaboration has been foremost in helping the construction industry meet the challenges posed by fragmentation of skills, processes and people. The contribution of cloud-based collaboration is because the concept of cloud-operation adheres to most of the design principles of Industry 4.0, such as interoperability, virtualization, decentralization, agility, service orientation and integrated business processes (Petri et al. 2017).

Cloud-based collaboration in the context of project management for construction has great potential. It has the ability to eliminate information latency and provide real-time information from the site to the head office. It has the potential to eliminate data entry duplication leading to significant productivity loss. It has the ability to collate information within the project and possibly across projects to provide data-driven analytics on various Key Performance Indicators (KPIs) for the industry. Ultimately, cloud-based collaboration technology has the potential to improve the reliability of project delivery within time and budget.

Following the above discussion on the evolution of the construction industry and project management, section 19.3 takes a critical look at current project management practices. Section 19.4 discusses the evolution of cloud-based collaboration and information sharing and proposes a technology platform for collaborative project management for Construction 4.0. Section 19.5 discusses what companies and organizations can do to transition and set a roadmap for themselves for Construction 4.0. It also presents a framework for decision makers to select technologies that are compliant with the demands of Construction 4.0 and discusses the benefits and challenges of cloud collaboration solutions. Section 19.6 and 19.7 conclude with suggestions for future directions and summarize the chapter with key takeaways for academics and practitioners.

19.3 Construction today – critical evaluation of current project management frameworks

19.3.1 Challenges faced in traditional project management to meet demands of the construction industry

Classical project management assumes that the scope of the project is clearly understood from the outset of the project and can be elaborated without ambiguity into a set of activities that are reasonably short in terms of duration and executability (PMI 2017). The theory is that the activities can be arranged into a logical network that can then be tracked from beginning to end. Managing the project is simply management of the networks, durations, and sequencing of activities. PERT and CPM were methodologies used to manage projects. These techniques were originally developed for the space programs and military applications where teams spent time and effort elaborating these activities, their relationships, durations etc. and monitored them regularly to ensure deviations were kept under control (Trietsch and Baker 2012). Earned value methodology (EVM) was developed for project performance measurement to monitor time and costs in an integrated manner (Bryde et al. 2018). EVM assumes that the schedules do not change, or changes are infrequent and significant events. Another inherent but unstated assumption in all of the above is that the project team is integrated and there is easy sharing of information needed to plan and execute the plan.

Literature on formal project management started emerging only in the late 1950s. Various professional associations like the Project Management Institute (PMI) and the Chartered

Institute of Building (CIOB) along with academic researchers have contributed significantly to the project management knowledge base. Over time, several practitioners and scholars began understanding the necessity of enhancing and expanding current knowledge and practices within the field, under the umbrella term – Rethinking Project Management, owing to the poor track record of projects and criticism of classical forms of project management (Cicmil et al. 2006, Svejvig and Andersen 2015, Dallasega et al. 2018).

As mentioned earlier, construction projects are increasingly managed as CSC. Stakeholders are not always involved in the project throughout its life cycle from design to procurement to construction. As shown in Figure 19.2, each stakeholder has a limited time presence during the life cycle of the project. Each has a period of intense involvement and a period where they are engaged in the project but more for answering queries and providing support. In Figure 19.2, the height of the rectangle indicates the effort and the length indicates the time duration of involvement with the overall time being from engineering through procurement, and construction phases of the project. For example, various design consultants work closely with each other and with the owner during the initial stages of the project. Although their involvement continues until the end of the project, it significantly tapers off during the construction period. The opposite is true for contractors, their involvement starts at the tail end of design and continues until handover. Even within construction, individual trade contractors join the project (and leave) at various times during the project. Thus, schedules are owned by distributed stakeholders and the schedule for the entire project is progressively elaborated as stakeholders participate. This complexity, combined with litigious environment, results in poor information sharing for effective project management (Thorpe and Mead 2001). All of these render classical project management methodologies (PERT, CPM, EVM) impractical or even useless in their current forms. Recent advances in lean construction, specifically the Last Planner System™ (LPS), aim to improve the culture of information sharing, increase reliability of planning, and reduce some of the uncertainty in the current construction operating environment (Ballard and Howell 1998).

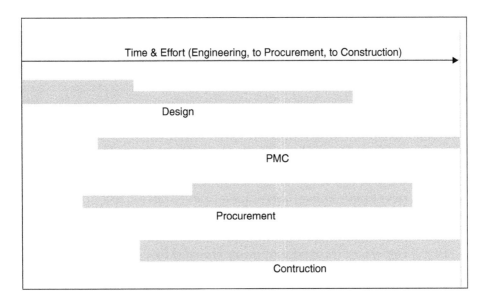

Figure 19.2   Timelines of stakeholder involvement in construction projects

The fragmentation of the stakeholders brings the second challenge to efficient project management. Each stakeholder holds significant information that has bearing on the project. Information sharing typically happens during the monthly progress review meetings that the Project Management Consultant (PMC) and/or Owner conducts to align all the stakeholders, review progress and discuss plans for the following time period. This, coupled with low trust among CSC stakeholders and the litigious environment of the industry, results in delayed and restricted information sharing among stakeholders. All of this causes latency in information sharing. In reality, projects can be efficiently managed only with information sharing and collaboration among stakeholders.

The third dimension of complexity comes from technology. Since stakeholders in the project belong to different organizations and tend to make their own technology choices that are useful to their individual organizations without due consideration of the consequences of their approach to the project(s) as a whole. Therefore, shared information is typically re-entered into the systems of the consuming stakeholders, resulting in a significant loss of productivity. In fact, the productivity loss is so large that additional personnel are hired and retained simply to enter data from one system to another. For example, daily progress reports by a contractor(s) are communicated by email, Excel spreadsheets, messages or phone calls from the field. This is re-entered into planning systems and/or accounting systems for internal updates within the stakeholder's organization and a second time with systems available at the Owner's for the same purpose of monitoring physical and financial progress. The construction project (as shown in Figures 19.1 and 19.2) that is at the intersection of these people, processes, and technologies has no common platform to store and exchange information across the CSC as well as foster collaboration. Inability to link all of this information implies that project management and controls are reactive, short term, and myopic. There is a lack of integrated project management and controls.

19.3.2 Need for a collaboration framework and process standards

A collaboration framework is a platform that allows people to share information. With the manner in which the construction industry is organized, construction project management can be efficient only if there is a collaboration framework that allows various stakeholders to share information. There is a need to evolve project management techniques that embrace these collaborative frameworks. The fragmentation of stakeholders as well as technology is here to stay, and hence any framework has to acknowledge and support this fragmentation of people, process, and technology. This applies not only to external stakeholders for an organization but also to silos of information that exist within different functions in the same organization. For example, procurement, legal, design, and site execution teams from the same stakeholder could use different technologies and hesitate to share information proactively. Each stakeholder (within and across the organization) would have its own economic incentives and profit motivations that will drive it and the stakeholder would only engage in ways and means that will bring in material benefit and more importantly not result in material or legal harm. What this means is that any collaboration framework must involve process enablers that will help them collaborate and operate cooperatively in the contractual and legal framework that binds them to the project.

19.3.3 Opportunities of improvements provided by Construction 4.0

With today's availability of technology and infrastructure, there is tremendous scope for improvement of construction project management (Oesterreich and Teuteberg 2016). Studies have shown that communication and information distribution costs account for nearly 5% of a project's overall cost and that is the value that can be unlocked with collaborative project management technologies. The opportunities offered today for collaboration come in a few different ways: 1) Information can be made available to all stakeholders on any device of their choice with the near ubiquitous availability of mobile devices; 2) Information can be stored and hence accessed anywhere with the advances that have been achieved in cloud storage solutions; 3) Data available in these systems and solutions can be analyzed by harnessing the cloud computing power available using Artificial Intelligence (AI)/Machine Learning (ML) techniques to provide information proactively to all stakeholders for real time decision making (Chong et al. 2014, Bloch and Sacks 2018).

However, in order to radically revamp project management using cloud-based collaboration, it is important to shift away from some of the current project management approaches and theories, and to modify them to achieve integrated project controls. What is needed is proactive information sharing in the CSC, elimination of information latency, and a set of forward-looking lead indicators into project performance that can be used to course correct projects in a near real-time basis to control time and costs of projects. Interconnections between data collected in the life cycle of projects are also important to deliver insights into people and process using technology so that reliability of delivering projects within time and on budget can be improved. All of these are opportunities for improvements provided by the Construction 4.0 framework (Oesterreich and Teuteberg 2016). These goals cannot be met with evolution of technology alone and there is a need for evolution of processes and culture in parallel. The next section will discuss specifics on the evolution of cloud-based solutions and its characteristics.

19.4 Construction 4.0 – cloud based collaboration and evolution of construction information supply chain solutions

This section discusses the construction information supply chain and its continuously evolving state. The evolution is caused by advancements in the technology and institutional mechanisms like contractual and organizational architecture as well as procedures. Their modes of addressing the various problems discussed in Section 19.3 are described and the means of achieving collaboration among construction industry stakeholders is postulated.

19.4.1 Evolution of collaborative information technology solutions for project management

The concept of collaboration has attracted the attention of practitioners and researchers alike in various areas, including intra- and inter-organizational collaboration. The popularity of this concept rests on the strong theoretical foundation provided by "Collaboration Theory". The seminal work by Wood and Gray (1991) describes collaboration as "a process of joint decision making among key stakeholders of a problem domain about the future of that domain". In the context of the construction industry and project management, the authors propose adapting Gray's definition for construction cloud-based collaboration as follows:

> The process of joint decision making among the stakeholders involved in a project across the construction supply chain by proactive sharing of information for meeting the ongoing and evolving challenges faced in the execution of the project in near real-time so as to proactively avoid time and cost overruns as well as value maximization for client without undue negative implications to the stakeholders sharing the information facilitated by a technology platform provided by cloud computing.

Cloud-based collaborative information technology (IT) solutions are of three types (Chong et al. 2014, Chen et al. 2016, Ko et al. 2016). The first is private cloud, in which, software and data are stored within the firewall of the organization. It is not available to anyone outside the organization. The second is the other extreme, in which, information is stored in a public network and is accessible to all (with relevant authentication and authorization). The third hybrid option is the midway point wherein some parts of the system and data are stored in the private network and some in the public network. In brief, the advantage of the private network is that the hardware and software required to run and operate the network is completely under the control of the stakeholder owning the network and they alone can determine who gets access to information, when, and for how long. But the cost of running and operating the network is high; it needs dedicated personnel and constant support to make sure the access control to the right personnel is provided at the right time and removed when not required. In this type of network, data is completely secure. In an environment where stakeholders in the CSC do not trust each other, this is a viable option. On the other hand, in the public network, the hardware and software are operated by a neutral service provider and each stakeholder pays to access the data and solutions. Here also, the data is secure in that it is only provided to personnel who have access, but the data is outside a corporate firewall and stakeholders may feel loss of control. However, the advantage is that the cost of this service is low. The personnel required to run and operate the service are provided by the cloud service provider. More importantly, the service is elastic. Additional computing power, storage, network etc. can easily be added or removed as demands change. Hence, each has its own advantages and disadvantages. The reasons for choosing one over the other of the above solutions is based on the company's strategic decisions of having data within or outside the corporate firewall as well as the associated cost. The hybrid option allows organizations to transition from private to public networks as trust in the networks and solutions grow. But in all three models, cloud-based solutions provide a networked environment that allows for electronic sharing of information.

Cloud computing refers to the ability to host an application and store data required for it in a networked remote computer. The application can be accessed locally on a computer or a mobile device using a local application or a browser. Cloud-based solutions are envisioned as a neutral network, in which, data across organizations working on a construction project, also known as the construction supply chain, is shared. Each of the participants inputs data based on their role and responsibility and each of them consumes data based on their need and requirements. For example, consultants upload drawings and changes to the drawings based on clarification requests from contractors. In this way, the consultant uploads data that the contractor consumes. Access control is provided to ensure that only people who can upload and "modify" the data are allowed to do so and only people who need to consume the data have access to download and consume it. For the smooth running of construction projects, the information required to run the project has to be effectively and efficiently shared with other stakeholders in a timely manner. Timely and efficient sharing of this information is key to efficient project management and would improve productivity.

Table 19.1 shows the various processes in the typical life cycle of a construction project that require sharing of information and lists some of the solutions available for each of the processes.

It should also be noted that the information is sourced from various stakeholders using technologies that each of them uses and might not always be the same even for the same process. For example, project schedules could be created and managed on Oracle's Primavera

Table 19.1 The various processes in the EPC phase of construction projects that involve collaboration and project management

Business Process	Tools Available*	Advantages	Limitations
Design & Analysis (3D/4D CAD & BIM)	Autodesk, Microstation, ArchiCAD, Trimble	Electronic rendering of drawings Easy to accommodate changes Easy to analyze and design various options/scenarios	Electronic document sharing not widespread across players in the CSC No integration to other tools like scheduling, estimation, and ERP functions BIM modeling and maintenance is time consuming and not easy
Estimation	Timberline, Trimble, CCE	Easy to compute costs, reduces scope for human errors in calculations	Largely lack of robust tools to automate quantity take offs from CAD drawings until the rise of BIM tools Lack of integration of material costs from supplier into estimation tools for updated costing information Lack of robust tools to integrate procurement across projects for cost savings
Drawing and document collaboration	McLaren, 4Projects, BIW, Oracle Aconex & Unifier, Autodesk BIM 360, Nadhi nPulse, Trimble, CMiS	Manage all project document flow Manage email and message flow Provide workflow capabilities	Typically not linked to ERP functions Typically not linked to scheduling functions Nadhi nPulse is an exception, but solution is still maturing
Project Planning/Scheduling	Microsoft Project, Oracle Primavera, Nadhi nPulse	Electronically create and manage schedules, track progress, generate reports including earned value management etc.	Schedules are static and rarely re-planned during execution Schedules are not capacity and material constrained Impact of changes to material supply, drawings delays to the plan are not easy to evaluate
Project coordination, issue management, risk management	Autodesk BIM 360, Oracle Primavera, Nadhi nPulse, Procore	Electronically manage project coordination, raise notifications etc.	Typically issues are not linked to schedules Nadhi nPulse and Oracle Primavera are exceptions
Project meeting management	Nadhi nPulse, Procore	Electronically manage project agenda items, meeting minutes etc.	Typically agenda items are not linked to schedules. Nadhi nPulse is an exception
Field progress tracking & monitoring	Oracle Mobility, Trimble, Nadhi nPulse, Procore	Use mobiles, RFID, biometrics to capture field progress, material stock, and labor mobilization	Data and workflow integration to ERP is lacking Largely lacks integration to scheduling

(Continued)

Table 19.1 (Cont.)

Business Process	Tools Available*	Advantages	Limitations
Procurement & supplier collaboration	Oracle, SAP, Nadhi nPulse	Use economic order quantity principles to provide procurement decision support Electronic exchange of information like POs, RFP, bids etc.	Construction procurement is yet to evolve into a mature industry Electronic supplier collaboration is an area that is in its early stages
Subcontractor collaboration (productivity)	Trimble, Oracle, SAP, Nadhi nPulse	Capture labor deployment at site and calculate productivity	Largely solutions are still evolving and productivity impact to scheduling is not done Integration to ERP functions is also largely lacking
Accounting, Finance, HR (ERP)	Intuit, Microsoft, Oracle, SAP, CMiS,	Manage accounting, taxation, finance, billing, HR, payroll etc.	Lack of integration of the various ERP functionality Lack of integration between planning and estimation to accounting, procurement, and billing
Quality control, assurance, and audit	Autodesk BIM 360, Procore, Nadhi nPulse	Digitally manage quality inspections Digitally manage material inspections Digitally manage NCs, and observations	Typically not linked to schedules Analytics of quality issues is evolving
Safety control, assurance, and audit	Autodesk BIM 360, Procore, Nadhi nPulse	Digitally manage safety audit inspections	Typically not linked to schedules Analytics of safety issues is evolving
Snagging/ de-snagging, punch list items	Autodesk BIM 360, Procore, Nadhi nPulse	Digitally manage punch list items	Typically not linked to schedules Analytics of punch list items is evolving
Digital twin, creation and managing BIM models with field updating for collaboration	Trimble, Autodesk BIM 360, Bentley	Electronically manage 3D parametric models Clash detection is easy Increases constructability Planning construction sequencing, vehicle movement etc. is feasible Reduces re-work	Model updating and maintenance across trades is cumbersome Model typically not integrated to ERP functions and scheduling functions for synchronous updates

* These are meant to be representative rather than exhaustive. There are several other vendors not mentioned here that offer solutions to each of these business processes

by one of the stakeholders while another might use Microsoft Project. A third might simply use Excel spreadsheets. They might all share progress of their activities to the PMC or Owner using output from their individual technologies using email, Microsoft Excel, or Adobe pdf.

All of this data might be electronically stored in a cloud-based document management system (DMS). In this scenario, although data is now centrally located in one cloud solution, it is not very useful in providing a holistic picture of where the project stands. In a sense, although the data is common for the project, the lack of the use of either the same technology or the lack of interoperability between the technologies renders the data ineffective for real-time analysis that supports decision making.

First generation collaboration solutions are about the need to simply share information among the stakeholders in a faster and more efficient manner. These solutions did not understand the business problems and were used for merely electronic transmission of information. Evolution in collaborative solutions came with the realization that, if the information is interlinked intelligently, the impact of the data on time and costs of the project can be understood as the power of fourth generation solutions that will help realize the potential of technology in a Construction 4.0 scenario. If this potential can be realized by ensuring that the business process allows for timely and proactive sharing of information by stakeholders, then the true benefits of collaborative project management can be realized. Then, project management would have transitioned into a more proactive, integrated project controls solution that generates forward-looking indicators where the objective is to reduce if not eliminate delays rather than capture (data) and monitor delays that can be pegged onto one stakeholder.

The following section discusses process enablers that provide the trust and confidence that allow stakeholders share data proactively.

19.4.2 *Evolution of process and legal enablers for project controls*

The interplay of process and technology has always been well understood in all industries and the construction industry is no exception to the rule. Lean construction as a process enabler to plan and execute collaboratively as a team has been evolving for nearly 25 years. While an exhaustive discussion on lean construction for collaborative planning is beyond the scope of this chapter, the founding principles of lean are worth stating:

* An approach for impeccable coordination and collaboration.
* A process to think of the project as a production system.
* A mentality of treating stakeholders as collective enterprise.

Research on lean construction has shown the benefits of collaborative planning, early sharing of site or execution constraints, measuring productivity and improving them on a continuous basis (Howell 1999, Ballard et al. 2002, IGLC 2019). Research has shown that more collaborative working and planning that transcend organizational boundaries, and having the project as the central focus, enables more reliable completion of projects within projected timelines and costs. Although lean has a number of tools that can be used at projects, experienced practitioners of lean will attest to the emphasis that lean places on people and culture. It emphasizes respect for individuals for their role in the project. Successful adoption of the tools helps creating a culture of trust and transparency across all the CSC stakeholders to collaborate and share information proactively without fear of blame and repercussions. It emphasizes creating a learning organization that aims for continuous improvement. Hence, lean, as a process enabler forms the core to create a collaborative environment. Enabling technology that supports collaborative working and supports the lean process across the stakeholders in the CSC acts as the glue that helps with the process adoption. And finally, supporting contractual structures bring together the various CSC stakeholders into the virtual project organization to work together

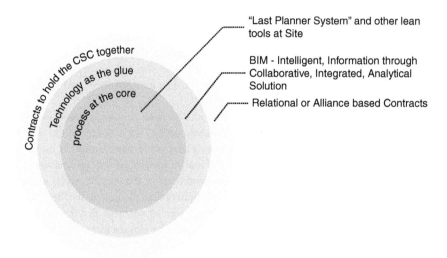

"Last Planner System" and other lean tools at Site

BIM - Intelligent, Information through Collaborative, Integrated, Analytical Solution

Relational or Alliance based Contracts

Contracts to hold the CSC together

Technology as the glue

process at the core

Figure 19.3   Lean enabled integrated project controls

and deliver the project collaboratively. Figure 19.3 illustrates the interlinking of processes, technologies, and contracts as discussed above.

The various contracting structures that support a more collaborative and transparent working environment are the alliance-based contracting, relational contracting, integrated project delivery (IPD), and consensus documents (Bygballe et al. 2015, TRB 2015, Rowlinson 2017). The core intent of each of these is to address the challenging issue of information sharing across organizational boundaries without fear of repercussion, particularly early sharing of constraints or delays that have dependencies on other agencies. The idea is to improve trust among the CSC stakeholders.

The entire field of technology evolution under the umbrella of "BIM" (Building Information Modeling) is also a relevant discussion in this context. In keeping with the lean principles of maximizing value and minimizing waste, BIM espouses virtual construction ahead of real construction. BIM involves the creation of three-dimensional (3D) parametric models of the construction project using BIM authoring tools. The benefits of doing this is that it helps eliminate or remove drawing coordination issues (or clashes), and re-work by providing a constructible drawing, and creating a digital twin that is representative of the real building. However, the actual practice of BIM in the field has shown that teams have not been able to proceed beyond clash detection using 3D BIM models. Issues stated as reasons include the lack of process (of sharing models, progress, early mobilization of contractors) and until recently, lack of technology. In fact, the current state of BIM adoption has only added an additional layer in technology fragmentation. Now, apart from the scheduling system, Enterprise Resource Planning (ERP) systems, and document management systems (DMS), CSC stakeholders have to update the 3D BIM models as well. The practical difficulty of this is that models are not linked to schedules, quantities (ERP), documents, issues and quality. But the promise of BIM relies on collaborative project management and integrated project controls. Should BIM models be enabled with lean processes and collaborative project management technologies, the potential of BIM can be realized.

Lean construction is an example of a collaboration framework that has been developed and evolved as a need to address the challenges in construction project management. Although a thorough discussion on lean construction and the Last Planner System™ is beyond the scope of this chapter, a deep understanding of the cultural implications and process enablers it brings into construction project management needs to be understood for readers to adopt technology for readiness for Construction 4.0. One of the underpinnings of the lean methodology is worth mentioning. The LPS methodology emphasizes that all stakeholders perform lookahead constraint analysis that essentially allows them to coordinate their work more seamlessly and proactively and eliminate the constraint at planning stage rather than at site. This helps in eliminating delays due to project coordination leading to a more reliable and smooth flow of construction. The LPS methodology measures PPC (percent plan complete) as the KPI metric to understand reliability of planning and documents root cause delay reasons for non-adherence to plan. This understanding of reasons for deviation from plan along with the concurrent use of VSM (value stream mapping) methodology are built on the principles of measuring performance and their concurrent and constant improvement during the life cycle of the project.

19.4.3 Vision for collaborative project management technology for integrated project controls

The authors believe that a Construction 4.0-ready collaborative solution for integrated project controls needs to be built on some fundamental ideas and assumptions. One is that the fragmentation in the industry is here to stay and has to be considered. The fallout of the fragmented information CSC is that every stakeholder belongs to a different organization and hence would have differing economic incentives that cannot easily be streamlined. The second idea is that the data that each stakeholder has or needs for a project strongly correlates to their role in the project and consequently determines data requirements from the solution. For instance, the data that an ERP system, scheduling system, BIM model, or document management system captures is different for different stakeholders (owner and contractor). In short, there will be multiple technology solutions that each CSC stakeholder will use for creating and sharing information as part of project planning and delivery and so the technology fragmentation is also real and permanent.

In this environment, we propose a neutral technology platform (Figure 19.4) into which all the stakeholders plug-in for efficient collaboration. Each would share the data (that they have to share) and would receive data from others (that they require). The technology platform is envisaged as a utility, much like the internet and/or a mobile enabled networks.

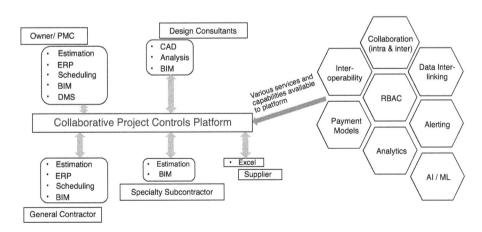

Figure 19.4   Collaborative project controls platform

In the proposed platform, the authors put forward a technology solution that is stakeholder neutral. This, in itself, is the strength of the solution. It ensures that technology ownership is distributed and is for the project or CSC. Such distributed ownership builds trust in that the platform is not biased for any individual stakeholder of the CSC. This ensures that the trust element (or lack of it) among stakeholders is not a barrier to adoption of the platform. Some of the other key characteristics of the platform are described below.

19.4.3.1 Intra- and inter-organization collaboration

The platform is envisioned as a collaborative platform that allows for all stakeholders in the CSC to have one version of the truth of the project. For instance, contractors would initiate requests for information (RFIs) that will be addressed by consultants (or PMC etc.) with the owner being aware of the communication as a subscriber. Consultants would provide drawings that are sent to relevant contractors with PMC and owner being subscribers without waiting for manual coordination. The site based information like material receipts and consumption, daily labor reports, physical, and quantity progress, site issues etc. are input into the solution by the onsite execution supervisors – people who create tasks for direct workers, of various contractors and suppliers. The platform automatically generates relevant management information system (MIS) reports and distributes them to all concerned, based on their business needs. All information entered by the contractors above are approved by the PMC prior to being used for billing. The system proactively follows up with all stakeholders for input and provides them with output as needed. Plan deviations are notified to the planning engineer as detected for them to take relevant corrective actions. Negative time impacts due to delays in issues, drawings, communication etc. are communicated to the planning engineer for re-planning as needed so that the time impact of delays in decision making is contained. The platform will complement in-house technology choices of individual stakeholders and provide end-to-end visibility to project issues for all. With this level of dynamic and real-time collaboration, we envision that the schedule for the project is kept in realistic and dynamic equilibrium. Any surprises in cost and time are kept to a minimum, if at all. As one can envision, such a solution can be a technology enabler for lean construction if it is practiced, although it does not require the same.

19.4.3.2 Interoperability

The platform itself would provide interoperability with existing construction point solutions. Financial progress from the onsite execution supervisors is written to various source ERP systems of the vendors for financial reconciliation while physical progress is written back to scheduling systems. The proposed platform is also built with robust security features so that data from one stakeholder is not shared with another, unless authorized. In addition, it would be easily configurable so that receiving and putting data from various stakeholders in a "new" project is not time consuming and is "plug and play". The data entry duplication that exists today is eliminated leading to significant productivity gains.

19.4.3.3 Intelligent interlinking and analytics

The transactional capabilities of the platform now capture all the information and actions performed by every stakeholder for the life cycle of the project from design to commissioning. The platform will also allow for data from the various processes and stakeholders to be interconnected. Since the platform is built for construction, it can take advantage of the domain knowledge to suggest and infer these interconnections much in the same way social media solutions suggest connections and links based on user profiles and usage patterns. Once these data level connections are established, as discussed above, the platform will be able to monitor and proactively notify individual stakeholders on the impact to them due to action or inaction by another.

This transactional data, along with interlinking, implies that the solution can provide the basis for performing domain specific analytics. Analytics can be performed at the process level (engineering, procurement etc.) or at the stakeholder level. Such analytics can help proactively identify strong and weak processes (and stakeholders) and provide lead indicators into project delays before they happen, in order to enable decision makers to make timely interventions.

19.4.3.4 Payment models

One of the key considerations to the adoption of such a platform is its perceived cost and benefits. In the author's experience, stakeholders place a heavy emphasis on the implementation of ERP systems to give themselves tight financial controls and are willing to spend for the same. But when it comes to collaboration, schedule control, and performance analytics, they are not willing to invest as much. The stakeholders also are not too forthcoming to spend for inter-organizational collaboration (yet).

In this model, the authors propose that everyone pays for the usage of the solution. As described above, the solution is to be treated like a common or shared utility. The authors propose a payment model that is built using the guiding principles of IPD or relational contracting with a fee for the technology platform being shared in proportion to the risk taken by individual stakeholders. Alternatively, a simpler payment model can be based on a transactional basis. Each stakeholder can pay for the number of transactions and the amount of data that is downloaded into their in-house systems (but not for uploading data).

19.4.3.5 Lead indicators as KPIs

The platform would be designed to generate forward-looking KPIs. The platform would compute time and cost delay of the end (or intermediate) milestones based on real time progress captured by the various stakeholders. Inaction or lack of progress would trigger notifications and alerts reminding them of the time and cost delays of inaction to future milestones.

The platform would generate forward-looking constraints that need to be acted on to avoid delays. The constraints would be computed to identify potential mis-alignments in document deliverables, project coordination issues that need resolution, material delivery schedules, labor mobilization plans, payment schedules, procurement plans with the construction execution plan. Inaction or delay in resolution in any of them could potentially lead to delay in onsite execution and project delays.

The platform would also generate production-based metrics, i.e. it would compute milestone completion based on productivity rather than simple physical progress. The platform would have the ability to correlate physical, financial, and virtual progress for various project types and contract types, to compute in real time, project progress with data captured from the field.

19.4.3.6 AI/ML enabled learning on process, stakeholder, project, CSC, and industry

The platform would have the ability to provide metrics. Metrics would be such that the data can be analyzed on various parameters including but not limited to cycle time (or productivity) for individual activities, processes, and stakeholders, and would measure other metrics like percent planned complete (PPC) etc.

Without losing privacy of information, the platform would have the ability to collate such metrics within the same project, for the same organization working across multiple projects, and across multiple projects, across multiple organizations for the industry, etc. All of these would be used to generate benchmarks that can then be used to provide guidelines for future estimation of projects.

The authors believe that such a platform would provide an agile and adaptive basis for construction projects to be run as agile CSCs that are adaptive, in near real time, to changes in site (and project) conditions to minimize waste and maximize value to clients. Such a dynamic platform would also allow for evolution of adaptive CPM and EVM, reflecting the dynamic nature of construction projects.

As a stakeholder in the CSC, looking to lay out a long-term corporate strategy for Construction 4.0 readiness, two questions must be answered: (1) How can the organization get the people and process ready for Construction 4.0? (2) How can the organization select a technology that is Construction 4.0-ready? The following section discusses these as organizations transition to Construction 4.0

19.5 Transitioning to Construction 4.0

19.5.1 Collaborative project management maturity model for transition

As a stakeholder in the Architecture, Engineering and Construction (AEC) market, looking to lay out a corporate long-term strategy to adopt technology solutions and evolve processes to improve operational efficiency and readiness to Construction 4.0, two questions need to be answered: (1) What is the current benchmark of the firm? (2) What steps must be taken to improve on the status quo? Maturity models have been a way to assess organization readiness to change in a structured manner. Their origins lie in the field of quality management, which describes quality improvement through a five-level maturity grid (Crosby 1979). Research and surveys have shown that processes have a maturity life cycle and that there is a correlation between improving process maturity and business performance (Lockamy and McCormack 2004). Business process maturation has also been shown to reduce conflict and encourage greater cooperative behavior while improving performance. Construction Supply Chain Maturity Model (CSCMM) was first proposed in (Vaidyanathan and Howell 2007) as a direct need to identify a structured mechanism to have organizations develop a strategy to change process and associated technology to achieve maturity, in order to achieve their operational efficiency goals. Subsequent studies on lean construction have proposed a lean construction maturity model (LCMM) (Nesensohn 2014). The LCMM methodology identified 50 parameters to assess the maturity level of an organization and ways to transition from one level to another. Another maturity model that has evolved in the space is the BIM maturity model (Liang et al. 2016).

CSCMM was proposed as a four level maturity model – Ad-hoc, Defined, Managed, and Controlled. Transition from one maturity level to another happened based on metrics measured for Process, Technology, Strategy, and Value. Details on how to assess an organization

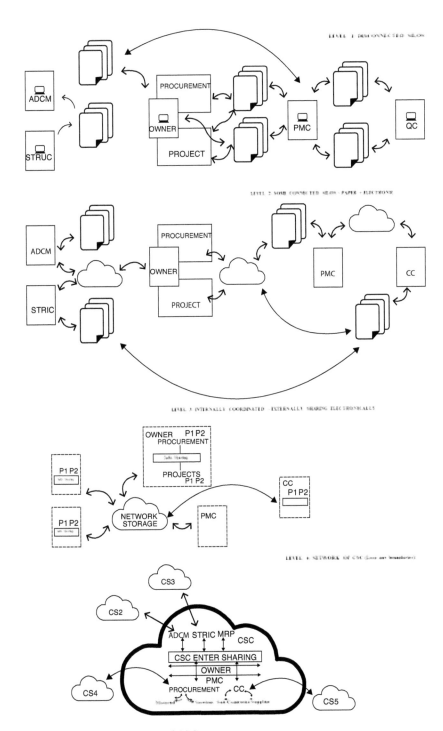

Figure 19.5 Levels of development of CSC

at each stage for each of the assessment parameters remains as a research topic. Adapting the same for collaborative project management, the authors propose a four level maturity model (Figure 19.5). Transition from one level to another can be based on technology parameters, and data parameters as described below.

The various maturity levels are defined in brief as below:

- *Level 1*: At this level, there are little to no processes. There are some processes and/or automation for functions within a company but no standards. Each project runs independent of each other. Within a project, collaboration across firms is paper based, ad-hoc, needs based and done more on a reactive than a proactive basis. There is little to no planning and there is information flow in all directions as shown in Figure 19.5. There is no visibility across the CSC, end to end of the project. There is no cross-functional cooperation and hence automation. Targets, if defined are missed. Heroics by people are required to get projects done. All of this leads to high collaboration and information sharing costs and low customer satisfaction.
- *Level 2*: At this level, there is some recognition to information sharing and collaboration across the CSC. Collaboration needs are more defined (like the lean project delivery process). Information sharing is both electronic (in a cloud storage) and paper based. Targets are defined, tracked, and met at times. Performance improves but is not consistently predictable. Collaboration and information sharing costs are still high, but customer satisfaction improves.
- *Level 3*: This is the breakthrough level. Organizational structures are aligned with CSC. Within each stakeholder, information sharing platforms, functions and processes are integrated. Firm level data entry duplication is eliminated. Information in the CSC is shared electronically. Across the CSC, data is stored in a common cloud system, but there is some data entry duplication. End-to-end visibility across the CSC for the project and project collaboration improves. Contract structures like IPD emerges and encourages information sharing. Targets are achieved more often and consistently. Collaboration and information sharing costs fall. There is no need for heroics to meet targets and complete projects. There is marked improvement in customer satisfaction.
- *Level 4*: At this level, CSC is complete, controlled, predictable, and managed. Boundaries of organizations within and across CSC are absent. Projects are planned and executed in a completely technology-enabled collaborative project management framework. Data entry duplication is completely eliminated. Advanced analytics, and forward-looking KPIs help projects avoid delays or quickly adapt to the same to get projects completed within costs and timelines. Competition is through networks of companies that form a supply chain. Firms are aligned through trust and mutual dependency beyond contractual and organizational boundaries. CSC competency is key to success. Targets are met as defined and customer satisfaction is high while collaboration and information sharing costs are kept low.

The authors propose the following guidelines for technology assessment parameters for firms to transition from one maturity level to another as discussed above.

- *Level 1*: At this level, most of the processes are not always digital and some are pen and paper. Collaboration between stakeholders occurs using hard copies.
- *Level 2*: Most individual processes are automated using technology; information sharing is electronic. There is no interoperability between the technologies; data entry is duplicated within an organization or across a project.

- *Level 3*: Within an organization, technology is integrated, more processes are automated, some interoperability exists between technologies; there is some early integration of information across stakeholders in a project.
- *Level 4*: All processes are seamlessly integrated through the project life cycle. Information is input once and used multiple times. Data from the systems are used to learn to further improve the processes.

Given the importance of data and data exchange, the authors propose the following data assessment parameters for data exchange and interoperability as firms transition across maturity levels.

- *Level 1*: No data standards.
- *Level 2*: Each stakeholder in the CSC has some data standards for data collection, collation, and aggregation. Standards are based on experience of individuals.
- *Level 3*: Stakeholder standards are evolved to project standards for the CSC. Projects begin to evolve common data environments to facilitate easier electronic data exchange that go beyond data storage in common networks.
- *Level 4*: Common data standards exist across the industry for all CSCs. This helps study of data to evolve industry KPIs and characteristics across process, stakeholders, project type, etc.

19.5.2 Selection and deployment of cloud-based collaboration tools

Given the number of solutions available for cloud-based project collaboration, choosing the right solution becomes critical. The proposed framework in Section 19.4.3 for a cloud-based collaboration platform by the authors offers an initial and broad set of parameters to consider while evaluating solutions for cloud collaboration for project management.

There are a variety of cloud-based project management tools available in the market and deployed across the construction project life cycle. These tools are often analyzed from the perspective of the features offered, like design drawing management, finance management, workflow management, sub-contractor management, schedule management, risk management, inventory and material management, etc. However, analyzing solutions based on their feature function capabilities indicate that a single application that provides the majority of required application features does not exist, which is self-explanatory considering the fact that these applications draw their competitive edge by focusing on a specific area and often offers integration with other relevant application areas. The need to manage this information across CSC is another challenge, something we have emphasized in this chapter. In this context, it is important to investigate if these tools are along the lines of the design principles of Industry 4.0 as outlined in Section 19.2. These design principles can be interpreted from the context of cloud-based project management tools as follows:

- *Interoperability*: ability of tools to communicate with different software using interfaces and standardized processes.
- *Virtualization*: ability to gather and monitor progress associated with construction projects through paperless electronic means. As technology evolves, data collection can be through acquiring, processing and using sensor data. This could be simple solutions like reading information from bar codes, quick response (QR) codes, radio frequency

identification (RFID) tags, or more advanced options like analyzing photos to infer progress etc. that can all be finally reported into various source systems.

*   *Decentralization*: enabling cloud-based tools to allow for actions to be delegated to its
    logical last mile across organizations at the CSC level and allow the relevant stakeholder
    to make decisions on their own. The solution should also allow for auto escalation of tasks
    to a higher level in case of inaction or failure by the stakeholder. All of this should respect
    data privacy and privacy of role. For instance, structural information should not be visible
    to mechanical, electrical and plumbing (MEP) trades and vice versa (unless the project/pro-
    cess requires it); financial information should only be visible to those who require it, etc.
*   *Real time capability*: collection and analysis of data in real time basis, which allows the
    construction processes to react spontaneously to progress or lack of it. In either case,
    the solution should provide planners with decision support options to modify plans, and
    publish to all stakeholders. At all times during the life cycle of the project, the plan should
    reflect actual site execution in near real time.
*   *Service orientation*: availability of tools to satisfy the customer/need requirements of both
    internal stakeholders as well as participants across CSC.
*   *Modularity*: ability of tools to adapt to changing requirements of stakeholders as well as
    construction projects by expansion or refinement of existing modules of software and
    hardware.

The authors recognize that readers are in different parts of the world and play various stakeholder
roles in the CSC, all in various stages of evolution and maturity. In order to make the selection of
a technology solution that is designed based on the principles and expectations outlined in Section
19.4.3, the authors are proposing a questionnaire that captures the intent of how a compliant solution can be chosen. Appendix 19.1 captures the above principles and their implications to project
collaboration and controls as a questionnaire. The authors propose that any stakeholders in the
CSC use that questionnaire as a starting point, adapt it to their individual needs and circulate it
among technology providers. A comparison of the responses should help an organization pick a
solution that is technically compliant with the organization strategy to adapt a solution as they
shift to a Construction 4.0 way of business. For solution providers and tool developers, the authors
recommend the questionnaire to be used as a guideline to develop a roadmap for their solutions
to make them more adaptable to the changing business environment towards Construction 4.0.

19.5.3 Benefits and challenges in adoption of cloud collaboration

Productivity improvement and automation is one of the fundamental drivers of technology
adoption for any business or process. With today's technology and processes, the time it takes
for planning engineers to analyze schedules, understand their time and cost impact and evaluate
alternates can be quite large. The more complex the project or more detailed the schedules, the
more difficult and time consuming the analysis is, but at the same time more important to be
done. Furthermore, unless the data collection for progress capture is decentralized, the planning
engineer does not have real-time information to analyze the schedules. Instead, they spend all of
their productive time collecting information with no time to analyze schedules. This defeats the
purpose of having any project management solution. But if a technology solution as proposed
in this chapter is adapted for a project and associated lean processes are adopted, what benefits
would the projects accrue? Beyond productivity, what are the other benefits of technology adoption? What are the challenges to technology adoption? We discuss the benefits and challenges of
adopting cloud-based collaboration technology for project management in this section.

The primary benefit of adapting collaborative project management solutions is to improve productivity. The time required to manage projects by planning engineers should be significantly reduced and planning engineers should have more time to analyze plans and develop course correction strategies since all of the data comes to the planning engineers from various stakeholders in real time.

Beyond productivity improvement, the authors believe that secondary benefits to adoption of collaborative project management solutions would be in helping people in project controls avoid blind spots. This will help in avoiding and eliminating delays, which is more powerful than monitoring delays and blaming individuals and stakeholders for the same. If a solution as proposed is adopted, the authors feel that projects can be managed more proactively and stakeholders would have end-to-end visibility into project progress beyond process and organization boundaries at the CSC level. Any decision made through such a technology would be objective and not subjective.

This leads to tertiary benefits. Data from the solution can be used to learn the performance of individual activities, processes, agencies, and industry in general. And with this learning benchmarks can be created that can be used for continuous improvement. While there are numerous benefits for cloud-based collaboration technology, adoption is critical for success. The process and maturity framework in Section 19.5.1, we believe, would assist the adoption process.

A challenge to adoption of technology is resistance to change. People tend to get comfortable with the current operating environment and open information sharing as required by collaborative project management described here can be seen as a risk. Not adopting commensurate processes with appropriate technology platforms would imply that stakeholders will not gain the benefits of the technology. Not obtaining a desired return on investment (ROI) can lead to premature giving up on the technology, which again is a challenge. Creating a collaborative project management platform as described requires technology vendors to develop interoperability standards. If the technology vendors do not cooperate to achieve that objective, the vision of the platform and its benefits cannot be realized.

As described in the chapter, collaborative project management and controls has great potential to transform the way construction projects are managed and significantly reduce information sharing costs. And collaborative project management and controls has the potential to radically improve reliability with which projects can be delivered on time and within budget. But as discussed, people, technology, and processes need to evolve in tandem for the value to be realized.

19.6 Conclusion

This chapter describes the technology solutions and the evolving hardware and software landscape that the construction industry possesses today for better managing projects. As companies and industries shift to Construction 4.0, decision makers need to understand the interplay of process and technology. Both must evolve for the potential to be realized. As practitioners in the industry, it is imperative that each realizes that all in the CSC need to evolve for maximum value to be realized. For decision makers and practitioners in the industry, there are a few insights from this chapter.

1. The operating system of projects must change from being silo-organization-based to being collaborative, team based, and sharing information enabled CSC. Technology and processes must go hand-in-hand. Technology solutions that are adopted by organizations must keep this in mind.

2. Stakeholders must understand that technological evolution is happening at a rapid pace. In the past five years, there has been a revolution in construction technology and in the evolution of solutions. Coupled with availability of connectivity, mobility, and data, this is changing the working environment and operating system for sites. But in order to take advantage of these technologies, processes must evolve as well. People and processes do not always change as fast as technology does. So, organizations have to pace the change in a manner that the organization can adopt. Else, the benefits of technology would not be realized.

3. There are a number of collaboration and project management solutions available with new ones being constantly developed. Selecting a solution that works for each organization will depend on their role in the CSC and the investments associated with the solution. However, while selecting the solution, organizations must keep in mind the parameters listed earlier in this chapter to ensure compatibility of the solution that they are selecting, that the larger collaborative CSC platform outlined in the chapter. This in turn, would ensure that organizations can seamlessly transition to the CSC cloud collaboration project management platform. The chapter lists key parameters to consider while selecting a solution for an organization as a questionnaire.

4. Each organization needs to evaluate critically where they stand on the maturity levels of technology and process. From there, they must draw a roadmap to evolve on the maturity scale. The context of each organization, the nature of projects that they work on, and the culture of the organization would determine the steps that each needs to evolve. This chapter has provided a few guidelines that should help in drawing that roadmap.

19.7 Summary

• This chapter described how project management and collaboration is evolving in a world that is shifting to Construction 4.0. It also discussed the interplay of processes and technology for effective and efficient technology-enabled collaboration and project management.

• The chapter proposes a collaborative project management platform and its characteristics to achieve efficient project management for construction projects. Then, the chapter presented a questionnaire to consider while selecting a technology for collaboration and project management.

• Finally, the chapter proposed a maturity framework that organizations need to consider before embarking on a technology adoption journey for Construction 4.0-ready collaborative project management.

References

Ballard, G. et al. 2002. Lean Construction Tools and Techniques. In Best, R. and De Valence, G. eds. *Design and Construction: Building in Value*. Oxford: Butterworth-Heinemann, 227–254.

Ballard, G. and Howell, G. 1998. Shielding Production: Essential Step in Production Control. *Journal of Construction Engineering and Management*, 124(1), 11–17.

Barbosa, F. et al. 2017. *Reinventing Construction – A Route to Higher Productivity*. London: McKinsey & Company.

Bloch, T. and Sacks, R. 2018. Comparing Machine Learning and Rule-based Inferencing for Semantic Enrichment of BIM Models. *Automation in Construction*, 91, 256–272.

Bryde, D., Unterhitzenberger, C. and Joby, R. 2018. Conditions of Success for Earned Value Analysis in Projects. *International Journal of Project Management*, 36(3), 474–484.

Bygballe, L. E., Dewulf, G. and Levitt, R. E. 2015. The Interplay Between Formal and Informal Contracting in Integrated Project Delivery. *Engineering Project Organization Journal*, 5(1), 22–35.

Chen, H.-M., Chang, K.-C. and Lin, T.-H. 2016. A Cloud-Based System Framework for Performing Online Viewing, Storage, and Analysis on Big Data of Massive BIMs. *Automation in Construction*, 71, 34–48.

Chong, H.-Y., Wong, J. S. and Wang, X. 2014. An Explanatory Case Study on Cloud Computing Applications in the Built Environment. *Automation in Construction*, 44, 152–162.

Cicmil, S. et al. 2006. Rethinking Project Management: Researching the Actuality of Projects. *International Journal of Project Management*, 24(8), 675–686.

Crosby, P. B. 1979. *Quality is Free: The Art of Making Quality Certain*. New York: McGraw-Hill.

Dallasega, P., Rauch, E. and Linder, C. 2018. Industry 4.0 as an Enabler of Proximity for Construction Supply Chains: A Systematic Literature Review. *Computers in Industry*, 99, 205–225.

DBIS. 2013. *UK Construction: An Economic Analysis of the Sector*. London: Department of Business Innovation and Skills.

Flyvbjerg, B. 2014. What You Should Know About Megaprojects and Why: An Overview. *Project Management Journal*, 45(2), 6–19.

Howell, G. A., ed. 1999. What is Lean Construction – 1999. *7th Annual Conference of the International Group for Lean Construction*, 1999/07/26 Berkeley, USA.

IGLC. 2019. About the International Group for Lean Construction. [online] Available from: http://iglc. net/Home/About [Accessed July 6 2019].

Ko, H. S., Azambuja, M. and Felix Lee, H. 2016. Cloud-based Materials Tracking System Prototype Integrated with Radio Frequency Identification Tagging Technology. *Automation in Construction*, 63, 144–154.

Lasi, H. et al. 2014. Industry 4.0. *Business & Information Systems Engineering*, 6(4), 239–242.

Liang, C. et al. 2016. Development of a Multifunctional BIM Maturity Model. *Journal of Construction Engineering and Management*, 142(11), 06016003.

Lockamy, A. and McCormack, K. 2004. The Development of a Supply Chain Management Process Maturity Model Using the Concepts of Business Process Orientation. *Supply Chain Management*, 9(4), pp. 272–278. https://doi.org/10.1108/13598540410550019.

Nesensohn, C. 2014. *An Innovative Framework for Assessing Lean Construction Matuirity*. (Doctor of Philosophy). Liverpool: Liverpool John Moores University.

Oesterreich, T. D. and Teuteberg, F. 2016. Understanding the Implications of Digitisation and Automation in the Context of Industry 4.0: A Triangulation Approach and Elements of a Research Agenda for the Construction Industry. *Computers in Industry*, 83, 121–139.

Oxford Economics. 2015. *Global Press Release on Global Construction 2030: A Global Forecast for the Construction Industry to 2030*. London, United Kingdom: Global Construction Perspectives and Oxford Economics.

Petri, I. et al. 2017. Coordinating Multi-site Construction Projects using Federated Clouds. *Automation in Construction*, 83, 273–284.

PMI. 2017. *A Guide to the Project Management Body of Knowledge (PMBOK Guide)*. Sixth Edition ed. Newtown Square, Pennsylvania: Project Management Institute, Inc.

Rowlinson, S. 2017. Building Information Modelling, Integrated Project Delivery and All That. *Construction Innovation*, 17(1), 45–49.

Schwab, K. 2017. *The Fourth Industrial Revolution*. New York: Crown Publishing Group.

Simchi-levi, D., Kaminsky, P. and Simchi-Levi, E. 2003. *Designing and Managing the Supply Chain: Concepts, Strategies, and Case Studies*. Boston: McGraw-Hill/Irwin.

Svejvig, P. and Andersen, P. 2015. Rethinking Project Management: A Structured Literature Review with a Critical Look at the Brave New World. *International Journal of Project Management*, 33(2), 278–290.

Thorpe, T. and Mead, S. 2001. Project-Specific Web Sites: Friend or Foe? *Journal of Construction Engineering and Management*, 127(5), 406–413.

TRB. 2015. *Alliance Contracting – Evolving Alternative Project Delivery*. Washington, DC: Transportation Research Board, National Academies of Sciences, Engineering, Medicine. The National Academies Press.

Trietsch, D. and Baker, K. R. 2012. PERT 21: Fitting PERT/CPM for Use in the 21st Century. *International Journal of Project Management*, 30(4), 490–502.

Vaidyanathan, K. and Howell, G., ed. 2007 Construction Supply Chain Maturity Model – Conceptual Framework. *15th Annual Conference of the International Group for Lean Construction*, 2007/07/18 East Lansing, MI, 170–180.

WEF. 2018. *Shaping the Future of Construction: Future Scenarios and Implications for the Industry.* Geneva: World Economic Forum.

Wood, D. J. and Gray, B. 1991. Toward a Comprehensive Theory of Collaboration. *The Journal of Applied Behavioral Science,* 27(2), 139–162.

Appendix

Appendix 19.1 Questionnaire for selection of Construction 4.0 compliant project collaboration and project management solution

Serial no	Parameter for Evaluation	Available	In Roadmap	Not Available	Remarks
1.0	**Project Management Section**				
1.1	Does the solution have ability to create and manage schedules?				
1.2	Does the solution have ability to bi-directionally integrate with Oracle Primavera?				
1.3	Does the solution have ability to bi-directionally integrate with MS Project?				
1.4	Does the solution have ability to bi-directionally integrate with any other scheduling system?				
1.5	Does the solution have native CPM capabilities?				
1.6	Does the schedule support ability to change schedules?				
1.7	Does the schedule support progressive elaboration of schedules?				
1.8	Does the solution have ability to update schedule using mobile devices?				
1.9	Does the solution have ability to provide access control to parts of the schedule for decentralized updates?				
2.0	**Project Collaboration and Coordination Section**				
2.1	Does it support collaboration with stakeholders within same organization?				
2.2	Does it support collaboration with other stakeholders in the CSC?				
2.3	Does the solution have ability to link schedules to project coordination issues?				

(Continued)

Serial no	Parameter for Evaluation	Available	In Roadmap	Not Available	Remarks
2.4	Does the solution have ability to link schedules to design deliverables?				
2.5	Does the solution have ability to link schedules to procurement schedules?				
2.6	Does the solution have ability to link material deliverables to schedules in real time?				
2.7	Does the solution have ability to link labor requirements to schedules in real time?				
3.0	**KPIs Section**				
3.1	Does the solution have the ability to compute milestone completion based on actual progress on the items above (schedule, issues, documents etc.)?				
3.2	Does it have the capability to compute EVM metrics based on actual progress?				
3.3	Does it support change to EVM metrics on change in schedule (see also 1.7 above)?				
3.4	Does it have ability to compute time cost of delay?				
3.5	Does it have the ability to visualize results in BIM authoring solutions?				
3.6	Does it have the ability to read data from BIM authoring solutions?				
4.0	**Real time capability**				
4.1	Does it have the ability to notify stakeholders on changes? How configurable? Please elaborate.				
4.2	Does it have the ability to notify stakeholders on missing future milestones?				
4.3	Does the solution have the ability to read data from sensor based devices to capture progress? Please elaborate.				
5.0	**Analytics**				

(Continued)

Serial no	Parameter for Evaluation	Available	In Roadmap	Not Available	Remarks
5.1	Does it have the ability to learn about delays based on patterns? Across projects, between similar processes, agencies, etc. Please elaborate.				
6.0	**Payment models**				
6.1	Does it have multiple payment models? Please elaborate.				
6.2	Does it have modular payment options? Please elaborate.				
7.0	**Others**				
7.1	Any other unique capabilities. Please elaborate.				

20

USE OF BLOCKCHAIN FOR ENABLING CONSTRUCTION 4.0

Abel Maciel

20.1 Aims

- To describe the basis and evolution of blockchain in the context of built environment activities and processes.
- To evaluate the implications and impact of blockchain technology upon the construction and property industries.
- To reflect upon the linkages between blockchain and Construction 4.0.

20.2 Introduction

The construction industry is a high impact sector and a key driver for all national economies, be it through employment of people, development and return on investment for the advancement of civilization or as a fundamental stimulator for any economic activity. Yet, the complexity of construction, in particular with regards to design, legal and financial aspects, causes recurrent and chronical inefficiencies. Emerging technologies related to the Fourth Industrial Revolution and specifically blockchain technology bring opportunities of overcoming these inefficiencies.

The Industry 4.0 emerged in the 2010s and is currently being defined as the merge of the 'physical and digital', the era where the 'bits' connect with 'atoms' and objects and environments become smart. There is a shift to almost instantaneous global distribution with popular YouTube videos being seen by billions of people and E-Scouters appearing simultaneously in many cities around the world.

The scope and magnitude of transformation is deep and wide, with, for example, software being distributed and update online and in real-time. This is an immense change and impacting billions of people. Never before we had to contemplate this scale of change.

The decomposition of the Industrial 4.0 can be describe in waves of changes (Case, 2017), with the first wave (1985–1999) charactering the development of *the building blocks of hardware and software* and the start of the broadcasting static information.

The second wave (2000–2015) was marked by the *software as a service* and the ability to search the world wide web, social networks and e-commerce. The second wave also introduced the use of mobile phones and the invention of the smartphone and the app economy. This increased the number of internet users to 3.2 billion by 2015 (BBC, 2015).

We are currently experiencing the third wave of the Industry 4.0 (2016–Now). Internet access is ubiquitous and new business models are 'born online'. This acceleration and degree of interconnectivity create the conditions for further development, bringing to us the rapid advances in the many different sciences in a similar fashion to the 1600's Renaissance. It might have an assimilation period of similar dimension (of about 200 years) or instead, it might match the accelerating times.

By 2025, it is expected that the world will have 75 billion things exchanging information in the IoT (Statista, 2019a). This information flow is becoming a detailed account of all and each one of us and how we interact with our environment. To prepare and respond to this new age of ubiquitous and embedded information, we have to consider technologies like blockchain.

Blockchain, also known as the '*Trust Protocol*' (Milutinovic et al., 2016; Tapscott, 2016; p. 4), can be described as the amalgamation of a few technologies for secure decentralized data management and value transaction. The interest in blockchain has been increasing since the idea of a decentralized cryptocurrency was developed in 2008 (Nakamoto, 2008) and Bitcoin was launched in 2009 (Wattles, 2017).

Blockchain has characteristics in its architecture that provide security, anonymity, provenance, immutability and purpose of data without any third-party organization in control of data transactions. Because of this automation, blockchain creates novel opportunities in the many digital applications and ecosystems out there. It becomes particularly disruptive in cyber-physical systems, the Internet of Things, cloud computing, cognitive computing and other automation and data exchange processes relevant for the manufacturing technologies and the Fourth Industrial Revolution (Hermann et al., 2016; Kagermann et al., 2013; Schwab, 2017).

The general field of Distributed Ledgers Technologies (DLT), with and without the blockchain component, is also highly relevant in application for Virtual Environments (VE), including Building Information Modelling (BIM) and digital construction management.

In this work we investigate the consequences of blockchain in relation to technologies associated with the construction sector, examining current case studies and discussing the prospects of the evolving technology itself.

We also discuss some known limitations and challenges in implementing the technology and how this reflects in prospective applications in the construction sectors.

20.3 Construction challenges in the era of BIM

Construction is manifested as a highly project-based sector, with tasks typically considered as non-repetitive activities where various professionals organized in firms come together to deliver housing, infrastructure of other construction projects. This delivery of projects is a complex undertaking, typically over-budget and over-schedule. The construction process is fragmented and coordination between the various stakeholders is often suboptimal, resulting in lost productivity, rework, delayed progress and consequentially, increased fees.

The complexity of construction projects starts from its onset, as these temporary endeavours are financed from multiple sources; from governmental backing of large infrastructure or public works, to private funding of commercial real estate. The driver for such financing is, generally, stimulus of economic activities. Additional interests that stimulate investment may include improvement of delivery of services, environmental resilience, transportation connectivity, efficient utilization of capital and for-profit activities.

In the UK, construction contributes more than £163 billion per annum and contributes 6.5% of GDP to the UK's Gross Domestic Product (GDP) between 2015 and 2020 (ONS, 2018, 2017). The importance and also the complexity of the construction sector can be outlined by a national macro-level that contributes 6% to UK GDP, or approximately £113 billion in 2017 by value. In total, 2.4 million jobs in Q3 2018 (Rhodes, 2018) are employed by the UK sector and 10.7 million jobs by the USA construction (Statista, 2019c).

On the other hand, at a sectorial micro-level, construction projects vary significantly by size and economic value. By market segment, housing accounts for 25% value of construction sector orders, infrastructure for 12%, repairs and maintenance for 34%, other private orders 22% and other public 7% (Rhodes, 2018).

In a typical construction project there are numerous stakeholders' organizations involved throughout the supply chain. This includes sponsors (client, funders), the design team, building contractors, the supply chain (manufacturers and product suppliers), the operation team, and the users, including tenants, residents, customers.

All parties have invested interests that are impacted by adverse project events often initiated by poor information and communication.

Construction is also a highly fragmented industry. This fragmentation can be described as informational and organizational, with over 99% of businesses being comprised of disperse Small-Medium Enterprises (SMEs) (White, 2015). This is consistent with the European average of construction industry structure (White, 2015).

The fragmentation of information in construction projects are a persistent problem characterized by a disconnect between design and construction (Bryde et al., 2013). This takes place mainly because of the lack of open and trustworthy distribution of coherent, cohesive information across all stakeholders. Blockchain technology has the potential to subvert some of these effects through open, transparent transactions, particularly if associated with BIM processes.

For a *Construction 4.0* in line with current emergent technologies, managerial supervision from all stakeholders on every aspect of a project has to crosslink all design, procurement, statutory compliance and operations in a framework that creates information symmetry in a project's structures, roles and responsibilities.

Currently, in construction procurement, the parties involved have to go to great lengths to have certainty regarding identity, reputability and a price guarantee. In a very near future, once the Internet of Things (IoT) has gathered and compiled enough data, blockchain is likely the mechanism that will govern how it is distributed. An immutable ledger of transactions offers great potential to manage and benchmark infrastructure and construction assets efficiently.

As the systems around us, including transportation, infrastructure, energy, waste and water, become more interconnected, a trusted system such as blockchain can offer a greater value yield. The technology has the potential to efficiently manage the ever increasing demand for infrastructure nodes and energy supply (Arup, 2019). BIM platforms and protocols are the natural choice to accelerate this digital transformation. BIM brings together architecture, engineering and construction professionals under a Computer Supported Collaborative Work (CSCW) process to produce a federated digital design model and therefore create a coherent dataset for the construction and management of a new asset.

20.3.1 Delivering projects in BIM

BIM projects start with the development of an Employer's Information Requirements (EIR) for a new asset. This is often developed with the assistance of the architect and design team and is later translated into the BIM Execution Plan (BEP or BXP) (BSI, 2014).

The BEP aims to define roles and responsibilities, activities and level of development of the digital model for each stage of the project. There are a number of cross-mapping references between BEPs and project contractual protocols, including the RIBA Plan of Work (Sinclair, 2012) and more recently, construction contracts (JCT, 2019).

EIRs and BEPs are often ignored, misunderstood or underestimated. This might happen because of general lack of understanding on the benefits of adopting the BIM process. There is still little education and training available on what is necessary as a BIM requirement. On the other hand, it is hard to justify the extra overheads to deliver a useful BIM 'layer' with the project. The processes to validate the model according to the EIRs and BEPs are also very labour intensive and prone to errors.

The delivery of a BIM is also very front-loaded process, with a consultant having to develop a great deal of work that will not immediately benefit their contribution to the project. Surely BIM is very useful for the design of the building, but the extra labour in delivering a high *Level of development* is considered currently to be excessive. For models with a high Level of Detail (LOD) and Level of Information (LOI), the latency and performance of the system is also a problem.

One of the main strategies for BIM platform interoperability is the Industry Foundation Class (IFC) open file format (OpenBIM, 2017). However, for complex geometries, other file formats are used, and extensive geometry and data scripting often the only viable way to avoid geometrical distortion and data loss in the interoperability and agile of federated models (Autodesk, 2019; Davidson, 2019). Nevertheless, BIM has an immense impact on the construction industry and has created noticeable efficiencies in the sector.

20.3.2 Procurement and integrated teams

As BIM is positioned by governments and construction professionals as a key solution to persistent problems in the construction industry, more research is needed to measure the benefits of BIM and how these benefits might be augmented with blockchain.

The effectiveness of BIM as a medium for communicating information within a construction team can indeed promote considerably more accurate, on-time and appropriate exchange of information. In that sense, it is possible to quantify some of its benefits in relation to information management in construction (Demian and Walters, 2014).

BIM solutions helps to foster earlier creation of critical information relating to design detailing, programming, logistics and coordination that help to generate significant value during the later production phases. These underlying trends highlight the core potential of BIM to foster better collaboration between project participants, thus placing considerable emphasis on its role in 'a human activity that ultimately involves broad process changes in construction' (Eastman and Rafael, 2008; Eastman et al., 2011).

There is some evidence to suggest that these benefits are observed not only on large scale projects but also projects of a relatively small scale (Aranda-Mena et al., 2009; Eastman and Rafael, 2008).

The application of BIM has resulted in a fast-changing landscape for construction contracts. Lawyers struggle to keep up with the pace of innovation and the need to provide legal solutions and accommodate new approaches. Intelligent contracts appear as a logical extension to BIM whereby the contractual performance itself becomes automated.

Smart contracts, as proposed today, are of short-term execution or of instantaneous effect. This is at odds with the complex and long-running nature of construction projects. Storage constraints, compatibility and reliability issues together with confidentiality and the long-term nature of distributed ledgers pose additional problems (Mason, 2016).

As new legal applications emerge, 'LawTech' or 'LegalTech' technologies are becoming part of the practices of law, including processes in courts. LawTech refers to the use of technology and software leveraging blockchain to provide legal services where advice is given both before the transaction commences and after disputes break out. This refers to the application of technology and software to help law firms with practice management, documentation, storage, billing, accounting and electronic discovery.

In relation to cash flow in construction, the Construction Act 2009 amended the Construction Act 1996 by incorporation of detailed notice requirements (LDEDCA, 2009, Part 8, SCCR, 1998, Part 2, s.2). The parties to a construction contract were no longer free to agree notice periods not otherwise in accordance with those specified in the legislation, or on failure to agree such notice periods under the contract those prescribed by the Scheme (as amended). This progress in legislation resolution and the advent of smart contracts could jointly promote unforeseen levels of development the construction sectors.

Blockchain in procurement can be used to effect if the parties agree to abide and produce an outcome within days rather than years in the most complex of disputes. They can be more easily incorporated in anticipatory disputes, such as in construction contracts.

The concept of 'algorithmic regulation', modelled on 'algorithmic trading systems' (Treleaven et al., 2013), is to stream compliance, social networks data, and other kinds of information from different sources to a platform where compliance reports are encoded using DLT and regulations are 'codifiable' and 'executable' as Decentralized Applications (DApps) in the blockchain.

Generally, five areas of impact have been highlighted and include 'intelligent regulatory advisor' as a front-end to the regulatory handbook; 'automated monitoring' of online and social media to detect consumer and market abuse; 'automated reporting' using online compliance communication and big data analytics; 'regulatory policy modelling' using smart contract technology to codify regulations and assess impact before deployment; and 'automated regulation' employing blockchain technology to automate monitoring and compliance (Treleaven and Batrinca, 2017).

20.3.3 Supply chains and circular economy

The structure of resource consumption in the built environment is shifting from a fragmented and untraceable workflow. Where once consumption took a linear form, from sourcing to use to disposal – sometimes termed the *take-make-use-dispose* model – we are now on the crossover of a different culture. The circular economy in the built environment is becoming a priority for developers looking to curb the consumption of natural resources, prevent waste and increase efficiency through the recycling and responsible sourcing of building resources (Accenture, 2014; Arup, 2016). This model aims to save on negative externalities such as carbon emission, unsustainable landfill and water extraction and widespread ecosystem pollution. In addition, it aims to create opportunities for the industry across the supply chain.

Better Supply Chain Management (SCM) software is in high demand (van der Meulen, 2018) and the effect of blockchain in circular economies should not be underestimated. It is very likely that blockchain will enable the effective and reliable tracking of materials, components and whole products throughout the supply and reuse chain, although the exact nexus between the physical and the digital record is still problematic. If this is resolved, there is great potential to create a perpetual cycle of use of units, elements and products, opening the industry to new economic models of construction processes where manufacturer, recycler and consumer can confidently assess the circularity of their products.

Some rudimentary application platform has been developed for open-source distributed communications protocol for a circular economy (Circularise, 2018). This allows information to be exchanged throughout the value chain, creating transparency around product histories and destinations of materials.

20.3.4 Health and safety

Since the fire of London's Grenfell Tower in 2018, there has been perplexity and intense debate in relation to the UK's statutory and compliance frameworks. How can such an accident like this happen in our time? There are a number of reports showing a mismatch of information, roles and responsibilities (Marrs, 2018). The Grenfell Tower accident was characterized by multiple and systemic failure point including professional services, design specification, component manufacturing, material performance, statutory compliance checks and tests, facilities management processes, etc.

Traceability of design and construction, from early EIR for BIM to unified operations and statutory compliance framework can now be envisioned with the advent of blockchain.

The immutability of blockchain has the potential to create a fully traceable and trusted *accountability chain*, also capturing decisions and interactions of all agents on the creation and maintenance of built assets. The very existence of such a record can not only have a persistent effect on the industry's behaviours but also provide the level of granularity necessary to derive accurate '*macro-behavioural economic models*' to better management risk in the industry and prevent incidents such as Grenfell Tower's. Currently, macro-behavioural economic modelling is a very difficult challenge to say the least (Baddeley, 2017).

20.4 Context and aspects of blockchain

Blockchain technology can improve the quality, traceability and security of live data collection in the construction sector. A digital immutable ledger allows a project to be mapped and tracked at every stage. During the design phase, this is useful for establishing ownership of models and tracking project progress reliably, benefiting stakeholders by reducing the opportunity for corruption, inefficiencies and contractual disputes.

Blockchain is central to cyber-physical systems and from its beginning, complexity beyond currency and payments was envisioned. the possibilities for programmable money and computational contracts were considered into the protocol of the earliest cryptocurrencies like Bitcoin (Nakamoto, 2008). This '*tokenized economy*' offers a large variety of possible transaction and applications yet to be discovered and exploited.

The technology has evolved from decentralization of money and payments into the decentralization of markets and contemplates the transfer of many other kinds of assets beyond cryptocurrency: from the creation of a unit of value through every time it is transferred or divided (Swan, 2015; Tapscott, 2016).

New developments in blockchain since 2014 are leveraging novel categories, distinctions, and understandings of value allocation, and new standard classifications and definitions are still emerging. Some of the terminology that broadly refers to the blockchain 2.0 space can include Smart Contracts, Smart Property, Decentralized Applications and Decentralized Autonomous Organizations (DAOs) (Choi, 2017; Dino et al., 2016; Swan, 2015).

It is also important to note that not all processes need an economy or a payments system, or peer-to-peer exchange, or decentralization, or robust public record keeping. The scale of operations is a relevant factor, because it might not make sense to have every tiny microtransaction recorded on a public blockchain and every node of the ledger to contain all data. Transactions could be batched into *sidechains* (Back et al., 2014) in which one overall daily transaction is recorded and nodes can be *sharded* to speed up ledgers synchronization (Kelly, 2014).

There is ample opportunity to explore more expansively the idea of the blockchain as an information technology converging the design and construction activities and the built environment, including what a consensus models might it mean and enable in the design process (Maciel, 2014). A key question is what is consensus-derived information; that is, what are its properties and benefits vis-à-vis other kinds of information?

Blockchain technology helps to distinguish at least three different levels of information. The first level being plain, unmodulated data. A second level can be described as data elements enriched by social network peer recommendation. The quality of the information is denser because of this social recommendation layer. On the newly proposed level three, blockchain consensus-validated data, peer recommendations are further formalized using a consensus mechanism about the quality and accuracy of the data (Swan, 2015). Blockchain technology thus is thought to produce a third tier of information modulated with quality attributes.

20.4.1 Basic principles

Blockchain is a growing list of records, called blocks. These blocks are linked using cryptography and each block contains a cryptographic hash of the previous block, a timestamp and transaction data.

A cryptographic hash is generated by a function that has certain properties which make it suitable for use in cryptography of blockchains. It is a mathematical algorithm that maps data of arbitrary size to a bit string of a fixed size (a hash) and is designed to be a one-way function, that is, a function which is infeasible to invert (Halevi and Krawczyk, 2006). The Bitcoin blockchain uses the SHA-256 (Secure Hash Algorithm) standard typically looking like the following in Table 20.1:

Table 20.1 SHA-256 example

'hash': '0000000082b5015589a3fdf2d4baff403e6f0be035a5d9742c1cae6295464449',

In Figure 20.1, we can find a basic diagram of the composition of a block in a Bitcoin blockchain. As blocks are created, each block has a header containing a number of key data.

A *timestamp* is added, registering when blocks are found. The block also contains a hash referring to parent block. This *Prev_Hash* of the previous block header ties each block to its parent, and therefore by induction to all previous blocks. This chain of references is the eponymic concept for the blockchain.

The Merkle Root or *Tx_Root*, is a reduced representation of the set of transactions that is confirmed with the block. The transactions themselves are provided independently, forming the body of the block. In the case of Bitcoin, there must be at least one transaction, called the *Coinbase*. This is a special transaction coining a program that may create new Bitcoins and collect the transactions fees. Other transactions are optional. An arbitrarily picked number or a *Nonce* is used to add entropy to a block header without rebuilding the Merkle tree.

The *target* corresponds to the difficulty of finding a new block. In the Bitcoin blockchain, the target is updated every 2016 blocks when the difficulty reset occurs.

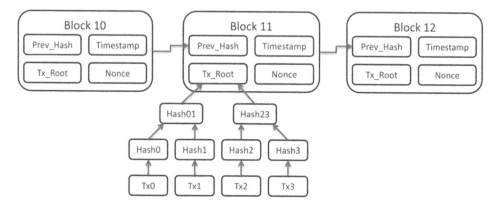

Figure 20.1   Bitcoin block data (Wander, 2015)

All of the above header items (i.e. all except the transaction data) get hashed into the block hash, which for one is proof that the other parts of the header have not been changed, and then is used as a reference by the succeeding block.

20.4.1.1 Synchronizing ledgers

As many miners compete to find the next block, often there will be more than one valid next block discovered. This is resolved as soon as one of the two forks progresses to a greater length, at which any client that receives the newest block knows to discard the shorter fork. These discarded blocks are referred to as *extinct blocks*.

This and other consensus mechanisms are necessary to create tolerance to failure in the given distributed system resilience. This is normally refed to as Byzantine Fault Tolerance (Lamport et al., 1982) and it can be summarized as any fault presenting different symptoms to different observers, or any loss of service due to a Byzantine consensus fault (Driscoll et al., 2004). The development of new consensus mechanisms is one of the main avenues to improve performance in blockchains.

20.4.2 Disruptive aspects of blockchain

Blockchain represents a significant leap in the way we digitally manipulate value, identity and automation of processes. There are numerous blockchain applications and use cases, and they have been previously structured into six categories (Strategic Registry, Identity, Smart Contracts, Dynamic Registries, Payment Infrastructures, and Others) across its two fundamental functions – record keeping and transacting (Carson et al., 2018).

20.4.2.1 A network of value

One of the primary concerns of a digital value network is the issue of double-spending. This refers to the incidence of an agent spending the same balance of a unit of value more than once, effectively creating a disparity between the spending record and the amount of that unit available. The issue of double-spending is a problem that cash does not have. A transaction using a digital currency like Bitcoin, however, occurs entirely digitally.

Bitcoins and other blockchains solve this double-spending problem with no central authority by using one or more consensus mechanisms. In the Bitcoin case, the waiting for confirmations when receiving payments leads to the transactions becoming more irreversible as the number of confirmations rises (BitcoinWiki, 2019; Hoepman, 2008; Osipkov et al., 2007). This reliable digital network of value with no central authority opens up a number of opportunities in a wide range of fields.

20.4.2.2 Data ownership and control

Decentralized Identifiers (DIDs) are a type of identifier for verifiable, 'self-sovereign' digital identity on a blockchain. DIDs are fully under the control of the DID owner, independent from any centralized registry, identity provider, or certificate authority. DIDs are URLs that relate a DID owner to means for trustable interactions with that owner. DIDs resolve to DID Documents, which are simple documents with instructions for the specific DID. Each DID Document may contain at least three things: proof purposes, verification methods, and service endpoints. For example, a DID Document can specify that a particular verification method, such as a cryptographic public key or pseudonymous biometric protocol, can be used to verify a proof that was created for the purpose of authentication. Service endpoints enable trusted interactions with the DID controller (W3C, 2019a; Windley, 2019).

Opensource blockchain projects like Hyperledger Indy (Hyperledger, 2018) enable purpose-built decentralized identity. It provides a toolkit for creating and using independent digital identities rooted on blockchains or other DLTs so that they are interoperable across administrative domains, applications, and any other 'silo'.

Generally speaking, because blockchains cannot be altered after the fact, it is essential that use cases for ledger-based identity are carefully consider in relation to its foundational components, including performance, scale, trust model, and privacy. In particular, 'privacy by design' and 'privacy-preserving' technologies are critically important for a public identity ledger where correlation can take place on a global scale.

For all these reasons, open source initiatives emerged to developed specifications, terminology, and design patterns for decentralized identity along with an implementation of these concepts that can be leveraged and consumed universally (Hyperledger, 2019). Other Hyperledger projects are infrastructure and tools to enable the exchange of blockchain-based data, support peer-to-peer messaging in various scenarios, and facilitates interoperable interaction between different blockchains and other DLTs (George et al., 2019).

20.4.3 Technical challenges and limitations

Blockchain limitations include those related to technical issues with the underlying technology as well as ongoing industry thefts and scandals, government regulation, and the mainstream adoption of the technology.

The issues are in clear sight of developers, with different answers and solutions suggested. Many are building different and new blockchains to circumvent limitations. One central challenge with the underlying blockchain technology is scaling up from the current maximum limit of transactions per second, especially if there were to be mainstream adoption of blockchains like Bitcoin (Lee, 2013). Some of the other issues include increasing the block size, addressing blockchain bloat, countering vulnerability to 51% mining attacks, and implementing hard forks (changes that are not backward compatible) to the code, as summarized here (Spaven, 2014).

20.4.3.1 *Throughput and latency*

The Bitcoin network has a potential issue with throughput in that it is processing only one transaction per second (tps), with a theoretical current maximum of 7 tps. The Ethereum blockchain supports 15 tps (Etherscan, 2019; Hertig, 2018).

For reference, metrics in other transaction processing networks are VISA (2,000 tps typical; 10,000 tps peak), Twitter (5,000 tps typical; 15,000 tps peak), and advertising networks (>100,000 tps typical) (Swan, 2015). A higher performance would be necessary and core Bitcoin and Ethereum developers are working to raise limits for when it becomes necessary (GitHub, 2019). One way that Bitcoin could handle higher throughput is if each block were bigger, though right now that leads to other issues with regard to size and blockchain bloat (Swan, 2015).

Average confirmation time of Bitcoin transactions from in 2019 was about 10 minutes (Statista, 2019b). For sufficient security, it is advised to wait longer for larger transfer amounts as the confirmation number must outweigh the cost of a double spend attack.

20.4.3.2 *Size and bandwidth*

The Bitcoin blockchain already takes a long time to download. If throughput were to increase by a factor of 2,000 to VISA standards, for example, that would be 1.42 PB/year or 3.9 GB/day. At 150,000 tps, the blockchain would grow by 214 PB/year. The Bitcoin community calls the size problem 'bloat'. Scaling to mainstream use may not require necessarily more nodes, but a more efficient access. This might motivate centralization, as more resources are needed to run a full node, and about 50% of all nodes are in only three countries (Bitnodes, 2019). It is an ongoing discussion whether location discrimination for running full nodes should be a factor for automated rewards by the network.

Innovation to address blockchain bloat and make the data more accessible are APIs that facilitate automated calls to the full Bitcoin blockchain. Some of the operations are to obtain address of balances and balances changes and notify user applications when new transactions or blocks are created on the network. There are also web-based block explorers (Blockchain, 2019), middleware applications allowing partial queries of blockchain data, and frontend customer-facing mobile e-wallets with greatly streamlined blockchain data.

20.4.3.3 *Security*

From the potential security issues with blockchain, the most troublesome is the possibility of a 51% attack, in which one mining entity could take control of the blockchain and double-spend previously transacted coins into his own account (Valkenburgh, 2018). The issue is the centralization tendency in mining where the competition to record new transaction blocks in the blockchain has meant that only a few large mining pools control the majority of the transaction recording. At present, the incentive is for them to be good players, and some have stated that they would not take over the network in a 51% attack, but the network is insecure (Rizzo, 2014). Double-spending might also still be possible in other ways – for example, spoofing users to resend transactions, allowing malicious coders to double-spend coins.

Another security issue is that the current cryptography standard that Bitcoin uses, Elliptic Curve Cryptography, might be crackable. Financial cryptography experts have proposed potential upgrades to address this weakness (Wang et al., 2019).

20.4.3.4 Resources

Some consensus mechanisms and mining draw an enormous amount of energy, all of it wasted. The earlier estimate cited was US$15 million per day, and other estimates are higher (O'Dwyer and Malone, 2014). Bitcoin's annual electricity consumption adds up to 45.8 TWh. The corresponding annual carbon emissions range from 22.0 to 22.9 MtCO2. This level sits between the levels produced by the nations of Jordan and Sri Lanka (Stoll et al., 2019). The wastefulness proof of work (PoW) and mining is what makes blockchains like Bitcoin trustable, but such a huge waste indicates the consensus mechanism needs to be improved and become sustainable.

20.4.3.5 Infrastructure and usability

Many technical issues in blockchain have to do with their infrastructure. One issue is the proliferation of blockchains, and that with so many different blockchains in existence, it is becoming easy to deploy the resources to launch a 51% attack on smaller chains as happened with the Ethereum Classic (ETC) in January 2019 (Nesbitt, 2019).

Another issue is that when chains are split for administrative or versioning purposes, there is no easy way to merge or cross-transact on forked chains.

From a different perspective, APIs for working with blockchains like Bitcoin, Ethereum and Hyperledger projects are far less user-friendly than the current standards of other easy-to-use modern APIs, such as widely used RESTful APIs (RESTfulAPI, 2019; W3C, 2019b).

20.4.3.6 Ecosystem

Another critical technical challenge and requirement is the creation of a complete ecosystem of solutions, particularly in service delivery. There are needs for secure decentralized storage (MaidSafe, 2019; Storj, 2019) messaging, transport, communications protocols, namespace and address management, network administration, and archival to name a few.

The blockchain industry may develop similarly to the cloud-computing industry, for which standard infrastructure components were defined and implemented at the beginning to allow for the development of value-added services instead of the core infrastructure. This is important in the blockchain economy due to the cryptographic engineering aspects of decentralized networks.

20.4.3.7 Innovations

Many of the innovations to the technical issues include *Offline Storage* for storing the bulk of consumer cryptocurrencies (Bonneau et al., 2015).

The proposition of *Dark Pools* envisions a more granular value chain such that big crypto-exchanges operate their own internal databases of transactions, and then periodically synchronize a summary of the transactions with the blockchain – an idea borrowed from the banking industry (Peters and Vishnia, 2016).

Hashing Algorithms like the one proposed for Litecoin and other cryptocurrencies are at least slightly faster than Bitcoin (Christidis and Devetsikiotis, 2016). As new, faster hashing algorithms are being developed, the prospect of new applications of blockchain become feasible.

There are many novel *Consensus Models* being proposed with lower latency, requiring less computational power, wasting fewer resources, and improving security for smaller chains (Zheng et al., 2017). Methods for consensus without mining or proof of work are also appearing and evolving. One example is the Practical Byzantine Fault Tolerance algorithm (Hyperledger, 2018).

To coordinate transactions between blockchains and create blockchain interoperability using several *Sidechains* is also being investigated and developed (Hwang et al., 2018).

20.5 Application considerations

Blockchain can play a significant role in facilitating new business models and the new transactional requirements in the construction industry. It makes possible for BIM platforms to be operated via smart contracts, signalling the team when to initiate a particular transaction and further automating construction procurement processes from the outset.

Here we discuss a few considerations on blockchain application development.

20.5.1 Levels of distribution and access

Over the past few years, blockchains have evolved in different builds and configurations. The content stored on the blocks of the blockchain and the activities performed by the various participants on the blockchain networks can be controlled depending upon how the blockchain is configured, and how it is expected to fulfil the desired business purpose.

Public and private blockchains are the two most common types used in various cryptocurrency networks and the private enterprises. A third category, permissioned blockchains, has also gained traction in recent times.

For instance, Bitcoin, the most popular cryptocurrency blockchain, allows anyone to participate in the network in the capacity of a full node, or a contributing miner. Anyone can take a read-only role or make legit changes to the blockchain like adding a new block or maintaining a full copy of the entire blockchain. Such blockchains, which allow equal and open rights to all participants, are called open, public, or un-permissioned blockchains. Recently, Bitcoin, Ethereum and EOS (2019) have been ranked as the most popular public blockchains and a number of performance indicators for public blockchains have been proposed (Tang et al., 2019).

If one needs to run a blockchain that allows only a selected entry of verified participants, like those for a private business, an internal private blockchain can be implemented. A participant can join such a private network only through an authentic and verified invitation, and a validation is necessary either by the network operator(s) or by a clearly defined set protocol implemented by the network (Seth, 2019). An incentive strategy, like a payment for mining for blocks, mainly aims to promote resource sharing, to stimulate group intelligence, and to promote collaborative communication. Private blockchain platforms are set to a specific group, rather than to everyone. Thus, incentive mechanisms or mining activities are not mandatory (Lu, 2019).

Permissioned blockchains can be nested in public blockchains and maintain an access control layer to allow certain actions to be performed only by certain identifiable participants. These blockchains differ from public as well as private blockchains.

The intrinsic configuration of such blockchains controls the participants' transactions and defines their roles in which each participant can access and contribute to predetermined parts in the blockchain. It may also include maintaining the identity of each blockchain participant on the network. An example of such blockchains include Ripple (2019), which determines roles for a select number of participants who can act as transaction validators on their network (Frankenfield, 2018).

The versatility of permissioned blockchains are popular in industry-level enterprises and businesses, for which security, identity and role definition are important. However, the transactions that occur on such a blockchain may also involve logistics partners, financing banks, and other vendors involved in the supply and financing process. These external parties, though part of the whole network, don't have to know the price at which the manufacturer supplies the products to various clients. Use of permissioned blockchains allows such role-limited implementations.

Some of the features of public, private and permissioned blockchains can be combined in hybrid, more exotic blockchain architectures (Foley, 2018).

20.5.2 Tokenized economies and alternatives

As previously mentioned, the *tokenization* of assets refers to the process of issuing blockchain tokens (specifically, a security token) that digitally represents a real tradable asset. This is in many ways similar to the traditional process of securitization (Laurent et al., 2018).

This new *token economy* offers some efficiencies by reducing friction associated with the creation and treading of securities. Some of the key advantages are greater liquidity, faster and cheaper transaction, more transparency and more accessibility (Laurent et al., 2018).

Some obstacles are the regulatory framework around the decentralized nature of the blockchain infrastructure. Security regulations are typically technology agnostic, and security tokens, depending on their exact features, can be implicated by the full scope of relevant security regulations and variations from different jurisdictions (HMT, 2018).

Tokenization represents a novel and powerful way to coordinate assets and value distribution in all verticals, including real estate, finance and naturally, in the construction sector.

20.5.3 Energy, speed and resilience

A *consensus mechanism* is a fault-tolerant device used in distributed computation systems to achieve the necessary agreement on a single data value or a single state of the network among distributed processes or multi-agent systems, such as with blockchains. It is useful in record-keeping, among other things (Frankenfield, 2019).

Bitcoin was originally described as a purely peer-to-peer version of electronic cash allowing online payments to be sent directly from one party to another without going through a financial institution (Nakamoto, 2008). It is one of the original blockchains and probably one of the most power-hungry platforms because of its proof of work (PoW) system to compile blocks in the ledger.

Ethereum describes itself as a decentralized platform that runs smart contracts (Ethereum Foundation, 2018) as Decentralized Applications (DApps). The Ethereum platform uses a Proof-of-Stake (PoS) involving the allocation of responsibility in maintaining the public ledger to a participant node in proportion to the number of virtual currency tokens held by it. This is a low-energy, low-cost methods to create consensus in the ledger. However, this method has been criticized for promoting cryptocoin saving, instead of spending.

R3 Corda is a blockchain and smart contract platform initially developed for financial applications but now being used in various domains (R3, 2019a). Corda can handle complex agreements. This capability has broad applications across industries including finance, supply chain and health care. The consensus mechanism in Corda is has two combined ties aiming to determining whether a proposed transaction is a valid ledger. They are *Validity* consensus – this is checked by each required signer before they sign the transaction, and *Uniqueness* consensus – this is only checked by a notary service (R3, 2019b).

Hyperledger is an open source collaborative effort created to advance cross-industry blockchain technologies. It is a global collaboration, hosted by Linux Foundation, developing enterprise grade blockchain technology (private and beyond) in finance, banking, Internet of Things, supply chains, and manufacturing (Hyperledger, 2018). Hyperledger architectures of the consensus layer in their many projects use a number of algorithms, including *Proof of Elapsed Time* (PoET), *Proof of Work* (PoW) and the use of voting-based methods including *Redundant Byzantine Fault Tolerance* (RBFT) and *Paxo* (Hyperledger, 2019). This creates flexibility to customize blockchains for particular applications and sectors.

Many new blockchains are appearing and some have creative propositions for their consensus mechanism. It is important to take into consideration the properties of the blockchain when selecting one for deploying a new application.

20.6 Blockchain and the construction sector

Construction brings together large teams to design and shape the built environment. Technologies and workflows like BIM create openness to collaboration across the industry. This momentum could be leveraged to bring the use of blockchain technology to the fore (Hughes, 2017).

Reports published since 2017 indicate that there is much progress to be made with the education, implementation and deployment of services based on blockchain technology. They point to a number of prospects and opportunities posed by using the technology in the sector, including efficiency (payments and transactions), cost savings, transparency and the augmentation and enhancements of IoT, AI and BIM (Arup, 2019; ICE, 2018; Reuters, 2018).

The fact that many surveys have all shared a number of the above points is encouraging as it suggests a willingness to critique and thereby improve understanding of the technology, not solely to appear 'ahead of the curve' with industry trends.

Significant progress is needed in the development of standards before adoption is achieved. There have not been many concrete indications of a future timeline for this, however it is expected that cities, energy and transport use cases will likely see adoption from 2025 (Arup, 2019, p. 18). Earlier reports such as by Thomson Reuters (2017, p. 12), Digital Catapult (2018) and the CDBB (Lamb, 2018) suggest that there is still a high degree of uncertainty about the future of blockchain. Indeed, it has been asserted that, by 2023, 90% of blockchain-based supply chain initiatives will suffer 'blockchain fatigue' due to a lack of strong use cases, according to Gartner (Omale, 2019). This conclusion was reached from data gathered between November 2017 and February 2018, including the finding that only 19% of respondents ranked blockchain as a very important technology for their business, and only 9% have invested in it (Omale, 2019).

Now more than a year later, it could be that these attitudes have improved following the further industry analysis and education that has taken place through these reports and continuing investment in piloting the technology. There has also been a doubling of tech investment in construction in the past decade (ENR, 2018) and there are a number of well-established firms providing blockchain solutions. A further detailed study to confirm or refute any changes in attitudes would be welcome.

In order to accelerate the understanding and development of blockchain technologies, more investment and collaboration is necessary. The University College London based Construction Blockchain Consortium (CBC, 2019) exists as a neutral platform to further these objectives through its series of white papers, providing knowledge transfer and assisting the development of use cases.

20.6.1 Blockchain and building information modelling

Blockchain technology is ideal for total life cycle management of an asset, from design to delivery to operation and reuse. The technology can act as a bridge between all stakeholders, allowing each party to track progress with the option to set up automatic payments, address responsibilities quickly and protect intellectual property.

Blockchain can help to better manage the construction progress monitoring stage, as well as solve the cash flow problem often experienced by companies.

As BIM evolve and further the 3D Design description model, adding extra dimensions of Time, CapEx, carbon cost and life cycle costing, and eventually include more complex dimensions of risk, financing and change control performance, asset management and project control performance, blockchain seems to embody the data architecture necessary to enable the deployment of these new dimensions. This evolution leading to highly integrated work-flows and closer collaboration will demand for more and better professional transdiscipli-nary in order deal with future challenges and to future-proof design and construction (Brown et al., 2010).

Virtual Environments, including VR, AR and XR are key coadjutants on the process of cap-turing and 'blockchaining' data at source, also visualizing this back to integrated teams. Open source real-time graphics engines (Unity, 2019; Unreal, 2018) are opportunities to further develop Construction 4.0 and enable some of the interface ergonomics necessary for block-chain adoption.

From the many types of AI and ML, novel methods like *Big Generative Adversarial Net-works* (BigGAN) are proving to be far more effective in yielding complex automation by unsupervised learning (Zhang et al., 2019). In an age of 'Fake News' and uncertainty of facts in an ever-increasing digitalized world, immutability and provenance of data it's not only timely, it is also absolutely paramount. This data forms the basis for realistic AI/ML applica-tions on predictions, operations and benchmarking of the built environment.

20.6.2 Construction supply chains

Another area in which blockchain is predicted to feature heavily in the next decade is the trans-actions and security mechanisms associated with the IoT. Blockchain can also pride 'trustless' interactions between users and machines and add value to the supply chains in a similar way it can add value to BIM.

In the context of the property sector, IoT is the method of collection and exchange of data processed by smart contracts. This has potential to disrupt supply chains, leading to live track-ing of goods and materials, allowing the circular economy to flourish, dramatically automating the 'metabolic' systems in the built environment and conserving energy (Arup, 2016).

20.6.3 Computational legal contracts and procurement

Currently, most construction projects work towards BIM Level 2 (BSI, 2014; 2013, 2007). Blockchain could accelerate the development and application of BIM Level 3 and realize the end-to-end transparent collaboration.

Computational contracts or smart contracts are currently described as applications that run exactly as programmed without any possibility of downtime, censorship, fraud or third-party interference (Ethereum Foundation, 2018). Smart contracts have the potential to become fully developed computational legal contracts and improve dramatically all aspects of construction project administration and payment systems in the sector. Computational legal contracts can become the engine for smart infrastructure and the combine circular economy.

The use of collaborative BIM has created a necessity for the use of single project insurance. Smart contracts operating with BIM processes offer a far more proactive method for delivering projects. The interoperability of smart contracts and BIM can leverage efficiencies in the allocation of accountability in seconds instead of days or weeks. Blockchain can address project complexity and in doing so reduce late payments, remediations and disputes that place companies under cash flow risk (ICE, 2018).

At the same time, collaborative procurement is a decisive enabler of digital transformation and are under intensive study. All the chronical issues identified in the sector, like frequent change requests, cost overruns and late delivery, failure to administer contracts correctly and/ or effectively, failure to manage projects, mishandling of delay can bring cash flow restrictions, such as late payments and other cash flow issues. To address these, enhanced project management and control, transparency and availability of accurate information in project governance are needed. A sector-generic and project-bespoke blockchain-enabled solution can alleviate these repercussions and optimize cash flow restrictions (ICE, 2018).

20.7 Challenges in the adoption of blockchain

Recent reports published in the period 2016 to 2019 BIM (Arup, 2019; Digital Catapult, 2018; ICE, 2018; Reuters, 2018) note that the full-scale adoption of blockchain could take years because the majority of use cases are in test phases and concerns about profits and the vested interests underpinning these may hinder adoption.

In general, there is still a lack of understanding and trust in the technology, which is slowing investment in, and adoption of, the technology.

The assumption that blockchain technology is inherently secure is wrong. Security vulnerabilities have been uncovered, with potentially more arising in future, and need to be addressed rapidly on an ongoing basis. Failure of firms to ensure that blockchain is appropriately integrated into their processes and business model – and with necessary interoperability – could result in any potential cost savings being nulled or reversed.

A system of globally compatible governance needs to be created in order for blockchain adoption to effectively address global challenges and work seamlessly across regions and borders.

It is expected the blockchain will become an invisible layer and will not represent a great deal of difference for the user but there are energy, scalability and flexibility issues. Consensus mechanisms have to be studied carefully and become more sustainable, faster and scalable.

Ergonomics and Users eXperience (UX) on the application of blockchain is also a factor. As BIM become more complex and map/represent wider phenomena in the built environment, this new knowledge has to be presented eloquently. Effective UX will have to be implemented to handle this new knowledge and some evidence supports that integrated team favour graphic and immersive UX systems (Demian and Walters, 2014).

This is a key aspect in facilitating BIM adoption and accelerating maturity of both technologies in the construction sectors.

20.8 The fourth wave: what might happen next

The Decentralized Autonomous Organization (DAO) represents an innovation in the design of organizations, in its emphasis on computerized rules and contracts, but the DAO's structures and functions also raise issues of governance (Chohan, 2017).

One of the first incarnations of a DAO was 'The DAO' (stylized Đ). This was form of investor-directed venture capital fund (Waters, 2016).

The DAO had an objective to provide a new decentralized business model for organizing both commercial and non-profit enterprises (Allison, 2016; Rennie, 2016). It was instantiated on the Ethereum blockchain (Ethereum Foundation, 2018), and had no conventional management structure or board of directors (Allison, 2016). The code of the DAO is open-source (Brady Dale, 2016).

The DAO was stateless, and not tied to any particular nation state. As a result, many questions of how government regulators would deal with a stateless fund were yet to be dealt with (The Economist, 2016). It was crowdfunded via a token sale in May 2016. It set the record for the largest crowdfunding campaign in history (Waters, 2016).

In June 2016, users exploited a vulnerability in The DAO code to enable them to siphon off one-third of The DAO's funds to a subsidiary account. In July 2016 the Ethereum community decided to hard-fork the Ethereum blockchain to restore virtually all funds to the original contract (Buterin, 2016).This was controversial, and led to a fork in Ethereum, where the original unforked blockchain was maintained as Ethereum Classic. The DAO was delisted from trading on major exchanges in late 2016.

Despite this incident, the concept of a global automated legal persona is very attractive. If we consider superhuman computational processes yielding better outcomes, as in algorithm trading, can a DAO eventually fix national health systems? Can they strategically invest globally to remediate global challenges? Perhaps this is far-fetched, but let's consider for a moment the interoperability issues in the construction sector: Could a DAO help to deliver BIM Level 4 or 5 – fully coordinate and interoperate BIM models, leveraging IFCs and CoBie ontologies to deliver better projects and protect IP in the industry?

The possibilities are many and fundamentally disruptive. They bring a new set of problematics that need to be investigated thoroughly.

20.9 Conclusion

Data and information asymmetry represent one of the major challenges for all agents in the construction sector. Supervising every aspect of a construction project is currently an opaque, fragmented and inefficient process.

Blockchain is a way of organizing transactions or information so that they can be viewed independently by relevant parties and cannot be changed after the fact.

To fully grasp the potential impact of blockchain in the built environment, it is worth taking a holistic view of Industry 4.0. Previous industrial revolutions were erratically distributed across the world, with change often taking significant time before affecting communities.

In today's industrial revolution we have almost instantaneous knowledge-driven dissemination. This is a new socio-technological phenomenon with effects yet to be understood and, needless to say, controlled. It is expected that the Fourth Industrial Revolution will transform the world with a far higher speed, scope and impact than any previous technological revolutions we have experienced.

In this new context of ubiquitous and embedded computing, we can still duplicate and tamper with data. We can fake news, manipulate imagery in real time and double-spend money. How can construction projects' participants prove their identity, stakes and actions? How can different communities have an effect on how their cities are developed and how can they interact with the new emerging intelligent environments they live in?

Blockchain offers much more than a secure, immutable and resilient DLT. It offers the means and opportunity to rethink financial, social and political relationships informing the built environment. It does so by providing digital assets with some of the properties and behaviours of physical objects. This represents one of the key technologies enabling the cyber-physical convergence in the Fourth Industrial Revolution.

The 2018 'Gartner Hype Cycle' displays a decline in the hype of blockchain (Gartner, 2018). Blockchain has now entered the 'implementation phase' and will probably become an invisible but important layer in an interconnected world.

DLT and blockchain also provides a new foundation for machines and humans to interact and exchange information. As a consequence, we may see disruption in infrastructure management, energy and real estate to autonomous transport and water management.

Blockchain can be used as an ID for asset for a circular economy, from design to delivery to operation and reuse. For its resilience, it is the favoured technology to connect people, assets and environments over long periods of time. It can create an '*automation of trust*' with parties having certainty regarding identity, reputability and a price guarantee.

Decentralized Smart Legal Contracts are emerging as the LegalTech to automate contract administration, avoiding chronic industry problems and optimizing construction cash flow. This represents an immense opportunity for the industry and can further improvements on areas such as IP protection, accountability, building management systems, operations and energy conservation.

Blockchain is only one piece of the puzzle in the Construction 4.0. It will very likely be the common thread connecting many technologies. It will require careful consideration on its design and implementation when creating the new global digital trust ecosystem in order to fulfil its promises.

20.10 Summary

- Construction projects rely on effective use of complex data and information, which often exist in different forms.
- Lack of consistency and symmetry of data can result in an opaque, fragmented and inefficient process.
- Blockchain allows organization of transactions or information so that they can be viewed independently and cannot be changed after the fact.
- Unlike previous industrial revolutions, which developed erratically across the world with a time-lag before affecting communities Industrial Revolution 4.0 features almost instantaneous knowledge-driven dissemination.
- The Fourth Industrial Revolution is likely to impact the world with higher speed, scope and impact than any previous technological revolution.
- This carries with it opportunities and risks with ubiquitous, embedded computing that can still allow data to be duplicated and tampered with.
- Construction project participants need a way to prove their identity, inputs and actions, whilst communities and stakeholders seek to have an effect on how their cities and emerging intelligent environments are developed.
- Blockchain offers a secure, immutable and resilient DLT, but also the means and opportunity to rethink financial, social and political relationships informing the built environment by providing digital assets with some of the properties and behaviours of physical objects.
- Blockchain is one of the key technologies enabling the cyber-physical convergence in the Fourth Industrial Revolution.

- Blockchain will probably become an invisible but important layer in an interconnected world, providing a foundation for machines and humans to interact and exchange information.
- As a core technology of IR4.0, it will cause disruption in infrastructure management, energy and real estate to autonomous transport and water management.
- Blockchain can create an '*automation of trust*' with parties having certainty regarding identity, reputability and a price guarantee.
- Smart Legal Contracts are emerging as the LegalTech to automate contract administration, avoiding chronic industry problems and optimizing construction cash flow.
- Blockchain is very likely be the common thread connecting many technologies.

References

Accenture, 2014. Circular Advantage Innovative Business Models and Technologies to Create Value in a World without Limits to Growth [WWW Document]. URL www.accenture.com/t20150523t053139__w__/us-en/_acnmedia/accenture/conversion-assets/dotcom/documents/global/pdf/strategy_6/accenture-circular-advantage-innovative-business-models-technologies-value-growth.pdf (accessed 7.27.19).

Allison, I., 2016. Ethereum Reinvents Companies with Launch of the DAO [WWW Document]. *International Business Times UK*. URL www.ibtimes.co.uk/ethereum-reinvents-companies-launch-dao-1557576 (accessed 3.9.19).

Aranda-Mena, G., Crawford, J., Chevez, A. and Froese, T., 2009. Building Information Modelling Demystified: Does It Make Business Sense to Adopt BIM? *International Journal of Managing Projects in Business*. https://doi.org/10.1108/17538370910971063.

Arup, 2016. Circular Economy in the Built Environment [WWW Document]. URL www.arup.com/perspectives/publications/research/section/circular-economy-in-the-built-environment (accessed 3.3.19).

Arup, 2019. Blockchain and the Built Environment [WWW Document]. URL https://static1.squarespace.com/static/58b6047520099e545622d498/t/5c7565a54192029977bc451f/1551197641233/Blockchain+and+the+built+environment.pdf (accessed 3.3.19).

Autodesk, 2019. Dynamo BIM. URL http://dynamobim.org/ (accessed 5.24.14).

Back, A., Corallo, M., Dashjr, L., Friedenbach, M., Maxwell, G., Miller, A., Poelstra, A., Timón, J. and Wuille, P., 2014. Enabling Blockchain Innovations with Pegged Sidechains [WWW Document]. URL https://blockstream.com/sidechains.pdf (accessed 3.3.19).

Baddeley, M., 2017. *Behavioural Economics: A Very Short Introduction*. OUP Oxford, Oxford.

BBC, 2015. Internet Used by 3.2 Billion in 2015. *BBC News*. URL https://www.bbc.co.uk/news/technology-32884867 (accessed 7.27.19).

BitcoinWiki, 2019. Double-spending Problem. *All about cryptocurrency – BitcoinWiki [WWW Document]*. URL https://en.Bitcoinwiki.org/wiki/Double-spending (accessed 7.27.19).

Bitnodes, 2019. Global Bitcoin Nodes Distribution [WWW Document]. URL https://bitnodes.earn.com/nodes/ (accessed 7.27.19).

Blockchain, 2019. Blockchain Explorer | BTC | ETH | BCH [WWW Document]. blockchain.com. URL www.blockchain.com/explorer (accessed 7.20.19).

Bonneau, J., Miller, A., Clark, J., Narayanan, A., Kroll, J. A. and Felten, E. W., 2015. SoK: Research Perspectives and Challenges for Bitcoin and Cryptocurrencies, In: 2015 IEEE Symposium on Security and Privacy. Presented at the 2015 IEEE Symposium on Security and Privacy, pp. 104–121. https://doi.org/10.1109/SP.2015.14.

Brown, V. A., Harris, J. A., and Russell, J. Y., 2010. *Tackling Wicked Problems through the Transdisciplinary Imagination*. Earthscan, London.

Bryde, D., Broquetas, M. and Volm, J. M., 2013. The Project Benefits of Building Information Modelling (BIM). *International Journal of Project Management 31*, 971–980. https://doi.org/10.1016/j.ijproman.2012.12.001.

BSI, 2007. BSI PAS 1192: 2007 Collaborative Production of Architectural, Engineering and Construction Information. Code of practice.

BSI, 2013. BSI PAS 1192-2:2013 Specification for Information Management for the Capital/delivery Phase of Construction Projects Using Building Information Modelling.

BSI, 2014. BSI PAS1192-3: Specification for Information Management for the Operational Phase of Assets Using Building Information Modelling: Incorporating Corrigendum No. 1.

Buterin, V., 2016. Hard Fork Completed. URL https://blog.ethereum.org/2016/07/20/hard-fork-completed/ (accessed 3.9.19).

Carson, B., Romanelli, G., Walsh, P. and Zhumaev, A. 2018. The Strategic Business Value of the Blockchain Market [WWW Document]. URL www.mckinsey.com/business-functions/digital-mckinsey/our-insights/blockchain-beyond-the-hype-what-is-the-strategic-business-value (accessed 6.4.19).

Case, S., 2017. *The Third Wave: An Entrepreneur's Vision of the Future.* Simon and Schuster, New York.

CBC, 2019. Construction Blockchain Consortium [WWW Document]. *CBC.* URL www.constructionblockchain.org (accessed 7.25.19).

Chohan, U. W., 2017. The Decentralized Autonomous Organization and Governance Issues (SSRN Scholarly Paper No. ID 3082055). *Social Science Research Network*, Rochester, New York.

Choi, E., 2017. *Beyond the Bitcoin – Smart Decentralised Financial Contracts of the Future.* New Zealand: Actuaries. URL https://actuaries.org.nz/wp-content/uploads/2016/07/13-Beyond-the-Bitcoin.pdf (accessed 11.17.19).

Christidis, K. and Devetsikiotis, M., 2016. Blockchains and Smart Contracts for the Internet of Things. *IEEE Access. 4*, 2292–2303. https://doi.org/10.1109/ACCESS.2016.2566339.

Circularise, 2018. Circularise [WWW Document]. URL www.circularise.com/ (accessed 3.9.19).

Dale, B., 2016. The DAO: How the Employeeless Company Has Already Made a Boatload of Money, Observer [WWW Document]. URL https://observer.com/2016/05/dao-decenteralized-autonomous-organizatons/ (accessed 3.9.19).

Davidson, S., 2019. Grasshopper [WWW Document]. URL www.grasshopper3d.com/ (accessed 7.28.19).

Demian, P. and Walters, D., 2014. The Advantages of Information Management through Building Information Modelling. *Construction Management and Economics 0*, 1–13. https://doi.org/10.1080/01446193.2013.777754.

Digital Catapult, 2018. Blockchain in Action: State of the UK Market [WWW Document]. URL www.digicatapult.org.uk/news-and-views/publication/blockchain-in-action-state-of-the-uk-market/ (accessed 3.3.19).

Dino, M., Zamfir, V. and Sirer, E. G., 2016. *A Call for A Temporary Moratorium on 'the DAO'.* Switzerland: Ethereum Foundation.

Driscoll, K., Hall, B., Paulitsch, M., Zumsteg, P. and Sivencrona, H., 2004. The Real Byzantine Generals, In: The 23rd Digital Avionics Systems Conference (IEEE Cat. No.04CH37576). Presented at The 23rd Digital Avionics Systems Conference (IEEE Cat. No.04CH37576), pp. 6.D.4–61. https://doi.org/10.1109/DASC.2004.1390734.

Eastman, C., Teicholz, P., Sacks, R. and Liston, K., 2011. *BIM Handbook: A Guide to Building Information Modeling for Owners, Managers, Designers, Engineers and Contractors*, 2nd ed. John Wiley & Sons, Hoboken, NJ.

Eastman, C. M. and Rafael, S., 2008. Relative Productivity in the AEC Industries in the United States for On-Site and Off-Site Activities. *Journal of Construction Engineering and Management 134*, 517–526. https://doi.org/10.1061/(ASCE)0733-9364(2008)134:7(517).

The Economist, 2016. The DAO of Accrue. *The Economist*, 19 May.

ENR, 2018. Tech Investment in Construction Doubles in past Decade |2018-06-13| Engineering News-Record [WWW Document]. URL www.enr.com/articles/44674-tech-investment-in-construction-doubles-in-past-decade?v=preview (accessed 6.3.19).

EOS, 2019. EOSIO – Blockchain Software Architecture [WWW Document]. *EOSIO*. URL https://eos.io/ (accessed 7.25.19).

Ethereum Foundation, 2018. Ethereum Project [WWW Document]. URL www.ethereum.org/ (accessed 4.4.17).

Etherscan, 2019. Ethereum (ETH) Blockchain Explorer [WWW Document]. URL https://etherscan.io/ (accessed 7.20.19).

Foley, I., 2018. Blockchains and Enterprise: The Hybrid Approach to Adoption [WWW Document]. NASDAQ.com. URL 1532703173 (accessed 7.24.19).

Frankenfield, J., 2018. Permissioned Blockchains [WWW Document]. *Investopedia*. URL www.investopedia.com/terms/p/permissioned-blockchains.asp (accessed 7.21.19).

Frankenfield, J., 2019. Consensus Mechanism (Cryptocurrency) [WWW Document]. *Investopedia*. URL www.investopedia.com/terms/c/consensus-mechanism-cryptocurrency.asp (accessed 7.25.19).

Gartner, 2018. Gartner Hype Cycle for Emerging Technologies. URL www.gartner.com/en/documents/3885468 (accessed 11.17.19).

George, N., CTO, Sovrin Foundation, sponsor and Hyperledger Aries, sponsor and contributor, 2019. Announcing Hyperledger Aries, infrastructure supporting interoperable identity solutions! *Hyperledger*. URL www.hyperledger.org/blog/2019/05/14/announcing-hyperledger-aries-infrastructure-supporting-interoperable-identity-solutions (accessed 5.29.19).

GitHub, 2019. The Ethereum Wiki. Contribute to Ethereum/Wiki Development by Creating an Account on GitHub. Ethereum. URL https://github.com/ethereum/wiki (accessed 11.17.19).

Halevi, S. and Krawczyk, H., 2006. Strengthening Digital Signatures Via Randomized Hashing. In: Dwork, C. (Ed.), *Advances in Cryptology – CRYPTO 2006, Lecture Notes in Computer Science*. Springer, Berlin Heidelberg, pp. 41–59.

Hermann, M., Pentek, T. and Otto, B., 2016. Design Principles for Industrie 4.0 Scenarios, In: 2016 49th Hawaii International Conference on System Sciences (HICSS). Presented at the 2016 49th Hawaii International Conference on System Sciences (HICSS), pp. 3928–3937. https://doi.org/10.1109/HICSS.2016.488.

Hertig, A., 2018. How Will Ethereum Scale? CoinDesk. URL www.coindesk.com/information/will-ethereum-scale (accessed 7.20.19).

HMT, 2018. Cryptoassets Taskforce: Final Report.

Hoepman, J.-H., 2008. Distributed Double Spending Prevention. ArXiv08020832 Cs.

Hughes, D., 2017. The Impact of Blockchain Technology on the Construction Industry. Medium. URL https://medium.com/the-basics-of-blockchain/the-impact-of-blockchain-technology-on-the-construction-industry-85ab78c4aba6 (accessed 6.4.19).

Hwang, G.-H., Chen, P.-H., Lu, C.-H., Chiu, C., Lin, H.-C. and Jheng, A.-J., 2018. InfiniteChain: A Multi-chain Architecture with Distributed Auditing of Sidechains for Public Blockchains. in: Chen, S., Wang, H. and Zhang, L.-J. (Eds.), *Blockchain – ICBC 2018, Lecture Notes in Computer Science*. Seattle: Springer International Publishing, pp. 47–60.

Hyperledger, 2018. Hyperledger Wiki [WWW Document]. URL https://wiki.hyperledger.org/ (accessed 10.28.18).

Hyperledger, 2019. Hyperledger Architecture, Volume 1 [WWW Document]. URL www.hyperledger.org/wp-content/uploads/2017/08/Hyperledger_Arch_WG_Paper_1_Consensus.pdf (accessed 7.25.19).

ICE, 2018. Blockchain Technology in Construction [WWW Document]. URL www.ice.org.uk/ICEDevelopmentWebPortal/media/Documents/News/Blog/Blockchain-technology-in-Construction-2018-12-17.pdf (accessed 3.3.19).

JCT, 2019. BIM and JCT Contracts [WWW Document]. URL www.jctltd.co.uk/product/bim-and-jct-contracts (accessed 7.18.19).

Kagermann, H., Wahlster, W. and Helbig, J., 2013. Recommendations for Implementing the Strategic Initiative INDUSTRIE 4.0 (final Report of the INDUSTRIE 4.0 Working Group). *Acatech – National Academy of Science and Engineering*, München.

Kelly, E., 2014. Introducing Elastic Scale Preview for Azure SQL Database [WWW Document]. URL https://azure.microsoft.com/en-us/blog/introducing-elastic-scale-preview-for-azure-sql-database/ (accessed 11.21.18).

Lamb, K., 2018. Blockchain and Smart Contracts: What the AEC Sector Needs to Know. *CDBB* 16.

Lamport, L., Shostak, R. and Pease, M., 1982. The Byzantine Generals Problem. *ACM Transactions on Programming Languages and Systems (TOPLAS)* 4, 382–401.

Laurent, P., Chollet, T., Burke, M. and Seers, T., 2018. The Tokenization of Assets is Disrupting the Financial Industry 6. URL www2.deloitte.com/content/dam/Deloitte/lu/Documents/financial-services/lu-tokenization-of-assets-disrupting-financial-industry.pdf (accessed 6.3.19).

LDEDCA, 2009. Local Democracy, Economic Development and Construction Act 2009 [WWW Document]. URL www.legislation.gov.uk/ukpga/2009/20/contents (accessed 6.3.19).

Lee, T. B., 2013. Bitcoin Needs to Scale by a Factor of 1000 to Compete with Visa. Here's How to Do it. *The Washington Post*, 12 November.

Lu, Y. 2019. The Blockchain: State-of-the-art and Research Challenges. *Journal of Industrial Information Integration*. https://doi.org/10.1016/j.jii.2019.04.002.

Maciel, A., 2014. *Co-present Collaborative Concept Formation: Participant Observation on Architectural Design*. London: The Bartlett School of Graduate Studies, UCL.

MaidSafe, 2019. About Us | MaidSafe [WWW Document]. URL https://maidsafe.net/about_us (accessed 7.20.19).

Marrs, C., 2018. Plans Emerge for Future of Grenfell Tower Site [WWW Document]. *Archit. J.* URL www.architectsjournal.co.uk/news/plans-emerge-for-future-of-grenfell-tower-site/10028723.article (accessed 1.5.19).

Mason, J., 2016. Intelligent Contracts and the Construction Industry. *Journal of Legal Affairs and Dispute Resolution in Engineering and Construction 9*, 3.

Milutinovic, M., He, W., Wu, H. and Kanwal, M., 2016. Proof of Luck: An Efficient Blockchain Consensus Protocol, in: Proceedings of the 1st Workshop on System Software for Trusted Execution.

Nakamoto, S., 2008. Bitcoin: A Peer-to-Peer Electronic Cash System. URL https://bitcoin.org/bitcoin.pdf (accessed 1.5.19).

Nesbitt, M., 2019. Ethereum Classic (ETC) is Currently Being 51% Attacked [WWW Document]. *Medium.* URL https://blog.coinbase.com/ethereum-classic-etc-is-currently-being-51-attacked-33be13ce32de (accessed 7.20.19).

O'Dwyer, K. J. and Malone, D., 2014. Bitcoin Mining and its Energy Footprint.

Omale, G., 2019. Gartner Predicts 90% of Blockchain-Based Supply Chain Initiatives Will Suffer 'Blockchain Fatigue' by 2023 [WWW Document]. *Gartner.* URL www.gartner.com/en/newsroom/press-releases/2019-05-07-gartner-predicts-90–of-blockchain-based-supply-chain (accessed 6.3.19).

ONS, 2017. Construction Statistics: Number 18, 2017 Edition: A Wide Range of Statistics And Analysis on the Construction Industry in Great Britain in 2016. [WWW Document]. URL www.ons.gov.uk/businessindustryandtrade/constructionindustry/articles/constructionstatistics/number182017edition (accessed 5.29.19).

ONS, 2018. Construction Statistics: Number 19, 2018 Edition: A Wide Range of Statistics and Analysis on the Construction Industry in Great Britain in 2017. [WWW Document]. URL www.ons.gov.uk/businessindustryandtrade/constructionindustry/articles/constructionstatistics/number192018edition#international-comparisons (accessed 5.29.19).

OpenBIM, 2017. OpenBIM [WWW Document]. URL www.openbim.org/ (accessed 1.24.17).

Osipkov, I., Vasserman, E. Y., Hopper, N. and Kim, Y., 2007. Combating Double-Spending Using Cooperative P2P Systems, In: 27th International Conference on Distributed Computing Systems (ICDCS '07). Presented at the 27th International Conference on Distributed Computing Systems (ICDCS '07), p. 41. https://doi.org/10.1109/ICDCS.2007.91.

Peters, G. and Vishnia, G., 2016. Overview of Emerging Blockchain Architectures and Platforms for Electronic Trading Exchanges (SSRN Scholarly Paper No. ID 2867344). *Social Science Research Network*, Rochester, NY.

R3, 2019a. Getting Set up for CorDapp Development – R3 Corda V3.3 Documentation [WWW Document]. URL https://docs.corda.net/getting-set-up.html (accessed 1.25.19).

R3, 2019b. Consensus – R3 Corda Master Documentation [WWW Document]. URL https://docs.corda.net/key-concepts-consensus.html (accessed 7.25.19).

Rennie, E., 2016. The Radical DAO Experiment [WWW Document]. URL www.swinburne.edu.au/news/latest-news/2016/05/the-radical-dao-experiment.php (accessed 3.9.19).

RESTfulAPI, 2019. REST Principles and Architectural Constraints – REST API Tutorial [WWW Document]. URL https://restfulapi.net/rest-architectural-constraints/ (accessed 7.20.19).

Rhodes, C., 2018. *Construction Industry: Statistics and Policy*. London: House of Commons Library 15.

Ripple, 2019. Ripple – One Frictionless Experience To Send Money Globally [WWW Document]. *Ripple.* URL www.ripple.com/ (accessed 7.21.19).

Rizzo, P., 2014. Ghash.io: We Will Never Launch a 51% Attack Against Bitcoin. *CoinDesk.* URL www.coindesk.com/ghash-io-never-launch-51-attack (accessed 7.20.19).

SCCR, 1998. The Scheme for Construction Contracts (England and Wales) Regulations 1998 [WWW Document]. URL www.legislation.gov.uk/uksi/1998/649/schedule/made (accessed 6.3.19).

Schwab, K. 2017. *The Fourth Industrial Revolution*, 1st ed. Portfolio Penguin, New York.

Seth, S., 2019. Public, Private, Permissioned Blockchains Compared [WWW Document]. *Investopedia.* URL www.investopedia.com/news/public-private-permissioned-blockchains-compared/ (accessed 7.21.19).

Sinclair, D., 2012. BIM Overlay to the RIBA Outline Plan of Work. URL www.architecture.com/-/media/gathercontent/riba-plan-of-work/additional-documents/bimoverlaytotheribaoutlineplanofworkpdf.pdf (accessed 7.21.19).

Spaven, E., 2014. The 12 Best Answers from Gavin Andresen's Reddit AMA. *CoinDesk*. URL www.coindesk.com/12-answers-gavin-andresen-reddit-ama (accessed 7.20.19).

Statista, 2019a. IoT: Number of Connected Devices Worldwide 2012–2025 [WWW Document]. *Statista*. URL www.statista.com/statistics/471264/iot-number-of-connected-devices-worldwide/ (accessed 7.28.19).

Statista, 2019b. Bitcoin Transaction Confirmation Time 2019 [WWW Document]. *Statista*. URL www.statista.com/statistics/793539/Bitcoin-transaction-confirmation-time/ (accessed 7.20.19).

Statista, 2019c. Construction sector employees United States 2017 Statistic [WWW Document]. *Statista*. RLwww.statista.com/statistics/187412/number-of-employees-in-us-construction/ (accessed 5.29.19).

Stoll, C., Klaaßen, L. and Gallersdörfer, U. 2019. The Carbon Footprint of Bitcoin. *Joule 3*, 1647–1661. https://doi.org/10.1016/j.joule.2019.05.012.

Storj, 2019. Decentralized Cloud Storage – Storj [WWW Document]. *Decentralized Cloud Storage – Storj*. URL https://storj.io (accessed 7.20.19).

Swan, M. 2015. *Blockchain: Blueprint for a New Economy*, 1st ed. O'Reilly, Beijing and Sebastopol, CA.

Tang, H., Shi, Y. and Dong, P., 2019. Public Blockchain Evaluation Using Entropy and TOPSIS. *Expert Systems with Applications 117*, 204–210. https://doi.org/10.1016/j.eswa.2018.09.048.

Tapscott, D., 2016. *Blockchain Revolution: How the Technology Behind Bitcoin is Changing Money, Business, and the World*. Portfolio, New York.

Thomson Reuters, 2017. Blockchain for Construction/Real Estate. URL https://mena.thomsonreuters.com/en/articles/blockchain-for-construction-and-real-estate.html (accessed 3.3.19).

Thomson Reuters, 2018. Blockchain for Construction Whitepaper [WWW Document]. URL https://mena.thomsonreuters.com/content/dam/openweb/documents/pdf/mena/white-paper/Blockchain_for_Construction_Whitepaper.pdf (accessed 3.3.19).

Treleaven, P. and Batrinca, B., 2017. Algorithmic Regulation: Automating Financial Compliance Monitoring And Regulation Using AI And Blockchain. *Journal of Financial Transformation 45*, 14–21.

Treleaven, P., Galas, M. and Lalchand, V., 2013. Algorithmic Trading Review. *Communications of the ACM. 56*, 76–85. https://doi.org/10.1145/2500117.

Unity, 2019. Unity [WWW Document]. *Unity*. URL https://unity.com/frontpage (accessed 7.25.19).

Unreal, 2018. VR Mode Controls | Unreal Engine [WWW Document]. URL https://docs.unrealengine.com/latest/INT/Engine/Editor/VR/Controls/index.html (accessed 1.10.18).

Valkenburgh, P. V., 2018. What is a 51% Attack, and What Can a Successful Attacker Do? [WWW Document]. *Coin Cent*. URL https://coincenter.org/entry/what-is-a-51-attack-and-what-can-a-successful-attacker-do-1 (accessed 7.20.19).

van der Meulen, R., 2018. Gartner Says Worldwide Supply Chain Management Software Revenue Grew 13.9 Percent in 2017 [WWW Document]. *Gartner*. URL www.gartner.com/en/newsroom/press-releases/2018-07-18-gartner-says-worldwide-supply-chain-management-software-revenue-grew-13-percent-in-2017 (accessed 7.27.19).

W3C, 2019a. Decentralized Identifiers (DIDs) V0.13 [WWW Document]. URL https://w3c-ccg.github.io/did-spec/ (accessed 5.29.19).

W3C, 2019b. Web Services Architecture: 3.1.3 Relationship to the World Wide Web and REST Architectures [WWW Document]. URL www.w3.org/TR/2004/NOTE-ws-arch-20040211/#relwwwrest (accessed 7.20.19).

Wander, M., 2015. Bitcoin Block Data Illustration [WWW Document]. URL https://i.stack.imgur.com/HrKX0.png (accessed 4.6.19).

Wang, L., Shen, X., Li, J., Shao, J. and Yang, Y., 2019. Cryptographic Primitives in Blockchains. *Journal of Network and Computer Applications 127*, 43–58. https://doi.org/10.1016/j.jnca.2018.11.003.

Waters, R., 2016. Automated Company Raises Equivalent of $120M in Digital Currency [WWW Document]. URL www.cnbc.com/2016/05/17/automated-company-raises-equivalent-of-120-million-in-digital-currency.html (accessed 3.9.19).

Wattles, J., 2017. Bitcoin Jumps after Futures Trading Begins [WWW Document]. *CNNMoney*. URL https://money.cnn.com/2017/12/10/technology/Bitcoin-futures-trading/index.html (accessed 11.21.18).

White, S., 2015. Business Population Estimates for the UK and Regions 2015 [WWW Document]. *Department for Business Innovation and Skills*, London. URL https://assets.publishing.service.gov. uk/government/uploads/system/uploads/attachment_data/file/467443/bpe_2015_statistical_release. pdf. (accessed 11.17.19).

Windley, P., 2019. Decentralized Identifiers [WWW Document]. URL www.windley.com/ archives/2019/02/decentralized_identifiers.shtml (accessed 7.27.19).

Zhang, C., Patras, P. and Haddadi, H., 2019. Deep Learning in Mobile and Wireless Networking: A Survey. IEEE Communications Survey. Tutor. 1. https://doi.org/10.1109/COMST.2019.2904897.

Zheng, Z., Xie, S., Dai, H., Chen, X. and Wang, H., 2017. An Overview of Blockchain Technology: Architecture, Consensus, and Future Trends, In: 2017 IEEE International Congress on Big Data (BigData Congress). Presented at the 2017 IEEE International Congress on Big Data (BigData Congress), pp. 557–564. https://doi.org/10.1109/BigDataCongress.2017.85.

PART III

Practical aspects of construction 4.0 including case studies, overview of start-ups, and future directions

21

CONSTRUCTION 4.0 CASE STUDIES

Cristina Toca Pérez, Dayana Bastos Costa, and Mike Farragher

21.1 Aims

- To demonstrate application of selected Construction 4.0 principles through case study examples from Brazil, USA and UK.
- To provide commentary from industry stakeholders and innovators regarding future directions of Construction 4.0.
- To reflect on the perceived efficacy of various innovations illustrated within the case studies.

21.2 Case study 1: 4D BIM for logistics purposes

21.2.1 Overview

The use of 4D Building Information Models (BIM) for planning construction site logistics has increased in recent years. According to Jongeling and Olofsson (2007) activity-based scheduling methods, such as the Critical Path Method (CPM) and the Gantt chart, used as basis for 4D models are efficient due to their high applicability in showing details about the relationship between activities. However, the simple translation of the output of a CPM network that contains only transformation activities implies that flow activities, mainly transportation activities, are neglected (Bortolini et al., 2015). In other words, those simulations do not represent constructive details required for planning the processes at an operational level. Hence, the aim of this case study is to show and evaluate an application of 4D BIM simulation to reduce logistic transportation waste at construction sites. Such reduction would enable the minimization of time spent in logistics tasks, and consequently it allows the efficiency improvement at the construction and building stage.

21.2.2 Methodology and results

An empirical study was performed in a residential social assistance housing project located in Petrolina city in the northeast of Brazil. This project is one of a number of social interest projects built under the Brazilian government's public housing program "Minha Casa, Minha Vida" (My Home, My Life in Portuguese). Steel panel formwork systems were used to build the reinforced cast-in-place concrete wall structure. That technology mainly involves four sub processes: reinforcement bars erection, electrical installation, formwork assembly and concrete placement. As a main project feature, the modular steel formworks used were made of large steel panel frames, which require heavy lifting equipment such as a hydraulic telescopic boom truck crane and must be handled by skilled laborer. The jobsite studied occupies 92.05m², with 184 one-story buildings totaling 368 units (see Figures 21.1 and 21.2).

This research was conducted by the Research Group in Construction Management and Technology of the Federal University of Bahia, Brazil (GETEC-UFBA) team in partnership with a local construction company (Company A). At the beginning of the study, logistics problems related to the truck which transported the massive steel formworks were identified. Thus, the planning of material flows at the operational level plays a key role in avoiding unnecessary transportation activities. The study lasted seven months, from May to November 2017, and it was performed in three stages: (1) Data collection; (2) BIM development; and (3) Outcome analysis.

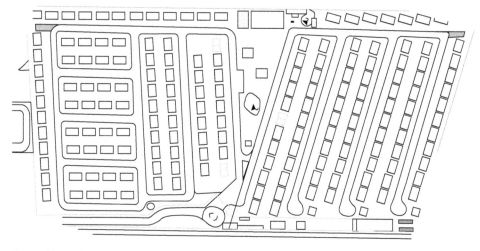

Figure 21.1   Floor plan of buildings

Figure 21.2   Example of a one-story building with two units

21.2.2.1 Stage 1: data collection by the research team

The aim of this stage was to identify the critical construction process that affects on-site logistics efficiency. To achieve this aim, the following information was collected during four site visits of 8 hours each:

- Photos, field notes, non-structured interviews with field engineers were used as source of evidence.
- Material storage area constraints were identified using Layout and Process Flow Diagram techniques. Examples of constrains identified were material storage workspace and delivery or transport, paths not available.
- The amount of productive, contributory and non-contributory work, especially the time wasted in transportation activities were collected using the Work Sampling technique. The findings of the work sampling revealed the steel formwork assembly as the critical process at the construction site, because it was the most time-consuming process with 51% of the observation made on labor force. Moreover, the steel formwork assembly was the process which wasted more time in non- value adding activities, 67% of contributory tasks and 11% of non-contributory task.
- The cycle time of steel formwork assembly process was measured. The cycle time was reduced from 16 hours to 9 hours. These improvements occurred after the research team provided feedback to site personnel based on the data collected on site. One of the proposed improvements suggested was the need to use appropriate equipment to transport the panels instead of transporting them by hand. That improvement was designated as first intervention.
- The transportation waste events were identified. In the study, the lack of information related to a sequence and transportation plan of the formwork panels was the main cause of transportation waste events. For example, the installation and the removal of the formwork was improvised every day and workers stocked the formwork in an incorrect manner (Figure 21.3), unsafe working conditions were frequently created because workers did not know which formwork panel was supposed to be installed, and consequently transported (Figure 21.4), and access and mobility problems were consequences of the presence of formwork panels stocked in the material transport paths (Figure 21.5).

Figure 21.3 Improvised anchors are used for keeping up the steel formwork panels

Figure 21.4 Formwork is transported above a worker

Figure 21.5 A formwork panel is stocked in the main door

21.2.2.2 Stage 2: BIM development at UFBA

The aim of this stage was to optimize the execution sequence of steel formworks panels.
    To achieve the stated aim, the following activities were performed in the lab:

- The formwork nomenclature was proposed.
- The 3D model of the main processes involved using Revit Architecture was developed.
- Suitable sequence plans were study and the most suitable sequence was proposed.
- The sequence proposed was scheduled using MS Project software, considering real times of the value-adding and non-value adding activities to perform the steel formwork panels installation, which were collected during the jobsite visits.
- A color coded legend was created to indicate the various kinds of activities performed.
- The sequence was simulated using the 4D BIM capabilities of Navisworks. The main challenges during the creation of the animations in Navisworks were the representation of activities performed by the equipment such as: the transportation of the formworks (Figure 21.6), the movement of the equipment empty to perform a new transport, and the use of equipment for performing assembly tasks (Figure 21.6). The solution adopted in

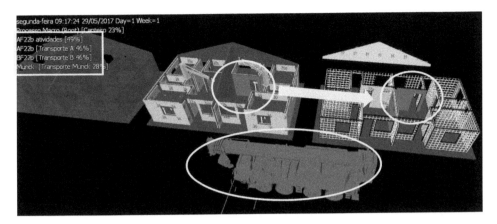

Figure 21.6   Transportation activity made by the telescopic handler

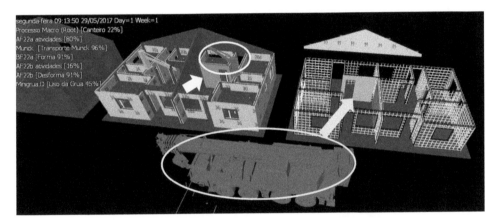

Figure 21.7   Crane being used for the formwork disassembly and telescopic handler working in the formwork assembly

this study for representing non-value adding activities, such as a transportation activity, that cannot be allocated in a specific place was to color the original point of departure of the element, the final point of installation and the equipment used for the transport with the same color (Figure 21.7).

21.2.2.3 Stage 3: outcome analysis

The aim of this stage was to identify the main contributions and limitations of the 4D BIM for simulating physical flows of the steel formwork assembly process in order to reduce transportation waste.

To achieve that goal, the following activities were performed:

- The 4D BIM model was presented, discussed and refined in a meeting with construction managers (Figure 21.8) and in a workshop on site involving the workers (Figure 21.9).

- Another round of site visits was performed aiming to observe the implementation of the new sequence plan proposed and discussed with the workers; This was the second intervention of the study.
- The new cycle time steel formworks assembly process was measured, and it took an average of 6.5 hours, which represents approximately a 2-hours reduction. The use of the 4D model allowed the time reduction due to the reduction of transportation waste consequences. The sequence plan proposed allowed the complete removal of improvisation activities once provided the correct information at the correct time (Figure 21.10). The sequence plan proposed enabled formwork were not disassembled until the exact time that they would be lifted and transported, avoiding work-in-progress and unsafe work conditions (Figure 21.11). In addition, transportation waste caused by access problems was remove because after the sequence implementation, the transport paths were always clear.
- The contributions of the 4D BIM for transportation waste reduction were identified.

Figure 21.8   Seminar with managers

Figure 21.9   Workshop with workers at work site

Figure 21.10 Workers are aware the sequence plan and improvisation activities are eliminated

Figure 21.11 Work-in-progress is reduced

21.2.3 Case study 1 conclusion

This case study pointed out the importance of having operational knowledge of transportation activities. That means that in order to guarantee simulation efficiency, information related to the piece of equipment used, the number of workers required for performing the transportation activities, the support workspace and support time for non-value adding activities must be known.

The practical results obtained showed, firstly, a considerable reduction of the cycle time (25% of reduction). This reduction was due to the information exchange allowed by the 4D BIM model with the stakeholders. Second, through the 4D model, the formwork assembly sequence could be shown in detail to managers and field engineers and thereby possible constraints or difficulties previously found could be discussed. Third, the screenshots from the 4D model were a very useful tool to show illustrate the sequence in the construction site. The screenshots assisted with the coordination of expected time and space flow of trades on site as well as the coordination of work in smaller areas. Finally, this way of communication appeared more effective than a traditional Gantt chart. However, the workers and engineers were not familiar with BIM tools; the understanding of the 4D model was not a challenge. Indeed, some very good feedback was obtained from the labor force.

21.3 Case study 2: the WikiHouse project

21.3.1 Overview

The development of open-source systems allows projects to be realized through public collaboration because the author uses a license that permits anyone to modify their source code. There are numerous open-source products available on the web; the plans or instructions for which are downloadable most times for free. Open-source products are not new. The term "open-source" was not actually used in the free software community before February 1998. It was created in part to clarify that the word "free" in free software referred to a "freedom to use" rather than "freedom from price" (Stallman, 1998). One way to describe open-source software is to say that it is s software created by communities of programmers who share their source code, that part of a program that is readable by humans, with anyone who might find the program, or a variation of the program, useful (Stallman, 1998). This collaboration model has been highly successful in the field of software and has begun to be used in other industries such as architecture. An example of one of the most famous initiative in open-source architecture is the WikiHouse project.

The WikiHouse project is maintained by Open Systems Lab, a non-profit company registered in England and Wales. WikiHouse is a digital construction system created in 2012 that enables anyone to download and "print" the parts for low-cost houses, using Computer Numerically Controlled (CNC) cutting machines (Castle, 2018). Parvin, the co-founder of WikiHouse, states that the emerging trend of the WikiHouse is welded to the word "empower" that stands out as a keyword for this technology (Parvin, 2014). According to Glancey (2014), to empower the people is related to democratize by working for many small companies, individuals and projects. The intent for any digital paradigm such as Construction 4.0 surely should be to bring together people and companies in an environment of collaboration (Gartner, 2017). The Digital Ecosystem is inherently "digital" although this aspect is a mechanism in which to empower people to build or even self-build. This idea is not new; historically vernacular domestic architecture pre-16th century was carried out without the use of an architect and master builders produced houses in simple forms or from plans purchased from pattern books.

Numerous later examples of self-build projects such as the work of Walter Segal and Buckminster Fuller were born out of the lightness of the structures of the modern movement. Kronenburg cites that Fuller saw the building industry as outdated and being made up of a series of wasteful unrelated processes that produced expensive products. An example of this could possibly be seen in the high decoration of the Victorian era; the "battle of the styles" (Kronenburg, 1995). It is from the prefabrication vein of thought that Parvin's WikiHouse has emerged. Take the self-build, kit format of Segal and Fuller's work and catalyze this with the use of information exchange and there appears a new empowerment through knowledge and ideas about how to build.

It is when a product such as a building, which is big, needs to be sourced that there appears to be a conundrum to be solved. Parvin is concerned with the affordability of architects and engineer's fees. Parvin cites that the benchmark for affordability of architect-designed buildings comes with a minimum wage of £24,000. Anybody below this threshold cannot really afford to employ an architect. This indirectly leads to the fact that architects therefore design and build for 1% of the world's population (TED, 2013). Parvin thinks this is wrong both on a social and also a logical standpoint.

Through WikiHouse, people can now freely access a library of building plans or 3D digital models for simple houses that are based on a modular system that have been designed using software such as SketchUp, which is free to download from the web (TED, 2013). Once a design is decided upon or changed to suit the searcher, a template file can be downloaded

The delivery chain

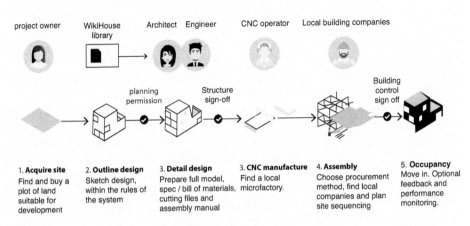

| 1. **Acquire site** | 2. **Outline design** | 3. **Detail design** | 3. **CNC manufacture** | 4. **Assembly** | 5. **Occupancy** |
| Find and buy a plot of land suitable for development | Sketch design, within the rules of the system | Prepare full model, spec / bill of materials, cutting files and assembly manual | Find a local microfactory. | Choose procurement method, find local companies and plan site sequencing | Move in. Optional feedback and performance monitoring. |

Figure 21.12   Typical procurement route. Source: WikiHouse Guide 1.0

with the click of a button to use in cutting the panels that make that house. This file can be programmed to cut the panels and joints out to form a kit for the house. What the user then receives from any operator locally who has a CNC machine is kind of IKEA flat pack for a house. Figure 21.12 shows the delivery chain for building a WikiHouse.

The construction of the WikiHouse system itself is technologically ingenious in that it uses low skill, dry assembly in its wedge and peg mechanisms. This reaches out to the majority of people in that the construction can be achieved without power tools. Even the mallet is CNC routed from the plywood sheet. Most designs within the Wren (Figure 21.13) and Blackbird range are based on a double 19 mm thick plywood sheet portal frame made into sections (Edward, 2018). Using the Wren system was highly beneficial to the construction phase as the build could be assembled by non-professionals and because it could be erected quickly without the use of heavy-lifting equipment.

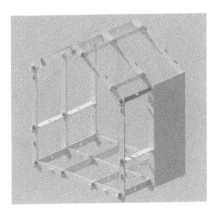

Figure 21.13   Wren system
Source: Wikihouse.cc

The Wren system has been developed utilizing BIM software such as Grasshopper, Rhino and SketchUp Pro. The project comprises a 3D virtual model from which working drawings, assembly instructions and building information can be extracted. Wren's source code falls within the Mozilla Public Licence (MPL) version 2.0, which is a worldwide, royalty-free, non-exclusive license not intended to limit any rights. WikiHouse uses standard sizes of commonly used materials to make the system cost-effective. For example, Wren's primary structure is composed of structural-grade Oriented Strand Board (OSB) or plywood panels that fit together like a large 3D jigsaw (Edward, 2018).

Most joints have two-piece interlocking wedges (Figure 21.14) instead of any nail or screw connections. The WikiHouse also has highly sustainable attributes in that materials are sourced locally, and panels fabricated with the use of nearby CNC machines. The houses boast high levels of airtightness with 1 to 1.5 air-changes per hour at 50 Pascal and low U-values; up to 1.4 W/m2K for wall panels. The structural performance of frames is consistent thanks to the uniformity of timber panels and high level of precision of CNC cutters. The result is that the current Wren system can achieve spans of 6500 mm and be used for double-story buildings (Edward, 2018)

21.3.2 The case of WikiFarmHouse

This section presents as an example of WikiHouse the case of the world's first two-story Wiki-House, known as the WikiFarmHouse (Figure 21.15 and Figure 21.16) with the use of the Wren system.

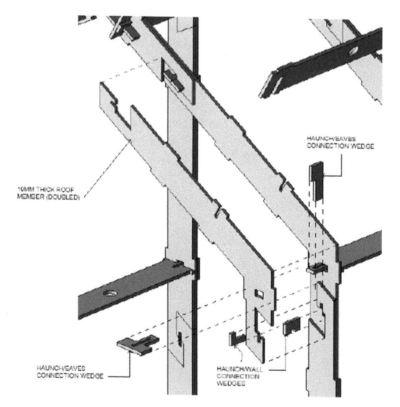

Figure 21.14 Typical joint assembly in the Wren system
Source: Wikihouse.cc

Figure 21.15 WikiFarmHouse completed
Source: Awikifarmhouse (n.d.)

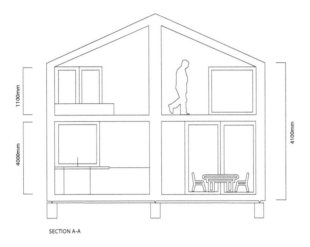

Figure 21.16 Drawing section
Source: Awikifarmhouse (n.d.)

The WikiFarmHouse is located about 20 km from Coventry, United Kingdom. The building was designed by Architecture 00 for an agricultural worker. The main activities performed at the field were: (a) Concrete foundation by professional workers (Figure 21.17); (b) CNC-milled plywood was cut in a nearby off-site workshop, it consists of over 500 sheets of plywood that took over a month to cut (Figure 21.18); (c) the sheets were transported to site by truck; (d) the sheets were assembled by a large group of volunteers, each component was filled with polystyrene insulation before being erected (Figure 21.19); (e) cassettes were lifted onto the foundation before connection pieces were hammered into place using open-source mallets (Figure 21.20); (f) the plywood structure was covered in a water-resistant breather membrane (Figure 21.21); (g) roof and cladding were installed using scaffolding (Figure 21.22).

Using the WikiHouse method has given owners a choice about how much of the labor to do themselves, enabled them to develop the project at their own pace and produce a building that meets their quality requirements (Edward, 2018). During some interviews, one of the owners pointed out that the total cost of the project was under £110,000, which includes everything

Figure 21.17 Concrete foundation
Source: Awikifarmhouse (n.d.)

Figure 21.18 CNC-milled plywood was cut in an offsite workshop
Source: Awikifarmhouse (n.d.)

Figure 21.19 Structural cassettes
Source: Awikifarmhouse (n.d.)

Figure 21.20   Cassettes were lifted onto the foundation
Source: Awikifarmhouse (n.d.)

Figure 21.21   Water-resistant membrane application
Source: Awikifarmhouse (n.d.)

Figure 21.22   Roof and cladding installation
Source: Awikifarmhouse (n.d.)

from planning and architects' fees to fit out costs as carpet fitting, etc. At £800 per m², the WikiFarmhouse is much more cost-effective than most self-build houses, which often vary between £1,500 and £2,000 per m² (GreenCoreConstruction, 2018).

21.3.3 Case study 2 conclusion

WikiHouse's development so far suggests that open-source architecture will have significant economic, social and environmental effects, and is therefore likely to disrupt the construction industry. The most advanced technology to date, Wren, has already been used to realize con-structed projects in the United Kingdom and abroad, including the double-story WikiFarm-House. The case study demonstrates that a high-performance, adaptable WikiHouse can be achieved for £800 per m², about half the cost of a typical site-built home of similar size. In addition to economic advantages, owners also benefit from having greater control over their building because they can make future modifications more easily.

21.4 Case study 3: the innovation lab

21.4.1 Overview

Oracle Construction and Engineering launched its Innovation Lab, a simulated work site, on August 23, 2018 (Figure 21.23), where visitors could interact with drones (Figure 21.24), autonomous vehicles (Figure 21.25), connected tools, remotely controlled pieces of equipment (Figure 21.26), sensors, and other technologies as well as related data integrations.

The vision for the Innovation Lab, which is located in Deerfield, Illinois, was shaped in part by input from several Oracle customers as well as other technology providers. The Innovation Lab is a model of an outdoor construction site, with girders, concrete-block walls, industrial fencing, gravel underfoot and a doublewide trailer for the simulated construction office (Mur-phy, 2018) that can be visited by owners, contractors, subcontractors and academia.

The Innovation Lab provides their visitors with several high level immersive experiences with technologies such as: (1) technologies used for material tracking; (2) technologies that

Figure 21.23   Innovation Lab
Source: (Oracle, 2018)

Figure 21.24 Drones used at Innovation Lab
Source: (Oracle, 2018)

Figure 21.25 Autonomous vehicle
Source: (Oracle, 2018)

Figure 21.26 Worker sensors and alerts
Source: (Oracle, 2018)

allow worker tracking and safety; (3) technologies that facilitate virtual work progress reporting; and (4) technology to demonstrate how information could be transmitted to a jobsite command center.

The facility's immersive experiences enable visitors to interact with technology in a realistic jobsite environment. According to Kaplanoglu (interview, 2019), executive director and innovation officer at Oracle Construction and Engineering, "immersive experience has been a tremendous success in terms of helping people understand what the solutions can do, visitors can understand what the real value of the solutions is. These are not PowerPoint presentations. These are physical experiences". One of the experiences that visitors can have outside the trailer is the use of RFID tags (Figures 21.27 and 21.28). Visitors can scan the material to track its location at the simulated work site.

The Innovation Lab experiences feature technologies that integrate with Oracle solutions to enable collaboration and unlock critical project intelligence to enhance outcomes

Figure 21.27   RFID tag scanner
Source: (Oracle, 2018)

Figure 21.28   RFID tag
Source: (Oracle, 2018)

Figure 21.29 Workers sensors and alerts
Source: (Oracle, 2018)

and drive continuous improvement (Oracle, 2018). Oracle has selected technologies that present high success individually and brings them together in a common platform. These technologies, as well as the data they yield or analyze, offer significant opportunities to transform project delivery by improving productivity, quality, safety, and standardization as well as by enabling continuous improvement across operations (Oracle, 2018). For that, Oracle works with several technology partners, referred to in this chapter as Company A, B, C, D, E, F, and G, to embed their innovations into the Lab, as well as its customers, who brought terrific energy and insights into their own needs and challenges, enabling the development of a collective vision of what the Lab should be. Some of the Oracle partners for the introduction of the Construction and Engineering Innovation Lab provide solutions for:

- 4D and 5D Building Information Modeling (BIM), Company A provides a solution that enables construction professionals to condition, query and connect BIM data to key work-flows across bid management, estimating, scheduling, site management and finance.
- Connected tools, Company B is empowering more productivity on the jobsite through their connected tools and asset solutions.
- Augmented reality, Company C empowers workforce by linking digital content to the real world to accelerate productivity, communication, and key business processes.
- Site mapping, Company D enables people, businesses and cities to harness the power of location. By making sense of the world through the lens of location.
- Material readiness application developed by Company E for the construction market, keeps crews productive with accurate and real-time information about material availability relative to the construction plan and schedule.
- Virtual progress reporting. Company F provides a 3D timeline that tracks visual progress, labor productivity, and predictive analytics that empower executives and their project teams to take actions to stay on time and on budget.
- Worker and equipment tracking, A platform developed by Company G connects the construction jobsite and provides real-time visibility into workers, safety, equipment, and asset management.

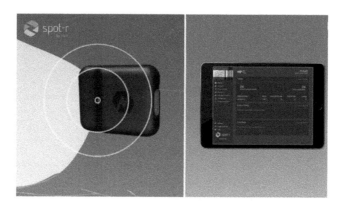

Figure 21.30 Spot-r Clips
Source: (Oracle, 2018)

Moreover, the Innovation Lab also works with a different business areas of Oracle, Oracle Communications, which provides solutions to communications customers, and Oracle Utilities, enabling synergies across industry verticals, among other benefits.

One technology in use at the Innovation Lab is worker tracking. If a worker falls on a construction site (Figure 21.29), a clip on his or her belt can send an alert (Figure 21.30): Worker A fell from a height of about 12 feet and is located in this spot on the site. Wearable devices can provide location by zone and floor, letting a company find an injured worker or collect time and attendance, without creating a feeling that it's watching workers' every move or knowing where they are when they leave the construction site (Murphy, 2018).

21.4.2 Case study: scheduling solution for adding value

The Innovation Lab has worked with a startup named Company H, which has been part of Oracle's startup accelerator program. Company H's solution aims to link reality capture information collected at work sites automatically with the project schedule.

Company H had its own platform and, working with Oracle, they bi-directionally linked their solution with Oracle's scheduling solution to create a 4D model (Figure 21.31). Reality capture

Figure 21.31 Worker data in BIM environment
Source: (Oracle, 2018)

Figure 21.32 Cloud-based data collected by drone
Source: (Oracle, 2018)

from photos, laser scans, videos from drone flights are used to create models (Figure 21.32). All this real data collected can be overlaid with the model, enabling all the information to be in a centralized platform and project participants to visualize the work progress linked to the schedule.

- Company H's solution also allows collection of real-time information that can be leveraged for predictive insights. Doing so requires moving from manual data collection to automated data collection.

21.4.3 Case study 3 conclusion

The Innovation Lab aims to provide a realistic view into the future of work sites. The feedback that the Innovation Lab has received from the construction industry has been very positive. Having a physical site where visitors can try new technologies seems to be very useful and interesting. However, to implement pilot tests in real work sites remains a challenge. The Innovation Lab is focusing on process improvements and how to enable and scale improvements with the use of technologies at the work site. There are many technologies that are readily available that can be applied in the construction and engineering domain.

21.5 Conclusion

This chapter presented three case studies to represent some of the most promising and prominent efforts performed across the world for demonstrating how Construction 4.0 principles, previously discussed, in this book can be applied to real construction projects. Each case study begins by providing an overview of the main principles and technologies applied in each case study. After that, the case studies are described and discussed. Finally, each chapter incorporates a reflective conclusion presenting the main benefits and limitations identified during the adoption of some Construction 4.0 concepts. In addition, the perception from construction stakeholders of each case study, such as field engineers, labor force and final users are presented.

21.6 Summary

- Construction companies' owners, labor force, research institutions and final users of the construction projects across the world are already benefiting from the implementation of Construction 4.0 principles and technologies in different ways. The feedback obtained from the AEC industry stakeholders has been extremely positive.
- Case study 1 discussed the main benefits and limitations of the use of 4D BIM for controlling and planning transportation activities at construction sites. This section presents the use of 4D BIM for control and planning the formwork assembly process in a residential construction project in Brazil. In this case study, although workers and engineers were not familiar with BIM tools; the understanding of the 4D BIM was not a challenge. The information exchange between researchers and workers using the 4D BIM model developed allowed a considerable reduction of the cycle time of the process study.
- Case study 2 presented and discussed the open-source concept through the presentation of the WikiHouse project, one of the most famous initiatives in open-source architecture. This section presents as an example the case of the world's first two-story Wiki-House, known as the WikiFarmHouse, located in UK. The final users of the building were extremely glad of having the opportunity to develop the project at their own pace and produce a building that meets their quality requirements.
- Case study 3 presented an innovation outdoor site located in the USA that can be visited by construction owners, contractors, subcontractors and academia, providing its visitors several high-level immersive experiences. Visitors provided very good feedback of this physical experience of having a realistic idea about the future of work sites.

References

Awikifarmhouse (n.d.). https://awikifarmhouse.wordpress.com/.

Bortolini, R., Shigaki, J. S.-I. and Formoso, C. (2015). "Site Logistics Planning and Control Using 4D Modeling: A Study in a Lean Car Factory Building Site". *Proceedings of the 23rd Annual Conference of the International Group for Lean Construction*, Perth, 28–31 July 2015. 361–370.

Castle, H. (2018). "Going from Zero to 00: Indy Johar on Shifting the Focus of Practice from Objects to Outcomes". *Architectural Design*, *88*(5), September/October 2018, 78–85.

Edward, D. (2018). "Building Open-Source. To What Extent does WikiHouse Apply the Open-Source Model to Architecture?" Dissertation. University of Kent.

Gartner. (2017). CIO Agenda 2017. Retrieved 20/03/2019, from www.gartner.com/en/search?keywords=cio%20agenda%202017.

Glancey, J. (2014). "*Alistair Parvin – His Open–sourced Platform, WikiHouse, Promises to Remake the Field. Is it a Threat to the Primacy of the Architect, or a Glimpse into Our Digital Design Future?*" Metropolis, 66–88.

GreenCoreConstruction. (2018). Retrieved 17/11/2019, www.greencoreconstruction.co.uk.

Jongeling, R. and Olofsson, T. (2007). "A Method for Planning of Work-flow by Combined Use of Location-based Scheduling and 4D CAD". *Automation in Construction*, *16*(2), 189–198, 10.1016/j.autcon.2006.04.001.

Kronenburg, Robert H. 1995. *Houses in Motion: The Genesis, History and Development of the Portable Building*. London: Academy Editions.

Murphy, C. (2018) "Construction Sites a Lift from Emerging Technology". *The Wall Street Journal*. September 4.

Oracle. (2018). www.oracle.com/corporate/pressrelease/construction-and-engineering-innovation-lab-082318.html.

Parvin, A. (2014). "WikiHouse Guide 1.0." www.wikihouse.cc/About.

Stallman, R. (1998). "Selling Free Software. The Philosophy of the GNU Project". Available: www.fsf.org/philosophy/selling.html.

TED. (2013). "TED: Architecture for the People by the People". Retrieved 12/07/2017, from www.ted.com/talks/alastair_parvin_architecture_for_the_people_by_the_people.

22

CYBER THREATS AND ACTORS CONFRONTING THE CONSTRUCTION 4.0

Erika A. Pärn and Borja Garcia de Soto

22.1 Aims

- Provide a review of cyber-space and cyber-physical attacks – looking at cases of cyber-attacks extracted from the Repository of Industrial Security Incidents (RISI) to identify the motivations for hacking and to identify the various types of hackers (also known as actors).
- To provide a richer understanding of the established, but fragmented, topic of cyber-crime.
- To report on innovative cyber-deterrence techniques – an explanation of how 'blockchain' can be successfully used to protect ensuing cyber threats (when compared to encryption and firewalls).
- Demonstrate the inextricable link between cyber threats and the overreliance upon networked systems and devices of the Construction 4.0 vision.

22.2 Introduction

Two decades into the 21st century and globally, smart cities can now provide fully integrated and networked connectivity between digital infrastructure assets and physical infrastructure to form seamless digital economies. However, this advancement has not been without its problems as industrial espionage, cyber-crime and politically driven cyber-interventions threaten to disrupt or physically damage the critical infrastructure that supports the generation of national wealth. The fact that this infrastructure also preserves the health, safety, and welfare of the population, matters not to those who would disrupt it. So what is the current state of affairs? What are the specific cyber threats that confront critical infrastructure asset management? This chapter highlights the very specific vulnerabilities of infrastructure that is reliant upon a common data environment (CDE) whilst implementing Building Information Modeling (BIM) at a project level.

With the rapid technological innovations of methods of construction and its materials we are now faced with the problems of the 21st-century construction industry, ranging from poor efficiency, waste of resources and a diminishing need for manually skilled workers as more machinery is introduced along with computing power into Construction 4.0. This is an attempt to replicate a sterile manufacturing environment driven by a need to standardize, improve profit margins and demonstrate work efficiency in the construction sector. The flux of cheaper technologies and breakthroughs in computing processing powers are opening the door to curious technology venture capitalists who seek to inject new funding into unchartered territories lured by the promise of Construction 4.0. But, is the rapid advancement of technology adoption in construction blinded by a conceit of knowing how to solve issues faced by construction, while faced with the possibility of creating new ones simultaneously? One such issue is that of cybersecurity of the digital assets that will be developed in Construction 4.0.

Through looking at the work being done in the field and analyzing case studies of cyber-physical attacks, we are looking at identifying the different types of hackers and what their motivation might be. We will also look at their reconnaissance techniques.

Does the direction for future research work lie in utilizing innovative blockchain technology as a potential risk mitigation measure for digital built environment vulnerabilities? While it is true that cybersecurity and digitalization of the built environment have attracted wide coverage in almost every domain and has been researched extensively, the fact is that there has been little research done that has adopted a holistic view of the perceived threats. Attention to deterrence applications and future developments in a digitized Architecture, Engineering, Construction and Owner-operated (Construction 4.0) sector has been scant.

This chapter is aimed at helping us all to be better informed of the threats and possible remedies, particularly from the perspective of the Construction 4.0 field.

22.2.1 *Cyber-physical connectivity*

Throughout history, buildings and infrastructure have provided us with a physically secure sanctuary, protecting us from theft and malicious attacks. Today's environment is no exception and offers the same practical physicality. However, contemporary operations and maintenance (O&M) works have become increasingly dependent upon an expansive web of cyber-physical connectivity (Karabacakn et al., 2016b) and this connectivity has been achieved via a union of smart sensor-based network technologies advanced computerization and computational intelligence techniques (Bessis and Dobre, 2014). Seen in the context of virtual assets, the data and information generated throughout a development's whole life cycle from the initial design, construction and eventual occupancy is the basis for the spread of knowledge, insightful business intelligence as well as an invaluable commercial commodity.

Intelligence on infrastructure asset performance can help in decision making via automated analytics that is geared towards driving economic prosperity, business profitability and environmental conservation (Lin et al., 2006); across the globe towards embedding digitalization throughout the Architecture, Engineering, Construction and Owner-operated (Construction 4.0) sector. For example, the UK government's mandated policy 'Digital Built Britain 2025' represents ambitious plans to merge digitized economies and infrastructure deployment (HM Government, 2015). This strategic vision has been realized via the BIM Level 2 mandate designed to extend the frontiers of digitized asset handover for building and infrastructure asset owners (HM Government, 2013). Many other EU member states (i.e. Ireland and Germany) and countries internationally (i.e. China, U.A.E., and Chile) have followed in suit by already mandating BIM or with the intention to do so by 2021 (McAuley et al., 2017).

What is not yet observable in Construction 4.0 has been well documented by manufacturing and financial businesses adopting digitalization and a heavy reliance on IT systems. Cybersecurity threats can disrupt at a scale formerly unknown to our industry. As the reliance on cyber-physical systems deepens, so does the impact of malicious cyber-physical attacks to Construction 4.0. In our attempt to mimic the efficiency and calibrated productivity of the manufacturing industry the construction sector is edging closer to the precipice and falling into harm's way from the threat of cyber-attacks. This is of particular importance in a movement towards the 'digital twin' modeling phenomenon of Construction 4.0.

While skeptics deny that a similar sterile environment like that of manufacturing could ever be replicated in construction, it is clear that many of the technologies currently being adopted in Construction 4.0 originate from the manufacturing setting where cyber-physical systems are being used to drive efficiency. As production moved to an increasing reliance on machinery and autonomous production lines, a growing number of cyber-physical systems have been introduced into manufacturing for real-time monitoring and processing capabilities. The untethered connectivity via the internet, offers the capacity to gather copious amounts of data for pattern recognition of production line productivity and quality. Similarly, digitalization of Construction 4.0 provides better means of information feedback to site workers, operators and the originators of the design both in terms of productivity and quality. From a cybersecurity perspective, anything operating with cyber-physical connectivity under the guise of 'smart' becomes vulnerable to cyber-attacks. Disruption could be caused to site activities and business operations but perhaps most importantly to the operation of our future buildings. We speculate on this because of the use of BIM in construction. BIM analogous to an umbilical cord supplies the data-rich content of the 'digital twin' to be used during building operations. Building operations are heavily intertwined with the businesses that reside and operate within it, and this cyber-physical connectivity threatens to disrupt business operations. This growing affinity with the digital twin puts building operations at jeopardy of cyber-physical attacks most commonly referred to in an Industry 4.0 context as affecting critical infrastructures. Construction 4.0 will require a period of introspection of all the necessary legislative, educational and business model changes needed to protect its businesses and that of its collaborating small and medium-sized enterprises (SMEs).

BIM has changed the way that information is managed, exchanged and transformed stimulating greater collaboration between stakeholders who network within a common data environment (CDE) (Eastman et al., 2011). Adaptation of a CDE for critical infrastructure developments – the processes, systems, technologies, and assets that are essential to economic security as well as public safety is a key facet of effective asset digitalization and offers potential 'long-term' life cycle savings for projects funded both by the government and the private sector. In the short term, front-loaded government expenditure was rather swiftly earmarked to augment operations management and that meant that a concerted effort had been made to develop accurate BIM asset information models (AIM) for large infrastructure asset managers that could be utility companies, Highways England, Network Rail or the Environment Agency (BSI, 2014a), and Deutsche Bahn AG (DB, 2019).

Government policy will continue to transform the standard way of developing and maintaining buildings and infrastructure within the smart built environment. However, the spread of cyber-physical connectivity that is part of CDE has created opportunities for hackers and terrorists, and as a result, there is a constant threat of cyber-crime (Boyes, 2013a). Surprisingly however, existing studies seem upbeat about the benefits that have been accrued from digitalization (BSI, 2014a, 2014b, 2014c; HM Government, 2015). The stakeholders in the infrastructure arena, including clients, project managers, designers, and coordinators find themselves being targeted by cyber-assailants attacking vital infrastructures through a digital portal facilitated by the CDE's integral networked systems that support O&M activities (Boyes, 2013a; Ficco et al., 2017). Strange to discover, therefore that much of what is written on the topic contains many examples of public policy considerations that evaluate critical infrastructure exposed to intentional attacks, natural disasters or physical accidents (Reggiani, 2013; Mayo, 2016). The discussion, however, says little about the substantial cyber-physical security risks posed by a wholesale digital shift within the construction industry (Kello, 2013). These significant risks could disrupt the stream of virtual data produced and have a profound detrimental impact upon a virtually enabled built environment. This, in turn, could lead to physical interruption as well as even the possible destruction of infrastructure assets. One example that illustrates this point is a possible attack on electricity generation with the obvious risk from that to members of the public (Stoddart, 2016).

With this worldwide menace, we want to take a comprehensive look at cyber threats that have the capability of impacting upon the built environment and specifically critical infrastructure. Case studies of cyber-physical attack help us to understand the different types of hackers, what motivates them and the investigation techniques that have been used. Useful also is some exploration of innovative blockchain technology for its potential mitigation of the risk for digital built environment vulnerabilities.

22.2.2 Specific challenges in the construction industry

The construction sector possesses large amounts of data, e.g. all the information generating during the bidding process, engineering designs, calculations and specifications, pricing, profit/loss data, employee information (including intellectual property), and banking records. In most cases, this data contains highly confidential or proprietary information; yet, construction companies are significantly vulnerable and should be proactive in implementing strategies and educate employees in an effort to secure data. However, the reality is that awareness and investment in high-level security in the industry is still very low, making this industry susceptible and particularly attractive to hackers. Therefore, a key element to be considered for the successful transition into digitalization of the industry is cybersecurity.

The challenges that construction companies are encountering due to the adoption of Construction 4.0 are not significantly different to those faced by other industries that have already implemented new technologies and are at a more advanced level of digitalization. However, some cyber risks are specific to the built environment, due to the number of stakeholders, the long supply chains, and the peculiarities of the different phases of the life cycle of construction projects (with a significant amount of data being generated).

22.2.2.1 Stakeholders and supply chain

The exposure to cyber-attacks in the construction industry is amplified by the number of stakeholder and the long supply chains, mostly consisting of SMEs with limited resources devoted to IT. While most general contractors and large subcontractors have cyber-security policies, many smaller subcontractors do not. The risks of cyber-attack also extend to the different project phases, as indicated below.

22.2.2.2 Pre-design/bidding phase

For example, during the tendering phase or the bidding process electronic tendering is becoming the standard, as digital procurement platforms save time and money; however, highly confidential or proprietary information such as project specifications, pricing, profit/loss data, employee information, and banking records could be exposed.

22.2.2.3 Planning and design phase

During the planning and design phases, an attack on the BIM could compromise essential project information, including personal data. It could also prevent access to the model or corrupt the project information, which might lead to construction issues in subsequent project phases (e.g. construction, operation and maintenance).

22.2.2.4 Construction/execution phase

New technology is also allowing the creation of smart and automated construction sites. They might include sensors equipped on the construction equipment or materials, a network of cameras to monitor construction progress in real-time, wearable technology to minimize safety hazards or robotic systems (connected to sensors to capture information that is fed to a control system) to assist workers or conduct construction activities autonomously. In addition, to promote transparency and improve communication, digital platforms are used to allow different project participants to access project data at the same time from different locations (currently using the combination of BIM and common data environments).

22.2.2.5 Operation & Maintenance (O&M) phase

This phase accounts for a significant portion of the cost of a built asset/infrastructure, and critical data is needed during this phase to ensure that the facility is operated and maintained correctly. During the operation and maintenance phase, new technology allows the possibility to move from rigid BMS to more flexible ones using sensors that interconnect different elements through the IoT. BMSs are particularly vulnerable and can compromise not only the performance of the building or infrastructure being managed, but also the corporation or owner, as in the case of US based retailer Target back in 2013 when hackers gained entry through a vendor-access HVAC (heating, ventilation, and air conditioning) building control system (Krebs, 2015).

22.3 The digital uprising

Across the world, rural communities are increasingly striving for economic migration to cities, and as a result, the upsurge in urbanization continues, a trend fueled by a forecasted 9.7 billion population growth by 2050 (UN, 2014a, 2015). For both developed and developing countries, this relentless drive towards urbanization presents a complex socio-economic challenge as well as raising important political issues such as deficiencies in health care provisions (UN, 2014b); lack of resources and malnutrition (UN, 2015); and environmental degradation and pollution (UN, 2015).

These challenges, however, could be addressed through a shrewd allocation of resources via social limitation measures (UN, 2014b). However, politicians worldwide also realize, it seems, that conservative attitudes and resistance to change could be an obstruction to government reform (c.f. Mokyr, 1992). Policies that have been developed have responded by mandating advanced technologies within smart city development as a cure-all to these challenges within the construction sector – a sector that has often been criticized for its reluctance to innovate (Eastman et al., 2011; Kiliccote et al., 2011; BSI, 2014a).

Despite a reluctance to change, the Construction 4.0 sector is widely thought of as being a typical economic stimulus (Eastman et al., 2011) – significantly contributing to gross domestic product. For example, the construction sector was a prime candidate for the UK government's BIM Level 2 mandate that aimed to immerse it within a digital economy. Specifically, the Digital Built Britain report (HM Government, 2015) says that:

> The UK has the potential to lead one of the defining developments of the 21st century, which will enable the country to capture not only all of the inherent value in our built assets, but also the data to create a digital and smart city economy to transform the lives of all.

Within this digital uprising, critical infrastructures lead the UK government's strategic agenda. Computerization has widened the capability of resolving infrastructure challenges such as: optimizing planning and economic development (Kshetri, 2013; Ryan, 2016); ensuring clean air, water and food supply (Ryan, 2016); as well as safeguarding integrated data and security systems (BSI 2014a). Through the various stages of an infrastructure asset's life cycle, this is further strengthened by BIM technology that can improve both information and performance management. BIM technology is essential during the asset's operational phase, which constitutes up to 80% of the whole life-cycle expenditure (Bosch et al., 2014; Liu and Issa, 2014) while the McNulty (2011) report predicts that the potential savings associated with digital asset management and supply chain management may reach £580m between 2018/2019 and facilitated through:

- Effective communication.
- Speed of action.
- Focus on detail and change.
- Incentives and mechanisms for cost reduction.

22.4 Smart cities and digital economies

There is no standardized commonly accepted definition for smart city. In October 2015, the Focus Group on Smart Sustainable Cities from the international telecommunication union (ITU-T Study Group 5) agreed on the following definition of a Smart Sustainable City after analyzing over 100 definitions related to smart cities:

> A smart sustainable city is an innovative city that uses information and communication technologies (ICTs) and other means to improve quality of life, efficiency of urban operation and services, and competitiveness, while ensuring that it meets the needs of present and future generations with respect to economic, social, environmental as well as cultural aspects.

(ITU, 2015)

Other definitions include the one from the British Standards Institute (BSI, 2014a), which defines smart cities as:

> The effective integration of physical, digital and human systems in the built environment to deliver a sustainable, prosperous and inclusive future for its citizens.

The term 'smart' sums up the fully integrated and networked connectivity. A wise hive is embedded within smart city philosophy serving to enhance intelligent analysis of real-time data and information generated to aid decision making and cost-effectiveness. Smart cities within the digital built environment form a cornerstone of a digital economy that will assist in maximizing resources and reducing costs and carbon emissions (whole life cycle). It should enable significant domestic and international growth and also ensure that an economy remains internationally competitive (HM Government, 2015).

The worldwide pace of digitization is set to continue with an expected $400bn investment allocated for smart city development by 2020. Smart infrastructure will be circa 12% of the cost (DBIS, 2013) but despite this substantial forecast expenditure, there has been scant academic attention to the complex array of the many strands of infrastructural asset management like roads, ports, rail, aviation, and telecommunications that will all provide an essential gateway to global markets.

22.4.1 *The ever presence of cyber-espionage and crime*

Smart city technology has increased the risk of cyber-attack that comes through expansive networked systems. However, the threat of cyber-crime has mainly been overlooked within the built environment. Academic studies agree that a gap exists between the state of security in practice and the level of mature security in standards that have been achieved (Korber, 2013; Markets and Markets, 2014; Vijayan, 2014).

The EU commission sees cybersecurity as one of seven building blocks for Construction 4.0 and supports efforts to develop this building block via Public Private Partnership (PPP) research projects and investments. For large bureaucratic institutions such as the European Commission cybersecurity is part and parcel of the legislative and legal framework for the Industry 4.0.

However, if we dig deeper into the understanding of cybersecurity, we see it has the ability to transform other areas of Construction 4.0 beyond the new legislation, namely new skills, new job, new digital innovations, and new partnerships and platforms. BIM, artificial intelligence, 5G, robotics, and IoT are earmarked as promising new facets of Construction 4.0, yet to ensure that widespread disruption does not halt significant contribution to global GDP, cybersecurity of such systems will have to become an inherent aspect to the changing demands of construction.

To better understand the potential threats to Construction 4.0 it is important to elaborate and understand the motivations and types of cyber attackers or hackers. The changing nature of cyber attackers has meant a transition into a black market and the business of information for many cyber assailants. What started with humble origins of hobbyists and teenage enthusiasts has quickly flourished into organized crime and even government-affiliated attacks while many cybersecurity experts hype the fear of cyber-terrorism and war.

Practitioners involved with smart buildings, grids, and infrastructures operations are said to operate in a way that does not reflect current thinking in academic research and discourse (Wendzel, 2016). The fact is that academic and policy attention has focused upon either hypothesized scenarios within international security studies (e.g. the protection of military, industrial and commercial secrets) (Rid, 2012; Segal, 2013); or on policy planning for cyber-warfare (McGraw, 2013) or ensuring the safety of computer systems or networks rather than preventing and dealing with cyber-physical attack. The obvious dangers in this include an impact on nuclear enrichment, hospital operations, public building operation and maintenance, and traffic management (Stoddart, 2016). These threats from cyber-crime are chiefly because of the increased adoption rate of networked devices but also as a result of industry's operational dependency upon IT systems (Boyes, 2013b).

Cyber-criminals have made it their business to be able to harness the basic and real value of digital assets (BSI, 2015) and they can decipher the digital economy and its intricacies more astutely than those in the areas that are under attack (Kello, 2013). The most recent 'WannaCry' ransomware attack in May 2017 was a worldwide cyber-attack by the Wanna-Cry ransomware cryptoworm that targeted computers that were running Microsoft Windows operating systems. The attacked data was encrypted, and ransom payments were demanded in the Bitcoin cryptocurrency. This showed the sophisticated measures that can be deployed by cyber-criminals in identifying, extracting and monetizing data found. While the value of digital assets to their owners and creators can vary, cyber-criminals manipulate data and information to encrypt, ransom or sell it piecemeal (Nicholson et al., 2012; Marinos, 2016).

There have been several prominent instances of unsecured critical infrastructure assets being physically damaged by cyber-crime, and these have been widely reported upon (Peng et al., 2015).

These include: the STUXNET worm that disarmed Iranian industrial and military assets at a nuclear facility (Lindsay, 2013); and the malware 'WannaCry' mentioned earlier that caused significant damage to the UK's National Health Service (NHS) patient database, German railway operations as well as businesses around the world (Clarke and Youngstein, 2017).

Cyber-attacks remain a national security threat for the prosperity of a digital economy and the functionality and safety of the digital built environment (Karabacak et al., 2016b). There have been many reports on the many different threats posed and these present real challenges, as cyber-attacks often enjoy anonymity as the malicious activity they are (Fisk, 2012; Kelly, 2015).

22.5 Cyberspace, cyber-physical attacks and critical infrastructure hacks

In the UK, security analysts from MI5 and MI6 have warned that industrial cyber-espionage is increasing in occurrence, sophistication and maturity, and could potentially lead to an entire shut down of critical infrastructure and services in sectors like power, transport, food and water supplies (Hjortdal, 2011). A number of politically driven infrastructure angles support this assertion and show that the prediction of a global pandemic can become a reality. These could include state-led cyber-attacks on digital infrastructures (banking, news outlets, electronic voting systems) similar to the one in Estonia in 2007 (Lesk, 2007; Pipyros et al., 2016) or the hacking of the US electricity network in 2009 (Hjortdal, 2011) or even the interference in Iranian nuclear plant facilities in 2005 (Denning, 2012).

Cyber-space is described as the global, virtual, computer-based and networked environment, consisting of 'open' and 'air gapped' internet, which directly or indirectly interconnects systems, networks, and other infrastructure and that is critical to society's needs (European Commission, 2013). Within the vast expanse of cyber-space, three partially overlapping territories exist – the world wide web of nodes accessible via URL, the internet consisting of interconnected computers and the 'cyber-archipelago' of computer systems that exist in isolation from the internet within a so-called air gap (Kello, 2013). A CDE hosted on any of these territories can be considered to be exposed to cyber-physical attack.

During a cyber-attack, codes are utilized to interfere with the function of a computer system either for strategic, ambiguous, experimental or political purposes (Nye, 2017). Expanding on this definition, it can be said that cyber-attack constitutes:

> any act by an insider or an outsider that compromises the security expectations of an individual, organization, or nation.
>
> *(Gandhi et al., 2011)*

Actual cyber-attacks can take many forms. They may be in the form of publicized web damages, information leaks, denial-of-service attacks (DoS), and other cyber actions that can be related to national security or military affairs. Cyber-physical attacks can cause disruption or damage to physical assets and pose serious threats to public health and safety or the destruction of the environment.

One of the earliest cyber-physical attacks to be publicly exposed took place in the Cold War when a Soviet oil pipeline exploded due to a so-called logic bomb (Reed, 2007).

The NIST (2017) framework for enhancing the ability of critical infrastructures to withstand cyber-physical attacks depends on two distinct domains being secured, information technologies (IT) and industrial control systems (ICS) (Rittinghouse and Hancock, 2003).

Common threats that can be delivered via IT and ICS include the theft of intellectual property, massive disruption to existing operations and the destruction, degradation or disablement of physical assets and operational ability (Szyliowicz, 2013). The European Union Agency for Network and Information Security (ENISA) outlines multiple common sources of attacks in its malware classification, including viruses, worms, trojans, botnets, spywares, scarewares, roguewares, adwares, and greywares.

Where around every 2,500 lines of code represent a potential vulnerability, cyber-attack targets are identified by a hacker's reconnaissance (Nye, 2017). This reconnaissance is the first and most important stage for a successful cyber-attack and determines a likely strategy for the attack. Strategies vary, but prominent methods include scanning, fingerprinting, footprinting, sniffing, and social engineering.

22.6 What motivates a cyber-attacker? Actors and incident analysis

One can get a comprehensive record of cyber-physical attack incidents categorized as either confirmed or likely, by querying the RISI database (RISI, 2015). However, those in the know think that attacks are more common than reports suggest and that the victims of attacks are often reluctant to disclose malicious cyber-attacks against themselves due to a fear that their reputation will be damaged. Cyber-physical attacks are therefore often shrouded in secrecy by states and private companies, and many states have already accepted the current digital arms race against an array of cyber-actors (or 'hackers') including hacktivists, malware authors, cyber-criminals, cyber-militias, cyber-terrorists, patriot hackers and script kiddies.

Cyber-actors are commonly classified within three categories either White Hats, Grey Hats or Black Hats, with the color of the hat indicating intention.

'White Hats' are mostly legitimately employed security researchers who perform simulated penetration testing hacks to assess the robustness of an organization's cyber-enabled systems (Cavelty, 2013). They do not have malevolent intentions but instead act on behalf of security companies and public interest (F-Secure, 2014).

Contemporary cyber-Robin Hood(s) (or hacktivists) are in the 'Grey Hat' category and act as vigilantes to challenge power structures (such as government) by embarrassing them with distributed denial of service (DDoS) attacks, web defacements, malware, ransomware, and Trojans. These hacktivists will often dabble with illegal means to hack, but they will believe that they are addressing a social injustice or supporting a good cause.

'Black Hats' often come from a criminal fraternity or have other malicious intent (Cavelty, 2013). These criminals use the same tools that the grey and white hat hackers use, but with the deliberate intention to cause harm, to vandalize, sabotage, shut down websites, commit fraud or any number of other illegitimate activities. Many states have focused increasingly on the activities of Grey Hats who have become the new uncontrolled source of hacking (Betz and Stevens, 2013).

22.6.1 Blurred lines: governments and civilians

State and non-state actors pose a two-pronged source of potential malicious attack or threats that face the Construction 4.0 sector. Motivations for these actors may be prompted by various things including patriotism, liberal activism, political ideology, criminal intent and hobby interests. We refer to 'A State' in the context of a political entity ('government') that has sovereignty over an area of territory and the people within it (Hjortdal, 2011; Rahimi, 2011).

Within this, state actors are people who are authorized to act on its behalf and are subject to regulatory control measures (Betz and Stevens, 2013). The roles of a state actor are multiple but often is focused on striving to create positive policy outcomes through approaches such as social movement coalitions (cf. Stearns and Almeida, 2004).

Non-state actors are persons or organizations who have sufficient political influence to act or participate in international relations to influence or cause change even though they are not part of a government or an established institution (Betz and Stevens, 2013).

Since the millennium, governments all over the world have become increasingly aware of cyber-crime and threats that come from such non-state actors. Some of the more notable actors include Anonymous, Ghost Net, The Red Hacker Alliance, Fancy Bear, and the Iranian Cyber Army.

Now, however, the boundary between state actors and non-state actors engaging in cyber-physical attacks has become increasingly blurred (Betz and Stevens, 2013; Papa, 2013; Brantly, 2014; Karabacak et al., 2016a; Stoddart, 2016). This has wider implications for the national security of states and national responsibility for non-state actors who often act on behalf of the state, from nationalistic and ideological motivation.

22.7 Looking at the literature

A comprehensive review of current literature points to four prominent clusters of industrial settings that provide the contextual framework for the analysis of cybersecurity research done to date, and those were construction, transport and infrastructure, information technology, political science and international relations. These groups in the literature contained the majority of the published research on cyber-crime. Within the groups, the following recurrent themes were identified:

*   National and global security.
*   Smart cities.
*   Critical infrastructure.
*   Industrial control systems.
*   Mobile or cloud computing.
*   Digitalization of the built environment.

Across all four sectors, the most popular discussed topics were: mobile cloud computing; national global security and critical infrastructure; smart cities; industrial control systems; and digitization of the built environment.

Strangely, however, within the construction literature, while a considerable effort was put into research on mobile and cloud computing, as well as digitization of the built environment, far less attention was paid to critical infrastructure; and national and global security. Much of the literature seemed oblivious to the threat of cyber-crime posed via the vulnerable CDE portal and BIM.

A CDE is usually established during the feasibility or concept design phases of a development (BSI, 2014a, 2014b). An information manager will then manage and validate the processes and procedures for the exchange of information across a network for each key decision gateway stage (including work in progress (WIP), shared, published and archive stages). Cloud-based CDE platforms are ubiquitous, but common solutions include Project-Wise, Viewpoint (4P), Aconex, Asite, and SharePoint. The internal workflow and typical external information exchange in BIM will rely on the re-use and sharing of information in a CDE.

Integrating BIM (and other file databases, e.g. IFC[1], gbXML[2], CSV[3], DWG[4], XML[5]) within a CDE will ensure a smooth flow of information between all stakeholders and is specified through its levels of development or design. The level of design (LOD) is classified on a linear scale ranging from LOD 1 that covers a conceptual 'low definition' design through to LOD 7 that will be for an as-built 'high definition' model. With each incremental increase in LOD, the range and complexity of asset information within models built will begin to swell, and the data contained within will become accessible to more stakeholders (Figure 22.1). As a consequence, the potential for cyber-crime increases and so it becomes essential that effective cyber-security deterrence measures are put in place.

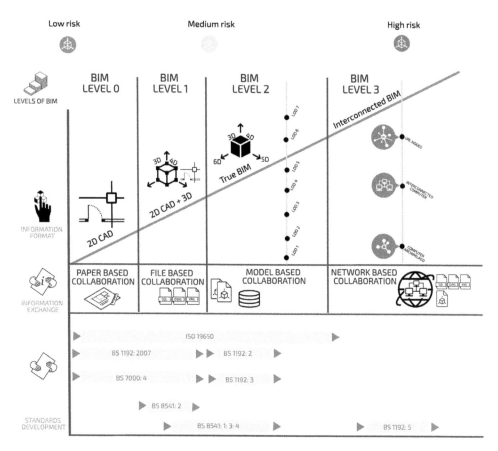

Figure 22.1   Adapted from Pärn and Edwards (2019): The cyber risks of networked CDE and levels of BIM

Perhaps the most crippling aspect of deterrence is the poor rate of acknowledgment (also known as tracebacking or source tracking); where this attribution seeks to determine the identity or location of an attacker or attacker's intermediary (Wheeler and Larsen, 2003; Brantly, 2014).

This connection increases attribution rates, for example, nefarious and malicious attacks on critical infrastructure by non-state 'patriot' actors who proclaim cyber-warfare in the name of nationalist ideologies can create ambiguity with state actors (Nicholson et al., 2012; Lindsay, 2015). What has been written on this subject does widely acknowledge that states actively recruit highly skilled hackers to counter-attack other state governed cyber-activities, in particular, critical infrastructure assets (Thomas, 2009; Pacek, 2012; Cavelty, 2013; Junio, 2013). The lack of identification or disclosure of attacker identities has made the hacking culture even more enticing for both non-state actors and state actors.

Network attribution or IP address traceability to a particular geographical region is possible and can lift the cyber veil to reveal the affiliation between the attacker and their government, but this remains difficult (Canfil, 2016). In the case of potential threats to the Construction 4.0 sector, acknowledgment of industrial cyber-espionage is still an imminent threat not only to the operational business but also for the nation-state security.

22.8 Cyber-deterrence

Cyber-deterrence measures rely mostly upon good practice adopted from standards ISO 27001 and ISO 27032 (ISO 27001, 2013; ISO 27032, 2012). In the context of the digital built environment (and specifically BIM), the recently published cyber-security good practice manual PAS 1198-Part 5 suggests deploying five measures of deterrence:

1. A built asset security manager.
2. A built asset security strategy (BASS).
3. A built asset security management plan (BASMP).
4. A security breach/incident management plan (SB/IMP).
5. Built asset security information requirements (BASIR).

Other guidance notes that can appear ambiguous and that refer to taking 'appropriate mitigation strategies' have not acknowledged the increased vulnerability of semantic and geometric information that is contained within a BIM (PAS 555, BSI, 2013; PAS 754, BSI, 2014). For example, a report by the Institute of Engineering and Technology (Boyes, 2013b) entitled 'Resilience and Cyber Security of Technology in the Built Environment', states that:

> Unauthorised access to BIM data could jeopardise security of sensitive facilities, such as banks, courts, prisons and defence establishments, and in fact most of the Critical National Infrastructure.
>
> *(Boyes, 2013b)*

The deterrence measures recommended in PAS 1192–5 has largely overlooked BIM data contained within a CDE and the onslaught of cyber-physical connectivity in critical infrastructures (Liu et al., 2012). Currently, the most common means employed to deter cyber-physical connectivity in critical BMS infrastructures is via network segregation (the firewall) and secure gateway protection (encryption) for securing against external threats complicit with ANSI/ISA-99 (ANSI, 2007; Nicholson et al., 2012).

- The CDE adopted with BIM in the Construction 4.0 sector acts as a springboard for the wider stakeholder engagement with networked data sharing in a centralized manner yielding such systems vulnerable for future cyber-physical attacks.
- To combat the threats posed, greater cooperation between stakeholders is urgently needed, and this should include governments internationally and private sector partners.
- Blockchain technology offers a secure approach to storing information, making data transactions, performing functions, and establishing trust, making it suitable for sensitive digital infrastructure data contained in BIM and CDE environments with high-security requirements.
- Future research is required to improve the understanding of motivations and to address the specific operational threats to critical infrastructure.

Notes

1 Industry Foundation Classes.
2 Green Building XML.
3 Comma-separated values.
4 AutoCAD Drawing Database (file extension).
5 Extensible Markup Language.

References

ANSI (2007) ISA-99.00.01-2007 Security for Industrial Automation and Control Systems; Part 1: Terminology, Concepts, and Models, ISA Available via: web.archive.org/web/20110312111418/www.isa.org/Template.cfm?Section=Shop_ISA&Template=%2FEcommerce%2FProductDisplay.cfm&Productid=9661 [Accessed: February 2018].

Bessis, N. and Dobre, C. (2014) *Big Data and Internet of Things: A Roadmap for Smart Environments.* London: Springer International Publishing.

Betz, D. J. and Stevens, T. (2013) Analogical Reasoning and Cyber Security, *Security Dialogue* Vol. 44, No. 2, pp. 147–164. DOI: https://doi.org/10.1177%2F0967010613478323.

Bosch, A., Volker, L. and Koutamanis, A. (2014) BIM in the Operations Stage: Bottlenecks and Implications for Owners, *Built Environment Project and Asset Management*, Vol. 5, No. 3, pp. 331–343. DOI: 10.1108/BEPAM-03-2014-0017.

Boyes, H. (2013a) Cyber Security of Intelligent Buildings: A Review, 8th IET International System Safety Conference incorporating the Cyber Security Conference 2013, Cardiff, UK. DOI: 10.1049/cp.2013.1698.

Boyes, H. (2013b) Resilience and Cyber Security of Technology in the Built Environment the Institution of Engineering and Technology, IET Standards Technical Briefing, London. Available via: www.theiet.org/resources/standards/-files/cyber-security.cfm?type=pdf [Accessed: February 2018].

Brantly, A. F. (2014) The Cyber Losers, *Democracy & Security*, Vol. 10, No. 2, pp. 132–155. DOI: https://doi.org/10.1080/17419166.2014.890520.

BSI (2013) PAS 555:2013 Cyber Security Risk. Governance and Management Specification, Available via: https://shop.bsigroup.com/ProductDetail/?pid=000000000030261972 [Accessed: February 2018].

BSI (2014a) PAS 180 Smart Cities. Vocabulary. British Standards Institution, London. Available via: www.bsigroup.com/en-GB/smart-cities/Smart-Cities-Standards-and-Publication/PAS-180-smart-cities-terminology/ [Accessed: February, 2018].

BSI (2014b) PAS 1192-3 Specification for Information Management for the Operational Phase of Assets Using Building Information Modelling, British Standards Institution, London. Available via: https://shop.bsigroup.com/ProductDetail/?pid=000000000030311237 [Accessed: February 2018].

BSI (2014c) PAS 754:2014 Software Trustworthiness. Governance and Management. Specification, Available via: https://shop.bsigroup.com/ProductDetail/?pid=000000000030284608 [Accessed: February 2018].

BSI (2015) PAS 1192-5 (2015) Specification for Security Minded Building Information Modelling, Digital Built Environments and Smart Asset Management, British Standards Institution, London. Available via: https://shop.bsigroup.com/ProductDetail/?pid=000000000030314119 [Accessed: February 2018].

Canfil, J. K. (2016) Honing Cyber Attribution: A Framework for Assessing Foreign State Complicity, *Journal of International Affairs*, Vol. 70, No. 1, pp. 217. Available via: www.questia.com/read/1G1-476843518/honing-cyber-attribution-a-framework-for-assessing [Accessed: February, 2018].

Carr, M. (2016) Public-private Partnerships in National Cyber-security Strategies, *International Affair*, Vol. 92, No. 1, pp. 43–62. Available via: www.chathamhouse.org/sites/files/chathamhouse/publications/ia/INTA92_1_03_Carr.pdf [Accessed: February 2018].

Cavelty, M. D. (2013) From Cyber-Bombs to Political Fallout: Threat Representations with an Impact in the Cyber-Security Discourse, *International Studies Review*, Vol. 15, pp. 105–122. DOI: 10.1111/misr.12023.

Clarke, R. and Youngstein, T. (2017) Cyberattack on Britain's National Health Service – A Wake-up Call for Modern Medicine. *New England Journal of Medicine*, Vol. 377, pp. 409–411. DOI: 10.1056/NEJMp1706754.

DB (2019) Implementation of Building Information Modeling (BIM) in the Infrastructure Division of Deutsche Bahn AG, Competence Center Major Projects 4.0, Frankfurt am Main, Germany. Available via: www.deutschebahn.com/de/bahnwelt/bauen_bahn/bim/BIM-1186016 [Accessed: June 2019].

DBIS (2013) Smart City Market: Opportunities for the UK, Department for Business, Innovation and Skills, BIS Research Papers Ref: BIS/13/1217, DBIS, London. Available via: www.gov.uk/government/publications/smart-city-market-uk-opportunities [Accessed: February 2018].

Denning, D. (2012) Stuxnet: What Has Changed? *Future Internet*, Vol. 4, No. 3, pp. 672–687. DOI: 10.3390/fi4030672.

Eastman, C., Teicholz, P., Sacks, R. and Liston, K. (2011). *BIM Handbook: A Guide to Building Information Modeling for Owners, Managers, Designers, Engineers and Contractors*. Hoboken, NJ: John Wiley & Sons.

Eom, S-J. and Paek, J-H. (2006) Planning Digital Home Services through an Analysis of Customers Acceptance, *Journal of Information Technology in Construction (ITcon)*, Vol. 11, Special issue IT in Facility Management, pp. 697–710, Available via: www.itcon.org/2006/49 [Accessed: February 2018].

European Commission (2013) Cybersecurity Strategy of the European Union: An Open, Safe and Secure Cyberspace, JOIN 1 Final, European Commission, Brussels. Available via: https://eeas.europa.eu/archives/docs/policies/eu-cyber-security/cybsec_comm_en.pdf [Accessed: February 2018].

Ficco, M., Choraś, M. and Kozik, R. (2017) Simulation Platform for Cyber-security and Vulnerability Analysis of Critical Infrastructures, *Journal of Computational Science*, Vol. 22, pp. 179–186. DOI: 10.1016/j.jocs.2017.03.025.

Fisk, D. (2012) Cyber Security, Building Automation, and the Intelligent Building, *Intelligent Buildings International*, Vol. 4, No. 3, pp. 169–181. DOI: 10.1080/17508975.2012.695277.

F-Secure Labs (2014) Havex Hunts for ICS and SCADA Systems, Available via: www.f-secure.com/weblog/archives/00002718.html [Accessed: February 2018].

Gandhi, R., Sharma, A., Mahoney, W., Sousan, W., Zhu, Q. and Laplante, P. (2011) Dimensions of Cyber-attacks: Cultural, Social, Economic, and Political, *IEEE Technology and Society Magazine*, Vol. 30, No. 1, pp. 28–38. DOI: 10.1109/MTS.2011.940293.

Hansen, L. and Nissenbaum, H. (2009) Digital Disaster, Cyber Security, and the Copenhagen School, *International Studies Quarterly*, Vol. 53, pp. 1155–1175. Available via: www.nyu.edu/projects/nissenbaum/papers/digital%20disaster.pdf [Accessed: February 2018].

Hjortdal, M. (2011) China's Use of Cyber Warfare: Espionage Meets Strategic Deterrence, *Journal of Strategic Security*, Vol. 4, No. 2, pp. 1–24. DOI: 10.5038/1944-0472.4.2.1.

HM Government (2013) Building Information Modeling Industrial Strategy: Government and Industry in Partnership, Government Construction Strategy, London. Available via: www.gov.uk/government/uploads/system/uploads/attachment_data/file/34710/12-1327-building-information-modelling.pdf [Accessed: February 2018].

HM Government (2015) Digital Built Britain: Level 3 Building Information Modelling – Strategic Plan, 26 February 2015, HM Publications, London. Available via: www.gov.uk/government/publications/uk-construction-industry-digital-technology [Accessed: February 2018].

ISO (2012) 27032 Information Technology – Security Techniques – Guidelines for Cybersecurity, International Organization for Standardization (ISO), Geneva, Switzerland. Available via: www.itgovernance.co.uk/shop/product/iso27032-iso-27032-guidelines-for-cybersecurity [Accessed: February 2018].

ISO (2013) 27001 The International Information Security Standard, International Organization for Standardization (ISO), Geneva, Switzerland. Available via: www.itgovernance.co.uk/iso27001 [Accessed: February 2018].

ITU (2015) Focus Group on Smart Sustainable Cities, International Telecommunication Union, 5–6 May 2015, Abu Dhabi, United Arab Emirates. Available via: www.itu.int/en/ITU-T/focusgroups/ssc/Pages/default.aspx [Accessed: June 2019].

Junio, T. J. (2013) How Probable is Cyber War? Bringing IR Theory back in to the Cyber Conflict Debate, *Journal of Strategic Studies*, Vol. 36, pp. 125–133. DOI: 10.1080/01402390.2012.739561.

Karabacak, B., Yildirim, S. O. and Baykal, N. (2016a) Regulatory Approaches for Cyber Security of Critical Infrastructures: The Case of Turkey, *Computer Law & Security Review*, Vol. 32, No. 3, pp. 526–539. DOI: 10.1016/j.clsr.2016.02.005.

Karabacak, B., Yildirim, S. O. and Baykal, N. (2016b) A Vulnerability-driven Cyber Security Maturity Model for Measuring National Critical Infrastructure Protection Preparedness, *International Journal of Critical Infrastructure Protection*, Vol. 15, October, pp. 47–59. DOI: 10.1016/j.ijcip.2016.10.001.

Kello, L. (2013) The Meaning of the Cyber Revolution: Perils to Theory and Statecraft, *International Security*, Vol. 38, pp. 7–40. DOI: 10.1162/ISEC_a_00138.

Kelly, A. (2015) Building Management Systems: the Cyber Security Blind Spot – QinetiQ White Paper, Available via: www.qinetiq.com/services-products/cyber/Pages/bms-the-cyber-security-blind-spot.aspx [Accessed: February 2018].

Kiliccote, S., Piette, M. A. and Ghatikar, G. (2011) Smart Buildings and Demand Response, AIP Conference Proceedings 1401, 328, American Institute of Physics, Doi: https://doi.org/10.1063/1.3653861.

Korber, S. (2013) Hackers' Next Target? Maybe Your Facility's Control Systems, Available via: http://goo.gl/d7QUnh. [Accessed: February 2018].

Krebs, B. (2015) Target Hackers Broke in Via HVAC Company, Available via: https://krebsonsecurity.com/2014/02/target-hackers-broke-in-via-hvac-company/ [Accessed: June 2019].

Kshetri, N. (2013) Cybercrime and Cyber-security Issues Associated with China: Some Economic and Institutional Considerations, *Electronic Commerce Research*, Vol. 13, No. 1, pp. 41–69. DOI: 10.1007/s10660-013-9105-4.

Lesk, M. (2007) The New Front Line: Estonia Under Cyber Assault, *IEEE Security & Privacy*, Vol. 5, No. 4, pp. 76–79. July–August 2007. DOI: 10.1109/MSP.2007.98.

Levy, Y. and Ellis, T. J. (2006) A Systems Approach to Conduct an Effective Literature Review in Support of Information Systems Research, *Informing Science*, Vol. 9, pp. 181–212. Available via: http://inform.nu/Articles/Vol9/V9p181-212Levy99.pdf [Accessed: February 2018].

Lin, S., Gao, J. and Koronios, A. (2006) Key Data Quality Issues for Enterprise Asset Management in Engineering Organisations, *International Journal of Electronic Business Management (IJEBM)*, Vol. 4, No. 1, pp. 96–110. Available via: http://ijebm.ie.nthu.edu.tw/IJEBM_Web/IJEBM_static/Paper-V4_N1/A10-E684_3.pdf [Accessed: February 2018].

Lindsay, J. R. (2013) Stuxnet and the Limits of Cyber Warfare, *Security Studies*, Vol. 22, No. 3, pp. 365–404. DOI: 10.1080/09636412.2013.816122.

Lindsay, J. R. (2015) The Impact of China on Cybersecurity: Fiction and Friction, *International Security*, Vol. 39, No. 3, pp. 7–47. DOI: 10.1162/ISEC_a_00189.

Liu, R. and Issa, R. R. A. (2014) Design for Maintenance Accessibility Using BIM Tools. *Facilities*, Vol. 32, No. 3/4, pp. 153–159, DOI: https://doi.org/10.1108/F-09-2011-0078.

Liu, J., Xiao, Y., Li, S., Liang, W. and Chen, C. P. (2012) Cyber Security and Privacy Issues in Smart Grids, *IEEE Communications Surveys & Tutorials*, Vol. 14, pp. 981–997. DOI: 10.1109/SURV.2011.122111.00145.

Marinos, L. (2016) ENISA Threat Taxonomy a Tool for Structuring Threat Information, European Union Agency for Network and Information Security. Available via: www.enisa.europa.eu/topics/threat-risk-management/threats-and-trends/enisa-threat-landscape/etl2015/enisa-threat-taxonomy-a-tool-for-structuring-threat-information/view [Accessed: February, 2018].

Markets and Markets (2014) Smart HVAC Controls Market by Product Type, Components, Application, Operation & Geography – Analysis and Forecast to 2014–2020, Available via: http://goo.gl/Ay2LjI. [Accessed: February 2018].

Mayo, G. (2016) Bas and Cyber Security: A Multiple Discipline Perspective, Proceedings of the American Society for Engineering Management 2016 International Annual Conference S. Long, E-H. Ng, C. Downing and B. Nepal eds. Available via: www.researchgate.net/publication/309480358_BAS_ AND_CYBER_SECURITY_A_MULTIPLE_DISCIPLINE_PERSPECTIVE [Accessed: February, 2018].

McAuley, B., Hore, A. and West, R. (2017) BICP Global BIM Study – Lessons for Ireland's BIM Programme Published by Construction IT Alliance (CitA) Limited, 2017, DOI: 10.21427/D7M049.

McGraw, G. (2013) Cyber War is Inevitable (Unless We Build Security In), *Journal of Strategic Studies*, Vol. 36, No. 1, pp. 109–119. DOI: 10.1080/01402390.2012.742013.

McNulty, R. (2011) Realising the Potential of GB Rail – Final Independent Report of the Rail Value for Money Study – Summary Report, Department for Transport, London, UK. Available via: www.gov. uk/government/uploads/system/uploads/attachment_data/file/4203/realising-the-potential-of-gb-rail-summary.pdf [Accessed: February 2018].

Metke, A. R. and /Ekl, R. L. (2010) Security Technology for Smart Grid Networks, *IEEE Transactions on Smart Grid*, Vol. 1, No. 1, pp. 99–107. DOI: 10.1109/TSG.2010.2046347.

Mokyr, J. (1992) Technological Inertia in Economic History, *The Journal of Economic History*, Vol. 52, No. 2, pp. 325–338. DOI: 10.1017/S0022050700010767.

National Institute of Standards and Technology (NIST) (2017) Framework for Improving Critical Infrastructure Cybersecurity, Draft Version 1.1, 10 January 2017. Available via: www.nist.gov/sites/ default/files/documents////draft-cybersecurity-framework-v1.11.pdf [Accessed: February 2018].

Nicholson, A., Webber, S., Dyer, S., Patel, T. and Janicke, H. (2012) SCADA Security in the Light of Cyber-warfare, *Computers & Security*, Vol. 31, No. 4, pp. 418–436. DOI: 10.1016/j.cose.2012.02.009.

Nye, J. S. (2017) Deterrence and Dissuasion in Cyberspace, *International Security*, Vol. 41, No. 3 (Winter 2016/17), pp. 44–71.

Pacek, B. (2012) Cyber Security Directed Activities, *Internal Security*, July–December 2012, pp. 119–137. Available via: https://search.proquest.com/openview/f4c2fab87dc19daeb2e0c56a6b601c36/1 ?pq-origsite=gscholar&cbl=1556339 [Accessed: February 2018].

Papa, P. (2013) US and EU Strategies for Maritime Transport Security: A Comparative Perspective. *Transport Policy*, Vol. 28, pp. 75–85.

Pärn, E. A. and Edwards, D. (2019) Cyber Threats Confronting the Digital Built Environment: Common Data Environment Vulnerabilities and Block Chain Deterrence, *Engineering, Construction and Architectural Management*, DOI: 10.1108/ECAM-03-2018-0101.

Peng, Y., Wang, Y., Xiang, C., Liu, X., Wen, Z. and Chen, D. (2015) Cyber-physical Attack-Oriented Industrial Control Systems (ICS) Modeling, Analysis and Experiment Environment, International Conference on Intelligent Information Hiding and Multimedia Signal Processing, pp. 322–326.

Pipyros, K., Mitrou, L., Gritzalis, D. and Apostolopoulos, T. (2016) Cyber Operations and International Humanitarian Law: A Review of Obstacles in Applying International Law Rules in Cyber Warfare, *Information & Computer Security*, Vol. 24, No. 1, pp. 38–52. DOI: 10.1108/ICS-12-2014-0081.

Rahimi, B. (2011) The Agonistic Social Media: Cyberspace in the Formation of Dissent and Consolidation of State Power in Postelection Iran, *The Communication Review*, Vol. 14, pp. 158–178. DOI: https:// doi.org/10.1080/10714421.2011.597240.

Reggiani, A. (2013) Network Resilience for Transport Security: Some Methodological Considerations. *Transport Policy*, Vol. 28, pp. 63–68.

Reed, T. (2007) *At the Abyss: An Insider's History of the Cold War*, New York: Random House Publishing Group.

Rid, T. (2012) Cyber War Will Not Take Place, *Journal of Strategic Studies*, Vol. 35, No. 1, pp. 5–32. DOI: 10.1080/01402390.2011.608939.

RISI (2015) The Repository of Industrial Security Incidents Database, Available via: www.risidata.com/ Database [Accessed: February 2018].

Rittinghouse, J. and Hancock, W. M. (2003) *Cybersecurity Operations Handbook*, Amsterdam, Netherlands: Elsevier Science. ISBN: 978-1-55558-306-4.

Ryan, D. J. (2016) Engineering Sustainable Critical Infrastructures, *International Journal of Critical Infrastructure Protection*, Vol. 15, pp. 47–59. DOI: http://dx.doi.org/10.1016/j.ijcip.2016.11.003.

Segal, A. (2013) The Code Not Taken: China, the United States, and the Future of Cyber Espionage, *Bulletin of the Atomic Scientists*, Vol. 69, No. 5, pp. 38–45. DOI: 10.1177/0096340213501344.

Stearns, L. B. and Almeida, P. D. (2004) The Formation of State Actor-Social Movement Coalitions and Favorable Policy Outcomes, *Social Policy*, Vol. 51, No. 4, pp. 478–504. DOI: 10.1525/ sp.2004.51.4.478.

Stoddart, K. (2016) Live Free or Die Hard: US-UK Cybersecurity Policies, *Political Science Quarterly*, Vol. 131, No. 4, pp. 803–842. DOI: 10.1002/polq.12535.

Szyliowicz, J. S. (2013) Safeguarding Critical Transportation Infrastructure: The US Case, *Transport Policy*, Vol. 28, pp. 69–74. DOI: https://doi.org/10.1016/j.tranpol.2012.09.008.

Thomas, N. (2009) Cyber Security in East Asia: Governing Anarchy, *Asian Security*, Vol. 5, pp. 3–23. DOI: 10.1080/14799850802611446.

Turk, Ž. and Klinc, R. (2017) Potentials of Blockchain Technology for Construction Management, *Procedia Engineering*, Vol. 196, pp. 638–645. DOI: https://doi.org/10.1016/j.proeng.2017.08.052.

UN (2014a) 2014 Revision of the World Urbanization Prospects, Available via: https://goo.gl/xwOSDS [Accessed: February 2018].

UN (2014b) World Urbanization Trends 2014: Key Facts, Statistical Papers – United Nations (Ser. A), Population and Vital Statistics Report. United Nations. DOI: 10.18356/685065dd-en.

UN (2015) World Population Projected to Reach 9.7 Billion by 2050, Available via: www.un.org/en/development/desa/news/population/2015-report.html [Accessed: February 2018].

Vijayan, J. (2014) With the Internet of Things, Smart Buildings Pose Big Risk, Available via: http://goo.gl/ONgwkN [Accessed: February 2018].

Wendzel, S. (2016) How to Increase the Security of Smart Buildings? Communications of the ACM, *May 2016*, Vol. 59, No. 5, pp. 47–49, DOI: 10.1145/2828636.

Wheeler, D. A. and Larsen, G. N. (2003) Techniques for Cyber-Attack Attribution, Institute For Defense Analyses, Alexandria, VA. Available via: www.researchgate.net/publication/235170094_Techniques_for_Cyber_Attack_Attribution [Accessed: February 2018].

23

EMERGING TRENDS AND RESEARCH DIRECTIONS

Eyuphan Koc, Evangelos Pantazis, Lucio Soibelman, and David J. Gerber

23.1 Aims

- Presenting a background on *diffusion of innovation* in the AEC industry and an expanded discussion on factors hindering and motivating innovation highlighting the clash between fragmentation and integration.
- Correlating a revised categorization of 4.0 concepts and technologies discussed in previous chapters in terms of their corresponding stage of diffusion across all project phases.
- Discussing how technological advancements revealed opportunities for novel and more collaborative project delivery methods such as IPD (Integrated Project Delivery), and how this has the potential to speed up integral 4.0 innovations in a reciprocal feedback manner.
- Introducing a broad vision on revising the design and construction curricula to allow for more cross-disciplinary information flow from disciplines that are traditionally left out from training and practice.
- Suggest future directions both in industry and academia for introducing interdisciplinary research methods which will allow engineers to take the leadership back in developing their own design and engineering tools (from software engineers) that will allow them to lead the AEC in the information age.

23.2 Introduction

"Industrial revolutions" are different periods in time signifying leaps in technology—through discoveries or revolutionary innovations—beyond what was or what could be imagined prior to their occurrence. The designations have naturally been "ex-post" where the so-called First Industrial Revolution resulted from the use of mechanization for production, second from the intensive use of electricity and the third associated with a widespread digitalization of building project delivery (Lasi et al. 2014). Figure 23.1 illustrates a timeline of industrial revolutions and lists paradigm shifting technological leaps associated with them.

Over the last decade, originating from the vision laid out by the German government for high-tech manufacturing, the term "Industry 4.0" was loosely coined to describe a Fourth

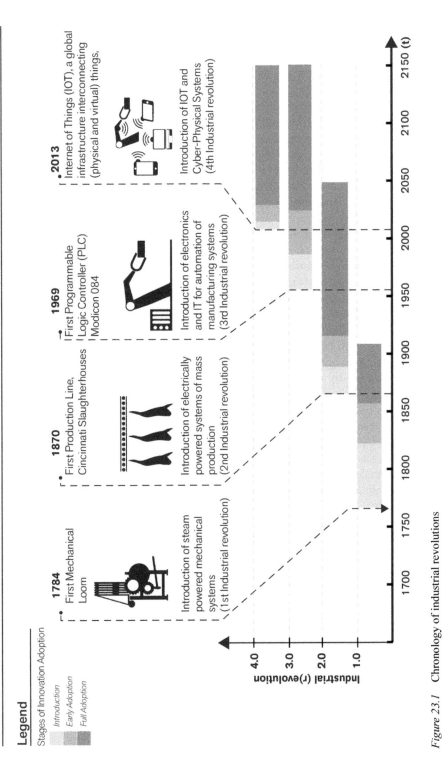

Figure 23.1 Chronology of industrial revolutions

Industrial Revolution. The so-called Fourth Industrial Revolution is characterized by the paradigm-shifting changes in manufacturing since the introduction of programmable logic controllers, underlining the concepts of *increased digitization and automation, modularization of production, mass customization and self-organization.* The AEC (architecture, engineering, construction) industry has been less perceptive of these developments than others such as the aerospace and automotive industries due to numerous factors stemming from the complexity of building design and delivery as well as the fragmented nature of construction industry combined with a technological risk aversion, all of which are factors that do not underpin innovation. As a consequence, the sector in many countries continues to suffer from low labor productivity, labor shortages due to aging population, low safety, time and cost overruns and logistic issues. In a similar vein, the construction industry has been criticized for being inefficient; often generating too much waste, emitting significant amounts of greenhouse gases (GHG) and consuming too much energy compared to other industries (Abanda et al. 2017).

Surveys in the building sector in North America suggests that complete deployment of green innovations can lead to a reduction of building energy consumption by 25–30%, a reduction in construction costs as much as 130 billion per year (Choi Granade et al. 2009) and returns as high as 20% of the total construction costs through the life cycle of a building (Kats et al. 2003). It is therefore not a secret that the AEC industry needs to revolutionize itself to develop holistic design and engineering solutions banking on life-cycle benefits that can compete with traditional solutions that are more cost-effective in isolation, i.e. when considered from the perspective of a particular discipline or phase. In this context, increased digitization and automation as well as a widespread adoption of concepts and tools under the Industry 4.0 umbrella could significantly improve the bottom-line of the AEC industry.

Against this backdrop, research in "Construction 4.0" is carried out in many directions including advancements in individual 4.0 technologies niched in the industry such as Building Information Modeling (BIM), adaptive building systems, robotics in construction, large-scale additive manufacturing, etc. Additionally, there is an increasing awareness of the potential benefits of technologies such as artificial intelligence and big data analytics, embedded sensing technologies, Virtual/Augmented reality (VR/AR), mobile/Cloud Computing as well as blockchain mostly stemming from the successful application and diffusion of such technologies outside the construction realm. Despite the potential benefits such as decreased design and construction durations, higher quality in projects delivered, and enhanced job safety; the industry is characteristically slow in adopting innovations which makes it challenging to achieve improvements in mentioned areas.

The importance of an innovative technology can be classified based on its effect on the existing supply chain, the design-construction-operation process or the actors involved (clients, stakeholders, designers, engineers, contractors, and managers). In the literature, various categorizations exist for innovations and researchers distinguish between autonomous vs. systemic, bounded vs. unbounded and integral vs. modular innovations (Sheffer 2011; Taylor et al. 2006; Teece 1996). In this chapter, the latter classification between *integral* and *modular innovations* is underlined. *Modular innovations* refer to disruptive technologies that fit within the existing divisions of work and do not cross conventional boundaries between disciplines (e.g. energy efficient smart bulbs and switches changing the outlook in the lighting industry). *Integral innovations* refer to technological advances that may introduce changes in the interaction of the modules or disrupt the overall system architecture. Such advances may introduce changes at the interfaces or design criteria between two or more modules, a change in the processes (construction sequence) of the overall system, or both. The authors perceive the developments in the industry through this lens as technologies and concepts classified to be under the umbrella of "Construction 4.0" are triggering both types of innovations for the industry.

Oesterreich and Teuteberg (2016) provide a thorough literature review of the subject matter, and a general taxonomy of key 4.0 technologies and concepts as they relate to the AEC industry. Specifically, "Construction 4.0" is divided into three categories, namely: (1) *smart construction site (smart factory)*, (2) *simulation and modeling*, (3) *digitization and virtualization*. The rapid adoption of technologies such as BIM, Digital Project Delivery, Radio Frequency Identification (RFID), and Cloud Computing have made them prevalent in the practice throughout the last decade. On the other hand, a number of technologies are advancing towards maturity (Artificial Intelligence and Big Data Analytics, Autonomous Robotic Construction, Additive Manufacturing and Virtual/Augmented reality) and are expected to impact the industry in the next decade (BCG 2016).

Within earlier chapters of this book, state-of-the-art in individual technologies is discussed in detail. In the background section, the objective of the authors is to (1) broadly discuss factors hindering innovation in the AEC industry together with factors motivating innovation, and (2) provide a snapshot of 4.0 concepts and technologies as categorized by Oesterreich and Teuteberg (2016) and analyze their respective stages of adoption from a *diffusion of innovations* viewpoint. These will set the stage for the following discussion on emerging trends in the AEC industry that are expected to accelerate the adoption of 4.0 technologies as well as the recommendations on future research directions. In this chapter, the authors' intention is to provide insights overarching individual technologies themselves.

23.3 Background

23.3.1 Innovation diffusion

"Diffusion" was characterized initially by French sociologist Gabriel Tarde in the late 19th century to refer to the spreading of social or cultural properties from one society or environment to another. The term was picked up by rural sociologists in the US investigating the rapidly expanding agriculture to understand how independent farmers were adopting hybrid seeds, equipment, and techniques. Initial efforts in rural sociology and agriculture helped construct the diffusion paradigm and *Diffusion of Innovations* theory was popularized by Roger's seminal work that synthesized hundreds of diffusion studies from various domains influencing the theory (Rogers 2003).

Beal and Bohlen (1957) in their early work titled "Process of Diffusion", assert that diffusion, defined as the process by which people accept new ideas, is a mental process meaning it happens through a series of complex acts rather than a single one. According to the authors, this mental process happens in five stages: awareness, interest, evaluation, trial, and adoption. An individual learns about the *idea*[1] during the *awareness* stage but may ignore the details thereof. During the *interest* stage, the individual becomes interested about details such as how the idea or product works and what is the associated potential. The *evaluation* stage is when the individual makes a mental trial of the idea or the product to ask questions on his or her capability of adopting it and if yes, what the impacts would be. Then the individual puts the idea or the product into *trial* and a successful trial leads to *adoption* where there is a large-scale, continued use of the idea which builds experience. Beal and Bohlen carefully warn that adoption of an idea does not necessitate constant application of it. It simply means that the idea or product is accepted mentally and there is an intention to include it in the practice.[2]

This broad conceptualization of the stages of how innovation diffuses into the practice as well as into supply and value chains is a useful one and will be adopted later in the chapter to demonstrate the status-quo of Industry 4.0 concepts and technologies in the AEC industry. It is essential to note that Beal's *stages of diffusion* differ from Roger's the *rate of adoption* where the former refers to the stages from the inception of the idea to its acceptance, and the latter is the relative speed with which an innovation is adopted by members of a social system.[3] In what follows, factors hindering the process of diffusion in the industry are briefly discussed.

23.3.2 Factors hindering diffusion of Construction 4.0

As mentioned, Industry 4.0 concepts and technologies mainly originated from the manufacturing industry where advancements in both infrastructures and operations found their initial test beds and enabled large-scale industrial improvements. The AEC industry possesses different types of challenges that are perceived as factors hindering the penetration and adoption of 4.0 concepts and technologies. Construction projects typically have higher numbers of interrelated processes, sub-processes and actors leading to higher complexity (Arayici and Coates 2012; Dubois and Gadde 2002). They are location-based, highly customized but temporary undertakings, factors all of which increase uncertainty and risk. In addition, the construction supply chain is highly fragmented with a high number of small and medium sized enterprises (SMEs) that lack the capacity (financial or otherwise) to invest in and benefit from new technologies (Kraatz et al. 2014). Specifically, fragmentation manifests itself in three dimensions; *vertical fragmentation* across building life cycle phases (design-construction-operations), *horizontal fragmentation* across trades and disciplines (mechanical, electrical, etc.) and *longitudinal fragmentation* across projects curtailing the sharing of knowledge. Figure 23.2 illustrates the three types of fragmentation in the AEC industry (Fergusson 1993; Sheffer 2011).

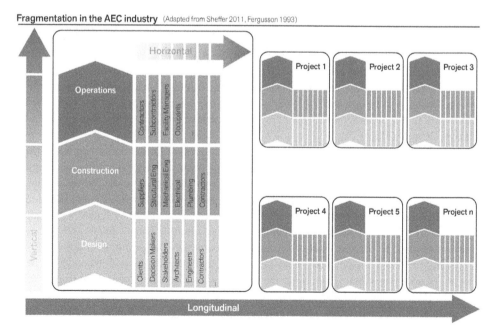

Figure 23.2   Three types of fragmentation in the construction industry

Other significant factors such as a long-standing risk aversion towards innovation focusing on *doing what works*, culture of low cost competitive bidding and a lack of computer savvy workforce in the industry are slowing down the wide adoption of 4.0 concepts and technologies. One less pronounced but highly impactful characteristic is conflicting incentives adversely affecting collaboration between the project stakeholders (owner, architect, contractor, subcontractor, material supplier etc.). In the wake of such characteristics, a slow adoption of technologies is currently observed despite clearly demonstrated benefits.

The idea of revolutionizing the AEC industry through automation has been tested before. High risk of injury, skilled labor shortages and an ageing workforce created a demand for research into automatic construction in Japan in the 90s. Japanese researchers realized that the development of single task human-machine construction systems could be an efficient and economical way to introduce automation in construction. Despite envisioned benefits, robotic construction did not prove to be viable for the industry at the time. Out of over 500 robotic construction platforms developed, less than ten made it to the industry (Bechthold 2010; Obayashi 1999). The main obstacles were reported to be (1) the high complexity and dynamic conditions of the construction sites were not providing a stable and structured operating environment for robots and (2) the lack of robotic control frameworks and high cost of sensors were prohibitive (Taylor et al. 2009). Today, enabling technologies including the advanced Modeling and Simulation Tools supported by Cloud Computing and Big Data, Digital Fabrication and Robotic Construction as well as the Internet of Things and Services (Oesterreich and Teuteberg 2016) are more mature and thus capable of facilitating the envisioned disruption across all sectors of AEC the industry.

23.3.3 Factors motivating and facilitating Construction 4.0

Above all, the factors motivating increased automation and digitization through 4.0 concepts and technologies are the Achilles heels of the AEC industry themselves that are mentioned above: low productivity, delays and overruns, quality, safety and waste issues. The vision is that a 4.0 version of the industry, if achieved, will not suffer from many of these problems (as shown in Figure 23.3).

A recent report by Boston Consulting Group, concludes that within ten years, full-scale digitalization in non-residential construction will lead to annual global cost savings of 13% to 21% in the engineering and construction phases and 10% to 17% in the operations phase (BCG 2016). The authors would like to discuss some other trends that are perceived as technological motivators and facilitators.

One facilitator is the fast-paced innovation in individual technologies themselves. Advancements in Internet of Things and Internet of Services have resulted in improvements that lowered the costs and enhanced the functionalities. Sensors with previously prohibitive costs became cheaper enabling their use in the construction and operation phases. Moreover, Cloud-Computing benefiting from advanced 5G network infrastructure has provided access to high performance computing for a much wider audience. On the other hand, BIM has emerged as the central platform that has the potential to combine all related 4.0 technologies that can largely benefit—both individually and collectively—from the creation of semantically rich digital instances of buildings. Despite being slow early on, the following waves of BIM adoption has paved the way for the digitization of various types of building related information virtually affecting all phases of projects (design, construction and operation).

In terms of project delivery, a much more collaborative setting implemented through Integrated Project Delivery (IPD) is promising to begin addressing the horizontal (fragmentation

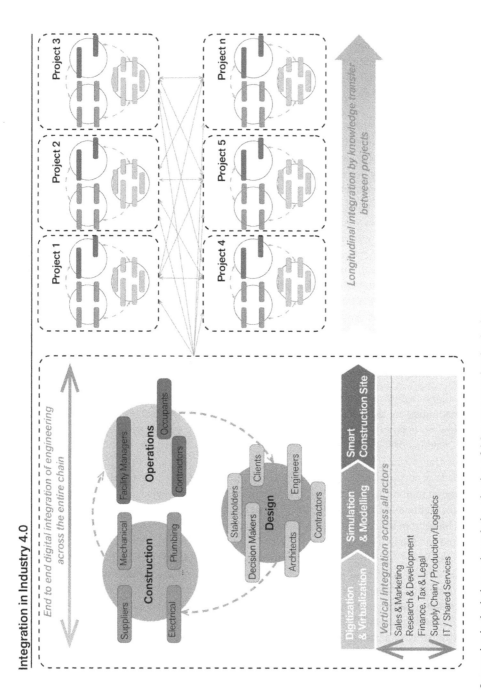

Figure 23.3 Integration in the industry through the adoption of 4.0 concepts and technologies

between disciplines such as electrical, mechanical, etc.) and vertical fragmentation (fragmentation between life cycle phases design, construction and operations) issues. IPD is a project delivery method that integrates people, systems, business, structures and practices into a process that collaboratively harness talents and insights of all participants to reduce waste and optimize efficiency through all phases of design, fabrication and construction of a building structure (Lahdenperä 2012).

All in all, an increased availability and accessibility of technology as well as the comprehension of the need for enhanced collaboration in project delivery are creating an innovation pull. In the next section, the authors begin their discussion by coming back to the diffusion concept to demonstrate the status quo regarding the 4.0 concepts and technologies in the AEC industry. It is asserted that a number of trends are relevant given this observation of the industry.

23.4 Emerging 4.0 trends in the AEC industry

In their review, Oesterreich and Teuteberg (2016) provide a comprehensive categorization of key Construction 4.0 concepts and technologies in three clusters: *Smart Factory, Modeling and Simulation, Digitization and Virtualization.* Here, the authors adopt their categorization and a) enrich it with a number of items regarded to be within the 4.0 realm that were missing from Oesterreich and Teuteberg's taxonomy, b) use the innovation diffusion concept and its five stages (awareness, interest, evaluation, trial and adoption) to demonstrate the stage of adoption of each concept and technology with respect to the different life-cycle phases. This is illustrated in Figure 23.4.

For example, it is asserted that *High Performance Computing (HPC)* is categorized under Modeling and Simulation and is at the stage of *awareness* for the construction phase whereas it is in the interest stage for design. This is because large databases enabling analytics tasks that would require HPC are virtually nonexistent for the construction phase. On the other hand, HPC has been used in a limited fashion for specific design applications which gave rise to a general interest for its potential, particularly for computational design approaches. Readers will also notice that some of the concepts and technologies that have very limited or virtually zero impact on a specific life-cycle phase have been grayed out on Figure 23.4 accordingly. Moreover, it is essential to note the non-rigid boundary between on-site and off-site concepts and technologies under the smart construction cluster. Almost all of these concepts and technologies were discussed in detail throughout the earlier sections of the book. The contribution here is the correlation of adoption with life-cycle phases as well as the connection made to the diffusion concept.

Hall et al. (2014) investigate the speed of diffusion for product and process innovations in the building industry and highlight that innovations that fit the current supply chain structure (modular innovations) diffuse up to three times faster than the innovations crossing disciplinary boundaries (integral innovations). Perhaps the most striking manifestation of this phenomenon has been BIM. Despite having a wide spectrum of functionalities in various contexts from supporting collaborative design through early involvement of stakeholders to enabling building performance simulations and smarter facility management, BIM diffusion started from design by not necessarily disturbing any existing interface in the construction supply chain. Its promise of improvements upon many of the shortcomings of the earlier CAD methods diffused BIM into design, and over time it has proven to create tangible benefits in terms of key project objectives (cost, quality, time). Consequently, the diffusion to other life-cycle phases happened to the point that BIM is currently defined as an nD-based methodology, designed to integrate the entire building information along the life cycle of

Emerging Concepts and Technologies in Construction 4.0

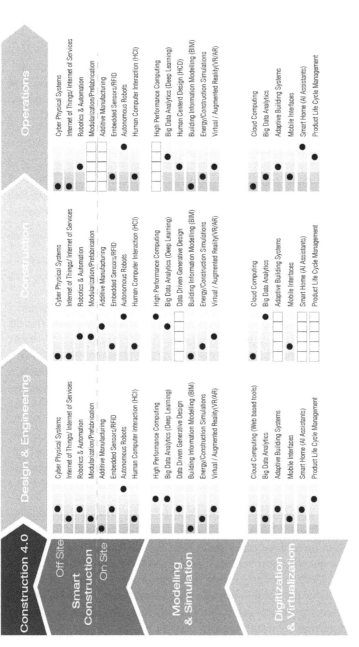

Figure 23.4 Categorization of concepts and technologies in Construction 4.0 into the clusters they belong in (e.g. Smart Construction Site) and corresponding stages of diffusion within different phases of the building life cycle

buildings, from design to construction, to operation and maintenance, and to reuse or dem-olition (Cheng and Ma 2013). In other words, the modular innovation (BIM for design) triggered an integral innovation that is promising to advance the industry beyond some of the vertical (BIM for all project phases) and horizontal fragmentation (BIM for disciplines/trades) challenges. Despite the fact that BIM and other technologies have been known con-cepts in the AEC industry, the authors believe that other key elements that relate to project delivery need to accompany technologies to overcome the deeply entrenched fragmentation issues, leading to the desired integral innovations in construction. Recommending IPD (Inte-grated Project Delivery) with a similar motivation, Hall et al. argue that a) *legal strategies* such as incentivized multi-party contracts guaranteeing cost reimbursement and providing fiscal transparency and flexibility, b) *management strategies* such as direct owner involve-ment and vision, team idea generation and lean construction principles as well as c) *work-place strategies* such as colocation, collaborative decision making and team accountability are key elements which can promote innovative solution that at the same time are more cost efficient than traditional project delivery methods.

Many of these arguments are also supported by Dallasega et al. (2018) in their work on Industry 4.0 and its potential to be the enabler of *proximity* for construction supply chains complementing Hall et al. in terms of the envisioned symbiosis between collaborative project delivery methods (or enhanced collaboration in general) and Construction 4.0. Dallasega et. al. highlight that one decisive characteristic of construction supply chains (CSCs) is that all involved actors have different distances, both physical and cognitive, to the location of pro-duction. Building on earlier work on proximity,[4] they assert 4.0 concepts and technologies have the power to bridge objective geographic distance while also being promising in improv-ing inter-organizational collaboration among actors in the construction supply chain. They also project 4.0 induced changes in subjective distances relating to organizational, cognitive, social, cultural, institutional, and technological proximities.

Another important aspect arising from studying Figure 23.4 is the coupling between the adoption of various concepts and technologies. For instance, for the adoption of Big Data Analytics in the construction phase, first rich construction datasets need to become availa-ble through the enhanced adoption of data generators such as IoT/IoS, RIFD, etc. For the realization of the potential of data analytics in construction, the challenge is the lack of open source, structured construction data as well as the scalability of insights based on the lim-ited data available because *every project is unique*. In a recent collaborative effort with a 1000-employee general contractor firm in California with a successful track record of serving high profile clients, the authors tried to apply data-driven approaches on productivity data from over 40 projects. So far, finding patterns have been challenging with project level attrib-utes that store aggregated data from weekly progress reports. For example, two projects with very similar characteristics that are built by the same contractor in the same neighborhood demonstrate significantly different productivity performances. This raised the question about the scalability of envisioned results and the longitudinal transferability of knowledge between projects given the complex, site-based nature of each undertaking. This research effort is still ongoing where the authors are expanding the data collection effort in terms of resolution while digging deeper into tasks and building elements rather than staying at project level attributes. Even then, Big Data Analytics (Deep Learning) methods may never be unlocked given the total number of past projects delivered by the firm. In this context, it is also significant to remember that a sector-wide data sharing initiative is needed that will circumvent the lack of financial incentives for data-sharing in the highly fragmented industry where cost competitive bidding still determines the project awardee.

Another coupling of interest is the relationship between robotics and automation, and modularization/prefabrication. Advancements in robotics and automation are projected to boost the adoption of modularization/prefabrication making off-site manufacturing a more viable alternative. Bechthold (2010) emphasized that the advances in robotics lowered the costs of construction robots and enhanced them with customized fabrication capabilities that add significant value. This is already benefiting off-site manufacturing of prefabricated units as discussed by Weinreich (2017) in his podcast titled "Tesla of Homebuilding". Weinreich's interview with the German machinery company Weinmann reveals that there are over 5,000 Weinmann machines operating in 150 homebuilding factories worldwide that are able to produce timber parts with respect to individually customized home layouts. The off-site manufacturing trend also demonstrated itself in the initial undertaking of Sidewalk Labs in Toronto, an Alphabet Group company (parent company of Google), where they laid out a vision for a holistic, integrated modular construction approach supported by a standardized parts library (apart from volumetric units) to achieve cost savings and product improvements that have not been captured by prior modular construction (Sidewalk Labs 2018). Overall, Sidewalk Lab's vision document for the Toronto Quayside project describes a comprehensive 4.0 undertaking with objectives ranging from net-zero energy community development to a suite of smart transportation systems supported by autonomous technologies. This integrated *from the internet-up community development* perspective could shift the paradigm enabling much faster adoption of 4.0 concepts and technologies in the future. Next, the authors discuss a number of research directions that are considered as promising ways to address issues with slow adoption of innovative technologies in the AEC industry.

A macro trend relating to Construction 4.0 is the targets proposed by governments related to overall efficiency of the industry both for the construction and operation phases. One of the more ambitious documents is the UK Construction Strategy that require the industry to dramatically improve its performance in four key areas by 2025 including lowering greenhouse gas emissions in the built environment by 50%, reducing the initial cost of construction and the whole life cost of built assets by 33%, reducing the overall time, from inception to completion, for new-build and refurbished assets by 50% and improving exports by 50% (Abanda et al. 2017). Interestingly, BIM and modularization/prefabrication are precisely suggested as opportunities to reach at these very ambitious targets. In California, all state buildings are planned to be zero net energy by 2025 and the state has stringent codes in regard to energy efficiency for newly built homes. In general, such initiatives enforced or motivated by legislation should speed up the diffusion of 4.0 innovations since they are considered necessary tools and concepts for designing, constructing and operating more efficiently.

23.5 Research directions

In this section, a number of research directions are discussed at a higher level than the individual concepts and technologies since previous chapters went into detailed considerations with that purpose. Again, perspectives presented on concepts of diffusion and fragmentation give direction to the thinking summarized here.

First, there is an obvious need for research to expand on integrated project management and delivery approaches. Earlier in the chapter, IPD was mentioned as a new promising project delivery method addressing key horizontal and vertical fragmentation issues. Specifically in "Construction 4.0" context, the authors assert that—by enabling various types of proximities discussed in Section 23.2—the convergence of actors and incentives through IPD will help facilitate the sharing of benefits (through multi-party contracts and guarantees

(Henisz et al. 2012)) associated with the adoption of 4.0 innovations in the industry. For example, in their work on construction simulations, Abourizk et al. (2011) argue that construction simulations have largely been limited to academic applications despite the beneficial use cases such as using simulations for designing construction operations and processes or experimenting with different scenarios to allow facility managers optimize their processes and achieve much higher levels of efficiency. According to the Abourizk et al., as the gap between the real and virtual worlds is closing, the realization of a construction project may take place in a virtual world with all details of its scope and constraints defined before or in parallel to its realization in the real world. For this, they emphasize the necessity of modeling and simulation platforms spanning the full life cycle of construction projects from design to construction and even operation, and that can accommodate virtually all the modules (e.g. tools for design, energy simulations, scheduling, etc.). In this junction, BIM and IPD appear as natural partners that could unleash integration across project phases, project actors and specific 4.0 concepts and technologies leveraged in a project. However, there is a lack of interdisciplinary research efforts and in the development of data-driven integrated frameworks that look beyond the boundaries of specific disciplines. Such frameworks can enable seamless collaboration between stakeholders, clients, policy makers, architects, engineers and contractors (*end to end integration of engineering across the entire value chain via information technologies*) by critically evaluating the state of the art in each discipline. In Figure 23.5, the authors graphically show Construction 4.0 concepts and technologies in terms of their "paths" across stages of innovation diffusion in time starting from their initial conception into today and to the future. The graph suggests that innovation and integration in the construction supply chain will go hand-in-hand with the convergence of technologies onto integrated project delivery settings.

Borrowing from the example of the rapidly growing gaming industry, BIM software and other design platforms have the potential to be disruptive and thus to be successful when they are "simple to learn but impossible to master" (Montfort and Bogost 2009). On the contrary, if design tools are difficult to use from the start, users tend to be discouraged and find

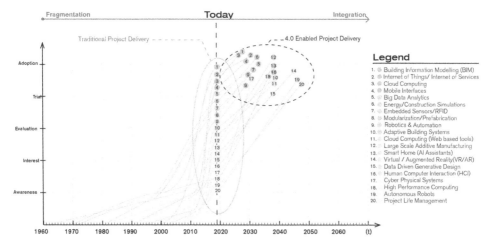

Figure 23.5 Construction 4.0 concepts and technologies in terms of their "paths" across stages of innovation diffusion in time, and the envisioned convergence of technologies onto integrated project delivery settings in the upcoming decades

it hard to engage with them as recent history has shown. In the last decade, there has been an increase in open source libraries and design tools which are developed from the bottom up (i.e. architects, engineers) by using open APIs of existing software as stepping stones. Such attempts have been successful because they have been created by computationally savvy architects and engineers within the industry. Moreover, they are structured in a way that their users can arrive at tangible results relatively quickly, albeit some results being not extremely accurate. Through continuous interaction, users delve into the tool to improve upon preliminary results while advancing their knowledge on the suite of tools and educating themselves in a particular area of expertise (i.e. environmental modeling). Overall, the developments associated with Industry 4.0 suggest a shift towards the development of an ecology of toolkits rather than a "one solution fits all approach" which is the current paradigm set forward by big software companies dominating the AEC industry. On the other hand, open access to data such as weather or building occupants' data are quickly changing the status quo of building modeling and simulation as data-driven approaches provide a more direct way of evaluating design intuition or the development of new design versions.

Figure 23.5 demonstrates the author's prediction of how innovation adoption in the AEC industry will progress in a symbiotic fashion with integration which will rapidly facilitate the appropriate circumstances for the wide application of Construction 4.0 concepts and technologies. The timeline envisioned may not happen exactly, but the change appears inevitable and the multiple facets of the transformation of the construction industry have been elaborately described in the previous chapters this book. However, one critical question arises: Who is going to be leading this change and what are the roles of designers and engineers therein? To answer this question, the importance of architecture and civil (construction) engineering education and the capacity of academic institutions to help students develop the diverse set of skills required to address the challenges of the 4.0 era becomes highly relevant and need to be seriously reconsidered. There is an obvious educational shortcoming that has not revealed itself yet in a significant manner as the industry has been reluctant in widely adopting a large-scale disruption until now. However, disruption waves are reaching the shores of the industry often coming from actors of other industries with different skillsets who are entering the AEC market with initiatives targeting large-scale development projects (e.g. Alphabet's Sidewalk Labs and its Toronto Quayside project) and partnerships that are based upon 4.0 concepts and technologies with the objectives of achieving more sustainable building designs, lower construction costs and improved productivity (e.g. IKEA and Skanska partnering to build low-cost, sustainable prefabricated homes[5]). By rapidly advancing towards digitizing construction, providing end-to-end building services (using outsourcing if necessary) and design automation/simulation, such actors already seem to be changing the industry. These early warnings of the mentioned large-scale disruption triggered by the forces of digitization, industrialization and globalization that is well under way. Moreover, "grand challenges" of the 21st century demand architects and engineers to cross boundaries between disciplines as they are solving problems and target holistic approaches to accommodate considerations of sustainability, resilience, cyber security, etc. The ability to develop a deeper understanding of theories, methods and tools outside their immediate domains (complexity theory, biomimetics, synthetic biology to name only a few) (Knippers et al. 2016) will increasingly be a distinguishing factor.

Although there are few institutes and graduate degree programs established in North America and Europe that are dedicated to the investigation of how aforementioned concepts are changing the way we design and construct in the digital age, the curricula do not satisfy these requirements (e.g. Santa Fe Institute, Wyss Institute/Harvard, Mediated Matter Group/ MIT, ITA/ETH Zurich, Institute for Computational Design/ITKE Stuttgart). In Figure 23.6,

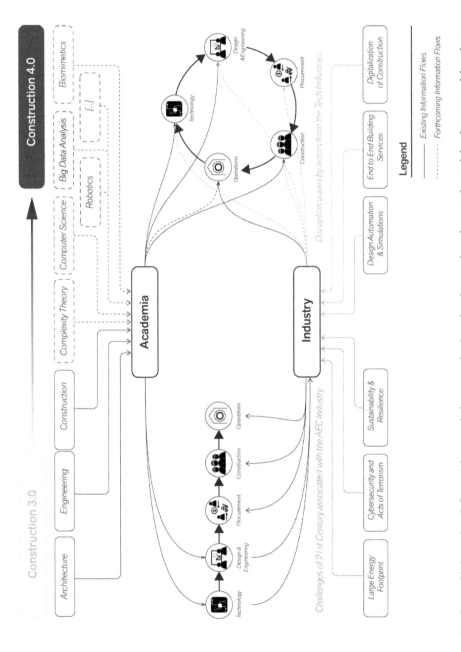

Figure 23.6 The flows of information (influence) between industry, academia and various project phases today and the changes envisioned to occur as the industry moves forward to a 4.0 state. Disruption waves originating in tech industries as well as the challenges of the 21st century are describing the disciplines to be integrated in the next generation curricula will be necessary

the authors visually tell the story of this interplay between industry of today and of the future as well as the desired advancements in academia. The idea is that today's design and engineering education need to allow for more cross disciplinary information flow and welcome sciences that are traditionally left out into the curricula to enable unique learning opportunities.[6] Otherwise, the threat is that architects and engineers could lose agency in project development to platforms developed by tech companies penetrating into the AEC industry and will become sidelined losing the opportunity to lead the 4.0 revolution. If disruptive enterprises grab market share and shake up existing business models, academics striving to prepare students to be employment-ready may be late in opening their eyes to the business models that are quite different from traditional construction companies (Hardie 2018).

The authors perceive the rapid transformation of ride-hailing services industry as an analogy and a warning. Taxi drivers were initially able to capitalize on telecommunications (radio) which expanded their business and made it easier for taxi riders to find rides. However, due to their slow adoption of further innovation required by the wide availability of smart phones, fully connected ride-hailing apps are threatening the bare existence of taxis. Capturing major market shares in hundreds of cities around the world, such apps are rendering taxi drivers obsolete. Stakeholders in the industry and the academia need to understand a potentially similar threat to reveal itself should the AEC industry miss the train on transforming itself.

23.6 Conclusion

"Construction 4.0" concepts and technologies will drive AEC innovation in the near future and the industry could finally face the large-scale disruption—from within or elsewhere—it has avoided. In this chapter, the authors intended to draw attention to the need of understanding the process of innovation diffusion in the AEC context. The industry has its characteristic challenges hindering innovation, however, there are also many factors creating an innovation pull. To track the emerging trends, there is a need to take *a multilevel, longitudinal perspective, and follow events implicating actors, artefacts, and institutions over time* (Garud et al. 2013). In light of the discussions in this chapter, innovation diffusion can be studied and perceived from a multiplicity of viewpoints: (1) from the interface between the "firm" and the industry it belongs to by understanding the dynamics and challenges of the diffusion in relation to technology standards and best practices, (2) from interface between the "firm" and its clients, and how their interaction influences technology related decision-making for individual projects, (3) from the interface between the "firm" and the technology providers to capture the advancements in individual technologies as well as higher level changes in paradigms. The methodology to understand the innovation diffusion process in construction should assume from the outset that digital innovation happens through a series of complex acts rather than a single one and that within the chronology of the innovation diffusion, nonlinearity and iteration may be present and should be accounted for (e.g. BIM as an idea being conceived decades before large-scale diffusion happened in practice). Lastly, the role of architects and engineers in the 4.0 revolution is a point of emphasis given the backdrop of disruption created by actors of other industries who are trying to gain market share by tackling the long-standing problems of the AEC industry. In this context, the authors emphasized that AEC education needs to open doors to theories, methodologies and tools from other domains. This is also necessitated by the grand challenges of this century highlighted as forces pushing the industry and academia towards change.

23.7 Summary

- Wider adoption of 4.0 concepts and technologies in the AEC industry in combination with legislation motivated by "grand challenges" are motivating faster diffusion of innovation. A vision on how this will progress was presented.

- With the described diffusion, the industry will move towards more integration—away from the inefficiencies of fragmentation—across project phases, disciplines and multiple projects. More collaborative project delivery methods are expected to play a significant role in this integration.

- The role of architects and engineers in "Construction 4.0" is a critical point of emphasis in terms of academic training. There is a need to open up curricula to teach "data" skills in a more enhanced fashion as well as to accommodate other fundamental sciences that are traditionally left out of architecture and engineering education. This diverse preparation is necessary for architects and engineers to lead the evolution of the construction industry and adapt to the challenges of this century.

Notes

1 Idea is loosely used for a new idea, product, or process.
2 In the case of BIM (Building Information Modeling), there is a general consensus in the industry on the benefits and there are many studies quantifying such benefits for various projects with different levels of BIM adoption. These are sufficient conditions for "adoption" under Beal and Bohlen's characterization of the stages and it does not require BIM to be used in the majority of the construction projects.
3 Rate of adoption is understood as a numeric indicator, e.g. number of heavy civil contractors adopting Building Information Modeling per year.
4 Proximity research identifies the distance (physical or cognitive) between two or more entities as a major determinant of knowledge transfer, innovation, and inter-organizational cooperation.
5 *BoKlok* is a housing concept, developed by Skanska and IKEA, www.boklok.com/. *Katerra* is another company revolutionizing the industry with off-site manufacturing, www.katerra.com/en.html.
6 An analogy could be made with "Taylorism" or "scientific management" approaches in the early 19th century when pioneers adopted a similar perspective to improve economic efficiency and labor productivity. For instance, the well-known Gannt chart was discovered and brought into the project management of large public infrastructure projects such as the Hoover Dam.

References

Abanda, F. H., Tah, J. H. M., and Cheung, F. K. T. (2017). "BIM in Off-site Manufacturing for Buildings." *Journal of Building Engineering*, 14(October), 89–102.

Abourizk, S., Halpin, D., Mohamed, Y., and Hermann, U. (2011). "Research in Modeling and Simulation for Improving Construction Engineering Operations." *Journal of Construction Engineering and Management*, 137(10), 843–852.

Arayici, Y., and Coates, P. (2012). "A System Engineering Perspective to Knowledge Transfer: A Case Study Approach of BIM Adoption." *Virtual Reality – Human Computer Interaction*, 179–206. IntechOpen, DOI: 10.5772/51052. Available at: www.intechopen.com/books/virtual-reality-human-computer-interaction/a-system-engineering-perspective-to-knowledge-transfer-a-case-study-approach-of-bim-adoption

Beal, G. M. and Bohlen, J. M. (1957). "The Diffusion Process." *Signal Processing*, 1(1), 1–6.

Bechthold, M. (2010). "The Return of the Future: A Second Go at Robotic Construction." *Architectural Design*, 80(4), 116–121.

Boston Consulting Group. (2016). "Digital in Engineering and Construction." The Boston Consulting Group.

Cheng, J. C. and Ma, L. Y. (2013). "A BIM-Based System for Demolition and Renovation Waste Estimation and Planning." *Waste Management*, 33(6), 1539–1551.

Choi Granade, H., Creyts, J., Derkach, A., Farese, P., Nyquist, S., and Ostrowski, K. (2009). *Unlocking Energy Efficiency in the U.S. Economy*. McKinsey Global Energy and Materials. Available at: www.mckinsey.com/~/media/mckinsey/dotcom/client_service/epng/pdfs/unlocking%20energy%20efficiency/us_energy_efficiency_exc_summary.ashx

Dallasega, P., Rauch, E., and Linder, C. (2018). "Industry 4.0 as an Enabler of Proximity for Construction Supply Chains: A Systematic Literature Review." *Computers in Industry*, 99(April), 205–225.

Dubois, A. and Gadde, L. E. (2002). "The Construction Industry as a Loosely Coupled System: Implications for Productivity and Innovation." *Construction Management and Economics*, 20(7), 621–631.

Fergusson, K. J. (1993). *Impact of Integration on Industrial Facility Quality*. Stanford, CA: Stanford University.

Garud, R., Tuertscher, P., and Van de Ven, A. H. (2013). "Perspectives on Innovation Processes." *Academy of Management Annals*, 7(1), 775–819.

Hall, D., Algiers, A., Lehtinen, T., Levitt, R. E., Li, C., and Padachuri, P. (2014). "The Role of Integrated Project Delivery Elements in Adoption of Integral Innovations." EPOC 2014 Conference, (July), 1–20.

Hardie, M. (2018). "Preparing Students for a Disruptive Construction Future." In M. Lamb (ed.), AUBEA 2017: Australasian Universities Building Education Association Conference 2017, vol. 1, pp. 70–77.

Henisz, W. J., Levitt, R. E., and Scott, W. R. (2012). "Toward a Unified Theory of Project Governance: Economic, Sociological and Psychological Supports for Relational Contracting." *The Engineering Project Organization Journal* (March–June 2012), 2, 37–55.

Kats, G., Alevantis, L., Berman, A., and Perlman, J. (2003). "The Costs and Financial Benefits of Green Buildings A Report to California's Sustainable Building Task Force." Capital E.

Knippers, J., Nickel, K. G., and Speck, T. (2016). *Biomimetic Research for Architecture and Building Construction*. Basel, Switzerland: Springer International Publishing, Springer.

Kraatz, J. A., Hampson, K. D., and Sanchez, A. X. (2014). "The Global Construction Industry and R&D." In J. A. Kraatz, K. D. Hampson, and A. X. Sanchez (eds), *R&D Investment and Impact in the Global Construction Industry*, London: Routledge, 4–23.

Lahdenperä, P. (2012). "Making Sense of the Multi-party Contractual Arrangements of Project Partnering, Project Alliancing and Integrated Project Delivery." *Construction Management and Economics*, 30(1), 57–79.

Lasi, H., Fettke, P., Kemper, H.-G., Feld, T., and Hoffmann, M. (2014). "Industry 4.0." *Business & Information Systems Engineering*, 6(4), 239–242.

Montfort, N., and Bogost, I. (2009). *Racing the Beam: The Atari Video Computer System*. Cambridge, MA: MIT Press.

Obayashi, S. (1999). *Construction Robot System Catalogue in Japan, Council for Construction Robot Research*. Tokyo: Japan Robot Association.

Oesterreich, T. D. and Teuteberg, F. (2016). "Understanding the Implications of Digitisation and Automation in the Context of Industry 4.0: A Triangulation Approach and Elements of A Research Agenda for the Construction Industry." *Computers in Industry*, 83, 121–139.

Rogers, E. M. (2003). *Diffusion of Innovations*, 5th Edition. New York: Free Press.

Sheffer, D. A. (2011). "Innovation in Modular Industries: Implementing Energy-Efficient Innovations in US Buildings." Dissertation.

Sidewalk Labs (2018). Sidewalk Toronto. www.sidewalktoronto.ca.

Taylor, J. E., Levitt, R. E., and Fellow, L. G. (2006). "Understanding and Managing Systemic Innovation in Project-based Industries." *Innovations: Project Management Research, 83–99.*

Taylor, M., Wamuziri, S., and Smith, I. (2009). "Automated Construction in Japan." *Proceedings of the Institution of Civil Engineers – Civil Engineering*, 156(1), 34–41.

Teece, D. J. (1996). "Firm Organization, Industrial Structure, and Technological Innovation." *Journal of Economic Behavior and Organization, 31(2), 193–224.*

Weinreich, A. (2017). "Predicting Our Future Episode 6: The Tesla of Homebuilding." *Stitcher*, stitcher.com/podcast/predicting-our-future/e/49745756?autoplay=true (June 5, 2019).

ACRONYMS

3D	Three Dimensional
3DCP	3D Concrete Printing
4D	Four Dimensional
4IR	Fourth Industrial Revolution
ABS	Acrylonitrile-butadiene-styrene
ADC	Analog-to-Digital Converters/Automated Data Collection
AEC	Architecture, Engineering & Construction
AEC/FM	Architectural, Engineering, Construction, and Facilities Management
AECO	Architecture, Engineering, Construction, and Operations
AECO-FM	Architecture, Engineering, Construction, Operations and Facility Management
AI	Artificial Intelligence
AIM	Asset Information Model
AIST	National Institute of Advanced Industrial Science and Technology
ALS	Airborne Laser Scanning
AM	Additive Manufacturing
ANAC	Agência Nacional de Aviação Civil (National Agency for Civil Aviation)
ANATEL	Agência Nacional de Telecomunicações (National Agency of Telecommunication)
ANSI	American National Standards Institute
API	Application Programming Interface
AR	Augmented Reality
ASCE	American Society of Civil Engineers
ASTM	American Society for Testing and Materials
AUVSI	Association for Unmanned Vehicle Systems International
AWP	Advanced Work Packaging
BAAM	Big Area Additive Manufacturing
BACnet	Building Automation and Control networks
BAS	Building Automation System
BCA	Building Construction Authority
BCF	BIM Collaboration Format

BCVTB	Building Controls Virtual Test Bed
BEMS	Building Energy Management Systems
BEP	BIM Execution Plan
BIM	Building Information Modeling
BLE	Bluetooth Low Energy
BMS	Building Management System
BOD	Building on Demand
BOT	Building Topology Ontology
BSI	British Standards Institution
BSRIA	Building Services Research and Information Association
C&D	Construction and Demolition
C4.0	Construction 4.0
CAA UK	Civil Aviation Authority-United Kingdom
CAD	Computer-Aided Design
CAM	Computer-Aided Manufacturing
CAVE	Cave Automatic Virtual Environment
CC	Cloud Computing/Contour Crafting
CDE	Common Data Environment
CI	Cincinnati Incorporated
CIM	Crane Information Models
CIOB	Chartered Institute of Building
CMES	Construction Methodology and Erection Sequence
CMMS	Computerized Maintenance Management System
CNC	Computer Numerically Controlled
CNN	Convolutional Neural Network
CO	Carbon Monoxide
COBie	Construction Operations Building Information Exchange
COBOD	Construction Building on Demand
CPCS	Cyber-Physical Construction Systems
CPM	Critical Path Method
CPPS	Cyber-Physical Production Systems
CPS	Cyber-Physical Systems
CRF	Conditional Random Fields
CSC	Construction Supply Chain
CSCMM	Construction Supply Chain Maturity Model
CSV	Comma-Separated Values
DAO	Decentralized Autonomous Organization
DCP	Digital Construction Platform
DDoS	Distributed Denial of Service
DED	Directed Energy Deposition
DfMA	Design for Manufacture and Assembly
DfS	Design for Safety
DfSK	Design for Safety Knowledge
DfSL	Design for Safety Language
DGAC	Direccion General de Aeronautica Civil (General Direction for Civil Aviation)
DIKW	Data, Information, Knowledge, and Wisdom
DMS	Document Management System

DMVR	Dense Multi-View 3D Reconstruction
DNS	Domain Name Server
DoS	Denial-of-Service
DWG	AutoCAD Drawing Database file extension
EASA	European Aviation Safety Agency
EER	Enhanced Entity – Relation
EIR	Equipment Interchange Receipt
EKF	Extended Kalman Filtering
EKF	Extended Kalman Filtering
EPC	Electronic Product Code
ER	Exchange Requirement
ERP	Enterprise Resource Planning
ES	Embedded Systems
EU	European Union
EVM	Earned Value Methodology
FAA	Federal Aviation Administration
FBR	Fast Brick Robotics
FDD	Fault Detection and Diagnosis
FDM	Fused Deposition Modelling
FM	Facilities Management
GA	Genetic Algorithm
gbXML	Green Building XML
GCP	Ground Control Points
GDP	Gross Domestic Product
GDPR	General Data Protection Regulation
GIS	Geographic Information Systems
GMAW	Gas-Metal Arc Welding
GNSS	Global Navigation Satellite System
GPS	Global Positioning Systems
HF	High Frequency
HMD	Head Mounted Display
HMI	Human and Machine Interaction
HPR	Humanoid Robot Prototype
HRC	Human-Robot Collaboration
HTML	Hyper Text Markup Language
HVAC	Heating, Ventilation, and Air Conditioning
I4.0	Industry 4.0
IAAC	Institute for Advanced Architecture of Catalonia
IAARC	International Association for Automation and Robotics in Construction
ICP	Iterative Closest Point
ICS	Industrial Control System
ICT	Information and Communication Technology
IDM	Information Delivery Manual
IEC	International Electrotechnical Commission
IFC	Industry Foundation Classes
IFD	International Framework for Dictionaries
IFR	International Federation of Robotics
IMU	Internal/Inertial Measurement Unit

IoP	Internet of People
IoS	Internet of Services
IoT	Internet of Things
IP	Internet Protocol
IPD	Integrated Project Delivery
ISA	International Society of Automation
ISARC	International Symposium on Automation and Robotics in Construction
ISO	International Organization for Standardization
IT	Information Technology
ITU	International Telecommunication Union
JSON	JavaScript Object Notation
k-NN	k-Nearest Neighbors
KPI	Key Performance Indicator
LADAR	Laser Distance and Ranging
LAR	Living Augmented Reality
LCCA	Life-cycle Cost Analysis
LCMM	Lean Construction Maturity Model
LF	Low Frequency
LiDAR	Light Detection and Ranging
LOD	Level of design/Linked (Open) Data
LOS	Line of Sight
LPDS	Lean Project Delivery System
LPS	Last Planner System™
LSTM	Long Short-Term Memory
M&E	Mechanical and Electrical
M2M	Machine-to-Machine
MAR	Mobile Augmented Reality
MCSC	Marine Corps Systems Command
MDN	Mixture Density Network
MEP	Mechanical Electrical and Plumbing
MI5	UK's Military Intelligence Service, Section 5
MI6	UK's Military Intelligence Service, Section 6
MIS	Management Information System
ML	Machine Learning
MLS	Mobile Laser Scanning
MmTSP	Multi-depot Multiple Travelling Salesman Problem
MNM	Marker Network Map
MR	Mixed Reality
MTOW	Maximum Take-off Weight
MTP	Multi-trade Prefabrication
mTSP	Multiple Travelling Salesman Problem
MVD	Model View Definition
MVS	Multi-View Stereo
NAS	National Airspace System
NCCR	Swiss National Centre of Competence in Research
NIBS	National Institute of Building Sciences
NIST	National Institute of Standards and Technology
NP	Nondeterministic Polynomial

O&M	Operations and Maintenance
OPC	Ordinary Portland Cement
ORNL	Oak Ridge National Laboratory
OWL	Web Ontology Language
PAS	Publicly Available Specification
PBF	Powder Bed Fusion
PC	Polycarbonate
PDN	Project Delivery Network
PERT	Program Evaluation and Review Technique
PLA	Polylactic Acid
PLM	Project Lifecycle Management
PMC	Project Management Consultant
PMI	Project Management Institute
PPC	Percentage Plan Complete
PPE	Personal Protective Equipment
PSD	Property Set Definition
PtD	Prevention through Design
QR	Quick Response
RANSAC	RANdom SAmple Consensus
RDF	Resource Description Framework
REST	Representational State Transfer
RFI	Request for Information
RFID	Radio Frequency Identification
RGB	Red, Green, and Blue
RISI	Repository of Industrial Security Incidents
ROI	Return on Investment
ROS	Robotic Operation System
RPA	Remote Pilot Aircraft
RTLS	Real-time Location Sensing System
SAM	Semi-Automated Mason
SCG	Siam Cement Group
SCM	Supply Chain Management
SD	System Dynamics
SDF	Signed Distance Fields
SfM	Structure-from-Motion
SHM	Structural Health Monitoring
SIFT	Scale Invariant Feature Transforms
SLAM	Simultaneous Localization and Mapping
SLM	Selective Laser Melting
SLS	Selective Laser Sintering
SME	Small and Medium-sized Enterprise
SoC	System on Chip
SRDBMS	Spatially Enhanced Relational Database Management System
SSH	Secure Socket Shell
STEP	Standard for the Exchange of Product
STR	Single-task Robot
sUAS	Small Unmanned Aerial System
SVM	Support Vector Machine

TC	Tower Crane
TCP/IP	Transmission Control Protocol/Internet Protocol
TLS	Terrestrial Laser Scanning
TOF	Time of Flight
TSA	Transportation Safety Administration
TSP	Travelling Salesman Problem
UA	Unmanned Aircraft
UAE	United Arab Emirates
UAS	Unmanned Aerial System
UAV	Unmanned Aerial Vehicles
UHF	Ultra High Frequency
UHPFRC	Ultra-High-Performance Fibre-Reinforced Concrete
UK	United Kingdom
UML	Unified Modeling Language
URI	Uniform Resource Identifier
URL	Uniform Resource Locator
USA	United States of America
USC	University of Southern California
UV	Ultraviolet
UWB	Ultra-Wide Band
VDC	Virtual Design and Construction
VLOS	Visual Line of Sight
VR	Virtual Reality
VRD	Virtual Retinal Display
VRP	Vehicle Routing Problem
VSM	Value Stream Mapping
VTOL	Vertical Take-off and Landing
WAAM	Wire and Arc Additive Manufacturing
WASP	World's Advanced Saving Project
WBS	Work Breakdown Structure
WEF	World Economic Forum
WIP	Work in Progress
WLAN	Wireless Local Area Network
WSN	Wireless Sensor Network
XML	Extensible Markup Language
XSD	XML Schema Definition

INDEX

2.4 G 354, **354**

360 Images/Videos/Camera 5, 241, *244*, 245, 248, 254, 257

3D: environments 132, 230; geometry 29, 164, 179, 227, 235, 245, 248–249, 252, 269; materials 157–164, 176; modeling 148, 150, 245; printing 5, 13, **18**, 25, 97, 155–156, 164–167, 172–173, 177–179, 192, 199, 204, 292, 307; robotics 159, 165, 167, 180–*181*, 189, 194, 198, 293; scanners *118*, 143–144, 150, 211, 226, 269

access information 210

active tag and passive tag 122, 353

activity recognition 251, 255

adaptation 9, 70, 73, 78, 80, 82, *86*, 144, 192, 327, 351, 443

additive manufacturing *14*, 16, **18**, 44, 106, 155–187, 189, 192, 198, 462–463

AEC/FM industry 245, 291, 301, 303, 326–328, 347

AECO 25, 131, 143–44, 147, 150

aerial construction 274, 279

agility 63, 66, 71, *86*, 188, 265, 372

airspace 265, **272**

alerts 32, 66, 70, *72*, 73–74, 76, 78–79, *86*, 357, 383, 435, 437

algorithms 10, 29, 38, 69, 76, 92, 95, 101, 113, 120, 124–125, 147–148, 150, 175, 198, 243, 246–249, 253–254, 279, 310–312, 315–321, 336, 351, 399, 401, 405–406, 408, 411

algorithmic image processing 143

analytics 6, 9, 13, *14*, 16, **18**, 23, 37, 49, 57, 69, 72, 77–82, 132, 240, 242–244, 249, *256*, 257–258, 307, 351, 355, 367, 372, **378**, 381–386, **393**, 399, 437, 442, 462–463, 467

appearance-based progress monitoring 252

AprilTags *313*, 314, 316

architecture, engineering & construction **18**, 75, 91–97, 99, 104, 107–109, 179, 188–190, 192–195, 199–201, 204–205, 222–226, 229–231, 237, 245, 291, 301, 326–328, 347, 355, 384, 440, 460, 462–467, 469–470, 475; and owner-operated 442

artificial intelligence 13, 32, 307, 375, 447, 462–463

as-built 93, 97, 114, 115, 127, 128, 146, 147, 150, 151, 211, 218, 245, 253, 255, 356, 451; models 114–115, 117, 248; BIM 142, 147–148; state 211, 218

as-designed 211, 218; information 150; models 148; states 211, 218

as-planned design 117

asset 5, **14**, 24, 47, 65, 69, **72**, 74, 80, **82**, 84, 85, 87, 150, 181, 255, 327, 357, 397, 407, 409, 412, 437, 441, 442, 443, 445, 446, 447, 451, 452, 453; information models (AIM) 443

augmented reality (AR) 23, 38–39, 131

automated progress monitoring 241, 254

automation 9, 11–13, *14*, 16, *18*, 25–28, 47, 55, 56, *57*, 97, 101, 107, 113, 131, 142, 148, 150, 156–158, 174–176, 189–194, 231, 249, 279, *289*, 290–295, 299, 301, 309, 322, 326, 350, 371, 386, 396, 402, 409, 412, 462, 465, 470, 472; in construction **18**, 101–102, 192, 290, *291*

autonomous 23–24, 92, 100–101, 104, 107–09, 146, 156, 165, 197, 265–268, **273**, 275, 279, 293, 297–299, 308–309, 312–13, *314*, 316–17, **320**, 322, 347, 400, 410–413, 434, **435**, 443, 462–463, 470

BEP 397, 398

bi-directional coordination 113, 117, *118*, *119*, 120, 123, 125, 127, 128, 284 big data 5, 6, 9, 13, *14*, 65, 81, *86*, 211, 264, 291, 322, 350–355, 367, 399, 455, 462, 465, 469

big visual data 240

BIM (Building Information Modeling) 475; Collaboration Format (BCF) 228; data 26, 76, 136, 147, 148, 225, 233, 237, 437, 452; implementation 69, 76, 209, *214*, 215, 218

Black Hats 449

blockchain 6, 13, *14*, 16, **18**, 66, 222, 237, 307, 395–413, 441, 444, 455, 462

bluetooth 98, 212, 312, 313, 354

building 5, 18; automation system (BAS) 28; elements 25, 159, 163, 180, 210, 211, 218, 226, 292, 469; energy management systems (BEMS) 28; information modeling (See BIM) 5, 25, 47, 57, 70, 135, 210, 212, 223, 240, 249, 328, 437, 441, 475; inspection 286; management system (BMS) 29, 412; managers 127; monitoring **18**, 27, 134, 139, 146–47, 151, 212, 217, 240, 253, 278–79, 356, 358, **361**

built environment 3, 5, 14–18, 23, 26, 30–32, 38, 71, 115, 150, 204, 240, 269, 302, 308–310, 314, 328, 395, 399, 401, 408–412, 442–448, 450, 470

camera 134, 150, 190, *191*, 200, 245, *247*, 248, 254, 257, 268, 270, 297, 313, 315, 317, 319, 321, 356

capture 6, 13, 25, 38, *54*, 55, *56*, 64, 69, 72, 80, 114–116, 133, 143, *145*, 146–150, 195, 211, 226, 241, *243*, 244–250, 254–257, 265, 269, 330, 347, 356, **377**, **378**, 382, 388, **393**, 438, 445, 474

carbon monoxide tunnel 359

case study 17, 80, 91, 101, 102, 120, 173, 183, 195, 199, 240, 242, 244, 255–259, 307, 316, 319, 321, 333, 337, 344, 350, 359, 367, 396, 421, 426, 434, 438–444

centralized information control system 294

circular economy 9, 399, 400, 409, 410, 412

Civil Engineering 106, 252, 290, *291*, 293

civil infrastructure systems 116

cloud based collaboration **18**, 370, 372, 375, 376, 387, 388, 389

cloud computing *5*, 6, 9, 13, 69, 95, 291, 328, 350, 355, 367, 375, 376, 396, 450, 462, 463, 465

collaboration framework 374, 380

collaborative project management 54, 370, 372, 379, 380, 381, 384, 386, 389, 390

collaborative virtual environments 123, 137, 139

common data environment (CDE) 13, *14*, 43, 58, 59, 70, 71, 77, 83, 387, 441, 443, 445

communication 16, **18**, 19, 23, 27, 38, 96, 100, 113, 116, 119, 127, 131, 136, 139, 202, 212, 228, 237, 242, 258, 268–271, 276, 317,

350–355, 358–360, *361*, *362*, 367, 374, 382, 397, 406, 427, 437, 445; inter-device 92, 95, 109, 223

competition 50, 83, 292, 386, 404

compliance checking processes 341

computer vision 143, 240, 241, 242, 244, 245, 248, 249, 250, 251, 252, 254, 255, 257, 258, 259

computerized maintenance management system (CMMS) 28

conditions and constraints 331, 333

constructability 26, *68*, *121*, 137, 181, 182, 193, 210, 218, 243, 257, **378**

construction automation 113, 116, 176, 192, 289, 290, 291, 292, 293, 294, 295, 299, 303

construction coordination 262

construction engineering management 113, 126, 156, 326

construction information supply chain 375

construction life-cycle phases 309, 464, 467

construction management *54*, 64, 357, 367, 396, 422

construction materials 114, 159, 252, 253, 255

construction monitoring 134, 278, 279

construction personnel 119

construction process *14*, 94, 96, 99, 101, 104, 106, 109, 113–119, 128, 131, 155, 170, 189, 196, 209, 252, 258, 264, 279, 292, 342, 350, 357, 388, 396, 399, 423

construction production control 71, 390

construction productivity 303

construction progress; management 114; monitoring 27, 38, 135, 139, 409

construction project; management 371, 374, 380, 381; planning **377**, 281

construction requirements 341, 342

construction resource tracking 250

construction risk management 330, *332*, 333, *335*, 336, 339, 442, 444, 452, 454

construction safety 27, 117, 137, 138, 327, 330, 350, 367

construction site 26, 101, 108, 123, 146, 169–171, 180, 197, 242, 245, 248, 252, 254, 257, 265–268, 274–280, 284, 289, 291, 293–295, 298, 322, 359, 365, 421, 440, 445; management 437

construction supply chain 76, 303, 328, 370, 371, 375, 376, 384, 409, 464, 467, 469, 471; maturity model 384

construction technology 70, 75, 241, 258, 291, 295, 390

construction value chain 328, 347

contracts 79, 399; computational 400, 409; construction 398–399; smart 18, 66, 398, 399, 400, 402, 406, 407, 409, 410, 453

contractor training 113, 127, 170

controlled factory environments 209

controls 53, *82*, 271, 374; closed-loop 92, 97, 106, 267; systems 32, 36, 231, 294, 445, 449, 450; technologies 113, 116
coordination 6, 16, **18**, 50, 67, 99, 113, 115, 117, *119*, 120, 123, 125, 127, 171, 179, 202, 204, 209, 212, 217, 229, 242, 255, *256*, 257, 278, 282, 284, **377**, 379, 380, 396, 398, 427
coordinator 121, 122, 258, 354, 360, **361**, *362*, 363
cost management 54, 66, 146
CPCS sensing system 116, 119
crane Information model 329, 341, 342, 343, 344, 345, 347
crane safety planning 329
critical; infrastructure 441, 443, 444, 446, 448, 449, 450, 452, 454, 455; path method 371, 421
culture 7, 62, 63, 64, 65, 69, 72, 74, 77, 85, *86*, 108, 115, 196, 327, 373, 375, 379, 390, 399, 452, 465
Cyber 9; actors 449; attack 127, 443, 444, 448, 449, 454; crime 441, 443, 447, 448, 450, 451, 454; deterrence 441, 452, 453, 454; physical ; reconnaissance 282, 283, 442, 449; security 9, 14, 17, 301, 302, 303, 308, 442, 443, 444, 447, 450, 451, 452, 453, 454, 457; space 441, 448; threats 18, 303, 441, 444, 454; vulnerabilities 322

DAO 410, 411, 413
dashboard *72*, 73, 82
data acquisition technologies 30, 97, 113, 117, 118, 119, 127
data: collection 16, 18, 28, 37, 147, 243, 248, 251, 254, 264, 268, 284, 293, 314–317, 321, 351, 355, 359, 361, 375, 387, 388, 400, 422, 439, 469; exchange 93, 94, 95, 96, 108, 142, 210, 218, 222, 226, 227, 231, 237, 387, 396; interoperability 32, 327; models 223, 224, 225, 226–229, 232, 233, 234, 235, 236, 237; standards 6, 13, 18, 32, 58, 59, 80, 83, 222, 223, 224, 229, 230, 231, 235, 237, 387; visualization 26–28, 36, 38, 70–71, 76, *86*, 104, 113, 132–136, 139, 212, 231, 243, 269, 274, 282, 327, 367
database 120–125, *126*, 203, 232, 250, 255, 258, 284, 337, 344, 360, *361*, 448, 455
decentralization 372, 388, 400
decision making 36, 53, 66, 78, 132, 134, 136, 139, 274, 306, 328, 329, 474; parameters 155, 160, 210, 386, 387; support 211, 290, **378**, 388
deep learning 27, 241, 242, 250, 253, 254, 291, 469
demand 77, 79, 85, 99, 143, 150, 165, 167, 173, 190, 209, 217, 236, 273, 281, 308, 397, 399, 409, 465, 472
demonstrator project 196, 205
deposition 97, 157, 160, 162, 163, 164, *169*, 293
design: element 290, 330, 331, 332, 333, 334, 336, 337, 338, 339; for safety 330, 336, 338;

review 134, 136, 137, 139; risk 331, 333, 336; solution 137, 171, 331, 332; team 25, 127, 137, 179, *181*, 397; to-production 193
DFAB House 195, 196, 197, 198, 199, 200, 205, 292, 293
DfS: language 336; required design features 331, *332*, 333; rule 331, *332*, 333, 334, 336, 337, 338, 339, 347
digital: construction 132, 150, 151, 158, 194, 396, 428; economies 441, 446, 447, 448, 453; ecosystem 1, 3, *4–5*, 6, 13, 17, **18**, 42–44, 47, 51–*52*, 53–*54*, 55, 58–60, 396, 412, 428, 453; fabrication *5*, **18**, 94, 96, 108, 188–205, 292, 293, 297, 465; platform 3, 42, 43, 50, 59, 202, 445; tools 11, 13, *14*, **18**, 43, 46, 51, 59, 189, 195, 202, 203, 327; transformation 5, 13, 26, 44, 46, 47, 54, 58, 59, 183, 240, 258, 284, 397, 410; twin 10, 11, 23, 24, 26, 37, 38, 44, 58, 66, 69, 70, 74, 76–80, 84, 85, 87, 92, 143, 146, 147, 341, *378*, 380, 443
digitalization/digitization 93, 109, 204, 291, 293, 301, 302, 303, 326, 327, 328, 350, 351, 355, 367, 442, 443, 444, 450, 460, 465
disaster management 282
DLT 396, 399, 412
DMVR 150
document management system 379, 380–*381*
domain knowledge **18**, 326, 328–29, 347, 383
drift: accumulation 318, 319, *320*; correction 313, 315, 319, 321
drone 100, 167, 195, *247*, 265, 271, 308, 439; fixed-wing 266, 356; hexacopters 266; quadcopter 266; rotary-wing 266
dust monitoring 280, 283, 300
dynamics 74, 114, 115, 117, 210, 212, 213, 215, 218, **296**, 474

early design stages 204, 210
earned value methodology 372–373, 384, **393**
earth moving 275, *276*
ecosystem 11, *45*, 47–48, *49*, 50, 55, 57, 399, 405; digital 3, *4–5*, 6, 13, 17, **18**, 42–44, 47, 51–*52*, 53–*54*, 55, 56, 58–60, 396, 412, 428, 453; physical 49; business 50
educational framework 126
EIRs 397, 398, 400
electronic product code 79, 352–353, **377**
endurance 271
energy analysis 69, 282, 355
engineering education 126, 472, 474–475
enterprise resource planning **377–378**, 380–383
environmental 13, 25, 79, 82, 87, 127, 167, 168, 170, 174, 180, 189, 199, 212, 227, 241, 250, 254, 268, 280, 299, 314, 334, 340, 350, 354, 356, 358–359, **361**, 363, 365, 371, 396, 434, 442, 445–446, 472; monitoring 314; perception 350

exchange requirement (ER) 227–229
explicit knowledge 329, 340, 347
EXPRESS 224, 225, 226, 229, 231, 232, 233, 235
eXtensible Markup Language (XML) 32, 223–224, 226, 228–29, 231, 232, **233**, 455n2, 455n5
extrusion 157, 158, 160, 161, 192, 198, 204, 227

facilities/facility: management 5, 36, 81, 131, 134, *135*, 136, 139, 226, 242–43, 245, 291, 301, 303, 326–28, 347, 357, 400, 467; managers 28–29, 36, 121–22, 136, 313, 471
Federal Aviation Administration (FAA) 265, 271, **272–273**, 284
fiducial markers 248, 313, 316, 317–319, *320*, 321
field personnel 127
flight 144, 266–268, 270–271, 273, 279, 284, 356; regulation *272*, 284; team 268, 284
foreshadow 68, 70
fourth industrial revolution 3, 7, 9, 19, 25, 43, 142, 301, 307, 350–351, 371–72, 395–396, 411–412, 462
fragmentation and inflexibility 19, 47, 209, 460
fusion 55, 157, 163
future 7, 9, 17, **18**, 19, 23–30, 32, 36, 42, 58, 63, 68, 72, 77, 99, 108, 113, 126, 131, 139, 143, 146, 155, 157, 164, 168, 175, 179–83, 188, 193, 205, 231, 254, 258, 264, 274, 278, 289, 292, 299, 301, 327, 337, 347, 359, 367, 369, 372, 375, 383, 392, 397, 408, 421, 434, 439, 440, 442, 446, 454, 460, 463, 470, 474

genetic algorithm 69, 125, 312
geographic information system (GIS) 25, 27, 38, 231, 274
geometric variability 106, 210, 218
geotagging 321
global positioning system (GPS) 98, 134, 265, 267–268, 270, 312, 356
global warming 358
graph database 344
graph theory 311
Grey Hats 449

hazardous gas 359, **361–362**, 363, 365, *366*
hazards 28, 138, 170, 211, 218, 241, 278, 330, 331, 336, 340, 342, 356, 357, **361–362**, 365, 445
head mounted displays 113, *137*
human: -computer Interaction 446; in-the-loop 102, 117; -machine interaction 99, 102–103, 202, 268, 292, 301–303, 465; -UAS interaction 268, 284

IAARC 290, 291, 292, 294, 303
IFC 59, 211, 223–24, 227, 327, 334, 336; limitation 229, 231, 398; schema architecture 222, 225–226, 228–229, 232, **233**, *234*, 235–236
image-based 142–143, 257; 3D Reconstruction 245, *246–247*, 252, 254; Sensing technologies 99, 115, 142–143
image processing 16, 143, 150, 241, 269, 282
immersive environments 138
improved quality 74, 196
in-situ fabricator (IF) 197, 293, 295, 297–298
indoor building environment applications 311
industrial/industrialized 5–6, 19, 60, 64, 97, 157, 162–163, 176, 194, 196, 199, 258–259, 265, 291, 293, 352, 358; automation 9, 100–101, 108, 164, 170, 189, 192, 197–198, 292, 294, 298, 301, 308, 310, 327, 353,449–450; construction 13, 16, 188, 192, 204; internet 24; revolution 3, 7, 9, 19, 43, 142, 301, 307, 350–51, 371–72, 395, 411–412, 460, *461*, 462; vision 143, 222, 264, 441, 448, 453
industry 4.0 3, 7, 9–10, *11*, 12, 15, 17, 26, 63, 80, 113–114, 131, 188, 199, 264, 293, 307, 327, 350–352, 367, 371–372, 387, 395–396, 411, 443, 447, 460, 464, 469, 472; adoption 191, 371; foundation class (IFC) 59, 211, 222–229, 231–232, **233–34**, 235–236, 327, 334, 336, 398
inertia measurement units IMU 123, 249, 312
inference mechanism 336
information: and communication technology (ICT) 113, 352, 446; delivery manual (IDM) 227–229; recognition 148; technology 46–47, 375–376, 401, 450
infrastructure 24, 28, 30, 32, 58, 96–97, 106, 114, 116, 123, 132, 181, 196, 224–225, 233, 235, 242, 245, 258, 281–282, 299, 302, 312–313, 316, 322, 327–328, 351–352, 355, 358–359, 371, 374, 396–397, 403, 407, 410, 412–413, 441–448, 450, 452–455, 465, 475
innovation 1, 3, 6–7, 11, 15–17, **18**, 19, 23–26, 28, 30, 32, 42–44, 46–47, 50, 51, 59–60, 62–76, 80–81, 83–*86*, 87, 108, 113, 156, 163, *181*, 182–183, 190, 192–193, 196–200, 202–205, 212, *214*, *216*, 237, 299, 301, 308, 350, 356, 398, 404–405, 410, 421, 434, *435*, 436–442, 444–447, 454, 460, 462–467, 469–472, 474–475
insights 65–66, 70, *72*, 74, 76–79, 83, 86, 104, 199, 328, 347, 375, 389, 437, 439, 442, 454, 463, 467, 469
in-situ fabrication 164, 168, 177, 199
inspection **18**, 27, 101, 134, 146–418, 150, 240–243, 248, 257, 264–269, 274, 278–729, 281–284, 289, 294, 308, 313, 345, 356, **378**
instance data 224
integrated 10, 13, 16, **18**, 23–24, 32, 37–39, 44, 48, 64, 66, 70–71, 76, 84–85, 87, 92–99, 106, 108–109, 113–114, 117, 122, 127, 137, 143, 148, 150, 156, 170–172, 180, 188–189, 192, 198–199, 210, 230, 237, 242, 256, 264–65,

267, 282, 294, 329–330, 336–337, 347,
351–352, 355, 357, 359–360, 372, 387, 398,
409–410, 441, 446, 470–471, *473*; practice/
manner 13, 372; project controls 223, 237,
374–375, 379–*381*; project delivery 64, 74, 76,
78, 93, 127, 188, 460, 465, 469–471
integration 6, 9–13, 16,**18**, 23–26, 29, 32, 37, 44,
47, 55, 57–60, 63–67, 70–74, 76, 81, 83–*86*,
91–99, 106, 108–109, 113–114, 117–118, 122,
127, 131, 137, 143, 148, 150, 156, 169–172,
177, 179–182, 188–189, 192, 198–199, 202,
210, 218, 223, 230, *232*, 240–243, 248–249,
256, 264–267, 270, 275, 282, 289, 292–94, 301,
326–37, 347, 350–52, 355, 359, **361**, 363–*64*,
367, 374–75, **377**–381, 386–387, 392, 398,
409–410, 436, 441, 446, 451, 460, 465–475
intelligent buildings 302, 358
intelligentization 350–351
international framework for dictionaries
(IFD) 228
internet of things (IoT) 6, 9–10, *11*, 13, *14*, 16,
18, 23–24, 30, 38–39, *43*, *54*, 66, 92, 109, 132,
146, 180–181, 211, 222–223, 233, 237, 264,
307, 326, 350–359, 367, 396–397, 408–409,
445, 465, 469
interoperability 6, 9, 12–13, **18**, 32, 47, 53, 94,
96, 104, 108–109, 223, 225–226, 229–231,
237, 327, 351–352, 367, 372, 379, *381*,382,
386–387, 389, 398, 406, 410–411
investment 63, 69, 71, 73, *75*, 78, 80, 82, 85, 168,
175, 181, 204, 213, 218, 241–242, 259, 290,
308, 322, 340, 371, 389, 395–396, 408, 410,
444, 447
ISARC 289–290, *291*, 292

JavaScript Object Notation (JSON) 223–224,
229, 231
jump factory 171

key performance indicator (KPI) 7, 65, 69, 78, 81,
86, 372, 381, 383, 386–387, **393**
knowledge 12, 13, 27, 59, 63, 66, 71–77, 83,
102, 116–17, 123, 138, 176, 203, 211, 214,
218, 233, 249, **272**, 328–331, 334, 340–342,
344, 347, 351–353, 373, 383, 410, 426, 442,
454, 464, 469, 472; enriched BIM **18**, 326,
329, 347; formalization 330; library 330–331,
334, 336–337, 339; representation 330, 341;
transfer 7, 210–211, 217–218, 408, 475

labor productivity 66, 70, 81, **82**, 370, 437, 462,
475
LADAR 143
laggard 44, 63, 64, 83
landmark based localization 316
laser: pulses 144, 269; scanning **5**, 6, 13, 23, 37,
39, 145, 146, 147, 148, 150, 252, 270, 356, 477

last planner 71, 373, **380**, 380, 480
LawTech 399
leadership 21, 44, 62, **68**, 73, 78, 79, 81,
86, 460
lean construction **67**, 71, 357, 367, 373, 379, 380,
382, 384, 469; maturity model 480
learning 13, 23, 27, 66, 69, **72**, 73, 74, 77, 80, 84,
87, 123, 134, 138, 215, 218, 241, 250, 253,
291, 379, 384, 389, 409, 469, 474
legal concerns 284
LegalTech 399, 412, 413
legislation 94, 109, 283, 399, 447, 470, 475
liability 71, 283, 284
light detection and ranging (LiDAR) 249, 267,
270, 275, 480
lighting 122, 137, 164, 172, 269, 280, 313, 314,
316, 462
lightweight translucent façade system 198
linked data 233, 235, 236
living augmented reality 136, 480
localization 29, 97, 98, 100, 106, 245–249, 297,
309, 312–317, 321, 322, 367, 481
logistics 70, 77, 81, 90, 176, 178, **181**, 182,
190, 203, 212, 242, 244, 353, 357, 367, 398,
407, 421

machine: learning 23, 28, 65, 66, 68, 76, 80, 83,
95, 101, 120, 242, 250–255, 258, 307, 375,
480; -to-Machine (M2M) 24, 480; vision 90,
143, 146, 147, 148, 150, 151, 290
malware 448, 449
management information system 382, 480
manipulating 282
manpower 291, 340, 341
manual 94, 136, 146, 147, 148, 156, 169, 170,
176, 230, 250, 267, 268, 326, 357, 382, **429**,
439, 452, 479
manufacturing 7, 9, 10–12, **14**, 16, **18**, 24, 26, 44,
63, 95–98, 102, 106, 108, 131, 142, 147, 150,
155, 157–166, 169–177, 179, 181, 183, 189,
192, 198, 202, 209, 264, 293, 327, 350, 353,
357, 367, 371, 396, 400, 408, 441, 460–464,
470, 475
marker detection 315, 318, 321
market: placement 314, 318; pressure 63, 64, 72,
73, 85
marketing 282
material systems 91, 92, 94, 100, 106, 194
maturity model 370, 384, 386, 391, 457, 478
mesh mould 197, 198, 200
mixed reality (MR) **18**, 25, 26, 38, **54**, 120, 131,
132, 133, 134, 135, 139, 480
MLS 146, 147, 480
mobile augmented reality 133, 480
mobile robots 308, 309, 316
model view definition (MVD) 228, 229, 236,
480

modeling 5, 13, 25, 47, 57, 66, 70, 135, 148, 150, 162, 189, 207, 210, 212, 218, 223, 227, 229, 237, 240, 245, 249, 293, 308, 328, 351, **377**, 380, 437, 441, 462, 465, 467, 471, 475, 478

modularity 53, 229, 235, 388

multi-depot 310, 480

multiple travelling salesman problem 310, 480

multispectral camera 268

multisystem platforms 195

multi-tier architecture analysis 360

multi-trade prefabrication 18, 209, 210, 212, 214, 215, 217, 480

National Institute of Standards and Technology (NIST) 24, 480

navigation 23, 249, 265, 267, 270, 279, 309, 310, 312–15, 316, 319–322, 479

NEST **107**, 196

network topology 355

networked systems 99

object: detection 250, 251; model 224, 228; recognition 143, 148, 255, 298; tracking 250

occupancy-based progress monitoring 252, 254

off-site construction 13, 192, 209, 210

on-demand building automation 28, 231, 232, 237

on-site: factory 98; robot 192, 194, 195; robotic fabrication 198, 199

operation and maintenance 26, 32, 307, 309, 330, 336, 358, 445, 469

optical laser scanners 142

optoelectronic devices 143, 146

optoelectric technology 142, 143

PAS 1192-5 302, 452

path planning 29, 316

pathway 75

pattern recognition 32, 74, 143, 443

payload capacity 271

phase shift 144, 145

photogrammetry 143, 145, 150, 227

platform product 44

point cloud data 143, 144, 147, 150, 249, 269, 356

pose estimation 250, 251, 255, 315

post-processing algorithms 147, 148, 150

post-construction 144, 148, 242, 264, 281

pre-construction 138, 242, 274

prefabrication 5, 13, 16, 25, 82, 84, 102, 104, 158, 164, 175, 180, 189, 192, 197–199, 207, 209, 212–218, 292, 299, 428, 470

prevention-through-design 330

privacy 83, 272, 283, 284, 308, 384, 388, 403, 457

process standard 7, 17, 374

processing 16, 94–96, 101, 104, 106, 109, 117, 119, 131, 134, 143–148, 150, 163, 189, 194, 198, 241, 269–271, 282, 322, 328, 352, 354, 357, 360, 387, 404, 442, 443

production network 209, 217, 218

productivity 7, 12, 26, 37, 42–44, 64–66, 70, 81–87, 96, 114, 125, 131, 139, 181, 199, 209, 217, 240, 243, 250, 254, 259, 274, 291, 295, 299, 303, 308, 322, 326, 351, 358, 370, 374, 376, 378, 382, 388, 396, 437, 443, 462, 465, 469, 475

program evaluation and review technique (PERT) 371, 372, 373

progress capture 147, 150, 383, 388

progress control 356

progress monitoring 19, 27, 38, 116, 135, 139, 151, 241, 245, 250, 254, 279, 308, 409

progress rate 209

project delivery networks 5, 62, 63, 64, 67, 69, 71, 78, 83, 84, 87

project evaluation 274, 275

project lifecycle management (PLM) 204

project management 16, 57, 120, 150, 204, 212, 276, 370–376, 379–381, 384, 386–390, 393, 410, 470, 475

project trades 210, 217, 218

publically-funded projects 215, 218

quality assessment 147, 148, 240, 241, 242, 257

quality control 26, 27, 94, 96, 101, 146, 147, 151, 196, 378

querying mechanism 231, 241, 449

radio frequency identification (RFID) tag 28, 37, 38, 98, 119–23, 125, 146, 270, 312, 313, 350, 351, 353, 357, 367, 377, 388, 436, 463

ransomware 448, 449

real-time capability 115

real-time monitoring 13, 16, 43, 113–116, 212, 284, 355, 443

reality capture 6, 13, 54–56, 66, 80, 143, 244, 248, 249, 255, 256, 257, 438

reinforcement learning algorithm 124

remote office 121

remote pilot certification 272, 273

representational state transfer (REST) 223

request for information 83

research and development 6, 7, 17, 95, 155, 167, 180, 183, 203, 242, 258, 289, 308, 309

resource availability 209, 217, 218

resource description framework 342

RGB Camera 268, 269, 319

risk management 170, 172, 302, 377, 387

risk register 330, 333, 334, 336, 337, 338, 339

risk review system 330, 336, 338

robot classification 308

robot swarms 291
robot system 268
robotic(s) construction 102, 292, 295, 463, 465
robotic fabrication 97, 189, 194, 198, 199
robotics operation system 317
robotics system 107, 108, 170, 181, 191, 289, 290, 293, 295, 300, 303, 307, 315, 322, 345
robotic technology 290, 291, 303, 308
root cause 54, 63, 66, 67, 77, 370, 381, 454
route planning 29, 309, 310, 312, 322
router 189, 354, 355, 360, 361, 362, 365
RTLS system 125, 126

safety breaches 211, 218
safety compliance 326, 340, 341, 342, 344, 347
safety management 29, 113, 115, 211, 212, 218, 251, 257, 278, 350, 356–359, 361, 365, 366, 367
scan-to-BIM 148, 149
scan-vs-BIM 148, 149
security and privacy models 283
security and surveillance 276
self-organized network 359
semantic information 29, 148, 150, 226, 237, 254, 328
semi-autonomous 92, 101, 107, 267, 268, 279, 309
sensing systems 115, 116, 119, 127
sensor feedback 92, 95, 97, 101, 104, 106
sensor integration 95
sensor node 350, 354, 359, 360, 361, 362, 363, 365
sensor technology 265
service orientation 9, 351, 372, 388
service robots 292, 312
SfM 150, 242, 245, 246, 247, 248, 249, 356
SHoP Architects 98, 137, 190, 191, 200, 201, 205
simulation(s) 5, 6, 13, 24, 29, 67, 69, 74, 76, 79, 81, 85, 91, 95, 106, 109, 116, 123, 132, 136, 139, 204, 211, 213, 215, 225, 251, 254, 268, 347, 355, 367, 421, 436, 463, 467, 471
single task robot 289, 293, 294, 197
site communication 276
site planning 242, 274, 275
site transportation 277
skill gaps 123
skilled labor 167, 183, 192, 291, 294, 422, 465
Sky factory 295, 299, 300
SLAM 245, 248, 249, 258, 316
slip-forming 198
small and medium-sized enterprises (SME) 397, 443, 444, 464
smart building 29, 30, 223, 237, 350, 352, 358
smart built Environment 30, 31, 32, 38, 358, 443
smart cities/City 9, 24, 30, 37, 79, 233, 302, 441, 445, 446, 447, 450

smart construction Sites 463, 468
smart dynamic Casting (SDC) 198, 200
smart factories 7, 351, 353, 463, 467
smart grid 24
smart helmets 69, 136
smart slab 198, 200
social engineering 449
spatial interference check 343, 344
spatial timber assemblies 198, 199, 200
specialized contractors 209
speckle noise 144
stakeholders 17, 19, 24–25, 27, 30, 38, 48, 50, 53, 59, 63, 66–69, 76–87, 104, 108, 115, 131, 137, 168, 174, 181, 225, 259, 268, 303, 326, 330, 370, 373–383, 386–390, 396, 409, 412, 421, 427, 439, 443, 451, 462, 465, 467, 471, 474
Standard for the Exchange of Product (STEP) 224, 227, 229
standards 6, 12, 13, 17, 24, 32, 59, 70, 72, 78–80, 83, 94, 109, 168, 170, 173, 177, 190, 196, 211, 222–225, 227–231, 233, 237, 279, 282, 327, 352, 374, 386, 389, 404, 408, 446, 453, 474
Structural Engineer 120, 121, 167
structural health monitoring 146, 147, 151, 350, 358
structural safety 358
structure from motion 150, 242, 245, 246, 247, 248, 249, 356
subcontracting 209, 371
superimposed 102, 210, 218, 252
supply chain management 113, 115, 170, 181
surveying 5, 104, 145, 147, 249, 250, 270, 283, 294, 356
Swiss National Centre of Competence in Research (NCCR) 195, 205
system architecture 113, 114, 118, 119, 341, 462
system dynamics 210, 213, 215
system integration 196, 361

tacit knowledge 329
take-off 480, 482
task allocation 308–310
taxonomy 50, 226, 251, 331, 463, 467
telecommunication 273, 302
temporary and multidisciplinary teams 211
thermal camera 268
threat Modeling 322
time of flight 144
TLS 146
tokenization 407
traditional construction 64, 66, 167, 169, 171, 180, 210, 474
training Requirements 284
transformation 3, 5, 16, 44, 64, 73, 240, 321, 421
transparency 63, 77, 242, 243, 327, 379

travelling Salesman Problem 310
trends 3, 5, 9, 13, 58, 63, 66, 192, 257, 302, 326, 460, 467
triangulation 144
trust 65, 79
twin 10, 23

UN Studio 190
unmanned aerial systems (UAS) 194, 265
unmanned aerial vehicle (UAV) 265

value stream mapping 381
variability 209, 210, 218
virtual design and construction 131, 182, 255
virtual environments 123, 137
virtual instructor 123
virtual reality (VR) 16, 23, 131, 139, 282
virtual site 125
virtualization 372, 387, 463

visual production management 255
visual line-of-sight (VLOS) 272
visualization 26, 36, 104, 132, 269, 367
vulnerability 83, 403

waste 16, 63, 64, 67, 137, 164, 167, 380, 421
wearable technology 445
web ontology 223
White Hats 449
wi-fi 98, 134, 268, 312, 316, 354
wireless sensor 113; network 350
work blockages 209, 217
work integration architectures 210
work packaging 66, 70
workforce training 113
work-related musculoskeletal disorders 123
workspace 309, 344, 423, 426

ZigBee 98, 354, 360